STUDENT'S SOLUTIONS MANUAL

EDUMEDIA SERVICES
WITH CONTRIBUTIONS FROM DEANA RICHMOND

BASIC MATHEMATICS

Brian Goetz

Kellogg Community College

Graham Smith

Kellogg Community College

John Tobey

North Shore Community College

www.pearsonhighered.com

Table of Contents

Chapter 1   Whole Numbers

1.1   Understanding Whole Numbers

GUIDED PRACTICE

1) There are two periods. 412 is in the thousands period. The number is written as four hundred twelve thousand, seven hundred six.

2) There are three periods. 12 is in the millions period. The number is written as twelve million, three.

3) There are two periods. 17 is in the thousands period. The number is written as seventeen thousand, four hundred.

4) 2 is in the millions period. The number in that period is 274. 2 is in the hundreds place of 274. 2 is in the hundred millions place.

5) 8 is in the ones period. The number in that period is 890. 8 is in the hundreds place of 890. 8 is in the hundreds place.

6)   451,386

$$
\begin{array}{r}
6 \\
80 \\
300 \\
1,000 \\
50,000 \\
400,000
\end{array}
$$

$451,386 = 400,000 + 50,000 + 1000 + 300 + 80 + 6$

7) $5,300,010 = 5,000,000 + 300,000 + 10$

8)

9)

10)

11) The digit 2 is in the tens place. $128 rounds up to $130 or rounds down to $120. 128 is closer to 130, so 130 is the better approximation of 128. $128 rounded to the tens place is $130.

12) 812 will round up to 900 or down to 800. 812 is closer to 800, so 812 rounds down to 800.

13) 5,632 will round up to 6,000 or down to 5,000. 5,632 is closer to 6,000, so 5,632 rounds up to 6,000.

14) 135 will round up to 140 or down to 130. Since 135 is halfway between 130 and 140, 135 rounds up to 140.

15) 9,423 will round up to 10,000 or down to 9,000. 9,423 is closer to 9,000, so 9,423 rounds down to 9,000.

16) Identify the digit in the thousands place. 3⃞4⃞,821; 34,821 will round to either 34,000 or 35,000. Since the hundreds digit is 8, we round up. 34,821 rounds to 35,000.

17) Identify the digit in the hundreds place. 17,⃞6⃞83; 17,683 will round to either 17,600 or 17,700. Since the tens digit is 8, we round up. 17,683 rounds to 17,700.

18) Identify the digit in the ten thousands place. ⃞5⃞2,908; 52,908 will round to either 50,000 or 60,000. Since the thousands digit is 2, we round down. 52,908 rounds to 50,000.

19) We must read the height from the top of the bar labeled "3 years." Mr. Descartes' car was worth $15,000 after 3 years.

20) The purchase price of the car is represented by the new bar. Mr. Descartes paid $33,000 for his car.

21) Descartes' car lost the most value between new and 1 year.

22) The last piece of data is shown above the bar labeled "4 years." Decartes' car was worth $13,000 after 4 years.

CONCEPT CHECKS AND PRACTICE EXERCISES

A1) The digits are 0, 1, 2, 3, 4, 5, 6, 7, 8, 9.

A2) There are three digits per period, so the largest place value of a number with five digits is in the second place of the second period, which is the ten thousands place.

A3) The first five periods are ones, thousands, millions, billions, and trillions.

A4) 153; There is one period. 153 is in the ones period. The number is written as one hundred fifty-three.

A5) 492; There is one period. 492 is in the ones period. The number is written as four hundred ninety-two.

A6) 7,005; There are two periods. 7 is in the thousands period. The number is written as seven thousand, five.

A7) 3,080; There are two periods. 3 is in the thousands period. The number is written as three thousand, eighty.

A8) 100,309; There are two periods. 100 is in the thousands period. The number is written as one hundred thousand, three hundred nine.

A9) 50,003; There are two periods. 50 is in the thousands period. The number is written as fifty thousand, three.

A10) 4,000,000,005; There are four periods. 4 is in the billions period. The number is written as four billion, five.

A11) 8,000,005,000; There are four periods. 8 is in the billions period. The number is written as eight billion, five thousand.

A12) 4 is in the thousands period. The number in that period is 234. 4 is in the ones place of 234. 4 is in the thousands place.

A13) 4 is in the thousands period. The number in that period is 564. 4 is in the ones place of 564. 4 is in the thousands place.

A14) 4 is in the ones period. The number in that period is 435. 4 is in the hundreds place of 435. 4 is in the hundreds place.

A15) 4 is in the thousands period. The number in that period is 435. 4 is in the hundreds place of 435. 4 is in the hundred thousands place.

A16) 4 is in the ones period. The number in that period is 456. 4 is in the hundreds place of 456. 4 is in the hundreds place.

A17) 4 is in the millions period. The number in that period is 435. 4 is in the hundreds place of 435. 4 is in the hundred millions place.

A18) 4 is in the thousands period. The number in that period is 945. 4 is in the tens place of 945. 4 is in the ten thousands place.

A19) 4 is in the billions period. The number in that period is 24. 4 is in the ones place of 24. 4 is in the billions place.

B1) a) yes

b) $505 = 500 + 5$
$550 = 500 + 50$

c) Answers may vary. In 550, the second 5 represents 5 tens, while in 505 it represents 5 ones.

B2) a) yes

b) $560 = 500 + 60$
$506 = 500 + 6$

c) Answers may vary. In 560, the 6 represents 6 tens, while in 506 it represents 6 ones.

B3) $1,534 = 1,000 + 500 + 30 + 4$

B4) $5,492 = 5,000 + 400 + 90 + 2$

B5) $6,002 = 6,000 + 2$

B6) $3,040 = 3,000 + 40$

B7) $205,100,309 = 200,000,000 + 5,000,000 + 100,000 + 300 + 9$

B8) $420,050,003 = 400,000,000 + 20,000,000 + 50,000 + 3$

B9) $4,000,000,005 = 4,000,000,000 + 5$

B10) $8,000,005,000 = 8,000,000,000 + 5,000$

C1) Answers may vary. If a tick mark were placed at every unit, the graph would be too large to fit on the paper.

C2) Answers may vary. The graphed number is already identified.

C3)

C4)

C5)

C6)

C7)

C8)

D1) a) 11,500 is halfway between 11,000 and 12,000.

b) Answers may vary. If the number is less than 11,500, then it is closer to 11,000.

D2) Answers may vary. If the number is greater than 500,000, then it is closer to 1,000,000.

D3) Answers may vary. Rounding is a way to approximate a number to a certain place value.

D4) 2,000 is a better approximation for 1,587.

D5) 1,000 is a better approximation for 1,298.

D6) 320 is a better approximation for 321.

D7) 170 is a better approximation for 172.

D8) 433,000 is a better approximation for 432,501.

D9) 237,000 is a better approximation for 237,499.

D10) Identify the digit in the tens place. 6|3; 63 will round to either 60 or 70. Since the ones digit is 3, we round down. 63 rounds to 60.

D11) Identify the digit in the tens place. 8|7; 87 will round to either 80 or 90. Since the ones digit is 7, we round up. 87 rounds to 90.

D12) Identify the digit in the hundreds place. 6|73; 673 will round to either 600 or 700. Since the tens digit is 7, we round up. 673 rounds to 700.

D13) Identify the digit in the hundreds place. 8|43; 843 will round to either 800 or 900. Since the tens digit is 4, we round down. 843 rounds to 800.

D14) Identify the digit in the thousands place. 1$5$|,732; 15,732 will round to either 15,000 or 16,000. Since the hundreds digit is 7, we round up. 15,732 rounds to 16,000.

D15) Identify the digit in the thousands place. 6$4$|,382; 64,382 will round to either 64,000 or 65,000. Since the hundreds digit is 3, we round down. 64,382 rounds to 64,000.

D16) Identify the digit in the hundred thousands place. 9|46,093; 946,093 will round to either 900,000 or 1,00,000. Since the ten thousands digit is 4, we round down. 946,093 rounds to 900,000.

D17) Identify the digit in the ten thousands place. 9|7,000; 97,000 will round to either 90,000 or 100,000. Since the thousands digit is 7, we round up. 97,000 rounds to 100,000.

D18) Identify the digit in the ten thousands place. $1$8|6,542; $186,542 will round to either $180,000 or $190,000. Since the thousands digit is 6, we round up. $186,542 rounds to $190,000. The cost of the new house is about $190,000.

D19) Identify the digit in the hundred thousands place. $3,$8|39,512; $3,839,512 will round to either $3,800,000 or $3,900,000. Since the ten thousands digit is 3, we round down. $3,839,512 rounds to $3,800,000. The cost of the airplane is about $3,800,000.

D20) Identify the digit in the hundreds place. $1,$4|83; $1,483 will round to either $1,400 or $1,500. Since the tens digit is 8, we round up. $1,483 rounds to $1,500. You have about $1,500 in the bank.

D21) Identify the digit in the thousands place. $1$2|,499; $12,499 will round to either $12,000 or $13,000. Since the hundreds digit is 4, we round down. $12,499 rounds to $12,000. The used car costs about $12,000.

E1) A bar graph can be used to show data that relates two pieces of information.

E2) Answers may vary. Data is a collection of facts that may be used to draw conclusions.

E3) The average winning speed in 1950 was about 120 mph.

E4) The average winning speed in 2000 was about 170 mph.

E5) The slowest winning speed from 1970 to 2000 was about 140 mph.

E6) The fastest winning speed from 1950 to 1980 was about 155 mph.

E7) The average winning speed increased over six decades.

E8) The average winning speed decreased over two decades.

SECTION 1.1   EXERCISES

For 1–15, refer to Concept Checks and Practice Exercises.

17) digits

19) rounding

21) 0, 1, 2, 3, 4, 5, 6, 7, 8, 9

23) 67,456; sixty-seven thousand, four hundred fifty-six

25) 3,127; three thousand, one hundred twenty-seven

27) 315; three hundred fifteen

29) 1,125,568; one million, one hundred twenty-five thousand, five hundred sixty-eight

31) 560,200,107; five hundred sixty million, two hundred thousand, one hundred seven

33) 12,005,000; twelve million, five thousand

35) 808,088

37) Two hundred one thousand, forty-one

39) $37 = 30 + 7$

41) $3,800 = 3,000 + 800$

43) $85,290 = 80,000 + 5,000 + 200 + 90$

45) $8,000,100,000 = 8,000,000,000 + 100,000$

47) $222 = 200 + 20 + 2$

49) $300 + 40 = 340$

51) $600 + 50 + 4 = 654$

53) $2,000 + 700 + 20 = 2,720$

55) $3,000 + 9 = 3,009$

57)

59)

61)

63) Identify the digit in the tens place. $\boxed{8}6$; 86 will round to either 80 or 90. Since the ones digit is 6, we round up. 86 rounds to 90.

65) Identify the digit in the hundreds place. $\boxed{8}68$; 868 will round to either 800 or 900. Since the tens digit is 6, we round up. 868 rounds to 900.

67) Identify the digit in the thousands place. $8\boxed{4},523$; 84,523 will round to either 84,000 or 85,000. Since the hundreds digit is 5, we round up. 84,523 rounds to 85,000.

69) Identify the digit in the hundred thousands place. $\boxed{6}89,512$; 689,512 will round to either 600,000 or 700,000. Since the ten thousands digit is 8, we round up. 689,512 rounds to 700,000.

71) 5,329 rounds to 5,300. A scrap dealer has loaded about 5,300 pounds of aluminum into a trailer.

73) 78 rounds to 80. It took about 80 hours for Zeus to complete the quilt for the charity auction.

75) a) Identify the digit in the hundreds place. $\$123,\boxed{5}64$; Since the tens digit is 6, we round up to $123,600.

   b) Identify the digit in the hundred thousands place. $\$\boxed{1}23,564$; Since the ten thousands digit is 2, we round down to $100,000.

   c) Identify the digit in the ten thousands place. $\$1\boxed{2}3,564$; Since the thousands digit is 3, we round down to $120,000.

77) a) Identify the digit in the hundreds place.
3,564,$\boxed{3}$26; Since the tens digit is 2, we
round down to 3,564,300.

b) Identify the digit in the hundred thousands
place. 3,$\boxed{5}$64,326; Since the ten thousands
digit is 6, we round up to 3,600,000.

c) Identify the digit in the millions place.
$\boxed{3}$,564,326; Since the hundred thousands
digit is 5, we round up to 4,000,000.

79) a) California had the most votes in the House.

b) California, 53; Texas, 32; Florida, 25; Ohio,
18; Michigan, 15; North Dakota, 1

c) North Dakota had more votes in the Senate
than in the House.

1.2 Adding Whole Numbers

GUIDED PRACTICE

1)

$9 + 3 = 12$

2) $6 + 5 = 11$
$5 + 6 = 11$
The order in which numbers are added does not
change the result.

3)
$$\begin{array}{r} 42 \\ +17 \\ \hline 59 \end{array}$$

4)
$$\begin{array}{r} {\scriptstyle 1} \\ 58 \\ +5 \\ \hline 63 \end{array}$$

5)
$$\begin{array}{r} {\scriptstyle 1\,1} \\ 73 \\ +38 \\ \hline 111 \end{array}$$

6)
$$\begin{array}{r} {\scriptstyle 1} \\ 92 \\ 23 \\ +11 \\ \hline 126 \end{array}$$

7)
$$\begin{array}{r} {\scriptstyle 1} \\ 3687 \\ +1032 \\ \hline 4719 \end{array}$$

8)
$$\begin{array}{r} {\scriptstyle 2\,1} \\ 482 \\ 813 \\ +992 \\ \hline 2287 \end{array}$$

9) Perimeter $= 1,520 + 730 + 1,250$
$= 2,250 + 1,250$
$= 3,500$ yards

10) $4 + y = 10$
$y = 6$
Use the basic fact $4 + 6 = 10$.

11) $35 + x = 68$
Starting at 35, count by tens. Don't go over 68.
35, 45, 55, 65
We are still 3 short of 68, so we need to add 3
ones. 35, 45, 55, 65, 66, 67, 68
3 tens + 3 ones $= 33$
$x = 33$

CONCEPT CHECKS AND PRACTICE EXERCISES

A1) Answers may vary. When you count on a
number line, you add 1 for each unit.

A2)
$4 + 2 = 6$

A3)
$6 + 2 = 8$

A4)
$7 + 8 = 15$

A5)
$9 + 6 = 15$

A6)

	7	3	9	8
5	12	8	14	13
7	14	10	16	15
8	15	11	17	16
6	13	9	15	14

A7)

	4	9	5	6
4	8	13	9	10
9	13	18	14	15
3	7	12	8	9
6	10	15	11	12

A8)
   a) $0+6=6$
   b) $1+2=3$
   c) $3+8=11$
   d) $7+2=9$
   e) $6+7=13$
   f) $8+5=13$
   g) $5+2=7$
   h) $4+9=13$
   i) $9+7=16$
   j) $8+8=16$

A9)
   a) $9+1=10$
   b) $6+9=15$
   c) $5+2=7$
   d) $2+8=10$
   e) $9+0=9$
   f) $7+6=13$
   g) $3+9=12$
   h) $8+4=12$
   i) $7+5=12$
   j) $8+7=15$

A10)
   a) $5+6=11$
   b) $8+5=13$
   c) $3+7=10$
   d) $4+9=13$
   e) $0+8=8$
   f) $9+5=14$
   g) $9+8=17$
   h) $1+4=5$
   i) $9+9=18$
   j) $6+8=14$

A11)
   a) $8+9=17$
   b) $5+1=6$
   c) $6+8=14$
   d) $2+6=8$
   e) $7+8=15$
   f) $8+1=9$
   g) $7+7=14$
   h) $6+5=11$
   i) $9+0=9$
   j) $9+4=13$

B1) If you have 13 ones, you can exchange them for 1 ten and 3 ones.

B2)
$$\begin{array}{r} 3\,2 \\ +\,5\,4 \\ \hline 8\,6 \end{array}$$

B3)
$$\begin{array}{r} 6\,1 \\ +\,3\,8 \\ \hline 9\,9 \end{array}$$

B4)
$$\begin{array}{r} 1 \\ 4\,2 \\ +\,3\,9 \\ \hline 8\,1 \end{array}$$

B5)
$$\begin{array}{r} 1 \\ 6\,5 \\ +\,2\,8 \\ \hline 9\,3 \end{array}$$

B6)
$$\begin{array}{r} 1\;1 \\ 9\,3 \\ +\;\;2\,8 \\ \hline 1\,2\,1 \end{array}$$

B7)
$$\begin{array}{r} 1 \\ 4\,8 \\ +\;\;8\,1 \\ \hline 1\,2\,9 \end{array}$$

B8)
$$\begin{array}{r} 1\;1 \\ 6\,5 \\ +\;\;7\,8 \\ \hline 1\,4\,3 \end{array}$$

B9)
$$\begin{array}{r} 1\;1 \\ 8\,3 \\ +\;\;5\,8 \\ \hline 1\,4\,1 \end{array}$$

B10)
```
   1 1
    1 7
    4 2
 +  8 3
 ──────
  1 4 2
```

B11)
```
    1
    6 5
    2 1
 +  8 3
 ──────
  1 6 9
```

C1) 2 + 3 is 5; 5 + 5 is 10; 10 + 7 is 17.

C2)
```
   3 4 5
 + 6 2 1
 ───────
   9 6 6
```

C3)
```
   8 9 3
 + 1 0 4
 ───────
   9 9 7
```

C4)
```
   1 1
   3 9 3
 + 4 7 8
 ───────
   8 7 1
```

C5)
```
   1 1
   2 6 3
 + 4 7 9
 ───────
   7 4 2
```

C6)
```
   1 1
   4 5 8 2
 + 3 9 8 7
 ─────────
   8 5 6 9
```

C7)
```
   1   1
   7 3 0 6
 + 1 8 3 7
 ─────────
   9 1 4 3
```

C8)
```
   1   1 1
   7 3 4 9 8
 + 1 8 3 9 5
 ───────────
   9 1 8 9 3
```

C9)
```
   1 1
   3 2 8 7 5
 + 4 5 9 5 0
 ───────────
   7 8 8 2 5
```

C10)
```
   1 1
   4 9 2
 + 6 3 4
 ───────
  1 1 2 6
```
The combined receipts on Wednesday and Thursday were $1,126.

C11)
```
    1
   2 3 4 2
 + 1 8 3 7
 ─────────
   4 1 7 9
```
The combined receipts on Friday and Saturday were $4,179.

FOCUS ON ADDING SEVERAL NUMBERS EFFICIENTLY PRACTICE EXERCISES

1) $6+4+9+1+8 = (6+4)+(9+1)+8$
    $= 10+10+8$
    $= 28$

2) $7+3+5+5+4 = (7+3)+(5+5)+4$
    $= 10+10+4$
    $= 24$

3) $1+7+4+2+6+5 = (1+7+2)+(4+6)+5$
    $= 10+10+5$
    $= 25$

4) $2+6+3+1+4+5 = (2+3+5)+(6+4)+1$
    $= 10+10+1$
    $= 21$

5) $4+7+6+2+5+1+4$
    $= (4+6)+(7+2+1)+5+4$
    $= 10+10+5+4$
    $= 29$

6) $5+6+3+2+4+1+3$
    $= (5+3+2)+(6+4)+1+3$
    $= 10+10+1+3$
    $= 24$

SECTION 1.2   EXERCISES

For 1–11, refer to Concept Checks and Practice Exercises.

13) Commutative Property of Addition

15) variable

17) sum

19)
$$5 + 2 = 7$$

21)
$$9 + 3 = 12$$

23)
$$3 + 19 = 22$$

25)
$$8 + 0 = 8$$

27)

	9	7	5	4
3	12	10	8	7
6	15	13	11	10
5	14	12	10	9
9	18	16	14	13

29)

	8	4	9	0
2	10	6	11	2
7	15	11	16	7
6	14	10	15	6
9	17	13	18	9

31) a) $9 + 1 = 10$
    b) $6 + 9 = 15$
    c) $5 + 2 = 7$
    d) $6 + 6 = 12$
    e) $9 + 8 = 17$
    f) $4 + 9 = 13$
    g) $5 + 8 = 13$

33) a) $8 + 9 = 17$
    b) $5 + 1 = 6$
    c) $6 + 8 = 14$
    d) $3 + 6 = 9$
    e) $9 + 6 = 15$
    f) $8 + 7 = 15$
    g) $7 + 9 = 16$

35)
```
   3 1
 + 3 8
   6 9
```

37)
```
   5 5
 + 2 4
   7 9
```

39)
```
     1
   4 8
 + 4 5
   9 3
```

41)
```
     1
   8 3
 +  7 6
 1 5 9
```

43)
```
     1
     5
     6
 +   3
   1 4
```

45)
```
     1
   1 5
   2 1
 + 1 7
   5 3
```

47)
```
   5 6 1
 +   3 8
   5 9 9
```

49)
```
       1
   2 6 5
 +   2 8
   2 9 3
```

51)
```
       1
   6 4 8
 + 2 8 1
   9 2 9
```

53)
```
   1 1 1
   5 8 3
 + 6 5 8
 1 2 4 1
```

55)
$$\begin{array}{r} {\scriptstyle 1\ 1} \\ 45 \\ 71 \\ +\ \ 37 \\ \hline 153 \end{array}$$
You need $153.

57)
$$\begin{array}{r} {\scriptstyle 1\ 1} \\ 75 \\ 41 \\ +\ \ 66 \\ \hline 182 \end{array}$$
She drove 182 miles.

59)
$$\begin{array}{r} {\scriptstyle 1\ 1\ 1} \\ 873 \\ +\ \ 158 \\ \hline 1031 \end{array}$$

61)
$$\begin{array}{r} {\scriptstyle 1\ 1} \\ 251 \\ +479 \\ \hline 730 \end{array}$$

63)
$$\begin{array}{r} {\scriptstyle 1\ 1} \\ 9406 \\ +\ \ 1733 \\ \hline 11139 \end{array}$$

65)
$$\begin{array}{r} {\scriptstyle 1\ 1} \\ 22875 \\ +44950 \\ \hline 67825 \end{array}$$

67)
$$\begin{array}{r} {\scriptstyle 1\ 1\ \ \ 1\ 1} \\ 975482 \\ +\ \ \ \ 81635 \\ \hline 1057117 \end{array}$$

69)
$$\begin{array}{r} {\scriptstyle 1\ 1\ 1\ 1\ \ \ 1} \\ 838738 \\ +\ 584539 \\ \hline 1423277 \end{array}$$

71) $7+4+3+6 = (7+3)+(4+6)$
$$= 10+10$$
$$= 20$$

73) $4+3+2+6+5+3 = (4+6)+(3+2+5)+3$
$$= 10+10+3$$
$$= 23$$

75) $4+7+4+2+1+2 = (4+4+2)+(7+1+2)$
$$= 10+10$$
$$= 20$$

77) Perimeter $= 10\,\text{in.}+12\,\text{in.}+10\,\text{in.}+12\,\text{in.}$
$$= 22\,\text{in.}+10\,\text{in.}+12\,\text{in.}$$
$$= 32\,\text{in.}+12\,\text{in.}$$
$$= 44\,\text{in.}$$

79) Perimeter $= 7\,\text{ft}+8\,\text{ft}+13\,\text{ft}$
$$= 15\,\text{ft}+13\,\text{ft}$$
$$= 28\,\text{ft}$$

81) Perimeter $= 120\,\text{ft}+75\,\text{ft}+120\,\text{ft}+75\,\text{ft}$
$$= 195\,\text{ft}+120\,\text{ft}+75\,\text{ft}$$
$$= 315\,\text{ft}+75\,\text{ft}$$
$$= 390\,\text{ft}$$
The race course is 390 feet long.

83) a) Total $= \$120+\$85+\$202+\$85$
$$= \$205+\$202+\$85$$
$$= \$407+\$85$$
$$= \$492$$
Pat spent $492.
  b) Since $492 is less than $500, Pat will not get the card.

85) $150+150+150$
$$= 300+150$$
$$= 450$$
The coastline of the upper peninsula is approximately 450 miles long.

87) Answers may vary. Example: The estimate is too low because the curve of the shoreline appears longer than the straight lines estimating its length.

89) Answers may vary.

91)
$$\begin{array}{r} {\scriptstyle 1} \\ 356 \\ +635 \\ \hline 991 \end{array}$$

93)
$$\begin{array}{r} {\scriptstyle 1\ 1} \\ 58 \\ 78 \\ +\ \ 53 \\ \hline 189 \end{array}$$

95) $\overset{1\ 1\ 1}{764}$
     $\underline{+\ 2\ 3\ 6}$
     $1\ 0\ 0\ 0$

97) $2\ 3\ 0\ 0\ 0\ 0\ 0\ 0$
     $\underline{+\ 2\ 3\ 5\ 0\ 0\ 0\ 0\ 0}$
     $\ \ 4\ 6\ 5\ 0\ 0\ 0\ 0\ 0$

The total sales for Jay-Z and Earth, Wind & Fire were \$46,500,000.

99) $94 + 50 + 94 + 50 = 144 + 94 + 50$
     $ = 238 + 50$
     $ = 288$

The distance of one lap is 288 feet.

$288 + 288 + 288 = 576 + 288$
$ = 864$

The distance each player runs in making three laps is 864 feet.

101) $10 + x = 17$
      $x = 7$

Use the basic fact $10 + 7 = 17$.

103) $7 + y = 16$
      $y = 9$

Use the basic fact $7 + 9 = 16$.

105) $11 = 4 + z$
       $4 + z = 11$
       $z = 7$

Use the basic fact $4 + 7 = 11$.

107) $38 = 29 + x$
       $29 + x = 38$
       $x = 9$

Use the basic fact $29 + 9 = 38$.

109) $24 + y = 45$

Starting at 24, count by tens. Don't go over 45.
24, 34, 44
We are "1 short of 45" so we need to add 1 one.
24, 34, 44, 45
$2\,\text{tens} + 1\,\text{one} = 21$
$y = 21$

111) $29 + z = 65$

Starting at 29, count by tens. Don't go over 65.
29, 39, 49, 59
We are "6 short of 65" so we need to add 6 ones.
29, 39, 49, 59, 60, 61, 62, 63, 64, 65
$3\,\text{tens} + 6\,\text{ones} = 36$
$z = 36$

1.3   Subtracting Whole Numbers

GUIDED PRACTICE

1) Start at 10 and move 6 units to the left.

$10 - 6 = 4$

2) $4\ 7$
     $\underline{-\ 1\ 2}$
     $\ \ 3\ 5$

3) $\overset{\ \ \ \ 4}{\cancel{5}^{1}3}$
     $\underline{-\ \ \ 8}$
     $\ \ 4\ 5$

4) $\overset{\ \ \ \ 8}{3\,\cancel{9}^{1}4}$
     $\underline{-\ 2\ 6\ 8}$
     $\ \ 1\ 2\ 6$

5) $\overset{5\ \ 13}{\cancel{6}\ \cancel{4}^{1}2}$
     $\underline{-\ 3\ 4\ 5}$
     $\ \ 2\ 9\ 7$

6) $\overset{\ \ \ \ 5}{3\,\cancel{6}^{1}3\ 5}$
     $\underline{-\ \ \ 4\ 6\ 2}$
     $\ \ 3\ 1\ 7\ 3$

7) $\overset{\ \ \ 4\ \ \ 9}{8\ \cancel{5}\ \cancel{0}^{1}7}$
     $\underline{-\ 3\ \ 3\ 1\ 9}$
     $\ \ 5\ \ 1\ 8\ 8$

8) $38 - 2 - 5 = 36 - 5$
    $ = 31$

9) $47 - 5 - 6 - 9 = 42 - 6 - 9$
    $ = 36 - 9$
    $ = 27$

10) $976 - 428 - 35 = 548 - 35$
     $ = 513$

11) Asia's highest point: 29,028 feet
Australia's highest point: 7,310 feet
$29,028 - 7,310 = 21,718$ feet
The largest difference between the highest mountain in Asia and the highest mountain is Australia is 21,718 feet.

12) Subtracting 12 from both sides of the equal sign will undo the addition of 12.
$$x + 12 = 15$$
$$x + 12 - 12 = 15 - 12$$
$$x = 3$$
Replace x with 12 and use basic facts to check the answer.
$$x + 12 = 15$$
$$3 + 12 = 15$$
$$15 = 15$$

13) Adding 15 to both sides of the equal sign will undo the subtraction of 15.
$$y - 15 = 19$$
$$y - 15 + 15 = 19 + 15$$
$$y = 34$$
Replace y with 34 and use basic facts to check the answer.
$$y - 15 = 19$$
$$34 - 15 = 19$$
$$19 = 19$$

14) Total order = items ordered + items needed
The number of ordered items is $190 + 75 = 265$.
$$\text{Total order} = \text{items ordered} + x$$
$$300 = 265 + x$$
$$300 - 265 = 265 - 265 + x$$
$$35 = x$$
The store must order 35 more items to earn the discount.

CONCEPT CHECKS AND PRACTICE EXERCISES

A1) Answers may vary. Start at the first number given and move to the left the number of tick marks to be subtracted.

A2)
$4 - 2 = 2$

A3)
$6 - 3 = 3$

A4)
$15 - 8 = 7$

A5)
$14 - 6 = 8$

A6)

	7	4	8	10
5	12	9	13	15
2	9	6	10	12
4	11	8	12	14
9	16	13	17	19

A7)

	2	6	7	3
5	7	11	12	8
6	8	12	13	9
7	9	13	14	10
8	10	14	15	11

A8) a) $11 - 6 = 5$
b) $13 - 5 = 8$
c) $4 - 1 = 3$
d) $9 - 2 = 7$
e) $13 - 7 = 6$
f) $10 - 8 = 2$
g) $7 - 2 = 5$
h) $8 - 4 = 4$
i) $16 - 9 = 7$
j) $14 - 8 = 6$

A9) a) $10 - 1 = 9$
b) $15 - 9 = 6$
c) $12 - 5 = 7$
d) $13 - 9 = 4$
e) $9 - 0 = 9$
f) $13 - 6 = 7$
g) $16 - 8 = 8$
h) $12 - 4 = 8$
i) $12 - 9 = 3$
j) $9 - 7 = 2$

A10) a) $11-5=6$
      b) $13-8=5$
      c) $7-4=3$
      d) $14-6=8$
      e) $8-8=0$
      f) $15-6=9$
      g) $12-8=4$
      h) $15-8=7$
      i) $18-9=9$
      j) $11-9=2$

A11) a) $17-8=9$
      b) $6-1=5$
      c) $14-5=9$
      d) $11-8=3$
      e) $15-7=8$
      f) $10-1=9$
      g) $14-7=7$
      h) $11-4=7$
      i) $9-7=2$
      j) $13-4=9$

B1) Answers may vary. Change one $10 bill for ten
     $1 bills. Changing the bills is the same as
     borrowing from the tens place.

B2) Answers may vary. Borrowing is unnecessary
     when subtracting a smaller digit from a larger
     digit.

B3) $\begin{array}{r} 3\ 9 \\ -\ 2\ 7 \\ \hline 1\ 2 \end{array}$

B4) $\begin{array}{r} 2\ 8 \\ -\ 1\ 5 \\ \hline 1\ 3 \end{array}$

B5) $\begin{array}{r} {}^3\!\!\!\not{4}\ {}^1 0 \\ -\ 1\ 5 \\ \hline 2\ 5 \end{array}$

B6) $\begin{array}{r} {}^2\!\!\!\not{3}\ {}^1 0 \\ -\ 1\ 7 \\ \hline 1\ 3 \end{array}$

B7) $\begin{array}{r} {}^3 \\ 3\ \not{4}\ {}^1 8 \\ -\ 2\ 2\ 9 \\ \hline 1\ 1\ 9 \end{array}$

B8) $\begin{array}{r} {}^1\quad {}^{11} \\ \not{2}\ \not{2}\ {}^1 3 \\ -\ 1\ 8\ 9 \\ \hline 3\ 4 \end{array}$

B9) $\begin{array}{r} {}^3 \\ \not{4}\ {}^1 5\ 3 \\ -\ 3\ 6\ 2 \\ \hline 9\ 1 \end{array}$

B10) $\begin{array}{r} {}^6 \\ \not{7}\ {}^1 8\ 5 \\ -\ 3\ 9\ 1 \\ \hline 3\ 9\ 4 \end{array}$

B11) $\begin{array}{r} {}^4 \\ \not{5}\ {}^1 0 \\ -\ 2\ 8 \\ \hline 2\ 2 \end{array}$

Oprah will have $22 left.

B12) $\begin{array}{r} {}^1 \\ \not{2}\ {}^1 0\ 0 \\ -\ \ \ 9\ 0 \\ \hline 1\ 1\ 0 \end{array}$

Maury will have $110 left on his card.

C1) Answers may vary. Example: If you remember
     this, you can write 100 directly as 9 tens and 10
     ones.

C2) Answer may vary. Example: We borrow the
     same number of times for each problem. We
     just subtract a few more columns in the second
     problem.

C3) $\begin{array}{r} {}^5 \\ \not{6}\ {}^1 0\ 7 \\ -\ 3\ 3\ 7 \\ \hline 2\ 7\ 0 \end{array}$

C4) $\begin{array}{r} {}^3\quad {}^9 \\ \not{4}\ \not{0}\ {}^1 2 \\ -\ 1\ 7\ 4 \\ \hline 2\ 2\ 8 \end{array}$

C5) $\begin{array}{r} {}^7 \\ 1\ \not{8}\ {}^1 4\ 8 \\ -\ \ \ 7\ 5\ 3 \\ \hline 1\ 0\ 9\ 5 \end{array}$

C6)
$$\begin{array}{r} {}^6\,\,{}^{1}\\ 2\,\cancel{7}\,8\,7\\ -\ 6\,9\,7\\ \hline 2\,0\,9\,0 \end{array}$$

C7)
$$\begin{array}{r} {}^4\ \ {}^9\\ \cancel{5}\,\cancel{0}\,{}^{1}0\\ -\ 3\,5\,6\\ \hline 1\,4\,4 \end{array}$$

C8)
$$\begin{array}{r} {}^5\ \ {}^9\\ \cancel{6}\,\cancel{0}\,{}^{1}0\\ -\ 2\,7\,8\\ \hline 3\,2\,2 \end{array}$$

C9)
$$\begin{array}{r} {}^1\ {}^9\ \ {}^9\\ \cancel{2}\,\cancel{0}\,\cancel{0}\,{}^{1}0\\ -\ \ \ 5\,2\,1\\ \hline 1\,4\,7\,9 \end{array}$$

C10)
$$\begin{array}{r} {}^2\ {}^9\ {}^9\ \ {}^9\\ \cancel{3}\,\cancel{0}\,\cancel{0}\,\cancel{0}\,{}^{1}0\\ -\ \ \ 2\,3\,3\,1\\ \hline 2\,7\,6\,6\,9 \end{array}$$

C11)
$$\begin{array}{r} {}^7\\ \cancel{8}\,{}^{1}7\,5\,0\\ -\ 4\,8\,0\,0\\ \hline 3\,9\,5\,0 \end{array}$$
The dealer will give him $3,950.

C12)
$$\begin{array}{r} {}^6\\ \cancel{7}\,{}^{1}3\,9\,5\\ -\ 6\,5\,0\,0\\ \hline 8\,9\,5 \end{array}$$
Alexis owes $895 more than her car is worth.

D1) Answers may vary. Subtract the first two numbers. From that answer, subtract the next value. Continue in this manner until all numbers have been subtracted.

D2) $20 - 5 - 2 = 15 - 2$
           $= 13$

D3) $30 - 7 - 3 = 23 - 3$
           $= 20$

D4) $69 - 4 - 3 = 65 - 3$
           $= 62$

D5) $48 - 5 - 1 = 43 - 1$
           $= 42$

D6) $732 - 258 - 309 = 474 - 309$
                 $= 165$

D7) $851 - 302 - 217 = 549 - 217$
                 $= 332$

D8) $490 - 20 - 2 = 470 - 2$
               $= 468$
The balance at the end of January 1 was $468.

D9)
$$\begin{array}{r} 468\\ -\ 350\\ \hline 118 \end{array}$$
The balance at the end of January 2 was $118.

D10)
$$\begin{array}{r} 118\\ +\ 250\\ \hline 368 \end{array}$$
The balance at the end of January 4 was $368.

D11) Since the balance on January 4 was $368, there was not enough money to write a check for $369 on January 5.

SECTION 1.3   EXERCISES

For 1–12, refer to Concept Checks and Practice Exercises.

13) borrow

15) difference

17) $9 - 7 = 2$

19) $14 - 6 = 8$

21)

	3	9	7	5
8	11	17	15	13
9	12	18	16	14
7	10	16	14	12
4	7	13	11	9

23)

	5	8	9	6
4	9	12	13	10
6	11	14	15	12
7	12	15	16	13
1	6	9	10	7

25) a) $12 - 6 = 6$
     b) $15 - 8 = 7$
     c) $14 - 6 = 8$
     d) $10 - 7 = 3$
     e) $11 - 5 = 6$
     f) $8 - 6 = 2$
     g) $18 - 9 = 9$

27) a) $12 - 8 = 4$
     b) $15 - 7 = 8$
     c) $16 - 9 = 7$
     d) $13 - 8 = 5$
     e) $11 - 4 = 7$
     f) $13 - 5 = 8$
     g) $17 - 9 = 8$

29)
$$\begin{array}{r} 3\,6 \\ -\,3\,4 \\ \hline 2 \end{array}$$

31)
$$\begin{array}{r} 4 \\ \cancel{5}\,{}^1 5 \\ -\;2\,8 \\ \hline 2\,7 \end{array}$$

33)
$$\begin{array}{r} 3 \\ \cancel{4}\,{}^1 8 \\ -\;1\,9 \\ \hline 2\,9 \end{array}$$

35)
$$\begin{array}{r} 7 \\ \cancel{8}\,{}^1 3 \\ -\;7\,6 \\ \hline 7 \end{array}$$

37) $25 - 6 - 3 = 19 - 3$
     $ = 16$

39) $55 - 21 - 17 = 34 - 17$
     $ = 17$

41)
$$\begin{array}{r} 4\;\;15 \\ \cancel{5}\,\cancel{6}\,{}^1 1 \\ -\;\;7\,8 \\ \hline 4\,8\,3 \end{array}$$

43)
$$\begin{array}{r} 1\;\;15 \\ \cancel{2}\,\cancel{6}\,{}^1 5 \\ -\;\;8\,8 \\ \hline 1\,7\,7 \end{array}$$

45)
$$\begin{array}{r} 5 \\ \cancel{6}\,{}^1 4\;8 \\ -\;2\,8\,1 \\ \hline 3\,6\,7 \end{array}$$

Johann has $367 left over.

47)
$$\begin{array}{r} 6\,5\,8 \\ -\,5\,2\,3 \\ \hline 1\,3\,5 \end{array}$$

The weight of the pumpkin increased by 135 pounds.

49) $75 - 11 - 37 = 64 - 37$
     $ = 27$

51) $175 - 41 - 22 = 134 - 22$
     $ = 112$

53) $455 - 65 - 45 = 390 - 45$
     $ = 345$

55) $500 - 32 - 37 = 468 - 37$
     $ = 431$

57)
$$\begin{array}{r} 7\;\;16 \\ \cancel{8}\,\cancel{7}\,{}^1 3 \\ -\;\;1\,8\,8 \\ \hline 6\,8\,5 \end{array}$$

59)
$$\begin{array}{r} 3\;\;9 \\ \cancel{4}\,\cancel{0}\,{}^1 0 \\ -\;\;2\,7\,9 \\ \hline 1\,2\,1 \end{array}$$

61)
$$\begin{array}{r} 8\;\;13\;\;9 \\ \cancel{9}\,\cancel{4}\,\cancel{0}\,{}^1 6 \\ -\;\;1\,7\,3\,8 \\ \hline 7\,6\,6\,8 \end{array}$$

63)
$$
\begin{array}{r}
{\scriptstyle 1\ \ \,11\ \ \ \ 6} \\
\not{2}\,\not{2}\,{}^{1}0\,\not{7}\,{}^{1}5 \\
-\ 1\ 4\ 9\ 5\ 7 \\
\hline
7\ 1\ 1\ 8
\end{array}
$$

65)
$$
\begin{array}{r}
{\scriptstyle 8\ \ \ \ 4\ \ \ \ 7} \\
\not{9}\,{}^{1}7\,\not{5}\,{}^{1}4\,\not{8}\,{}^{1}2 \\
-\ \ \ 8\ 1\ 6\ 3\ 5 \\
\hline
8\ 9\ 3\ 8\ 4\ 7
\end{array}
$$

67)
$$
\begin{array}{r}
{\scriptstyle 7\ \ \ \ \ \ \ 6\ \ 12} \\
\not{8}\,{}^{1}3\ 8\ \not{7}\,\not{3}\,{}^{1}8 \\
-\ 5\ 8\ 4\ 5\ 3\ 9 \\
\hline
2\ 5\ 4\ 1\ 9\ 9
\end{array}
$$

69)
$$
\begin{array}{r}
{\scriptstyle 2} \\
\not{3}\,{}^{1}1 \\
-\ \ 6 \\
\hline
2\ 5
\end{array}
$$

71)
$$
\begin{array}{r}
6\ 3 \\
-2\ 0 \\
\hline
4\ 3
\end{array}
$$

73)
$$
\begin{array}{r}
{\scriptstyle 5} \\
\not{6}\,{}^{1}4 \\
-3\ 6 \\
\hline
2\ 8
\end{array}
$$

75)
$$
\begin{array}{r}
7\ 8 \\
-5\ 4 \\
\hline
2\ 4
\end{array}
$$

77)
$$
\begin{array}{r}
7\ 6 \\
-3\ 6 \\
\hline
4\ 0
\end{array}
$$

79)
$$
\begin{array}{r}
{\scriptstyle 6\ \ \ \ 9} \\
1\ 2\ 4\ \not{7}\,\not{0}\,{}^{1}0 \\
-1\ 2\ 4\ 3\ 4\ 7 \\
\hline
3\ 5\ 3
\end{array}
$$
Tina has driven 353 miles.

81)
$$
\begin{array}{r}
{\scriptstyle 12\ 13} \\
\not{1}\,\not{3}\,\not{4}\,{}^{1}5 \\
-6\ 8\ 9 \\
\hline
6\ 5\ 6
\end{array}
$$
Amy needs to read 656 pages to finish the book.

83)
$$
\begin{array}{r}
{\scriptstyle 10\ \ \ \ \ 6} \\
\not{1}\,\not{1}\,{}^{1}6\,\not{7}\,{}^{1}0 \\
-\ \ 9\ 8\ 6\ 6 \\
\hline
1\ 8\ 0\ 4
\end{array}
$$
Tara will save \$1,804 if she buys from the private seller.

85)
$$
\begin{aligned}
57-18-21-3 &= 39-21-3 \\
&= 18-3 \\
&= 15
\end{aligned}
$$
It took the winning team 8 hours 15 minutes to finish the leg.

87)
$$
\begin{array}{r}
{\scriptstyle 1} \\
4\ 2\ 0 \\
+\ \ 8\ 2 \\
\hline
5\ 0\ 2
\end{array}
$$
The balance after the paycheck was deposited was \$502.

89)
$$
\begin{array}{r}
{\scriptstyle 4} \\
\not{5}\,{}^{1}0\ 2 \\
-\ \ 5\ 0 \\
\hline
4\ 5\ 2
\end{array}
$$
The balance after April 15 was \$452.

91)
$$
\begin{array}{r}
{\scriptstyle 3\ \ \ \ 2} \\
1\ \not{4}\,{}^{1}4\ \not{3}\,{}^{1}3 \\
-\ \ 3\ 5\ 0\ 6 \\
\hline
1\ 0\ 9\ 2\ 7
\end{array}
$$
The difference in altitude of the highest points in Colorado and North Dakota is 10,927 feet.

93)
$$
\begin{aligned}
x-16 &= 12 \\
x-16+16 &= 12+16 \\
x &= 28
\end{aligned}
$$

95)
$$
\begin{aligned}
y-28 &= 13 \\
y-28+28 &= 13+28 \\
y &= 41
\end{aligned}
$$

97)
$$
\begin{aligned}
8 &= z-22 \\
8+22 &= z-22+22 \\
30 &= z
\end{aligned}
$$

99)
$$
\begin{aligned}
23 &= x+4 \\
23-4 &= x+4-4 \\
19 &= x
\end{aligned}
$$

101)
$$
\begin{aligned}
34 &= y-8 \\
34+8 &= y-8+8 \\
42 &= y
\end{aligned}
$$

103) $64 = z + 12$
             $64 - 12 = z + 12 - 12$
             $52 = z$

1.4   Multiplying Whole Numbers

GUIDED PRACTICE

1) Product: 63
   Factors: 7, 9
   63 is the product of 7 and 9.
   7 and 9 are factors of 63.

2) $4 \cdot 5 = 20$
   $5 \cdot 4 = 20$
   The order in which factors are multiplied does not change the result.

3) $\begin{array}{r} 3\,2 \\ \times\ 3 \\ \hline 9\,6 \end{array}$

4) $\begin{array}{r} {\scriptstyle 2} \\ 5\,3 \\ \times\ 7 \\ \hline 3\,7\,1 \end{array}$

5) $\begin{array}{r} 9\,0 \\ \times\ 6 \\ \hline 5\,4\,0 \end{array}$

6) 10,000 has four zeros.
   $10{,}000 \cdot 63 = 630{,}000$

7) 10,000,000 has seven zeros.
   $6513 \cdot 10{,}000{,}000 = 65{,}130{,}000{,}000$

8) $\begin{array}{r} {\scriptstyle \not4\,\not1} \\ {\scriptstyle \not2\,\not1} \\ 2\,7\,3 \\ \times\ 6\,5 \\ \hline 1\,3\,6\,5 \\ \underline{1\,6\,3\,8\,0} \\ 1\,7\,7\,4\,5 \end{array}$ $273 \cdot 65 = 17{,}745$

9) $\begin{array}{r} {\scriptstyle \not1} \\ {\scriptstyle \not2\,\not4} \\ {\scriptstyle \not2\,\not4} \\ 9\,3\,6 \\ \times\,2\,8\,7 \\ \hline 6\,5\,5\,2 \\ 7\,4\,8\,8\,0 \\ \underline{1\,8\,7\,2\,0\,0} \\ 2\,6\,8\,6\,3\,2 \end{array}$ $936 \cdot 287 = 268{,}632$

10) $212 \approx 200$
    $285 \approx 300$
    $212 \cdot 285 \approx 200 \cdot 300$
    $ = 60{,}000$

11) $1{,}183 \approx 1{,}000$
    $299 \approx 300$
    $1{,}183 \cdot 299 \approx 1{,}000 \cdot 300$
    $\phantom{1{,}183 \cdot 299} = 300{,}000$

12) $5 \cdot 3 \cdot 7 \cdot 2 = 15 \cdot 7 \cdot 2$
    $ = 105 \cdot 2$
    $ = 210$

13) $2 \cdot 12 \cdot 3 \cdot 5 \cdot 2 = 24 \cdot 3 \cdot 5 \cdot 2$
    $ = 72 \cdot 5 \cdot 2$
    $ = 360 \cdot 2$
    $ = 720$

14) $A = l \cdot w$
    $ = 15 \cdot 16$
    $ = 240$ square inches

15) $8 \cdot x = 48$
    $x = 6$
    Basic Fact: $8 \cdot 6 = 48$

16) Repeatedly add 41 until we reach 205.
    41, 82, 123, 164, 205
    Five 41's add to 205.
    $41 \cdot y = 205$
    $y = 5$

CONCEPT CHECKS AND PRACTICE EXERCISES

A1) Answers may vary. Example: Multiplication is a quick way to perform repeated addition.

A2) $2 \cdot 5 = 10$
    $5 \cdot 2 = 10$

A3) Factors: 3, 4
    Product: 12

A4) $3 \cdot 5 = 5 + 5 + 5 = 15$

A5) $4 \cdot 7 = 7 + 7 + 7 + 7 = 28$

A6) $3 \cdot 6 = 6 + 6 + 6 = 18$

A7) $4 \cdot 8 = 8 + 8 + 8 + 8 = 32$

A8)

×	1	2	3	4	5	6	7	8	9	10
1	1	2	3	4	5	6	7	8	9	10
2	2	4	6	8	10	12	14	16	18	20
3	3	6	9	12	15	18	21	24	27	30
4	4	8	12	16	20	24	28	32	36	40
5	5	10	15	20	25	30	35	40	45	50
6	6	12	18	24	30	36	42	48	54	60
7	7	14	21	28	35	42	49	56	63	70
8	8	16	24	32	40	48	56	64	72	80
9	9	18	27	36	45	54	63	72	81	90
10	10	20	30	40	50	60	70	80	90	100

A9) a) $0 \cdot 5 = 0$
      b) $6 \cdot 8 = 48$
      c) $9 \cdot (6) = 54$
      d) $1 \times 3 = 3$
      e) $5 \cdot 6 = 30$
      f) $7 \cdot 6 = 42$
      g) $4 \cdot 7 = 28$
      h) $10 \times 4 = 40$
      i) $8(8) = 64$
      j) $4 \cdot 4 = 16$

A10) a) $4 \cdot 1 = 4$
       b) $6 \times 9 = 54$
       c) $3 \cdot 0 = 0$
       d) $7(7) = 49$
       e) $5 \times 10 = 50$
       f) $8 \cdot 7 = 56$
       g) $5 \cdot 9 = 45$
       h) $4 \cdot (5) = 20$
       i) $9 \cdot 8 = 72$
       j) $3 \cdot 2 = 6$

A11) a) $4 \cdot 3 = 12$
       b) $10 \cdot 3 = 30$
       c) $8 \times 9 = 72$
       d) $4 \cdot 8 = 32$
       e) $5 \cdot (7) = 35$
       f) $9 \cdot 7 = 63$
       g) $0 \cdot 10 = 0$
       h) $7(8) = 56$
       i) $3 \times 6 = 18$
       j) $6 \cdot 7 = 42$

A12) a) $3 \times 3 = 9$
       b) $8 \cdot 6 = 48$
       c) $5 \cdot 8 = 40$
       d) $0(0) = 0$
       e) $7 \cdot 9 = 63$
       f) $6 \times 6 = 36$
       g) $4 \cdot 10 = 40$
       h) $4 \cdot (9) = 36$
       i) $9 \cdot 9 = 81$
       j) $4 \cdot 2 = 8$

B1) Answers may vary. When multiplying $4 by 9, the answer is $36, or 3 tens and 6 ones. The 6 goes into the ones column and the three must be 'carried' into the tens column.

B2) Answers may vary. We may need to carry digits to the next higher place value.

B3)
$$\begin{array}{r} 41 \\ \times\ 2 \\ \hline 82 \end{array}$$

B4)
$$\begin{array}{r} 13 \\ \times\ 3 \\ \hline 39 \end{array}$$

B5)
$$\begin{array}{r} {}^{3} \\ 16 \\ \times\ 5 \\ \hline 80 \end{array}$$

B6)
$$\begin{array}{r} {}^{3} \\ 15 \\ \times\ 6 \\ \hline 90 \end{array}$$

B7)
$$\begin{array}{r} {}^{5} \\ 98 \\ \times\ 7 \\ \hline 686 \end{array}$$

B8)
$$\begin{array}{r} {}^{5} \\ 97 \\ \times\ 8 \\ \hline 776 \end{array}$$

B9)
$$\begin{array}{r} 70 \\ \times\ 6 \\ \hline 420 \end{array}$$

B10)
$$\begin{array}{r} 60 \\ \times\ 7 \\ \hline 420 \end{array}$$

B11)   ₆
       2 8
      × 8
      ‾‾‾‾
      2 2 4
Bianca will earn $224.

B12)   ₃
       1 4
      × 8
      ‾‾‾‾
      1 1 2
James will earn $112.

C1) Answers may vary. $3 \cdot 100$ gives three hundreds, so appending two zeros is a quick way to multiply by 100.

C2) The Commutative Property of Multiplication states that the order in which two numbers are multiplied does not change the product, so $38 \cdot 1,000 = 1,000 \cdot 38$.

C3) 10 has one zero.
     $14 \cdot 10 = 140$

C4) 10 has one zero.
     $23 \cdot 10 = 230$

C5) 1,000 has three zeros.
     $523 \cdot 1000 = 523,000$

C6) 100 has two zeros.
     $12 \cdot 100 = 1,200$

C7) 100 has two zeros.
     $100 \cdot 73 = 7,300$

C8) 1,000 has three zeros.
     $1,000 \cdot 879 = 879,000$

C9) 10,000 has four zeros.
     $10,000 \cdot 987 = 9,870,000$

C10) 100,000 has five zeros.
     $100,000 \cdot 87 = 8,700,000$

D1) Answers may vary. In multiplying 34 by 12, we need to multiply $1 \cdot 34$ and $2 \cdot 34$.

D2) Answers may vary. In multiplying 34 by 12, we must add the products of $1 \cdot 34$ and $2 \cdot 34$, making sure to align the place values correctly.

D3)   4 2 3
       × 1 2
       ‾‾‾‾‾‾
         8 4 6
       4 2 3 0
       ‾‾‾‾‾‾
       5 0 7 6

D4)   1 4 2
       × 2 1
       ‾‾‾‾‾‾
         1 4 2
       2 8 4 0
       ‾‾‾‾‾‾
       2 9 8 2

D5)   ₅ ₄
         3 6 5
       × 1 8
       ‾‾‾‾‾‾
       2 9 2 0
       3 6 5 0
       ‾‾‾‾‾‾
       6 5 7 0

D6)   ₁ ₁
         7 3 4
       × 2 4
       ‾‾‾‾‾‾
       2 9 3 6
       1 4 6 8 0
       ‾‾‾‾‾‾‾
       1 7 6 1 6

D7)   ₁
         ₁
         1 5 2
       × 3 1 2
       ‾‾‾‾‾‾‾
         3 0 4
       1 5 2 0
       4 5 6 0 0
       ‾‾‾‾‾‾‾
       4 7 4 2 4

D8)   ₄ ₂
         ₁ ₁
         1 6 4
       × 7 3 1
       ‾‾‾‾‾‾‾
         1 6 4
       4 9 2 0
       1 1 4 8 0 0
       ‾‾‾‾‾‾‾‾‾
       1 1 9 8 8 4

D9)   ₃ ₂
         ₁
         ₃ ₂
         9 4 3
       × 7 3 9
       ‾‾‾‾‾‾‾
       8 4 8 7
       2 8 2 9 0
       6 6 0 1 0 0
       ‾‾‾‾‾‾‾‾‾
       6 9 6 8 7 7

D10)
$$
\begin{array}{r}
\not{1}\ \not{1} \\
\not{6}\ \not{1} \\
\not{3}\ \not{2} \\
1\,8\,6 \\
\times 2\,7\,4 \\
\hline
7\,4\,4 \\
1\,3\,0\,2\,0 \\
\underline{3\,7\,2\,0\,0} \\
5\,0\,9\,6\,4
\end{array}
$$

D11)
$$
\begin{array}{r}
\not{2}\ \not{1} \\
\not{6}\ \not{6} \\
1\,9\,5 \\
\times\ 3\,7 \\
\hline
1\,3\,6\,5 \\
\underline{5\,8\,5\,0} \\
7\,2\,1\,5
\end{array}
$$

The contestant consumes 7,215 calories.

D12)
$$
\begin{array}{r}
\not{4}\ \not{2} \\
\not{2}\ \not{1} \\
1\,9\,5 \\
\times\ 5\,3 \\
\hline
5\,8\,5 \\
\underline{9\,7\,5\,0} \\
1\,0\,3\,3\,5
\end{array}
$$

Kobayashi consumed 10,335 calories.

FOCUS ON ESTIMATION PRACTICE EXERCISES

1) $65 \approx 70$
$73 \approx 70$
$65 \cdot 73 \approx 70 \cdot 70$
$\quad\quad = 4{,}900$
Answer of 4,745 appears to be correct.

2) $58 \approx 60$
$93 \approx 90$
$58 \cdot 93 \approx 60 \cdot 90$
$\quad\quad = 5{,}400$
Answer of 5,394 appears to be correct.

3) $1{,}258 \approx 1{,}000$
$\quad 68 \approx 70$
$1{,}258 \cdot 68 \approx 1{,}000 \cdot 70$
$\quad\quad\quad = 70{,}000$
Answer of 58,544 appears to be incorrect.

$1{,}258 \cdot 68 = 85{,}544$

$$
\begin{array}{r}
\not{1}\ \not{3}\ \not{4} \\
\not{2}\ \not{1}\ \not{6} \\
1\,2\,5\,8 \\
\times\ 6\,8 \\
\hline
1\,0\,0\,6\,4 \\
\underline{7\,5\,4\,8\,0} \\
8\,5\,5\,4\,4
\end{array}
$$

4) $1{,}893 \approx 2{,}000$
$\quad 42 \approx 40$
$1{,}893 \cdot 42 \approx 2{,}000 \cdot 40$
$\quad\quad\quad = 80{,}000$
Answer of 97,506 appears to be incorrect.

$1{,}893 \cdot 42 = 79{,}506$

$$
\begin{array}{r}
\not{3}\ \not{3}\ \not{1} \\
\not{1}\ \not{1} \\
1\,8\,9\,3 \\
\times\ 4\,2 \\
\hline
3\,7\,8\,6 \\
\underline{7\,5\,7\,2\,0} \\
7\,9\,5\,0\,6
\end{array}
$$

5) $359 \approx 400$
$\quad 34 \approx 30$
$359 \cdot 34 \approx 400 \cdot 30$
$\quad\quad = 12{,}000$
Answer of 21,306 appears to be incorrect.

$359 \cdot 34 = 12{,}206$

$$
\begin{array}{r}
\not{1}\ \not{2} \\
\not{2}\ \not{2} \\
3\,5\,9 \\
\times\ 3\,4 \\
\hline
1\,4\,3\,6 \\
\underline{1\,0\,7\,7\,0} \\
1\,2\,2\,0\,6
\end{array}
$$

6) $483 \approx 500$
$\quad 67 \approx 70$
$483 \cdot 67 \approx 500 \cdot 70$
$\quad\quad = 35{,}000$
Answer of 29,306 appears to be incorrect.

$483 \cdot 67 = 32{,}361$

$$
\begin{array}{r}
\not{4}\ \not{1} \\
\not{2}\ \not{2} \\
4\,8\,3 \\
\times\ 6\,7 \\
\hline
3\,3\,8\,1 \\
\underline{2\,8\,9\,8\,0} \\
3\,2\,3\,6\,1
\end{array}
$$

7) $4{,}863 \approx 5{,}000$
$\quad 192 \approx 200$
$4{,}863 \cdot 192 \approx 5{,}000 \cdot 200$
$\quad\quad\quad = 1{,}000{,}000$
Answer of 830,466 appears to be incorrect.

$4{,}863 \cdot 192 = 933{,}696$

$$
\begin{array}{r}
\not{2}\ \not{2}\ \not{1} \\
\not{1}\ \not{1} \\
4\,8\,6\,3 \\
\times\,1\,9\,2 \\
\hline
9\,7\,2\,6 \\
4\,3\,7\,6\,7\,0 \\
\underline{4\,8\,6\,3\,0\,0} \\
9\,3\,3\,6\,9\,6
\end{array}
$$

8) $7,412 \approx 7,000$
     $967 \approx 1,000$
     $7,412 \cdot 967 \approx 7,000 \cdot 1,000$
     $ = 7,000,000$
     Answer of 716,404 appears to be incorrect.

$$
\begin{array}{r}
{\scriptstyle \cancel{3}\,\cancel{1}\,\cancel{1}} \qquad 7,412 \cdot 967 = 7,167,404 \\
{\scriptstyle \cancel{2}\,\,\cancel{1}} \\
{\scriptstyle \cancel{2}\,\,\cancel{1}} \\
7\,4\,1\,2 \\
\times 9\,6\,7 \\
\hline
5\,1\,8\,8\,4 \\
4\,4\,4\,7\,2\,0 \\
6\,6\,7\,0\,8\,0\,0 \\
\hline
7\,1\,6\,7\,3\,0\,4
\end{array}
$$

E1) Answers may vary. Multiply the first pair of factors. Multiply this product by the next factor. Continue until you have multiplied by all the factors.

E2) $2 \cdot 9 \cdot 5 = 18 \cdot 5$
     $ = 90$

E3) $5 \cdot 7 \cdot 2 = 35 \cdot 2$
     $ = 70$

E4) $2 \cdot 9 \cdot 2 \cdot 3 = 18 \cdot 2 \cdot 3$
     $ = 36 \cdot 3$
     $ = 108$

E5) $8 \cdot 3 \cdot 7 \cdot 2 = 24 \cdot 7 \cdot 2$
     $ = 168 \cdot 2$
     $ = 336$

E6) $14 \cdot 6 \cdot 5 = 84 \cdot 5$
     $ = 420$

E7) $16 \cdot 7 \cdot 5 = 112 \cdot 5$
     $ = 560$

E8) $2 \cdot 8 \cdot 3 \cdot 3 \cdot 2 = 16 \cdot 3 \cdot 3 \cdot 2$
     $ = 48 \cdot 3 \cdot 2$
     $ = 144 \cdot 2$
     $ = 288$

E9) $3 \cdot 4 \cdot 2 \cdot 9 \cdot 3 = 12 \cdot 2 \cdot 9 \cdot 3$
     $ = 24 \cdot 9 \cdot 3$
     $ = 216 \cdot 3$
     $ = 648$

SECTION 1.4   EXERCISES

For 1–17, refer to Concept Checks and Practice Exercises.

19) factors

21) factor; factor; product

23) $2 \cdot 4 = 4 + 4 = 8$

25) $4 \cdot 2 = 2 + 2 + 2 + 2 = 8$

27)

×	7	8	4	3
5	35	40	20	15
8	56	64	32	24
6	42	48	24	18
9	63	72	36	27

29) a) $6 \cdot 6 = 36$
     b) $3 \cdot (6) = 18$
     c) $9 \times 4 = 36$
     d) $9 \cdot 7 = 63$
     e) $4(7) = 28$
     f) $8 \times 9 = 72$
     g) $6 \cdot 5 = 30$

31) a) $5 \cdot 4 = 20$
     b) $(4) \cdot 6 = 24$
     c) $7 \times 8 = 56$
     d) $6 \cdot 3 = 18$
     e) $8(4) = 32$
     f) $9 \times 9 = 81$
     g) $8 \cdot 5 = 40$

33) $\begin{array}{r} {\scriptstyle 1} \\ 2\,3 \\ \times\ 4 \\ \hline 9\,2 \end{array}$

35) $\begin{array}{r} {\scriptstyle 1} \\ 4\,2 \\ \times\ 7 \\ \hline 2\,9\,4 \end{array}$

37) $\begin{array}{r} {\scriptstyle 2\ 1} \\ 5\,4\,3 \\ \times\ \ \,5 \\ \hline 2\,7\,1\,5 \end{array}$

39)
```
  1 1
  7 4 5
×     3
2 2 3 5
```

41)
```
  1 1
2 4 6 3
×     3
7 3 8 9
```
Her total ATV sales were $7,389.

43)
```
    2 3 1
  5 1 4 7 2
×         5
2 5 7 3 6 0
```
The total amount collected was $257,360.

45) 10 has 1 zero.
$84 \cdot 10 = 840$

47) 100 has 2 zeros.
$854 \cdot 100 = 85,400$

49) 1,000 has 3 zeros.
$561 \cdot 1,000 = 561,000$

51) 100 has 2 zeros.
$5,789 \cdot 100 = 578,900$

53)
```
  1
  4 5
×2 0
9 0 0
```

55)
```
        1
      9 2 6
×   3 0 0
2 7 7 8 0 0
```

57)
```
    4 0 0
×   3 0 0
1 2 0 0 0 0
```
The total amount given is $120,000.

59)
```
    5 0 0 0
×     5 2
  1 0 0 0 0
2 5 0 0 0 0
2 6 0 0 0 0
```
$260,000 was given to the scholarship fund.

61)
```
  ⁵
  8 4
×2 8
6 7 2
1 6 8 0
2 3 5 2
```

63)
```
  ⁵
  ⁵
  5 7
×7 8
4 5 6
3 9 9 0
4 4 4 6
```

65) $8 \cdot 7 \cdot 4 = 56 \cdot 4$
$= 224$

67) $12 \cdot 14 \cdot 13 = 168 \cdot 13$
$= 2,184$

69)
```
  1 1
    1
  3 3 2
×   5 4
  1 3 2 8
1 6 6 0 0
1 7 9 2 8
```

71)
```
    1
    6 1
  7 4 2
×   3 9
  6 6 7 8
2 2 2 6 0
2 8 9 3 8
```

73)
```
  2 1 1
  4 1 1
  1 5 2 2
×   5 8 1
  1 5 2 2
1 2 1 7 6 0
7 6 1 0 0 0
8 8 4 2 8 2
```
A total of $884,282 was donated.

75)
```
    5 3 4
    4 4 4
  3 6 9 8
×   8 0 0 5
  1 8 4 9 0
2 9 5 8 4 0 0 0
2 9 6 0 2 4 9 0
```

77)
```
      ⅄⅄
      ⅄⅄
      ⅄⅄
      465
    ×657
    3255
   23250
  279000
  305505
```

79)
```
      ⅄⅄
      ⅄⅄
      ⅄⅄
      765
    ×478
    6120
   53550
  306000
  365670
```

81)
```
     ⅟⅟
      ⅟
      245
    ×321
      245
     4900
    73500
    78645
```

83)
```
      ⅄⅄
      ⅄⅄
      293
    ×405
     1465
   117200
   118665
```

85) $A = l \cdot w$
    $= 10 \cdot 8$
    $= 80$ square inches

87) $A = l \cdot w$
    $= 2,029 \cdot 167$
    $= 338,843$ square feet

89)
```
      ⅟
      ⅟
      35
    ×46
     210
    1400
    1610
```

91)
```
      54
    ×53
     162
    2700
    2862
```

93)
```
     1 1
     123
    ×  6
     738
```

95)
```
      31
    ×15
     155
     310
     465
```
Juan earned $465.

97)
```
     3
     850
    ×  6
    5100
```
Mia will receive $5,100 from the rental.

99)
```
      90
    × 90
    8100
```
The area of the infield is 8,100 square feet.

101)
```
      42
    ×13
     126
     420
     546
```
Terrell can travel 546 miles.

103) a)
```
      250
    × 14
     1000
     2500
     3500
```
The total of all payments was $3,500.

$9,000 - 3,500 = 5,500$
$5,500 was still owed on the car.

b) Since $5,200 is less than $5,500, the car is worth less than the amount owed on it.

105) a)
$$\begin{array}{r} 9\,4 \\ \times\ 5\,0 \\ \hline 4\,7\,0\,0 \end{array}$$

The area of the court is 4,700 square feet.

b) $400 \cdot 10 = 4{,}000$ square feet

c) 4,000 square feet is less than 4,700 square feet, so 10 gallons is not enough.

107) $8 \cdot 45 + 2 \cdot 35 + 6 \cdot 70 = 360 + 70 + 420$
$$= 430 + 420$$
$$= 850$$
Herman will charge $850.

109) $5 \cdot x = 15$
$$x = 3$$
Basic Fact: $5 \cdot 3 = 15$

111) $8 \cdot y = 32$
$$y = 4$$
Basic Fact: $8 \cdot 4 = 32$

113) $80 = 10 \cdot z$
$$10 \cdot z = 80$$
$$z = 8$$
Basic Fact: $10 \cdot 8 = 80$

115) Repeatedly add 220 until we reach 660.
220, 440, 660
Three 220's add to 660.
$$660 = 220 \cdot x$$
$$220 \cdot x = 660$$
$$x = 3$$

1.5   Dividing Whole Numbers

GUIDED PRACTICE

1) $10 \div 5 = \boxed{?} \ \Rightarrow \ \boxed{?} \cdot 5 = 10$

2) $24 \div 6 = \boxed{?} \ \Rightarrow \ \boxed{?} \cdot 6 = 24$
$$4 \cdot 6 = 24$$
Therefore, $24 \div 6 = 4$.

3) $12 \div 3 = 4 \Rightarrow 3\overline{)12}$ with quotient 4
The dividend is 12.
The divisor is 3.
The quotient is 4.

4) 6 goes into 29 four times.
$$\begin{array}{l} 6 \cdot 4 = 24 \\ 6 \cdot 5 = 30 \end{array} \to 29 \ \Rightarrow \ 6\overline{)29}$$

$$\begin{array}{r} 4 \\ 6\overline{)29} \\ -24 \\ \hline 5 \end{array}$$

Because 6 does not go into 5, the remainder is 5. To show that 6 goes into 29 four times with a remainder of 5, we write $29 \div 6 = 4\mathrm{R}5$.

5)
$$\begin{array}{r} 2\ 4\ \mathrm{R}1 \\ 3\overline{)7\ 3} \\ -6 \\ \hline 1\ 3 \\ -1\ 2 \\ \hline 1 \end{array}$$

6)
$$\begin{array}{r} 7\ 6\ \mathrm{R}3 \\ 6\overline{)4\ 5\ 9} \\ -4\ 2 \\ \hline 3\ 9 \\ -3\ 6 \\ \hline 3 \end{array}$$

7) For $17 \div 3 = 5\mathrm{R}2$, identify the quotient, remainder, and divisor.
Quotient = 5
Remainder = 2
Divisor = 3
Write the mixed number.
$$\text{quotient}\ \frac{\text{remainder}}{\text{divisor}} = 5\frac{2}{3}$$

8) Divide the dividend by the divisor.
$$25 \div 6 = 4\mathrm{R}1$$
Quotient = 4
Remainder = 1
Divisor = 6
Write the mixed number.
$$\text{quotient}\ \frac{\text{remainder}}{\text{divisor}} = 4\frac{1}{6}$$

9) $11 \approx 10$ (Rounded to the tens place.)
$193 \approx 200$ (Rounded to the hundreds place.)
$$\begin{array}{r} 20 \\ 10\overline{)200} \\ -20 \\ \hline 00 \end{array}$$
$$193 \div 11 \approx 20$$

10) $31 \approx 30$ (Rounded to the tens place.)
       $571 \approx 600$ (Rounded to the hundreds place.)

$$
\begin{array}{r}
20 \\
30)\overline{600} \\
-60 \\
\hline
00
\end{array}
$$

$571 \div 31 \approx 20$

11)

$$
\begin{array}{r}
1\,7 \;\; \text{R15} \\
2\,1)\overline{3\,7\,2} \\
-2\,1 \\
\hline
1\,6\,2 \\
-1\,4\,7 \\
\hline
1\,5
\end{array}
$$

12)

$$
\begin{array}{r}
3\,8 \\
1\,2)\overline{4\,5\,6} \\
-3\,6 \\
\hline
9\,6 \\
-9\,6 \\
\hline
0
\end{array}
$$

13)

$$
\begin{array}{r}
3\,8\,5 \;\; \text{R2} \\
2\,2)\overline{8\,4\,7\,2} \\
-6\,6 \\
\hline
1\,8\,7 \\
-1\,7\,6 \\
\hline
1\,1\,2 \\
-1\,1\,0 \\
\hline
2
\end{array}
$$

14) $42 \div y = 7 \;\Rightarrow\; 7 \cdot y = 42$
                              $y = 6$

15) $65 \div x = 13 \;\Rightarrow\; 13 \cdot x = 65$
       Multiples of 13 are 13, 26, 39, 52, 65.
       Five 13's add to 65.
       $x = 5$

CONCEPT CHECKS AND PRACTICE EXERCISES

A1) $\boxed{?} \cdot 9 = 45$

A2)   Answers may vary. There is no multiplication
       problem for which $0 \cdot \boxed{?} = 6$.

A3) $30 \div 3 = \boxed{?} \;\Rightarrow\; \boxed{?} \cdot 3 = 30$
                              $10 \cdot 3 = 30$
       Therefore, $30 \div 3 = 10$.

A4) $40 \div 4 = \boxed{?} \;\Rightarrow\; \boxed{?} \cdot 4 = 40$
                              $10 \cdot 4 = 40$
       Therefore, $40 \div 4 = 10$.

A5) $60 \div 10 = \boxed{?} \;\Rightarrow\; \boxed{?} \cdot 10 = 60$
                              $6 \cdot 10 = 60$
       Therefore, $60 \div 10 = 6$.

A6) $48 \div 8 = \boxed{?} \;\Rightarrow\; \boxed{?} \cdot 8 = 48$
                              $6 \cdot 8 = 48$
       Therefore, $48 \div 8 = 6$.

A7) $63 \div 7 = \boxed{?} \;\Rightarrow\; \boxed{?} \cdot 7 = 63$
                              $9 \cdot 7 = 63$
       Therefore, $63 \div 7 = 9$.

A8) $54 \div 6 = \boxed{?} \;\Rightarrow\; \boxed{?} \cdot 6 = 54$
                              $9 \cdot 6 = 54$
       Therefore, $54 \div 6 = 9$.

A9) $80 \div 20 = \boxed{?} \;\Rightarrow\; \boxed{?} \cdot 20 = 80$
                              $4 \cdot 20 = 80$
       Therefore, $80 \div 20 = 4$.

A10) $100 \div 50 = \boxed{?} \;\Rightarrow\; \boxed{?} \cdot 50 = 100$
                              $2 \cdot 50 = 100$
       Therefore, $100 \div 50 = 2$.

A11)

×	7	6	8	2
5	35	30	40	10
9	63	54	72	18
6	42	36	48	12
4	28	24	32	8

A12)

×	3	5	9	6
9	27	45	81	54
7	21	35	63	42
4	12	20	36	24
8	24	40	72	48

A13) a) $10 \div 5 = 2$
       b) $48 \div 8 = 6$
       c) $\dfrac{54}{6} = 9$
       d) $3 \div 3 = 1$
       e) $6 \overline{)30}$ with 5 above
       f) $42 \div 6 = 7$
       g) $\dfrac{28}{7} = 4$
       h) $40 \div 0$; undefined
       i) $64 \div 8 = 8$
       j) $4 \overline{)16}$ with 4 above

A14) a) $4 \div 1 = 4$
       b) $54 \div 9 = 6$
       c) $\dfrac{0}{3} = 0$
       d) $49 \div 7 = 7$
       e) $0 \overline{)7}$; undefined
       f) $56 \div 7 = 8$
       g) $\dfrac{54}{9} = 6$
       h) $20 \div 5 = 4$
       i) $72 \div 8 = 9$
       j) $2 \overline{)6}$ with 3 above

A15) a) $12 \div 3 = 4$
       b) $30 \div 3 = 10$
       c) $\dfrac{72}{9} = 8$
       d) $32 \div 8 = 4$
       e) $7 \overline{)35}$ with 5 above
       f) $63 \div 7 = 9$
       g) $\dfrac{0}{7} = 0$
       h) $56 \div 8 = 7$
       i) $6 \div 0$; undefined
       j) $7 \overline{)42}$ with 6 above

A16) a) $9 \div 3 = 3$
       b) $48 \div 6 = 8$
       c) $\dfrac{40}{8} = 5$
       d) $25 \div 5 = 5$
       e) $9 \overline{)63}$ with 7 above
       f) $36 \div 6 = 6$
       g) $\dfrac{40}{10} = 4$
       h) $36 \div 9 = 4$
       i) $81 \div 9 = 9$
       j) $0 \overline{)8}$; undefined

B1) Answers may vary. Example: It will help keep the digits aligned.

B2) Answers may vary. Example: The remainder is what is left over after subtracting the divisor as many times as possible.

B3)
$$
\begin{array}{r}
1\ 8\ \text{R}3 \\
4\overline{)\ 7\ 5} \\
-\ 4 \\
\hline
3\ 5 \\
-\ 3\ 2 \\
\hline
3
\end{array}
$$

B4)
$$
\begin{array}{r}
4\ 1\ \text{R}1 \\
2\overline{)\ 8\ 3} \\
-\ 8 \\
\hline
0\ 3 \\
-\ 2 \\
\hline
1
\end{array}
$$

B5)
$$
\begin{array}{r}
8\ \text{R}3 \\
8\overline{)\ 6\ 7} \\
-\ 6\ 4 \\
\hline
3
\end{array}
$$

B6)
$$
\begin{array}{r}
6\ \text{R}6 \\
7\overline{)\ 4\ 8} \\
-\ 4\ 2 \\
\hline
6
\end{array}
$$

B7)
$$\begin{array}{r} 1\ 6\ 3\,\text{R2} \\ 4\overline{)\ 6\ 5\ 4} \\ \underline{-\ 4} \\ 2\ 5 \\ \underline{-\ 2\ 4} \\ 1\ 4 \\ \underline{-\ 1\ 2} \\ 2 \end{array}$$

B8)
$$\begin{array}{r} 2\ 8\ 1\,\text{R2} \\ 3\overline{)\ 8\ 4\ 5} \\ \underline{-\ 6} \\ 2\ 4 \\ \underline{-\ 2\ 4} \\ 0\ 5 \\ \underline{-\ 3} \\ 2 \end{array}$$

B9)
$$\begin{array}{r} 8\ 7\,\text{R1} \\ 6\overline{)\ 5\ 2\ 3} \\ \underline{-\ 4\ 8} \\ 4\ 3 \\ \underline{-\ 4\ 2} \\ 1 \end{array}$$

B10)
$$\begin{array}{r} 8\ 9\,\text{R2} \\ 7\overline{)\ 6\ 2\ 5} \\ \underline{-\ 5\ 6} \\ 6\ 5 \\ \underline{-\ 6\ 3} \\ 2 \end{array}$$

B11)
$$\begin{array}{r} 3\ 0\ 0 \\ 5\overline{)\ 1\ 5\ 0\ 0} \\ \underline{-\ 1\ 5} \\ 0\ 0\ 0 \end{array}$$

Each player will get $300.

B12)
$$\begin{array}{r} 2\ 5\ 0 \\ 6\overline{)\ 1\ 5\ 0\ 0} \\ \underline{-\ 1\ 2} \\ 3\ 0 \\ \underline{-\ 3\ 0} \\ 0\ 0 \end{array}$$

Each person will get $250.

FOCUS ON THE MEANING OF REMAINDERS PRACTICE EXERCISES

1) $14 \div 3 = 4\text{R2}$
    Quotient: 4
    Remainder: 2
    Divisor: 3
    $$\text{quotient}\,\frac{\text{remainder}}{\text{divisor}} = 4\frac{2}{3}$$

2) $21 \div 8 = 2\text{R5}$
    Quotient: 2
    Remainder: 5
    Divisor: 8
    $$\text{quotient}\,\frac{\text{remainder}}{\text{divisor}} = 2\frac{5}{8}$$

3) $26 \div 3 = 8\text{R2}$
    $$\text{quotient}\,\frac{\text{remainder}}{\text{divisor}} = 8\frac{2}{3}$$

4) $17 \div 8 = 2\text{R1}$
    $$\text{quotient}\,\frac{\text{remainder}}{\text{divisor}} = 2\frac{1}{8}$$

5) $32 \div 5 = 6\text{R2}$
    $$\text{quotient}\,\frac{\text{remainder}}{\text{divisor}} = 6\frac{2}{5}$$

6) $29 \div 7 = 4\text{R1}$
    $$\text{quotient}\,\frac{\text{remainder}}{\text{divisor}} = 4\frac{1}{7}$$

7) $41 \div 5 = 8\text{R1}$
    $$\text{quotient}\,\frac{\text{remainder}}{\text{divisor}} = 8\frac{1}{5}$$

8) $53 \div 7 = 7\text{R4}$
    $$\text{quotient}\,\frac{\text{remainder}}{\text{divisor}} = 7\frac{4}{7}$$

C1) Answers will vary. Estimating can help determine whether or not the answer is correct.

C2) a) $4{,}832 \approx 5{,}000$
     b) $92{,}387 \approx 90{,}000$
     c) $23 \approx 20$
     d) $850 \approx 900$

C3) $16 \approx 20$ (Rounded to the tens place.)
$387 \approx 400$ (Rounded to the hundreds place.)

$$\begin{array}{r} 20 \\ 20\overline{)400} \\ -40 \\ \hline 00 \end{array}$$

$387 \div 16 \approx 20$

C4) $12 \approx 10$ (Rounded to the tens place.)
$212 \approx 200$ (Rounded to the hundreds place.)

$$\begin{array}{r} 20 \\ 10\overline{)200} \\ -20 \\ \hline 00 \end{array}$$

$212 \div 12 \approx 20$

C5) $48 \approx 50$ (Rounded to the tens place.)
$973 \approx 1,000$ (Rounded to the hundreds place.)

$$\begin{array}{r} 20 \\ 50\overline{)1000} \\ -100 \\ \hline 00 \end{array}$$

$973 \div 48 \approx 20$

C6) $33 \approx 30$ (Rounded to the tens place.)
$912 \approx 900$ (Rounded to the hundreds place.)

$$\begin{array}{r} 30 \\ 30\overline{)900} \\ -90 \\ \hline 00 \end{array}$$

$912 \div 33 \approx 30$

C7) $17 \approx 20$ (Rounded to the tens place.)
$5,342 \approx 5,000$ (Rounded to the thousands place.)

$$\begin{array}{r} 250 \\ 20\overline{)5000} \\ -40 \\ \hline 100 \\ -100 \\ \hline 00 \end{array}$$

$5,342 \div 17 \approx 250$

C8) $23 \approx 20$ (Rounded to the tens place.)
$7,921 \approx 8,000$ (Rounded to the thousands place.)

$$\begin{array}{r} 400 \\ 20\overline{)8000} \\ -80 \\ \hline 000 \end{array}$$

$7,921 \div 23 \approx 400$

C9) $58 \approx 60$ (Rounded to the tens place.)
$6,185 \approx 6,000$ (Rounded to the thousands place.)

$$\begin{array}{r} 100 \\ 60\overline{)6000} \\ -60 \\ \hline 000 \end{array}$$

$6,185 \div 58 \approx 100$

C10) $95 \approx 100$ (Rounded to the tens place.)
$3,182 \approx 3,000$ (Rounded to the thousands place.)

$$\begin{array}{r} 30 \\ 100\overline{)3000} \\ -300 \\ \hline 00 \end{array}$$

$3,182 \div 95 \approx 30$

C11) $23 \approx 20$ (Rounded to the tens place.)
$9,400 \approx 9,000$ (Rounded to the thousands place.)

$$\begin{array}{r} 450 \\ 20\overline{)9000} \\ -80 \\ \hline 100 \\ -100 \\ \hline 00 \end{array}$$

Each person will move about 450 pounds of rock.

C12) $27 \approx 30$ (Rounded to the tens place.)
$9,400 \approx 9,000$ (Rounded to the thousands place.)

$$\begin{array}{r} 300 \\ 30\overline{)9000} \\ -90 \\ \hline 000 \end{array}$$

Each person will move about 300 pounds of rock.

D1) $31 \approx 30$ (Rounded to the tens place.)
$271 \approx 300$ (Rounded to the hundreds place.)
$300 \div 30 = 10$

D2) $31 \cdot 9 = 279$, which is greater than 271, so
31 must go into 271 less than 9 times.

D3) Answers may vary. Example The 24 represents
24 tens. Since $24 \div 12 = 2$, the 2 belongs in the
quotient above the 4.

D4) Answers may vary. Example: It would have
made it very clear where the 2 belonged.

D5)
```
           6 R3
    5 3) 3 2 1
       - 3 1 8
             3
```

D6)
```
             7 R49
    6 2) 4 8 3
       - 4 3 4
             4 9
```

D7)
```
           4 3 R10
    2 2) 9 5 6
       - 8 8
           7 6
         - 6 6
           1 0
```

D8)
```
             2 7
    3 1) 8 3 7
       - 6 2
         2 1 7
       - 2 1 7
             0
```

D9)
```
           1 5 9 R6
    1 6) 2 5 5 0
       - 1 6
         9 5
       - 8 0
         1 5 0
       - 1 4 4
             6
```

D10)
```
           1 8 9 R4
    2 1) 3 9 7 3
       - 2 1
         1 8 7
       - 1 6 8
           1 9 3
         - 1 8 9
               4
```

D11) a)
```
             6 3 2
    1 2) 7 5 8 4
       - 7 2
         3 8
       - 3 6
           2 4
         - 2 4
             0
```
Deana's monthly gas bill will be $632.

b)
```
             7 9 3
    1 2) 9 5 1 6
       - 8 4
         1 1 1
       - 1 0 8
             3 6
           - 3 6
               0
```
Deana's monthly gas bill will be $793.

SECTION 1.5   EXERCISES

For 1–16, refer to Concept Checks and Practice
Exercises.

17) quotient

19) divisor

21) dividend ÷ divisor = quotient

23)
$$\text{divisor} \overline{)\text{dividend}}^{\text{quotient}}$$

25) a) $16 \div 4 = x \implies x \cdot 4 = 16$
b) $y = 4$

27) a) $28 \div 7 = x \implies x \cdot 7 = 28$
b) $x = 4$

29) a) $54 \div 9 = z \implies z \cdot 9 = 54$
b) $z = 6$

31) a) $60 \div 12 = y \implies y \cdot 12 = 60$
b) $y = 60$

33)

×	6	8	9	3
5	30	40	45	15
4	24	32	36	12
7	42	56	63	21
8	48	64	72	24

35) a) $9\overline{)36}$ with quotient 4

b) $\dfrac{12}{4} = 3$

c) $8\overline{)64}$ with quotient 8

d) $\dfrac{24}{3} = 8$

e) $49 \div 7 = 7$

f) $5\overline{)45}$ with quotient 9

g) $0 \div 1 = 0$

37) a) $56 \div 8 = 7$

b) $\dfrac{18}{3} = 6$

c) $8\overline{)72}$ with quotient 9

d) $\dfrac{8}{0}$; undefined

e) $36 \div 6 = 6$

f) $3\overline{)21}$ with quotient 7

g) $63 \div 9 = 7$

39)
```
       1 3
  8 ) 1 0 4
     - 8
       2 4
     - 2 4
         0
```

41)
```
       1 2
  4 ) 4 8
     - 4
       0 8
     -   8
         0
```

43)
```
       1 6
  8 ) 1 2 8
     - 8
       4 8
     - 4 8
         0
```

45)
```
       5 4
  6 ) 3 2 4
     - 3 0
       2 4
     - 2 4
         0
```

47)
```
       5 R1
  6 ) 3 1
     - 3 0
         1
```

49)
```
       3 R3
  7 ) 2 4
     - 2 1
         3
```

51)
```
       2 1 R6
  9 ) 1 9 5
     - 1 8
       1 5
     -   9
         6
```

53)
```
       3 2
  9 ) 2 8 8
     - 2 7
       1 8
     - 1 8
         0
```

55)
```
       3 7
  5 ) 1 8 5
     - 1 5
       3 5
     - 3 5
         0
```
Each person will inventory 37 shelves.

57)
```
        2 1 6
    3 ) 6 4 8
      - 6
        0 4
      -   3
          1 8
        - 1 8
            0
```
The amount of each payment is $216.

59)
```
          6 1 R4
    7 ) 4 3 1
      - 4 2
          1 1
        -   7
            4
```

61)
```
        2 6 1
    9 ) 2 3 4 9
      - 1 8
          5 4
        - 5 4
            0 9
          -   9
              0
```

63)
```
        2 8 4 3 6
    2 ) 5 6 8 7 2
      - 4
        1 6
      - 1 6
          0 8
        -   8
              0 7
            -   6
                1 2
              - 1 2
                  0
```

65)
```
          1 9 3 3 R3
    8 ) 1 5 4 6 7
      - 8
        7 4
      - 7 2
          2 6
        - 2 4
            2 7
          - 2 4
              3
```

67)
```
          4 1 1
    4 ) 1 6 4 4
      - 1 6
          0 4
        -   4
            0 4
          -   4
              0
```

69) $\dfrac{76{,}483}{0}$; undefined

71)
```
              4 7
    4 2 ) 1 9 7 4
        - 1 6 8
            2 9 4
          - 2 9 4
                0
```

73)
```
            2 1
    2 8 ) 5 8 8
        - 5 6
            2 8
          - 2 8
              0
```

75)
```
              5 R30
    5 4 ) 3 0 0
        - 2 7 0
            3 0
```
Five boxes can be stacked.

77)
```
              9 3
    1 4 ) 1 3 0 2
        - 1 2 6
            4 2
          - 4 2
              0
```
He read an average of 93 pages each day.

79) a)
```
            2 0 0 0
    2 0 ) 4 0 0 0 0
        - 4 0
            0 0 0 0
```
b)
```
            2 0 8 6
    2 0 ) 4 1 7 2 0
        - 4 0
            1 7 2
          - 1 6 0
              1 2 0
            - 1 2 0
                0
```

81) a)
```
              2 0
    1 0 0 ) 2 0 0 0
          - 2 0 0
              0 0
```
b)
```
                1 6
    1 3 2 ) 2 1 1 2
          - 1 3 2
              7 9 2
            - 7 9 2
                  0
```

83) a)
```
            3 5 0 0
    2 0 ) 7 0 0 0 0
        - 6 0
          1 0 0
        - 1 0 0
            0 0 0
```

b)
```
            3 7 6 9 R1
    1 8 ) 6 7 8 4 3
        - 5 4
          1 3 8
        - 1 2 6
            1 2 4
          - 1 0 8
              1 6 3
            - 1 6 2
                  1
```

85) a)
```
                1 0 0 0
    1 0 0 ) 1 0 0 0 0 0
          - 1 0 0
              0 0 0 0
```

b)
```
                7 4 5
    1 4 2 ) 1 0 5 7 9 0
          - 9 9 4
              6 3 9
            - 5 6 8
                7 1 0
              - 7 1 0
                    0
```

87)
```
            3 2
    1 8 ) 5 7 6
        - 5 4
          3 6
        - 3 6
            0
```

89)
```
          8 2 R1
    8 ) 6 5 7
      - 6 4
        1 7
      - 1 6
          1
```

91)
```
              4 8
    1 1 ) 5 2 8
        - 4 4
          8 8
        - 8 8
            0
```

93)
```
              6 0
    2 0 ) 1 2 0 0
        - 1 2 0
            0 0
```
Dewey earned $60 per hour.

95)
```
            5 6 0 0
    1 1 ) 6 1 6 0 0
        - 5 5
          6 6
        - 6 6
            0 0 0
```
Each person invested $5,600.

97)
```
            3
    6 2 ) 1 8 6
        - 1 8 6
            0
```
They drove 3 hours.

99) a)
```
            3 2 8
    1 2 ) 3 9 3 6
        - 3 6
          3 3
        - 2 4
            9 6
          - 9 6
              0
```
His average monthly bill is $328.

b)
```
                5
    3 2 8 ) 1 6 4 0
          - 1 6 4 0
                  0
```
He can cover his heating bills for 5 months.

101) a)
$$
\begin{array}{r}
1\,4\,0\,0 \\
1\,5\,)\overline{\,2\,1\,0\,0\,0} \\
-\,1\,5 \quad\;\; \\
\hline
6\,0 \\
-\,6\,0 \\
\hline
0\,0\,0
\end{array}
$$
Each employee will receive $1400.

b)
$$
\begin{array}{r}
1\,5\,0\,0 \\
1\,4\,)\overline{\,2\,1\,0\,0\,0} \\
-\,1\,4 \quad\;\; \\
\hline
7\,0 \\
-\,7\,0 \\
\hline
0\,0\,0
\end{array}
$$
Each employee will receive $1500.

103) a)
$$
\begin{array}{r}
\overset{3}{3}\,7 \\
\times \quad 5 \\
\hline
1\,8\,5
\end{array}
$$
Jeanette spent $185.

b) $190 - 185 = 5$
Since $5 is less than $15, no, she does not have enough money to pay for a foam finger.

105) $24 \div x = 4$
$\quad\quad x = 6$

107) $7 = 42 \div y$
$\quad\quad y = 6$

109) $110 \div z = 11$
$\quad\quad\quad z = 10$

111) $800 = 4 \cdot x$
$\quad\quad x = 200$

1.6   Exponents, Groupings, and the Order of Operations

GUIDED PRACTICE

1) The base is 9.
The exponent is 8.

2) 2^4 has a base of 2 and an exponent of 4.
To evaluate 2^4, we must multiply four factors of the base, 2.
$$
\begin{aligned}
2^4 &= 2 \cdot 2 \cdot 2 \cdot 2 \\
&= 4 \cdot 2 \cdot 2 \\
&= 8 \cdot 2 \\
&= 16
\end{aligned}
$$

3) There are seven factors of 3.
The base is 3.
The exponent is 7.
$3 \cdot 3 \cdot 3 \cdot 3 \cdot 3 \cdot 3 \cdot 3 = 3^7$

4) $$
\begin{aligned}
3^4 &= 3 \cdot 3 \cdot 3 \cdot 3 \\
&= 9 \cdot 3 \cdot 3 \\
&= 27 \cdot 3 \\
&= 81
\end{aligned}
$$

5) $$
\begin{aligned}
12^2 &= 12 \cdot 12 \\
&= 144
\end{aligned}
$$

6) Any number raised to the power of zero is 1.
$3,589^0 = 1$

7) $$
\begin{aligned}
15 - (12 \div 4) &= 15 - 3 \\
&= 12
\end{aligned}
$$

8) $$
\begin{aligned}
3 \cdot [15 - 9] &= 3 \cdot 6 \\
&= 18
\end{aligned}
$$

9) $$
\begin{aligned}
\frac{42 - 32}{15 \div 3} &= \frac{42 - 32}{5} \\
&= \frac{10}{5} \\
&= 10 \div 5 \\
&= 2
\end{aligned}
$$

10) $3(\text{player 1's points} + \text{player 2's points})$
$$
\begin{aligned}
3(12 + 9) &= 3(21) \\
&= 63
\end{aligned}
$$
Candace scored 63 points.

11) $$
\begin{aligned}
12 + 6 \cdot 3 &= 12 + 18 \\
&= 30
\end{aligned}
$$

12) $$
\begin{aligned}
30 - (2 + 3) \cdot 4 &= 30 - (5) \cdot 4 \\
&= 30 - 20 \\
&= 10
\end{aligned}
$$

13) $30 + 2 \cdot 9 - 8 = 30 + 18 - 8$
$= 48 - 8$
$= 40$

14) $16 + 30 \div 3 \cdot 2 = 16 + 10 \cdot 2$
$= 16 + 20$
$= 36$

15) $5 + (2+3)^2 - 20 = 5 + (5)^2 - 20$
$= 5 + 25 - 20$
$= 30 - 20$
$= 10$

16) $27 \div (9-6)^3 + 8 \cdot 2 - 7 = 27 \div (3)^3 + 8 \cdot 2 - 7$
$= 27 \div 27 + 8 \cdot 2 - 7$
$= 1 + 8 \cdot 2 - 7$
$= 1 + 16 - 7$
$= 17 - 7$
$= 10$

17) a) The temperatures range from 55 to 83, so the average value must be above 55 and below 83. If you draw a line that goes through the middle of the data values, it appears that the average daily high temperature is about 70 degrees F.

b) average $= \dfrac{\text{the sum of the data values}}{\text{the number of data values}}$
$= \dfrac{55 + 58 + 72 + 83 + 62}{5}$
$= \dfrac{330}{5}$
$= 66$

The average high temperature is $66°$ F.

CONCEPT CHECKS AND PRACTICE EXERCISES

A1) The exponent is written above and to the right of the base.

A2) The exponent indicates how many of the base must be multiplied.

A3) The base is 3.
The exponent is 7.

A4) The base is 7.
The exponent is 2.

A5) The base is 13.
The exponent is 2.

A6) The base is 4.
The exponent is 10.

A7) $6^3 = 6 \cdot 6 \cdot 6$
$= 36 \cdot 6$
$= 216$

A8) $5^3 = 5 \cdot 5 \cdot 5$
$= 25 \cdot 5$
$= 125$

A9) $3^4 = 3 \cdot 3 \cdot 3 \cdot 3$
$= 9 \cdot 3 \cdot 3$
$= 27 \cdot 3$
$= 81$

A10) $3^5 = 3 \cdot 3 \cdot 3 \cdot 3 \cdot 3$
$= 9 \cdot 3 \cdot 3 \cdot 3$
$= 27 \cdot 3 \cdot 3$
$= 81 \cdot 3$
$= 243$

A11) $16^2 = 16 \cdot 16$
$= 256$

A12) $21^2 = 21 \cdot 21$
$= 441$

A13) $7^0 = 1$

A14) $9^0 = 1$

B1) Answers may vary. Example: A grouping indicates that the operations inside must be done first.

B2) Operations with numbers in parentheses are calculated before a number with exponents.

B3) $5 + (6-2) = 5 + 4$
$= 9$

B4) $13 - (2 \cdot 6) = 13 - 12$
$= 1$

B5) $28 \div [14 - 10] = 28 \div 4$
$= 7$

B6) $[5+4] \cdot 3 = 9 \cdot 3$
$= 27$

B7) $\dfrac{7+8}{8-7}=\dfrac{15}{8-7}$
$=\dfrac{15}{1}$
$=15$

B8) $\dfrac{4\cdot 4}{3+5}=\dfrac{16}{3+5}$
$=\dfrac{16}{8}$
$=2$

B9) $\dfrac{3+(4-2)}{6-1}=\dfrac{3+2}{6-1}$
$=\dfrac{5}{6-1}$
$=\dfrac{5}{5}$
$=1$

B10) $\dfrac{4-2}{12\div(5+1)}=\dfrac{4-2}{12\div(6)}$
$=\dfrac{4-2}{2}$
$=\dfrac{2}{2}$
$=1$

B11) $(893\cdot 12)-9{,}312=10{,}716-9{,}312$
$=1{,}404$
The theater group made \$1,404.

B12) $(7\cdot 8)-12=56-12$
$=44$
Danika ends up with \$44.

C1) Answers may vary. Example: When everyone follows the same order, the results will always be the same.

C2) When considering only the operations of multiplication and division, they are performed as encountered from left to right.

C3) When considering only the operations of addition and subtraction, they are performed as encountered from left to right.

C4) $6-2+3=4+3$
$=7$

C5) $12-6+3=6+3$
$=9$

C6) $20\div 5\cdot 2=4\cdot 2$
$=8$

C7) $30\div 5\cdot 3=6\cdot 3$
$=18$

C8) $12-4\cdot 2=12-8$
$=4$

C9) $12+8\div 4=12+2$
$=14$

C10) $3+4^2\div 8=3+16\div 8$
$=3+2$
$=5$

C11) $6^2-3\cdot 8=36-3\cdot 8$
$=36-24$
$=12$

C12) $25-(6-2)^2+9=25-(4)^2+9$
$=25-16+9$
$=9+9$
$=18$

C13) $8^2-(12+4)\cdot 2=64-(16)\cdot 2$
$=64-32$
$=32$

C14) $3\cdot(8-4)+2\cdot 3^2=3\cdot(4)+2\cdot 3^2$
$=3\cdot(4)+2\cdot 9$
$=12+2\cdot 9$
$=12+18$
$=30$

C15) $3\cdot 2^3-8(15-13)=3\cdot 2^3-8(2)$
$=3\cdot 8-8(2)$
$=24-8(2)$
$=24-16$
$=8$

SECTION 1.6   EXERCISES

For 1–9, refer to Concept Checks and Practice Exercises.

11) Exponents

13) grouping

15) When considering only multiplication and division, the operation appearing first when moving from left to right should be performed first.

17) $4 \cdot 4 \cdot 4 = 4^3$

19) $6 \cdot 6 \cdot 6 \cdot 6 \cdot 6 = 6^5$

21) $6^5 = 6 \cdot 6 \cdot 6 \cdot 6 \cdot 6$

23) $9^3 = 9 \cdot 9 \cdot 9$

25) a) $1^2 = 1$
    b) $3^2 = 9$
    c) $2^3 = 8$
    d) $4^2 = 16$
    e) $7^2 = 49$
    f) $6^2 = 36$
    g) $9^2 = 81$

27) a) $5^2 = 25$
    b) $2^2 = 4$
    c) $5^3 = 125$
    d) $3^2 = 9$
    e) $6^2 = 36$
    f) $3^3 = 27$
    g) $9^2 = 81$

29) $3^4 = 3 \cdot 3 \cdot 3 \cdot 3$
$$= 9 \cdot 3 \cdot 3$$
$$= 27 \cdot 3$$
$$= 81$$

31) $5^1 = 5$

33) $1^8 = 1$

35) $10^3 = 10 \cdot 10 \cdot 10$
$$= 100 \cdot 10$$
$$= 1,000$$

37) $2^4 = 2 \cdot 2 \cdot 2 \cdot 2$
$$= 4 \cdot 2 \cdot 2$$
$$= 8 \cdot 2$$
$$= 16$$

39) $5^2 = 5 \cdot 5$
$$= 25$$

41) $4^3 = 4 \cdot 4 \cdot 4$
$$= 16 \cdot 4$$
$$= 64$$

43) $2^6 = 2 \cdot 2 \cdot 2 \cdot 2 \cdot 2 \cdot 2$
$$= 4 \cdot 2 \cdot 2 \cdot 2 \cdot 2$$
$$= 8 \cdot 2 \cdot 2 \cdot 2$$
$$= 16 \cdot 2 \cdot 2$$
$$= 32 \cdot 2$$
$$= 64$$

45) $(16 + 4) \cdot 5 = 20 \cdot 5$
$$= 100$$

47) $8(24 - 15) = 8 \cdot 9$
$$= 72$$

49) $(24 \div 6) \cdot 7 = 4 \cdot 7$
$$= 28$$

51) $19 - (33 \div 11) = 19 - 3$
$$= 16$$

53) $\dfrac{21 - 3}{7 + 2} = \dfrac{18}{7 + 2}$
$$= \dfrac{18}{9}$$
$$= 2$$

55) $\dfrac{29 + 48}{3 + 4} = \dfrac{77}{3 + 4}$
$$= \dfrac{77}{7}$$
$$= 11$$

Each person received 11 morel mushrooms.

57) $12 \div 3 + 4 \cdot 9 = 4 + 4 \cdot 9$
$$= 4 + 36$$
$$= 40$$

59) $81 \div 3^2 - 6 = 81 \div 9 - 6$
$$= 9 - 6$$
$$= 3$$

61) $\dfrac{4+3\cdot 6}{7+4} = \dfrac{4+18}{7+4}$

$\qquad\qquad = \dfrac{22}{7+4}$

$\qquad\qquad = \dfrac{22}{11}$

$\qquad\qquad = 2$

63) $5\left(4+3^2\right) = 5\left(4+9\right)$

$\qquad\qquad = 5\cdot 13$

$\qquad\qquad = 65$

65) $8^2 - \left(4^2 - 6\right) = 64 - \left(16 - 6\right)$

$\qquad\qquad = 64 - 10$

$\qquad\qquad = 54$

67) $12 \div 3^2 = 12 \div 9$; d

69) $12 \div 2 \cdot 3 = 6 \cdot 3$; c

71) $8 - 2 + 4 = 6 + 4$; g

73) $3^3 + 4 = 27 + 4$; a

75) $40 - 10 \times 2 + 3^2 = 40 - 10 \times 2 + 9$

$\qquad\qquad = 40 - 20 + 9$

$\qquad\qquad = 20 + 9$

$\qquad\qquad = 29$

77) $5 - \left(9 - 7\right)^2 + 1 = 5 - 2^2 + 1$

$\qquad\qquad = 5 - 4 + 1$

$\qquad\qquad = 1 + 1$

$\qquad\qquad = 2$

79) $24 \div 8 \times 3 - \left(5 - 3\right)^2 = 24 \div 8 \times 3 - \left(2\right)^2$

$\qquad\qquad = 24 \div 8 \times 3 - 4$

$\qquad\qquad = 3 \times 3 - 4$

$\qquad\qquad = 9 - 4$

$\qquad\qquad = 5$

81) $17 \cdot 8,500 + 35 \cdot 6,100 = 144,500 + 213,500$

$\qquad\qquad = 358,000$

If all the tickets were sold, \$358,000 was earned.

83) average $= \dfrac{\text{the sum of the data values}}{\text{the number of data values}}$

$\qquad = \dfrac{9 + 16 + 23}{3}$

$\qquad = \dfrac{48}{3}$

$\qquad = 16$

85) average $= \dfrac{\text{the sum of the data values}}{\text{the number of data values}}$

$\qquad = \dfrac{100 + 200 + 240}{3}$

$\qquad = \dfrac{540}{3}$

$\qquad = 180$

87) average $= \dfrac{\text{the sum of the data values}}{\text{the number of data values}}$

$\qquad = \dfrac{0 + 90 + 100 + 110}{4}$

$\qquad = \dfrac{300}{4}$

$\qquad = 75$

89) average $= \dfrac{\text{the sum of the data values}}{\text{the number of data values}}$

$\qquad = \dfrac{120 + 240 + 180 + 100}{4}$

$\qquad = \dfrac{640}{4}$

$\qquad = 160$

91) a) Answers may vary. Example: 40°F

b) average $= \dfrac{\text{the sum of the data values}}{\text{the number of data values}}$

$\qquad = \dfrac{45 + 38 + 32 + 43 + 52}{5}$

$\qquad = \dfrac{210}{5}$

$\qquad = 42$

The average high temperature is 42°F.

93) average $= \dfrac{\text{the sum of the data values}}{\text{the number of data values}}$

$\qquad = \dfrac{13 + 13 + 10 + 11 + 13}{5}$

$\qquad = \dfrac{60}{5}$

$\qquad = 12$

The average number of top movies with a PG-13 rating between 2002 and 2006 is 12.

1.7   Properties of Whole Numbers

GUIDED PRACTICE

1) Neither 2, 3, 4, 5, 6 divide it evenly.
   Beacause 7 is not divisible by any number other than 1 and itself, 7 is a prime number.

2) 3 and 5 can divide 15 evenly.
   $15 \div 3 = 5$
   $15 \div 5 = 3$
   Because 15 is divisible by a whole number other than itself and 1, 15 is a composite number.

3) a) A number is divisible by 2 if it is even.
      915 is odd, so it is not divisible by 2.
      720 is even, so it is divisible by 2.
      837 is odd, so it is not divisible by 2.

   b) A number is divisible by 3 if the sum of its digits is divisible by 3.
      915: 9 + 1 + 5 = 15; 15 is divisible by 3, so 915 is divisible by 3.
      720: 7 + 2 + 0 = 9; 9 is divisible by 3, so 720 is divisible by 3.
      837: 8 + 3 + 7 = 18; 18 is divisible by 3, so 837 is divisible by 3.

   c) A number is divisible by 5 if it ends in 0 or 5.
      915's last digit is 5, so it is divisible by 5.
      720's last digit is 0, so it is divisible by 5.
      837's last digit is 7, so it is not divisible by 5.

4) 11 is odd, so 2 does not divide it.
   11 is not divisible by 3.
   11 ends in 1, so it is not divisible by 5.
   11 is not divisible by 7.
   11 does divide itself evenly.
   11 is prime because it is divisible only by itself and 1.

5) 35 is odd, so it is not divisible by 2.
   3 + 5 = 8, which is not divisible by 3. Therefore, 3 does not divide 35 evenly.
   The last digit of 35 is 5, so 5 does divide 35 evenly.
   35 is composite because it is divisible by a number other than itself or 1.

6) Check the primes whose square is less than 87.
   $2^2 = 4$, $3^2 = 9$, $5^2 = 25$, $7^2 = 49$, $11^2 = 121$,
   $13^2 = 169$
   We must check 2, 3, 5, and 7.

7) Check the primes whose square is less than 109.
   $2^2 = 4$, $3^2 = 9$, $5^2 = 25$, $7^2 = 49$, $11^2 = 121$,
   $13^2 = 169$
   We must check 2, 3, 5, and 7.

8) $2^2 = 4$, $3^2 = 9$, $5^2 = 25$, $7^2 = 49$, $11^2 = 121$
   Because 121 is more than 53, we must check through 7.
   2 doesn't divide 53 evenly because 53 is odd.
   3 doesn't divide 53 evenly because $5 + 3 = 8$ is not divisible by 3.
   5 doesn't divide 53 evenly because 53 does not end in 0 or 5.
   7 doesn't divide 53 evenly because $7 \cdot 7 = 49$ and $7 \cdot 8 = 56$.
   53 is prime.

9) $12 \div 1 = 1$ $12 \div 2 = 6$ $12 \div 3 = 4$
   $12 \div 12 = 1$ $12 \div 6 = 2$ $12 \div 4 = 3$
   The six factors of 12 are 1, 2, 3, 4, 6, and 12.

10) The three factor pairs of 12 are 1 and 12, 2 and 6, 3 and 4.

11) The two-number factorizations of 12 are
    $1 \cdot 12$, $2 \cdot 6$, $3 \cdot 4$.

12) Start with any two-number factorization of 12.
    $12 = 2 \cdot 6$
    This is not a prime factorization of 12 because 6 is composite. However, we can write 6 as a product of primes. $6 = 2 \cdot 3$
    $12 = 2 \cdot 6$
    $12 = 2 \cdot 2 \cdot 3$
    The prime factorization of 12 is
    $12 = 2 \cdot 2 \cdot 3$.
    Each factor is prime and the product is 12.

13)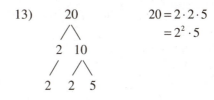
    $20 = 2 \cdot 2 \cdot 5$
    $\quad = 2^2 \cdot 5$

14)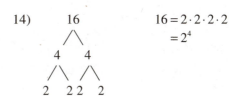
    $16 = 2 \cdot 2 \cdot 2 \cdot 2$
    $\quad = 2^4$

15) The first five multiples of 4 are as follows.
$1 \cdot 4 = 4$
$2 \cdot 4 = 8$
$3 \cdot 4 = 12$
$4 \cdot 4 = 16$
$5 \cdot 4 = 20$

16) The first five multiples of 12 are as follows.
$1 \cdot 12 = 12$
$2 \cdot 12 = 24$
$3 \cdot 12 = 36$
$4 \cdot 12 = 48$
$5 \cdot 12 = 60$

CONCEPT CHECKS AND PRACTICE EXERCISES

A1) A prime number is divisible only by 1 and itself.

A2) A composite number is divisible by a whole number other than 1 and itself.

A3) 12, 34 and 220 are divisible by 2 since they are even numbers.

A4) 12, 75, 87, 111 and 213 are divisible by 3 since the sum of the digits in each is divisible by 3.

A5) 75 and 220 are divisible by 5 since they end in 0 or 5.

A6) The largest prime number that must be checked is 3 because $5^2 = 25$, which is greater than 23.

A7) a) $2^2 = 4$, $3^2 = 9$, $5^2 = 25$, $7^2 = 49$, $11^2 = 121$
Since 121 is more than 63, we only have to check through 7. We must check 2, 3, 5, 7.

b) 2 doesn't divide 63 evenly because 63 is odd.
3 divides 63 evenly because $6 + 3 = 9$ is divisible by 3.
63 is composite.

A8) a) $2^2 = 4$, $3^2 = 9$, $5^2 = 25$, $7^2 = 49$, $11^2 = 121$
Since 121 is more than 75, we only have to check through 7. We must check 2, 3, 5, 7.

b) 2 doesn't divide 75 evenly because 75 is odd.
3 divides 75 evenly because $7 + 5 = 12$ is divisible by 3.
75 is composite.

A9) a) $2^2 = 4$, $3^2 = 9$, $5^2 = 25$, $7^2 = 49$, $11^2 = 121$
Since 121 is more than 93, we only have to check through 7. We must check 2, 3, 5, 7.

b) 2 doesn't divide 93 evenly because 93 is odd.
3 divides 93 evenly because $9 + 3 = 12$ is divisible by 3.
93 is composite.

A10) a) $2^2 = 4$, $3^2 = 9$, $5^2 = 25$, $7^2 = 49$, $11^2 = 121$
Since 121 is more than 97, we only have to check through 7. We must check 2, 3, 5, 7.

b) 2 doesn't divide 97 evenly because 97 is odd.
3 doesn't divide 97 evenly because $9 + 7 = 16$ is not divisible by 3.
5 doesn't divide 97 evenly because 97 does not end in 0 or 5.
7 doesn't divide 97 evenly because $7 \cdot 13 = 91$ and $7 \cdot 14 = 98$.
97 is prime.

A11) a) $2^2 = 4$, $3^2 = 9$, $5^2 = 25$, $7^2 = 49$,
$11^2 = 121$, $13^2 = 169$
Since 169 is more than 133, we only have to check through 11. We must check 2, 3, 5, 7, 11.

b) 2 doesn't divide 133 evenly because 133 is odd.
3 doesn't divide 133 evenly because $1 + 3 + 3 = 7$ is not divisible by 3.
5 doesn't divide 133 evenly because 133 does not end in 0 or 5.
7 does divide 133 evenly because $133 \div 7 = 19$.
133 is composite.

A12) a) $2^2 = 4$, $3^2 = 9$, $5^2 = 25$, $7^2 = 49$,
$11^2 = 121$, $13^2 = 169$
Since 169 is more than 141, we only have to check through 11. We must check 2, 3, 5, 7, 11.

b) 2 doesn't divide 141 evenly because 141 is odd.
3 does divide 141 evenly because $1 + 4 + 1 = 6$ is divisible by 3.
141 is composite.

A13) a) $2^2 = 4$, $3^2 = 9$, $5^2 = 25$, $7^2 = 49$, $11^2 = 121$

Since 121 is more than 89, we only have to check through 7. We must check 2, 3, 5, 7.

b) 2 doesn't divide 89 evenly because 89 is odd.

3 doesn't divide 89 evenly because $8 + 9 = 17$ is not divisible by 3.

5 doesn't divide 89 evenly because 89 does not end in 0 or 5.

7 doesn't divide 89 evenly because $7 \cdot 12 = 84$ and $7 \cdot 13 = 91$.

89 is prime.

A14) a) $2^2 = 4$, $3^2 = 9$, $5^2 = 25$, $7^2 = 49$, $11^2 = 121$

Since 121 is more than 57, we only have to check through 7. We must check 2, 3, 5, 7.

b) 2 doesn't divide 57 evenly because 57 is odd.

3 does divide 57 evenly because $5 + 7 = 12$ is divisible by 3.

57 is composite.

B1) Answers may vary. Example: A prime factorization is a way to write a number as a product of prime numbers.

B2) Answers may vary. Example: The use of exponents enables us to write prime factorizations more concisely. We can immediately see how many times each factor is repeated in the factorization.

B3) a) $36 \div 1 = 36$, $36 \div 2 = 18$, $36 \div 3 = 12$, $36 \div 4 = 9$, $36 \div 6 = 6$, $36 \div 9 = 4$, $36 \div 12 = 3$, $36 \div 18 = 2$, $36 \div 36 = 1$

The nine factors of 36 are 1, 2, 3, 4, 6, 9, 12, 18, 36.

b) The two-number factorizations of 36 are $1 \cdot 36$, $2 \cdot 18$, $3 \cdot 12$, $4 \cdot 9$, $6 \cdot 6$.

c)
```
        36                36 = 2·2·3·3
       /\                    = 2²·3²
      6   6
     /\  /\
    2 3 2  3
```

B4) a) $40 \div 1 = 40$, $40 \div 2 = 20$, $40 \div 4 = 10$, $40 \div 5 = 8$, $40 \div 8 = 5$, $40 \div 10 = 4$, $40 \div 20 = 2$, $40 \div 40 = 1$

The eight factors of 40 are 1, 2, 4, 5, 8, 10, 20, 40.

b) The two-number factorizations of 40 are $1 \cdot 40$, $2 \cdot 20$, $4 \cdot 10$, $5 \cdot 8$.

c)
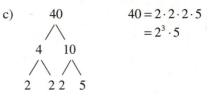

$$40 = 2 \cdot 2 \cdot 2 \cdot 5$$
$$= 2^3 \cdot 5$$

B5) $35 = 5 \cdot 7$

B6) $33 = 3 \cdot 11$

B7) $50 = 10 \cdot 5$
$$= 2 \cdot 5 \cdot 5$$
$$= 2 \cdot 5^2$$

B8) $45 = 9 \cdot 5$
$$= 3 \cdot 3 \cdot 5$$
$$= 3^2 \cdot 5$$

B9) $48 = 16 \cdot 3$
$$= 4 \cdot 4 \cdot 3$$
$$= 2 \cdot 2 \cdot 2 \cdot 2 \cdot 3$$
$$= 2^4 \cdot 3$$

B10) $100 = 2 \cdot 50$
$$= 2 \cdot 2 \cdot 25$$
$$= 2 \cdot 2 \cdot 5 \cdot 5$$
$$= 2^2 \cdot 5^2$$

B11) $72 = 8 \cdot 9$
$$= 2 \cdot 4 \cdot 3 \cdot 3$$
$$= 2 \cdot 2 \cdot 2 \cdot 3 \cdot 3$$
$$= 2^3 \cdot 3^2$$

B12) $56 = 8 \cdot 7$
$$= 2 \cdot 4 \cdot 7$$
$$= 2 \cdot 2 \cdot 2 \cdot 7$$
$$= 2^3 \cdot 7$$

C1) Answers may vary. Example: 10, 20, 30, 40

C2) 1, 2, 5, 10

C3) factors

C4) multiples

C5) 8, 16, 24, 32, 40

C6) 7, 14, 21, 28, 35

C7)   30, 60, 90, 120, 150

C8)   20, 40, 60, 80, 100

C9)   16, 32, 48, 64, 80

C10)   18, 36, 54, 72, 90

C11)   24, 48, 72, 96, 120

C12)   32, 64, 96, 128, 160

SECTION 1.7   EXERCISES

For 1–9, refer to Concept Checks and Practice
Exercises.

11)   prime number

13)   multiple

15)   divisible

17)   prime factorization

19)   A number is divisible by three if the sum of its
      digits is divisible by 3.

21)   17 is only divisible by itself and 1, so it is
      prime.

23) $32 \div 2 = 16$, so it is composite.

25) $39 \div 3 = 13$, so it is composite.

27) $121 \div 11 = 11$, so it is composite.

29)   Factors of 6: 1, 2, 3, 6

31)   Factors of 11: 1, 11

33)   Factors of 28: 1, 2, 4, 7, 14, 28

35)   Factors of 32: 1, 2, 4, 8, 16, 32

37)   Factors of 75: 1, 3, 5, 15, 25, 75

39)   Factors of 34: 1, 2, 17, 34

41)   a) $2^2 = 4$, $3^2 = 9$, $5^2 = 25$
           Since 25 is more than 10, we only have to
           check through 3. We must check 2, 3.

      b)   2 does divide 10 evenly because 10 is even.
           10 is composite.

43)   a) $2^2 = 4$, $3^2 = 9$, $5^2 = 25$
           Since 25 is more than 23, we only have to
           check through 3. We must check 2, 3.

      b)   2 doesn't divide 23 evenly because 23 is
           odd.
           3 doesn't divide 23 evenly because $2 + 3 = 5$
           is not divisible by 3.
           23 is prime.

45)   a) $2^2 = 4$, $3^2 = 9$, $5^2 = 25$, $7^2 = 49$
           Since 49 is more than 39, we only have to
           check through 5. We must check 2, 3, 5.

      b)   2 doesn't divide 39 evenly because 39 is
           odd.
           3 does divide 39 evenly because $3 + 9 = 12$
           is divisible by 3.
           39 is composite.

47)   a) $2^2 = 4$, $3^2 = 9$, $5^2 = 25$, $7^2 = 49$
           Since 49 is more than 43, we only have to
           check through 5. We must check 2, 3, 5.

      b)   2 doesn't divide 43 evenly because 43 is
           odd.
           3 doesn't divide 43 evenly because $4 + 3 = 7$
           is not divisible by 3.
           5 doesn't divide 43 evenly because 43 does
           not end in 0 or 5.
           43 is prime.

49)   a) $2^2 = 4$, $3^2 = 9$, $5^2 = 25$, $7^2 = 49$, $11^2 = 121$
           Since 121 is more than 73, we only have to
           check through 7. We must check 2, 3, 5, 7.

      b)   2 doesn't divide 73 evenly because 73 is
           odd.
           3 doesn't divide 73 evenly because
           $7 + 3 = 10$ is not divisible by 3.
           5 doesn't divide 73 evenly because 73 does
           not end in 0 or 5.
           7 doesn't divide 73 evenly because
           $7 \cdot 10 = 70$ and $7 \cdot 11 = 77$.
           73 is prime.

51)   a) $2^2 = 4$, $3^2 = 9$, $5^2 = 25$, $7^2 = 49$, $11^2 = 121$
           Since 121 is more than 93, we only have to
           check through 7. We must check 2, 3, 5, 7.

      b)   2 doesn't divide 93 evenly because 93 is
           odd.
           3 does divide 93 evenly because $9 + 3 = 12$
           is divisible by 3.
           93 is composite.

53)   18 $18 = 2 \cdot 3 \cdot 3$
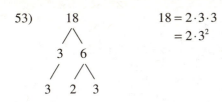
                      $= 2 \cdot 3^2$

55)   42 $42 = 2 \cdot 3 \cdot 7$

57) $12 = 4 \cdot 3$
     $= 2 \cdot 2 \cdot 3$
     $= 2^2 \cdot 3$

59) 31 is prime.

61) $18 = 2 \cdot 9$
     $= 2 \cdot 3 \cdot 3$
     $= 2 \cdot 3^2$

63) $42 = 6 \cdot 7$
     $= 2 \cdot 3 \cdot 7$

65) $72 = 8 \cdot 9$
     $= 2 \cdot 2 \cdot 2 \cdot 3 \cdot 3$
     $= 2^3 \cdot 3^2$

67) 17 is prime.

69) $100 = 4 \cdot 25$
     $= 2 \cdot 2 \cdot 5 \cdot 5$
     $= 2^2 \cdot 5^2$

71) $60 = 4 \cdot 15$
     $= 2 \cdot 2 \cdot 3 \cdot 5$
     $= 2^2 \cdot 3 \cdot 5$

73) 6, 12, 18, 24, 30

75) 25, 50, 75, 100, 125

77) 14, 28, 42, 56, 70

79) 17, 34, 51, 68, 85

81) 150, 300, 450, 600, 750

83) 35, 70, 105, 140, 175

85) 1, 2, 3, 6

87) 8, 12, 16, 20, 24, 80

89) 1, 2, 3, 6, 12

91) 5, 10, 20, 75, 80

93) 1, 2, 8

95) 20, 80

97) A box of 36 pencils can be divided evenly between 3 people because 3 is a factor of 36.

99) Stamps come in sheets of 20. Liam can sell 400 stamps to Sam because 400 is a multiple of 20.

1.8   The Greatest Common Factor and Least Common Multiple

GUIDED PRACTICE

1) Factors of 12: $\boxed{1}, \boxed{2}, \boxed{3}, 4, \boxed{6}, 12$
    Factors of 18: $\boxed{1}, \boxed{2}, \boxed{3}, \boxed{6}, 9, 18$
    The four common factors of 12 and 18 are 1, 2, 3 and 6.

2) From Guided Practice 1, the common factors of 12 and 18 and 1, 2, 3 and 6.
    The greatest of the common factors is 6.
    The GCF of 12 and 18 is 6.

3) Factors of 20: $\boxed{1}, \boxed{2}, \boxed{4}, 5, 10, 20$
    Factors of 8: $\boxed{1}, \boxed{2}, \boxed{4}, 8$
    1, 2, and 4 are common factors. 4 is the greatest of the common factors.
    GCF = 4

4) Factors of 14: $\boxed{1}, \boxed{2}, \boxed{7}, \boxed{14}$
    Factors of 42: $\boxed{1}, \boxed{2}, 3, 6, \boxed{7}, \boxed{14}, 21, 42$
    1, 2, and 7, and 14 are common factors. 14 is the greatest of the common factors.
    GCF = 14

5) Factors of 8: $\boxed{1}, 2, 4, 8$
    Factors of 27: $\boxed{1}, 3, 9, 27$
    1 is the only common factor. 1 is the greatest of the common factors.
    GCF = 1

6) The first six multiples of 4:
$$1 \cdot 4 = 4$$
$$2 \cdot 4 = 8$$
$$3 \cdot 4 = 12$$
$$4 \cdot 4 = 16$$
$$5 \cdot 4 = 20$$
$$6 \cdot 4 = 24$$

The first four multiples of 6:
$$1 \cdot 6 = 6$$
$$2 \cdot 6 = 12$$
$$3 \cdot 6 = 18$$
$$4 \cdot 6 = 24$$

The first two common multiples of 4 and 6 are 12 and 24.

7) From Guided Practice 6, the first two common multiples of 4 and 6 are 12 and 24. The least of the common multiples is 12. The LCM of 4 and 6 is 12.

8) Multiples of 5: 5, 10, 15, 20, 25, 30
Multiples of 6: 6, 12, 18, 24, 30
The LCM is 30.

9) Multiples of 12: 12, 24, 36, 48
Multiples of 16: 16, 32, 48
The LCM is 48.

10) a) The items on the turkey sub are written inside the left circle. The turkey sub has turkey, tomato, and swiss.

b) The items on the ham sub are written inside the right circle. The ham sub has ham, lettuce, peppers, and swiss.

c) The common items are written in the overlapping area. Swiss is the only common item.

11) $4 = 2 \cdot 2$
$6 = 2 \cdot 3$
The common factor is 2.

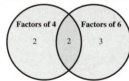

12) $20 = 2 \cdot 2 \cdot 5$
$12 = 2 \cdot 2 \cdot 3$
The common factors are 2 and 2.

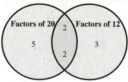

13) $50 = 2 \cdot 5 \cdot 5$
$20 = 2 \cdot 2 \cdot 5$
The common factors are 2 and 5.

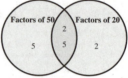

GCF $= 2 \cdot 5$
$\quad = 10$
LCM $= 2 \cdot 5 \cdot 2 \cdot 5$
$\quad\quad = 100$

14) $18 = 2 \cdot 3 \cdot 3$
$54 = 2 \cdot 3 \cdot 3 \cdot 3$

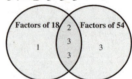

GCF $= 2 \cdot 3 \cdot 3$
$\quad = 18$
LCM $= 2 \cdot 3 \cdot 3 \cdot 3$
$\quad\quad = 54$

15) Multiples of 5: 5, 10, 15, 20, 25, 30
Multiples of 6: 6, 12, 18, 24, 30
The LCM is 30.
It will be 30 days until they need to borrow a car again.

CONCEPT CHECKS AND PRACTICE EXERCISES

A1) 8 is a factor of 16, but not of 18.

A2) Answers may vary. 5 is a common factor of 10 and 20, but so is 10. Because 10 is greater than 5, 10 is the greatest common factor (GCF) of 10 and 20.

A3) Answers may vary. For example, 12 and 18 have 3 as a common factor, but the greatest common factor (GCF) is 6.

A4) a) Factors of 6: 1, 2, 3, 6
         Factors of 12: 1, 2, 3, 4, 6, 12
     b) Common factors of 6 and 12: 1, 2, 3, 6
     c) The GCF is 6.

A5) a) Factors of 12: 1, 2, 3, 4, 6, 12
         Factors of 36: 1, 2, 3, 4, 6, 9, 12, 18, 36
     b) Common factors of 12 and 36: 1, 2, 3,
         4, 6, 12
     b) The GCF is 12.

A6) Factors of 6: 1, 2, 3, 6
     Factors of 4: 1, 2, 4
     Common factors of 6 and 4: 1, 2
     The GCF is 2.

A7) Factors of 10: 1, 2, 5, 10
     Factors of 15: 1, 3, 5, 15
     Common factors of 10 and 15: 1, 5
     The GCF is 5.

A8) Factors of 9: 1, 3, 9
     Factors of 12: 1, 2, 3, 4, 6, 12
     Common factors of 9 and 12: 1, 3
     The GCF is 3.

A9) Factors of 8: 1, 2, 4, 8
     Factors of 10: 1, 2, 5, 10
     Common factors of 8 and 10: 1, 2
     The GCF is 2.

A10) Factors of 6: 1, 2, 3, 6
     Factors of 5: 1, 5
     Common factors of 6 and 5: 1
     The GCF is 1.

A11) Factors of 7: 1, 7
     Factors of 10: 1, 2, 5, 10
     Common factors of 7 and 10: 1
     The GCF is 1.

A12) Factors of 28: 1, 2, 4, 7, 14, 28
     Factors of 56: 1, 2, 4, 7, 8, 14, 28, 56
     Common factors of 28 and 56: 1, 2, 4, 7,
                                         14, 28
     The GCF is 28.

A13) Factors of 32: 1, 2, 4, 8, 16, 32
     Factors of 48: 1, 2, 3, 4, 6, 8, 12,
                     16, 24, 48
     Common factors of 32 and 48: 1, 2, 4, 8, 16
     The GCF is 16.

B1) The least common multiple of two numbers is
     the smallest number that is divisible by the two
     numbers.

B2) The greatest common factor of two numbers is
     the largest number that divides the two numbers
     exactly.

B3) The LCM can be greater than the two numbers.

B4) The GCF can be less than the two numbers.

B5) Multiples of 4: 4, 8, 12, 16
     Multiples of 16: 16
     LCM: 16

B6) Multiples of 5: 5, 10, 15, 20, 25
     Multiples of 25: 25
     LCM: 25

B7) Multiples of 20: 20, 40, 60
     Multiples of 15: 15, 30, 45, 60
     LCM: 60

B8) Multiples of 9: 9, 18, 27, 36
     Multiples of 12: 12, 24, 36
     LCM: 36

B9) Multiples of 50: 50, 100, 150
     Multiples of 75: 75, 150
     LCM: 150

B10) Multiples of 30: 30, 60
     Multiples of 20: 20, 40, 60
     LCM: 60

B11) Multiples of 600: 600, 1,200,
     Multiples of 400: 400, 800, 1,200,
     LCM: 1,200

B12) Multiples of 800: 800, 1,600, 2,400
     Multiples of 600: 600, 1,200, 1,800, 2,400
     LCM: 2,400

B13) 10 is the LCM of 2 and 10.

B14) 15 is the GCF of 15 and 30.

B15) 9 is the GCF of 9 and 27.

B16) 12 is the LCM of 12 and 6.

C1) 14 must be the GCF because it is less than either
     number.

C2) 84 must be the LCM because it is greater than
     either number.

C3) a) If all of the numbers inside a Venn diagram are multiplied, the result will be the LCM.

b) If the numbers inside the overlapping portion of a Venn diagram are multiplied, the result will be the GCF.

C4) The factor in the overlapping section of the diagram is understood to be 1. The GCF of 6 and 5 is 1.

C5) a) $2 \cdot 3 \cdot 3 = 18$. The number represented by the circle on the right is 18.

b) $2 \cdot 3 = 6$. The GCF of 24 and 18 is 6.

c) $2 \cdot 2 \cdot 2 \cdot 3 \cdot 3 = 72$. The LCM of 24 and 18 is 72.

C6) a) $1 \cdot 11 = 11$. The factor in the right side of the circle is understood to be 1.

b) The GCF of 55 and 11 is 11.

c) $5 \cdot 11 = 55$. The LCM of 11 and 55 is 55.

C7) $6 = 2 \cdot 3$
$24 = 2 \cdot 2 \cdot 2 \cdot 3$
GCF: 6; LCM: 24
$GCF(7, 14) = 7$
$LCM(7, 14) = 14$

C8) $9 = 3 \cdot 3$
$36 = 2 \cdot 2 \cdot 3 \cdot 3$
GCF: 9; LCM: 36

C9) $15 = 3 \cdot 5$
$25 = 5 \cdot 5$
GCF: 5; LCM: 25

C10) $21 = 3 \cdot 7$
$35 = 5 \cdot 7$
GCF: 7; LCM: 105

C11) $22 = 2 \cdot 11$
$55 = 5 \cdot 11$
GCF: 11; LCM: 110

C12) $36 = 2 \cdot 2 \cdot 3 \cdot 3$
$48 = 2 \cdot 2 \cdot 2 \cdot 2 \cdot 3$
GCF: 12; LCM: 144

C13) $64 = 2 \cdot 2 \cdot 2 \cdot 2 \cdot 2 \cdot 2$
$14 = 2 \cdot 7$
GCF: 2; LCM: 448

C14) $34 = 2 \cdot 17$
$12 = 2 \cdot 2 \cdot 3$
GCF: 2; LCM: 204

C15) $9 = 3 \cdot 3$
$16 = 2 \cdot 2 \cdot 2 \cdot 2$
GCF: 1; LCM: 144

C16) $25 = 5 \cdot 5$
$12 = 2 \cdot 2 \cdot 3$
GCF: 1; LCM: 300

C17) $20 = 2 \cdot 2 \cdot 5$
$70 = 2 \cdot 5 \cdot 7$
GCF: 10; LCM: 140

C18) $50 = 2 \cdot 5 \cdot 5$
$90 = 2 \cdot 3 \cdot 3 \cdot 5$
GCF: 10; LCM: 450

SECTION 1.8 EXERCISES

For 1–9, refer to Concept Checks and Practice Exercises.

11) factor

13) least common multiple

15) greatest common factor

17) Factors of 6: 1, 2, 3, 6
Factors of 9: 1, 3, 9
Common factors of 6 and 9: 1, 3
GCF: 3

19) Factors of 30: 1, 2, 3, 5, 6, 10, 15, 30
Factors of 18: 1, 2, 3, 6, 9, 18
Common factors of 30 and 18: 1, 2, 3, 6
GCF: 6

21) Factors of 7: 1, 7
Factors of 42: 1, 2, 3, 6, 7, 14, 21, 42
Common factors of 7 and 42: 1, 7
GCF: 7

23) Factors of 40: 1, 2, 4, 5, 8, 10, 20, 40
Factors of 50: 1, 2, 5, 10, 25, 50
Common factors of 40 and 50: 1, 2, 5, 10
GCF: 10

25) Multiples of 5: 5, 10
Multiples of 10: 10
LCM: 10

27) Multiples of 33: 33
Multiples of 11: 11, 22, 33
LCM: 33

29) Multiples of 3: 3, 6, 9, 12, 15, 18, 21, 24, 27, 30
33, 36, 39, 42, 45, 48, 51, 54, 57, 60, 63, 66
Multiples of 22: 22, 44, 66
LCM: 66

31) Multiples of 30: 30, 60, 90
Multiples of 18: 18, 36, 48, 72, 90
LCM: 90

33) a) $1 \cdot 2 \cdot 7 = 14$. $\boxed{?}$ is equal to 1.

b) $2 \cdot 7 = 14$. The GCF of 210 and 14 is 14.

c) $2 \cdot 3 \cdot 5 \cdot 7 = 210$. The LCM of 210 and 14 is 210.

35) a)
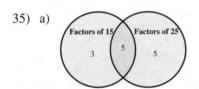

b) GCF: 5

c) LCM: $3 \cdot 5 \cdot 5 = 75$

37) a)
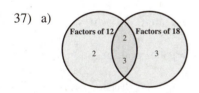

b) GCF: $2 \cdot 3 = 6$

c) LCM: $2 \cdot 2 \cdot 3 \cdot 3 = 36$

39) a)
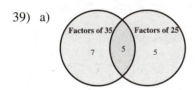

b) GCF: 5

c) LCM: $5 \cdot 5 \cdot 7 = 175$

41) a)
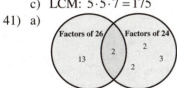

b) GCF: 2

c) LCM: $2 \cdot 2 \cdot 2 \cdot 3 \cdot 13 = 312$

43) a)
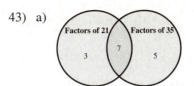

b) GCF: 7

c) LCM: $3 \cdot 5 \cdot 7 = 105$

45) a)
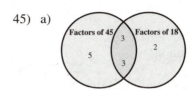

b) GCF: $3 \cdot 3 = 9$

c) LCM: $2 \cdot 3 \cdot 3 \cdot 5 = 90$

47) a)
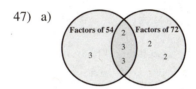

b) GCF: $2 \cdot 3 \cdot 3 = 18$

c) LCM: $2 \cdot 2 \cdot 2 \cdot 3 \cdot 3 \cdot 3 = 216$

49) $2 = 1 \cdot 2$
$3 = 1 \cdot 3$
LCM: $2 \cdot 3 = 6$
$5 + 6 = 11$
The next time they will work together is on the 11th.

51)

	9	7	4	8
7	63	49	28	56
5	45	35	20	40
6	54	42	24	48
3	27	21	12	24

53) a) $50 = 2 \cdot 5 \cdot 5$
$140 = 2 \cdot 2 \cdot 5 \cdot 7$
LCM: $2 \cdot 2 \cdot 5 \cdot 5 \cdot 7 = 700$

b) $700 \div 50 = 14$
C must make 14 revolutions before the green arrows on C and D line up again.

c) $700 \div 140 = 5$
D must make 5 revolutions before the green arrows on C and D line up again.

1.9   Applications with Whole Numbers

GUIDED PRACTICE

1) "Tripling" indicates multiplying by 3.
   The result $= 3 \cdot 40$

2) "Go into" indicates division.
   The answer $= 32 \div 4$

3) "Deposit" indicates addition.
   Balance $= \$50 + \$45$

4) "Withdrew" indicates subtraction.
   We must subtract 25 from $6.
   $$\begin{aligned} \text{new balance} &= \text{old balance} - \text{withdrawal} \\ &= 76 - 25 \\ &= 51 \end{aligned}$$
   The new balance is $51.

5) $\begin{aligned} \#\text{cakes} &= (\text{cakes per hour}) \cdot (\text{hours}) \\ &= 8 \cdot 4 \\ &= 32 \end{aligned}$
   Buggs can make 32 carrot cakes.

6) $\begin{aligned} \#\text{bags} &= (\text{lb carrots}) \div (\text{lb per bag}) \\ &= 45 \div 3 \\ &= 15 \end{aligned}$
   Daniel fills 15 bags of carrots.

7) $\begin{aligned} \text{avg temp} &= \frac{\text{sum of the daily temps}}{\text{number of days}} \\ &= \frac{64 + 61 + 72 + 59}{4} \\ &= \frac{256}{4} \\ &= 64 \end{aligned}$
   The four-day average of 64 degrees was lower than 65 degrees.

8) $\begin{aligned} \text{avg} &= \frac{\text{amount of water wasted}}{\text{number of hours}} \\ &= \frac{\text{change in meter}}{\text{number of hours}} \\ &= \frac{13{,}037 - 12{,}983}{2} \\ &= \frac{54}{2} \\ &= 27 \end{aligned}$
   The average amount of water wasted per hour was 27 gallons.

CONCEPT CHECKS AND PRACTICE EXERCISES

A1) a) Answers may vary. Examples include sum, add, total, plus, more than, increased by, deposited, greater than.

   b) Answers may vary. Examples include difference, subtract, minus, less, fewer than, less than, decreased by, loss of, withdraw.

   c) Answers may vary. Examples include product, times, of, double, tripled, multiplied by.

   d) Answers may vary. Examples include quotient, divided by, divided into, goes into, per, ratio.

   e) Answers may vary. Examples include is, was, equals, is the same as.

A2) Answers may vary. Example: A child has 10 pennies. She is given 20 more. How many pennies does she have?

A3) Answers may vary. Example: A soccer coach has $20 to spend on his 10 team members. How much can he spend on each member of the team?

A4) "Difference" indicates subtraction. $15 - 7$

A5) "Sum" indicates addition. $8 + 9$

A6) "Doubling" indicates multiplication by 2. $2 \cdot 15$

A7) "Divide into" indicates division. $42 \div 3$

A8) "Withdrawn" indicates subtraction. $80 - 40$

A9) "Of" indicates multiplication. $5 \cdot 6$

A10) "Product" indicates multiplication. $3 \cdot 4 \cdot 5$

A11) "Total" indicates addition. $3 + 4 + 5$

B1) Answers may vary. Example: Sarah pays $50 a month for her wireless service. How much does she pay for this service in one year?

B2) Answers may vary. Example: Simon has purchased a rug for his bedroom for $600 and was given the opportunity to purchase it for 12 months "same as cash". If he chose to take advantage of this plan, what were his monthly payments?

B3) $75 \div 15 = 5$
   It will take 5 hours to paint 75 bird houses.

B4) $5 \cdot 144 = 720$
720 pens were ordered.

B5) $\$84,312 - \$75,000 = \$9,312$
The repairs cost $9,312.

B6) $212 - 183 = 29$
The difference in the class sizes in 29 students.

B7) $\$45 + \$63 + \$21 + \$12 = \$141$
The total amount collected was $141.

B8) $\$835 - \$245 - \$120 = \$470$
$470 was left.

B9) $43,560 \cdot 23 = 1,001,880$
The area of 23 acres is 1,001,880 square feet.

B10) $\$176,000 \div 4 = \$44,000$
Their average income is $44,000.

C1) Answers may vary. Example: Writing an intermediate word equation may help the student focus on the words which lead directly to an equation.

C2) Answers may vary. Example: There is a great deal of difference between 1 inch and 1 mile. Writing the units gives greater clarity to the answer.

C3) $\text{avg} = \dfrac{\text{sum of life expectancies}}{\text{number of entries}}$
$= \dfrac{70 + 71 + 73 + 74}{4}$
$= \dfrac{288}{4}$
$= 72$
The average life expectancy of men is 72 years.

C4) $\text{avg} = \dfrac{\text{sum of life expectancies}}{\text{number of entries}}$
$= \dfrac{75 + 76 + 78 + 79}{4}$
$= \dfrac{308}{4}$
$= 77$
The average life expectancy of women is 77 years.

C5) $450 - 212 - 175 = 238 - 175$
$= 63$
63 freshman or sophomores entered the competition.

C6) $3 \cdot 3 + 4 \cdot 2 + 16 \cdot 2 = 9 + 4 \cdot 2 + 16 \cdot 2$
$= 9 + 8 + 16 \cdot 2$
$= 9 + 8 + 32$
$= 17 + 32$
$= 49$
Yanis's bill was $49.

C7) $(150 + 85 + 39 + 43) - 75 = (235 + 39 + 43) - 75$
$= (274 + 43) - 75$
$= 317 - 75$
$= 242$
The total bill was $242.

C8) $350 - (45 + 52 + 35) = 350 - (97 + 35)$
$= 350 - 132$
$= 218$
Janise spent $218 on the project.

SECTION 1.9   EXERCISES

For 1–9, refer to Concept Checks and Practice Exercises.

11) division

13) division

15) equality

17) addition

19) subtraction

21) subtraction

23) equality

25) "Multiplied" indicates multiplication.
$45 \cdot 19 = 855$

27) "Less than" indicates subtraction.
$38 - 27 = 11$

29) "Total" indicates addition.
$7 + 17 + 22 = 46$

31) "More than" indicates addition; "product" indicates multiplication.
$(6 \cdot 8) + 10 = 58$

33) "Withdrew" indicates subtraction.
$\$1,236 - \$245 = \$991$
The new balance is $991.

35) $76,473 + 5,500 = 81,973$
The next oil change chould occur at 81,973 miles.

37) $\$834 - \$350 = \$484$
The sale price is $484.

39) $12 \cdot 12 = 144$
There are 144 eggs in a case.

41) $43 + 16 - 7 = 59 - 7$
$ = 52$
The temperature at 5 P.M. was 52 degrees.

43) $(200 \cdot 85) \div 8 = 17,000 \div 8$
$ = 2,125$
It takes 2,125 gallons of water.

45) $\$940 \div 4 = \$235$
His average travel cost per year is $235.

47) $\text{avg low temp} = \dfrac{\text{sum of the daily low temps}}{\text{number of days}}$
$\phantom{\text{avg low temp}} = \dfrac{55 + 61 + 52}{3}$
$\phantom{\text{avg low temp}} = \dfrac{168}{3}$
$\phantom{\text{avg low temp}} = 56$
The average low temperature for the last three days is 56 degrees.

49) $43,560 \cdot 52 = 2,265,120$
There are 2,265,120 square feet on the farm.

51) $364 \div 14 = 26$
Kendra's car averaged 26 miles per gallon.

53) $2 \cdot 26 + 3 \cdot 14 = 52 + 42$
$ = 94$
The total cost is $94.

55) $(114,744 - 114,456) \div 9 = 288 \div 9$
$ = 32$
The gas mileage of this trip was 32 mpg.

57) $12(950 - 639) - 1968 = 12(311) - 1,968$
$ = 3,732 - 1,968$
$ = 1,764$
The yearly profit is $1,764.

CHAPTER 1   REVIEW EXERCISES

1) 7 is in the thousands period. The number in that period is 272. 7 is in the tens place of 272. 7 is in the ten thousands place.

2) 2 is in the millions period. The number in that period is 2. 2 is in the ones place of 2. 2 is in the millions place.

3) One million, ninety-eight thousand, three hundred forty-two

4) Thirty-four thousand, nine hundred eight

5) $30,402 = 30,000 + 400 + 2$

6) $105,030 = 100,000 + 5,000 + 30$

7) Identify the digit in the thousands place.
93⬚8⬚,412; 938,412 will round to either 938,000 or 939,000. Since the hundreds digit is 4, we round down. 938,412 rounds to 938,000.

8) Identify the digit in the hundreds place. ⬚87;
There isn't a digit written in the hundreds place, so we use 0. 87 will round to either 100 or 0. Since the tens digit is 8, we round up. 87 rounds to 100.

9) $\{16, 25, 32, 43\}$

10) $\{50, 80, 125, 175\}$

11) There are 70 McDonald's restaurants within 10 miles of Southfield, Michigan.

12) They would have to travel at most 50 miles to get to any one of 300 restaurants.

13) $\begin{array}{r} 4\,5 \\ +\,1\,3 \\ \hline 5\,8 \end{array}$

14) $\begin{array}{r} 2\,1 \\ +\,7\,6 \\ \hline 9\,7 \end{array}$

15)
```
      1
    127
  +  45
    172
```

16)
```
     11
    385
  +  87
    472
```

17)
```
      1
   1548
  + 531
   2079
```

18)
```
      1
   2632
  + 433
   3065
```

19)
```
    1   1
   8382
  + 7152
  15534
```

20)
```
    1   1
   7392
  + 9285
  16677
```

21)
```
      1
    420
  +285
    705
```
A ring with a 0.4 carat ruby and a 0.2 carat diamond will cost $705. Yes, he can afford this ring.

22)
```
      1
    725
  +125
    850
```
A ring with a 0.4 carat diamond and a 0.2 carat ruby will cost $850. Yes, he can afford this ring.

23)
```
    75
  − 43
    32
```

24)
```
    56
  − 23
    33
```

25)
```
      6
   1 7¹2
  −  38
   1 3 4
```

26)
```
      8
   4 9¹2
  −  55
   4 3 7
```

27)
```
      1
   1 2¹7 5
  −  183
   1 0 9 2
```

28)
```
    4   1
   5¹3 2¹6
  −  417
   4 9 0 9
```

29)
```
    7   9
   8 0¹5 7
  −  3265
   4 7 9 2
```

30)
```
    2   9
   5 3 0¹7
  −   289
   5 0 1 8
```

31) The difference in salaries was the greatest in the 4th year after graduating.
```
     3
    4¹3
  − 3 6
      7
```
Elaine earned $7,000 less than Blair that year.

32)
```
     2
    23
    28
    35
  + 36
   122
```
Elaine earned $122,000 in four years.
```
     2
    19
    27
    36
  + 43
   125
```
Blair earned $125,000 in four years.

33) Blair earned the most money in the four years.

$$
\begin{array}{r}
1\ 2\ 5 \\
-1\ 2\ 2 \\
\hline
3
\end{array}
$$

Blair earned $3,000 more over the four years.

34) When adding on a number line, start at the first number and move to the **right**. Sometimes, we need to **carry** a 1 to the next highest place value. To check an addition answer, we can **subtract**.

When subtracting on a number line, start at the first number and move to the **left**. Sometimes, we need to **borrow** a 1 from the next highest place value. To check a subtraction answer, we can **add**.

35) $21 \approx 20$ $21 \cdot 41 \approx 20 \cdot 40$
    $41 \approx 40$ $= 800$

$$
\begin{array}{r}
2\ 1 \\
\times\ 4\ 1 \\
\hline
2\ 1 \\
8\ 4\ 0 \\
\hline
8\ 6\ 1
\end{array}
$$

36) $32 \approx 30$ $32 \cdot 27 \approx 30 \cdot 30$
    $27 \approx 30$ $= 900$

$$
\begin{array}{r}
\not{1} \\
3\ 2 \\
\times\ 2\ 7 \\
\hline
2\ 2\ 4 \\
6\ 4\ 0 \\
\hline
8\ 6\ 4
\end{array}
$$

37) $67 \approx 70$ $67 \cdot 24 \approx 70 \cdot 20$
    $24 \approx 20$ $= 1,400$

$$
\begin{array}{r}
\not{1} \\
\not{1} \\
6\ 7 \\
\times\ 2\ 4 \\
\hline
2\ 6\ 8 \\
1\ 3\ 4\ 0 \\
\hline
1\ 6\ 0\ 8
\end{array}
$$

38) $38 \approx 40$ $38 \cdot 79 \approx 40 \cdot 80$
    $79 \approx 80$ $= 3,200$

$$
\begin{array}{r}
\not{8} \\
\not{7} \\
3\ 8 \\
\times\ 7\ 9 \\
\hline
3\ 4\ 2 \\
2\ 6\ 6\ 0 \\
\hline
3\ 0\ 0\ 2
\end{array}
$$

39) $184 \approx 200$ $184 \cdot 53 \approx 200 \cdot 50$
    $53 \approx 50$ $= 10,000$

$$
\begin{array}{r}
\not{4}\not{2} \\
\not{2}\not{1} \\
1\ 8\ 4 \\
\times\ 5\ 3 \\
\hline
5\ 5\ 2 \\
9\ 2\ 0\ 0 \\
\hline
9\ 7\ 5\ 2
\end{array}
$$

40) $285 \approx 300$ $285 \cdot 76 \approx 300 \cdot 80$
    $76 \approx 80$ $= 24,000$

$$
\begin{array}{r}
\not{8}\not{8} \\
\not{8}\not{8} \\
2\ 8\ 5 \\
\times\ 7\ 6 \\
\hline
1\ 7\ 1\ 0 \\
1\ 9\ 9\ 5\ 0 \\
\hline
2\ 1\ 6\ 6\ 0
\end{array}
$$

41) $2 \cdot 6 \cdot 7 \cdot 5 = 12 \cdot 7 \cdot 5$
    $ = 84 \cdot 5$
    $ = 420$

42) $8 \cdot 3 \cdot 7 \cdot 2 = 24 \cdot 7 \cdot 2$
    $ = 168 \cdot 2$
    $ = 336$

43) $A = l \cdot w$
    $ = 28 \cdot 15$
    $ = 420$ square feet

   $A = l \cdot w$
   $ = 21 \cdot 19$
   $ = 399$ square feet

$$
\begin{array}{r}
{}^{3}\ {}^{1}1 \\
\not{4}\ \not{2}\ {}^{1}0 \\
-3\ 9\ 9 \\
\hline
2\ 1
\end{array}
$$

The area of the short rectangle is 21 square feet larger than the area of the tall rectangle.

44) $15 \cdot \$372 = \$5,580$
    The total of her payments is $5,580.
    $\$12,856 - \$5,580 = \$7,276$
    She still owes $7,276.

45) $4 \cdot 3 + 3 \cdot 6 = 12 + 18$
    $ = 30$
    The bears scored 30 points.

46) $12(850 - 817) = 12 \cdot 33$
    $ = 396$
    They will save $396 in the first 12 months if they sign a 24-month lease.

47) $72 \div 9 = \boxed{?} \implies \boxed{?} \cdot 9 = 72$
    $\phantom{72 \div 9 = \boxed{?} \implies} 8 \cdot 9 = 72$
    Therefore, $72 \div 9 = 8$.

48) $56 \div 8 = \boxed{?} \implies \boxed{?} \cdot 8 = 56$
    $\phantom{56 \div 8 = \boxed{?} \implies} 7 \cdot 8 = 56$
    Therefore, $56 \div 8 = 7$.

49) $7 \approx 7$ (Rounded to the ones place.)
$73 \approx 70$ (Rounded to the tens place.)

$$
\begin{array}{r}
1\ 0 \\
7\overline{)\ 7\ 0} \\
-7\ 0 \\
\hline
0\ 0
\end{array}
\qquad
\begin{array}{r}
1\ 0\,\text{R}3 \\
7\overline{)\ 7\ 3} \\
-7\ 0 \\
\hline
3
\end{array}
$$

50) $3 \approx 3$ (Rounded to the ones place.)
$58 \approx 60$ (Rounded to the tens place.)

$$
\begin{array}{r}
2\ 0 \\
3\overline{)\ 6\ 0} \\
-6 \\
\hline
0\ 0
\end{array}
\qquad
\begin{array}{r}
1\ 9\,\text{R}1 \\
3\overline{)\ 5\ 8} \\
-3 \\
\hline
2\ 8 \\
-2\ 7 \\
\hline
1
\end{array}
$$

51) $14 \approx 10$ (Rounded to the tens place.)
$171 \approx 200$ (Rounded to the hundreds place.)

$$
\begin{array}{r}
2\ 0 \\
1\ 0\overline{)\ 2\ 0\ 0} \\
-2\ 0 \\
\hline
0\ 0
\end{array}
\qquad
\begin{array}{r}
1\ 2\,\text{R}3 \\
1\ 4\overline{)\ 1\ 7\ 1} \\
-1\ 4 \\
\hline
3\ 1 \\
-2\ 8 \\
\hline
3
\end{array}
$$

52) $18 \approx 20$ (Rounded to the tens place.)
$237 \approx 200$ (Rounded to the hundreds place.)

$$
\begin{array}{r}
1\ 0 \\
2\ 0\overline{)\ 2\ 0\ 0} \\
-2\ 0 \\
\hline
0\ 0
\end{array}
\qquad
\begin{array}{r}
1\ 3\,\text{R}3 \\
1\ 8\overline{)\ 2\ 3\ 7} \\
-1\ 8 \\
\hline
5\ 7 \\
-5\ 4 \\
\hline
3
\end{array}
$$

53) $21 \approx 20$ (Rounded to the tens place.)
$6{,}572 \approx 7{,}000$ (Rounded to the thousands place.)

$$
\begin{array}{r}
3\ 5\ 0 \\
2\ 0\overline{)\ 7\ 0\ 0\ 0} \\
-6\ 0 \\
\hline
1\ 0\ 0 \\
-1\ 0\ 0 \\
\hline
0\ 0
\end{array}
\qquad
\begin{array}{r}
3\ 1\ 2\,\text{R}20 \\
2\ 1\overline{)\ 6\ 5\ 7\ 2} \\
-6\ 3 \\
\hline
2\ 7 \\
-2\ 1 \\
\hline
6\ 2 \\
-4\ 2 \\
\hline
2\ 0
\end{array}
$$

54) $36 \approx 40$ (Rounded to the tens place.)
$7{,}540 \approx 8{,}000$ (Rounded to the thousands place.)

$$
\begin{array}{r}
2\ 0\ 0 \\
4\ 0\overline{)\ 8\ 0\ 0\ 0} \\
-8\ 0 \\
\hline
0\ 0\ 0
\end{array}
\qquad
\begin{array}{r}
2\ 0\ 9\,\text{R}16 \\
3\ 6\overline{)\ 7\ 5\ 4\ 0} \\
-7\ 2 \\
\hline
3\ 4\ 0 \\
-3\ 2\ 4 \\
\hline
1\ 6
\end{array}
$$

55) $4 \cdot 200 = 800$ two-by-fours

$$
\begin{array}{r}
9\,\text{R}35 \\
8\ 5\overline{)\ 8\ 0\ 0} \\
-7\ 6\ 5 \\
\hline
3\ 5
\end{array}
$$

The carpenter can build nine sheds.

56) $3 \cdot 100 = 300$ bags

$$
\begin{array}{r}
3\,\text{R}45 \\
8\ 5\overline{)\ 3\ 0\ 0} \\
-2\ 5\ 5 \\
\hline
4\ 5
\end{array}
$$

Three hours will pass before more bags are
needed.

57)   **First:** Perform operations within grouping
symbols.
Second: Evaluate any exponential expressions.
Third: Perform multiplication and division as
they occur from left to right.
Fourth: Perform addition and subtraction as
they occur from left to right.

58) $2^3 = 8$

59) $4^2 = 16$

60) $2 + 3^3 = 2 + 27$
$ = 29$

61) $100 - 7^2 = 100 - 49$
$ = 51$

62) $15 - 3 + 2 = 12 + 2$
$ = 14$

63) $21 \div 7 \cdot 3 = 3 \cdot 3$
$ = 9$

64) $(5 - 3)^2 + 5 = 2^2 + 5$
$ = 4 + 5$
$ = 9$

65) $25-(8-7)^3 = 25-1^3$
$= 25-1$
$= 24$

66) $15 \div (8-3) = 15 \div 5$
$= 3$

67) $42-(52-33)+5 = 42-19+5$
$= 23+5$
$= 28$

68) $\dfrac{7-(5-2)}{12 \div (8-2)} = \dfrac{7-3}{12 \div 6}$
$= \dfrac{7-3}{2}$
$= \dfrac{4}{2}$
$= 2$

69) $\dfrac{40-5 \cdot 6}{4^2-11} = \dfrac{40-5 \cdot 6}{16-11}$
$= \dfrac{40-30}{16-11}$
$= \dfrac{10}{5}$
$= 2$

70) Answers may vary. Example: 85

71) avg exam score $= \dfrac{\text{sum of the exam scores}}{\text{number of exams}}$
$= \dfrac{71+79+88+95+87}{5}$
$= \dfrac{420}{5}$
$= 84$

Alex's average exam score is 84. This corresponds to a grade of B.

72) The lowest grade Alex received on any one exam was 71, a C–. The average will not be lower than the lowest grade.

73) $2^2 = 4,\ 3^2 = 9,\ 5^2 = 25,\ 7^2 = 49$
Since 49 is more than 39, we only have to check through 5. We must check 2, 3, 5.
2 doesn't divide 39 evenly because 39 is odd.
3 does divide 39 evenly because $3+9 = 12$ is divisible by 3.
39 is composite.

74) $2^2 = 4,\ 3^2 = 9,\ 5^2 = 25,\ 7^2 = 49$
Since 49 is more than 41, we only have to check through 5. We must check 2, 3, 5.
2 doesn't divide 41 evenly because 41 is odd.
3 doesn't divide 41 evenly because $4+1 = 5$ is not divisible by 3.
5 doesn't divide 41 evenly because 41 does not end in 0 or 5.
41 is prime.

75) $20 \div 1 = 20 \qquad 20 \div 2 = 10 \qquad 20 \div 4 = 5$
$20 \div 20 = 1 \qquad 20 \div 10 = 2 \qquad 20 \div 5 = 4$
The three factor pairs of 20 are 1 and 20, 2 and 10, 4 and 5.

76) $36 \div 1 = 36 \qquad 36 \div 2 = 18 \qquad 36 \div 3 = 12$
$36 \div 36 = 1 \qquad 36 \div 18 = 2 \qquad 36 \div 12 = 3$
$36 \div 4 = 9 \qquad 36 \div 6 = 6$
$36 \div 9 = 4 \qquad 36 \div 6 = 6$
The five factor pairs of 36 are 1 and 36, 2 and 18, 3 and 12, 4 and 9, 6 and 6.

77)
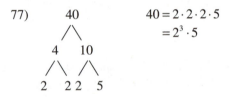
$40 = 2 \cdot 2 \cdot 2 \cdot 5$
$= 2^3 \cdot 5$

78)
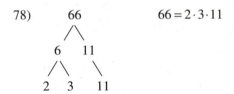
$66 = 2 \cdot 3 \cdot 11$

79) The first eight multiples of 6 are as follows.
$1 \cdot 6 = 6$
$2 \cdot 6 = 12$
$3 \cdot 6 = 18$
$4 \cdot 6 = 24$
$5 \cdot 6 = 30$
$6 \cdot 6 = 36$
$7 \cdot 6 = 42$
$8 \cdot 6 = 48$

80) The first six multiples of 8 are as follows.
$1 \cdot 8 = 8$
$2 \cdot 8 = 16$
$3 \cdot 8 = 24$
$4 \cdot 8 = 32$
$5 \cdot 8 = 40$
$6 \cdot 8 = 48$

81) 33 must be the greatest common factor, because the GCF must be less than or equal to the other numbers.

82) 198 must be the least common multiple, because the LCM must be greater than or equal to the other numbers.

83) Factors of 24: 1, 2, 3, 4, 6, 8, 12, 24
Factors of 16: 1, 2, 4, 8, 16
Common factors of 24 and 16: 1, 2, 4, 8
GCF: 8

84) Factors of 28: 1, 2, 4, 7, 14, 28
Factors of 21: 1, 3, 7, 21
Common factors of 28 and 21: 1, 7
GCF: 7

85) Multiples of 12: 12, 24, 36, 48, 60
Multiples of 15: 15, 30, 45, 60
LCM: 60

86) Multiples of 12: 12, 24, 36
Multiples of 9: 9, 18, 27, 36
LCM: 36

87) $20 = 2 \cdot 2 \cdot 5$
$24 = 2 \cdot 2 \cdot 2 \cdot 3$
LCM: $2 \cdot 2 \cdot 2 \cdot 3 \cdot 5 = 120$
GCF: $2 \cdot 2 = 4$

88) $32 = 2 \cdot 2 \cdot 2 \cdot 2 \cdot 2$
$28 = 2 \cdot 2 \cdot 7$
LCM: $2 \cdot 2 \cdot 2 \cdot 2 \cdot 2 \cdot 7 = 224$
GCF: $2 \cdot 2 = 4$

89) $20 = 2 \cdot 2 \cdot 5$
$40 = 2 \cdot 2 \cdot 2 \cdot 5$
LCM: $2 \cdot 2 \cdot 2 \cdot 5 = 40$
GCF: $2 \cdot 2 \cdot 5 = 20$

90) $14 = 2 \cdot 7$
$15 = 3 \cdot 5$
LCM: $2 \cdot 3 \cdot 5 \cdot 7 = 210$
GCF: 1

91) "Product" indicates multiplication.
$16 \cdot 24 = 384$

92) "Quotient" indicates division.
$256 \div 64 = 4$

93) "Quotient" indicates division; "increased by" indicates addition.
$(72 \div 8) + 17 = 9 + 17$
$\qquad\qquad\qquad = 26$

94) "Product" indicates multiplication; "decreased by" indicates subtraction.
$(8 \cdot 7) - 30 = 56 - 30$
$\qquad\qquad\qquad = 26$

95) $(97 - 76) \div 3 = 21 \div 3$
$\qquad\qquad\qquad = 7$
Angela was charged $7 for each box.

96) $(2,598 - 1,944) \div 3 = 654 \div 3$
$\qquad\qquad\qquad\qquad = 218$
The amount of each payment was $218.

97) avg output $= \dfrac{\text{sum of the outputs}}{\text{number of windmills}}$
$\qquad\qquad = \dfrac{115 + 118 + 117 + 126 + 99}{5}$
$\qquad\qquad = \dfrac{575}{5}$
$\qquad\qquad = 115$
The average output for each windmill is 115 kilowatt hours of electricity. Since 115 is greater than 110, the farmer is making money.

98)
```
      1
    1 2
  × 7 3
  ─────
    3 6
  8 4 0
  ─────
  8 7 6
```
876 kilowatt hours per day are needed in the subdivision.

```
    8 7 6
  − 6 0 0
  ───────
    2 7 6
```
The subdivision will need 276 kilowatt hours more per day.

CHAPTER 1   TEST

1) 8 is in the thousands period. The number in that period is 876. 8 is in the hundreds place of 876. 8 is in the hundred thousands place.

2) Sixty thousand, eight hundred nine

3) Identify the digit in the ten thousands place.
8$\boxed{2}$7,398; 827,398 will round to either 820,000
or 830,000. Since the next digit is 7, we round
up. 827,398 rounds to 830,000.

4) $107,060 = 100,000 + 7000 + 60$

5) $\{23, 30, 10, 18\}$

6) The largest gem Dante can purchase for $300 is
a 0.3 carat ruby.

7)
$$
\begin{array}{r}
1\,3\,7 \\
+\ \ 4\,2 \\
\hline
1\,7\,9
\end{array}
$$

8)
$$
\begin{array}{r}
2\,9\,3 \\
-\ \ 5\,2 \\
\hline
2\,4\,1
\end{array}
$$

9)
$$
\begin{array}{r}
{}^{3}\ {}^{16} \\
\cancel{4}\,\cancel{7}\,{}^{1}5 \\
-\,1\,8\,6 \\
\hline
2\,8\,9
\end{array}
$$

10)
$$
\begin{array}{r}
{}^{1}\ {}^{1} \\
3\,9\,8 \\
+\,2\,7\,5 \\
\hline
6\,7\,3
\end{array}
$$

11)
$$
\begin{array}{r}
{}^{1}\ {}^{1} \\
4\,2\,3\,5 \\
+\ \ 7\,8\,3 \\
\hline
5\,0\,1\,8
\end{array}
$$

12)
$$
\begin{array}{r}
{}^{6}\ {}^{9} \\
\cancel{7}\,\cancel{0}\,{}^{1}3\,2 \\
-\ \ \ 6\,7\,1 \\
\hline
6\,3\,6\,1
\end{array}
$$

13)
$$
\begin{array}{r}
{}^{4}\ {}^{9} \\
\cancel{5}\,\cancel{0}\,{}^{1}4\,5 \\
-\ \ \ 3\,5\,2 \\
\hline
4\,6\,9\,3
\end{array}
$$

14)
$$
\begin{array}{r}
{}^{1}\ {}^{1} \\
6\,9\,5\,6 \\
+\ \ 3\,7\,2 \\
\hline
7\,3\,2\,8
\end{array}
$$

15) $28 + 37 + 48 = 113$
The CEO earned a total of $113,000,000 in
those years.

16) $48 - 28 = 20$
The CEO's compensation increased
$20,000,000 from 2004 to 2006.

17) $83 - 48 = 35$
The new CEO earned $35,000,000 million
more.

18) $\text{avg earnings} = \dfrac{\text{sum of the earnings}}{\text{number of years}}$

$$
= \frac{28 + 37 + 48 + 83}{4}
$$

$$
= \frac{196}{4}
$$

$$
= 49
$$

The company paid its CEO an average of
$49,000,000 each of these four years.

19) $48 \approx 50$ $48 \cdot 72 \approx 50 \cdot 70$
$72 \approx 70$ $= 3,500$

$$
\begin{array}{r}
\cancel{8} \\
\cancel{1} \\
4\,8 \\
\times 7\,2 \\
\hline
9\,6 \\
3\,3\,6\,0 \\
\hline
3\,4\,5\,6
\end{array}
$$

20) $423 \div 11 \approx 400 \div 10$

$$
\begin{array}{r}
4\,0 \\
1\,0\,)\overline{4\,0\,0} \\
-4\,0 \\
\hline
0\,0
\end{array}
\qquad
\begin{array}{r}
3\,8\,\text{R}5 \\
1\,1\,)\overline{4\,2\,3} \\
-3\,3 \\
\hline
9\,3 \\
-8\,8 \\
\hline
5
\end{array}
$$

21) $392 \div 18 \approx 400 \div 20$

$$
\begin{array}{r}
2\,0 \\
2\,0\,)\overline{4\,0\,0} \\
-4\,0 \\
\hline
0\,0
\end{array}
\qquad
\begin{array}{r}
2\,1\,\text{R}14 \\
1\,8\,)\overline{3\,9\,2} \\
-3\,6 \\
\hline
3\,2 \\
-1\,8 \\
\hline
1\,4
\end{array}
$$

22) $162 \cdot 67 \approx 200 \cdot 70$

$$
\begin{array}{r}
2\,0\,0 \\
\times\ 7\,0 \\
\hline
1\,4\,0\,0\,0
\end{array}
$$

$$
\begin{array}{r}
1\,6\,2 \\
\times 6\,7 \\
\hline
1\,1\,3\cdot4 \\
9\,7\,2 \\
\hline
1\,0\,8\,5\,4
\end{array}
$$

23) $5{,}723 \cdot 492 \approx 6{,}000 \cdot 500$

$$
\begin{array}{r}
6\,0\,0\,0 \\
\times 5\,0\,0 \\
\hline
3\,0\,0\,0\,0\,0\,0
\end{array}
$$

$$
\begin{array}{r}
5\,7\,2\,3 \\
\times 4\,9\,2 \\
\hline
1\,1\,4\,4\,6 \\
5\,1\,5\,0\,7 \\
2\,2\,8\,9\,2 \\
\hline
2\,8\,1\,5\,7\,1\,6
\end{array}
$$

24) $6{,}834 \div 52 \approx 7{,}000 \div 50$

$$
\begin{array}{r}
1\,4\,0 \\
5\,0\,\overline{)\,7\,0\,0\,0} \\
-\,5\,0 \\
\hline
2\,0\,0 \\
-\,2\,0\,0 \\
\hline
0\,0
\end{array}
\qquad
\begin{array}{r}
1\,3\,1\,\text{R}22 \\
5\,2\,\overline{)\,6\,8\,3\,4} \\
-\,5\,2 \\
\hline
1\,6\,3 \\
-\,1\,5\,6 \\
\hline
7\,4 \\
-\,5\,2 \\
\hline
2\,2
\end{array}
$$

25)
$$
\begin{array}{r}
1\,9\,0 \\
\times 1\,5 \\
\hline
9\,5\,0 \\
1\,9\,0 \\
\hline
2\,8\,5\,0
\end{array}
$$

The force acting on the truck is 2,850 pounds.

26)
$$
\begin{array}{r}
1\,7\,2 \\
1\,8\,\overline{)\,3\,0\,9\,6} \\
-\,1\,8 \\
\hline
1\,2\,9 \\
-\,1\,2\,6 \\
\hline
3\,6 \\
-\,3\,6 \\
\hline
0
\end{array}
$$

The jack is making 172 square inches of contact.

27)
$$
\begin{array}{r}
3\,2 \\
2\,8\,\overline{)\,8\,9\,6} \\
-\,8\,4 \\
\hline
5\,6 \\
-\,5\,6 \\
\hline
0
\end{array}
$$

The author sold 32 books.

28)
$$
\begin{array}{r}
2\,8 \\
\times 1\,9 \\
\hline
2\,5\,2 \\
2\,8\,0 \\
\hline
5\,3\,2
\end{array}
$$

The author earned an income of $532.

29) $3^4 - 31 = 81 - 31$
$$= 50$$

30) $53 - 24 + 6 = 29 + 6$
$$= 35$$

31) $3 + 2 \cdot 5 = 3 + 10$
$$= 13$$

32) $40 \div 2 \cdot 5 = 20 \cdot 5$
$$= 100$$

33) $2 \cdot (5-3)^2 = 2 \cdot (2)^2$
$$= 2 \cdot 4$$
$$= 8$$

34) $\dfrac{24 - 4 \cdot 2}{2^3} = \dfrac{24 - 4 \cdot 2}{8}$
$$= \dfrac{24 - 8}{8}$$
$$= \dfrac{16}{8}$$
$$= 2$$

35) $40 \div 1 = 40 \qquad 40 \div 2 = 20$
$40 \div 40 = 1 \qquad 40 \div 20 = 2$

$40 \div 4 = 10 \qquad 40 \div 5 = 8$
$40 \div 10 = 4 \qquad 40 \div 8 = 5$

The four factor pairs of 40 are 1 and 40, 2 and 20, 4 and 10, and 5 and 8.

36)   42 $42 = 2 \cdot 3 \cdot 7$

37) Multiples of 25: 25, 50, 75
    Multiples of 15: 15, 30, 45, 60, 75
    LCM: 75

38) Factors of 30: 1, 2, 3, 5, 6, 10, 15, 30
    Factors of 20: 1, 2, 4, 5, 10, 20
    Common factors of 30 and 20: 1, 2, 5, 10
    GCF: 10

39) $42 = 2 \cdot 3 \cdot 7$
    $24 = 2 \cdot 2 \cdot 2 \cdot 3$
    LCM: $2 \cdot 2 \cdot 2 \cdot 3 \cdot 7 = 168$
    GCF: $2 \cdot 3 = 6$

40) $16 = 2 \cdot 2 \cdot 2 \cdot 2$
    $28 = 2 \cdot 2 \cdot 7$
    LCM: $2 \cdot 2 \cdot 2 \cdot 2 \cdot 7 = 112$
    GCF: $2 \cdot 2 = 4$

41) $583 - (20 \cdot 18) = 583 - 360$
    $ = 223$
    $223 is left in the account.

42) $75 \div 15 + 100 \div 10 = 5 + 10$
    $ = 15$
    His truck used 15 gallons of gas.

Chapter 2   Fractions

2.1   Visualizing Fractions

GUIDED PRACTICE

1) There are three shaded pieces.
   The numerator is 3.
   There are 5 pieces in all.
   The denominator is 5.

   The fraction is $\dfrac{\text{numerator}}{\text{denominator}}$ or $\dfrac{3}{5}$.

2) Since there are 6 equal-sized parts in all, the
   denominator is 6. Fraction $=\dfrac{}{6}$
   Since there are 4 shaded parts, the numerator is
   4. Fraction $=\dfrac{4}{6}$

3) The denominator is the total number of hours in
   a school day.
   Denominator $= 3+1+2+1 = 7$
   The numerator is the number of reading hours.
   Numerator $= 3$
   Fraction $=\dfrac{\text{numerator}}{\text{denominator}}=\dfrac{3}{7}$

4) Since the denominator is 8, we draw an object
   with 8 equal-sized parts.
   Since the numerator is 3, we shade 3 of those
   parts.

5) Proper Fractions
   Fraction a) is proper because the numerator is
   less than the denominator.
   Picture e) is a proper fraction because less than
   one whole object is shaded.

   Improper Fractions
   Fractions d) and f) are improper because the
   numerator is equal to or greater than the
   denominator.
   Pictures b) and c) show improper fractions
   because at least one whole object is shaded.

6) Since there are 3 equal-sized parts in each
   object, the denominator is 3. Fraction $=\dfrac{}{3}$
   Since there are 7 shaded parts, the numerator is
   7. Fraction $=\dfrac{7}{3}$

7) Since the denominator is 5, we draw objects
   with 5 equal-sized parts.
   The numerator is 11, so we must shade 11 parts.
   A total of 3 whole objects must be drawn to
   shade 11 parts.

8) Use the chart to create a visual prepresentation
   of each fraction.

   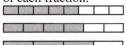

   Since $\dfrac{3}{4}$ represents a larger portion of a whole

   than $\dfrac{4}{7}$, $\dfrac{3}{4}$ is larger than $\dfrac{4}{7}$.

9) Use the chart to create a visual prepresentation
   of each fraction.

   Order the fractions from smallest to largest.
   $\dfrac{5}{8}, \dfrac{4}{6}$ and $\dfrac{3}{4}$
   Graph the fractions on the same number line.

10) a)   One-third is less than one-half.
    $\dfrac{1}{3} < \dfrac{1}{2}$

    b)   Two-sixths is greater than one-sixth.
    $\dfrac{2}{6} > \dfrac{1}{6}$

11) Since numerators are the same, the fracton with
    larger pieces is greater than the other.
    Since 5^{ths} are larger than 6^{ths}, $\dfrac{4}{5} > \dfrac{4}{6}$.

12) Since the denominators are the same, the
    fraction with fewer pieces is less than the other.
    $\dfrac{3}{5} < \dfrac{4}{5}$

13) 8 is the part of the cars that have been washed
    and 13 is the number of cars on the lot.
    $\dfrac{\text{part}}{\text{whole}} = \dfrac{\text{numerator}}{\text{denominator}} = \dfrac{8}{13}$

14) 25 is the part of the cars that are sedans.
    $25 + 38 = 63$ is the number of cars on the lot.
    $\dfrac{\text{part}}{\text{whole}} = \dfrac{\text{numerator}}{\text{denominator}} = \dfrac{25}{63}$

15)   To mark fourths, the ruler must be cut into 4
       equal-sized pieces.

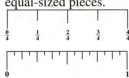

16)   Decide which marks represent halves and
       fourths.

       Count halves from the left and graph $\frac{1}{2}$.

       Count fourths from the left and graph $\frac{3}{4}$ and $\frac{4}{4}$.

17)   Decide which marks represent eighths and
       sixteenths.

       Count eights from the left and graph $\frac{3}{8}$.

       Count sixteenths from the left and graph $\frac{9}{16}$.

CONCEPT CHECKS AND PRACTICE EXERCISES

A1)   For the fraction $\frac{2}{7}$, the numerator is 2 and the
       denominator is 7.

A2)   When you represent a picture as a fraction, the
       number of equal-sized pieces in a whole
       becomes the denominator and the number of
       shaded pieces become the numerator.

A3)   Since there are five equal-sized parts, the
       denominator is 5.
       Since there are three shaded parts, the
       numerator is 3.
       Fraction $= \frac{3}{5}$

A4)   Since there are four equal-sized parts, the
       denominator is 4.
       Since there are three shaded parts, the
       numerator is 3.
       Fraction $= \frac{3}{4}$

A5)   Since there are two equal-sized parts, the
       denominator is 2.
       Since there is one shaded part, the numerator is 1.
       Fraction $= \frac{1}{2}$

A6)   Since there are two equal-sized parts, the
       denominator is 2.
       Since there are two shaded parts, the numerator
       is 2.
       Fraction $= \frac{2}{2}$

A7)   Since there are six equal-sized parts, the
       denominator is 6.
       Since there are four shaded parts, the numerator
       is 4.
       Fraction $= \frac{4}{6}$

A8)   Since there are four equal-sized parts, the
       denominator is 4.
       Since there are two shaded parts, the numerator
       is 2.
       Fraction $= \frac{2}{4}$

A9)   The denominator is the total items of clothing.
       Denominator $= 4 + 6 + 3 = 13$
       The numerator is the number of shoes.
       Numerator $= 4$
       Fraction $= \frac{\text{numerator}}{\text{denominator}} = \frac{4}{13}$

A10)   The denominator is the total bags of fruit.
       Denominator $= 5 + 1 + 2 = 8$
       The numerator is the number of bags of grapes.
       Numerator $= 5$
       Fraction $= \frac{\text{numerator}}{\text{denominator}} = \frac{5}{8}$

B1)   When representing a fraction as a picture, the
       numerator indicates the number of shaded parts.

B2)   When representing a fraction as a picture, the
       denominator indicates the total number of
       equal-sized parts in the whole.

B3)   Since the denominator is 6, we draw an object
       with six equal parts.
       Since the numerator is 3, we shade three of
       those parts.
       The corresponding picture is c.

B4)   Since the denominator is 6, we draw an object
      with six equal parts.
      Since the numerator is 5, we shade five of those
      parts.
      The corresponding picture is a.

B5)   Since the denominator is 5, we draw an object
      with five equal parts.
      Since the numerator is 3, we shade three of
      those parts.
      The corresponding picture is d.

B6)   Since the denominator is 7, we draw an object
      with seven equal parts.
      Since the numerator is 6, we shade six of those
      parts.
      The corresponding picture is b.

B7)   Since the denominator is 3, we draw an object
      with three equal-sized parts.
      Since the numerator is 2, we shade two of those
      parts.

B8)   Since the denominator is 8, we draw an object
      with eight equal-sized parts.
      Since the numerator is 5, we shade five of those
      parts.

B9)   Since the denominator is 4, we draw an object
      with four equal-sized parts.
      Since the numerator is 0, we shade zero of those
      parts.

B10) Since the denominator is 2, we draw an object
      with two equal-sized parts.
      Since the numerator is 0, we shade zero of those
      parts.

B11)

B12)

C1)   Since an improper fraction may represent more
      than one whole object, we may need to draw
      more than one object to represent the fraction as
      a picture.

C2)   An improper fraction that can be drawn with
      exactly one object is equal to one. The
      numerator and denominator are equal.

C3)   Since the denominator is 3, we draw objects
      with three equal-sized parts.
      The numerator is 6, so we must shade six parts.
      A total of two whole objects must be drawn to
      shade six parts.
      The corresponding picture is c.

C4)   Since the denominator is 5, we draw objects
      with five equal-sized parts.
      The numerator is 6, so we must shade six parts.
      A total of two whole objects must be drawn to
      shade six parts.
      The corresponding picture is b.

C5)   Since the denominator is 5, we draw objects
      with five equal-sized parts.
      The numerator is 5, so we must shade five parts.
      One whole object must be drawn.
      The corresponding picture is d.

C6)   Since the denominator is 1, we draw objects
      with one part.
      The numerator is 6, so we must shade six parts.
      A total of six whole objects must be drawn to
      shade six parts.
      The corresponding picture is a.

C7)   Since the denominator is 2, we draw objects
      with two equal-sized parts.
      The numerator is 4, so we must shade four
      parts.
      A total of two whole objects must be drawn to
      shade four parts.

C8)   Since the denominator is 5, we draw objects
      with five equal-sized parts.
      The numerator is 10, so we must shade ten
      parts.
      A total of two whole objects must be drawn to
      shade ten parts.

C9)   Since the denominator is 5, we draw objects
      with five equal-sized parts.
      The numerator is 7, so we must shade seven
      parts.
      A total of two whole objects must be drawn to
      shade seven parts.

C10)   Since the denominator is 3, we draw objects with three equal-sized parts.
The numerator is 5, so we must shade five parts. A total of two whole objects must be drawn to shade five parts.

D3) $\dfrac{4}{6} > \dfrac{1}{2}$

C11)   Since there are three equal-sized parts in each object, the denominator is 3.
Since there are a total of six shaded parts, the numerator is 6.

Fraction $= \dfrac{6}{3}$

Since the numerator is greater than or equal to the denominator, the fraction is improper.

D4) $\dfrac{3}{4} > \dfrac{5}{8}$

D5) $\dfrac{6}{8} > \dfrac{5}{7}$

D6) $\dfrac{1}{5} < \dfrac{1}{4} < \dfrac{1}{3}$

C12)   Since there are four equal parts in each object, the denominator is 4.
Since there are a total of six shaded parts, the numerator is 6.

Fraction $= \dfrac{6}{4}$

Since the numerator is greater than or equal to the denominator, the fraction is improper.

D7) $\dfrac{2}{3} < \dfrac{5}{7} < \dfrac{3}{4}$

D8) $\dfrac{3}{8} < \dfrac{3}{7} < \dfrac{3}{6} < \dfrac{3}{5}$

C13)   The denominator is the total packages of ice cream.
Denominator $= 3$
The numerator is the packages of vanilla ice cream.
Numerator $= 1$

Fraction $= \dfrac{\text{numerator}}{\text{denominator}} = \dfrac{1}{3}$

Since the numerator is less than the denominator, the fraction is proper.

D9) $\dfrac{2}{7} < \dfrac{2}{6} < \dfrac{2}{5} < \dfrac{2}{4}$

E1)   When comparing fractions that have the same denominator, a larger numerator indicates a larger fraction because there will more shaded parts.

C14)   The denominator is the total cartons of milk.
Denominator $= 4$
The numerator is the cartons of chocolate milk.
Numerator $= 3$

Fraction $= \dfrac{\text{numerator}}{\text{denominator}} = \dfrac{3}{4}$

Since the numerator is greater than or equal to the denominator, the fraction is proper.

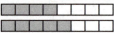

E2)   When comparing fractions that have the same numerator, a smaller denominator indicates a larger fraction because each individual part will be larger.

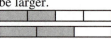

C15)   Answers may vary. Example: It is not possible to draw a whole with zero parts.

E3) $\dfrac{9}{7} < \dfrac{10}{7}$

D1)   Answers may vary. Example: Given that the "whole" pictures are the same size, the larger fraction picture will have more area shaded than the smaller fraction picture.

E4) $\dfrac{3}{3} > \dfrac{2}{3}$

E5) $\dfrac{15}{31} < \dfrac{15}{19}$

E6) $\dfrac{21}{52} < \dfrac{21}{25}$

D2) $\dfrac{1}{5} > \dfrac{1}{7}$

E7)　$\dfrac{6}{2} > \dfrac{6}{4}$

E8)　$\dfrac{7}{10} < \dfrac{7}{4}$

E9)　$\dfrac{8}{90} < \dfrac{10}{90}$

E10)　$\dfrac{70}{120} > \dfrac{20}{120}$

SECTION 2.1　EXERCISES

For 1–15, refer to Concept Checks and Practice Exercises.

17)　fraction

19)　proper fraction

21)　denominator

23)　Answers may vary. Example: The numerator tells you the number of shaded parts.

25)　Since there are four equal-sized parts, the denominator is 4.
Since there are seven shaded parts, the numerator is 7.
Fraction $= \dfrac{7}{4}$

27)　Since the denominator is 6, we draw an object with six equal-sized parts.
Since the numerator is 1, we shade one of those parts.

29)　Since the denominator is 2, we draw objects with two equal-sized parts.
The numerator is 4, so we must shade four parts.
A total of two whole objects must be drawn to shade four parts.

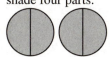

31)　The denominator is the total number of children.
Denominator = 6
The numerator is the number of children in pink shirts.
Numerator = 2
Fraction $= \dfrac{\text{numerator}}{\text{denominator}} = \dfrac{2}{6}$

32)　Since the denominator is 2, we draw an object with two equal-sized parts.
Since the numerator is 0, we shade zero of those parts.

33)　Since there are six equal-sized parts in the object, the denominator is 6.
Since there are a total of zero shaded parts, the numerator is 0.
Fraction $= \dfrac{0}{6}$

35)　Since there are four equal-sized parts in each object, the denominator is 4.
Since there are a total of three shaded parts, the numerator is 3.
Fraction $= \dfrac{3}{4}$

37)　a)

b)　$\dfrac{4}{8}$

39)　a)

b)　$\dfrac{3}{4}$

41)　a)

b)　$\dfrac{1}{3}$

43)　Subtraction; there are fewer shaded parts at the conclusion of the "problem" than at the start.

45) a)

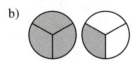

b) $\dfrac{1}{2} = \dfrac{2}{4}$

47) a)

b) $\dfrac{3}{4} = \dfrac{6}{8}$

49) $\dfrac{4}{7} > \dfrac{3}{7}$ With like denominators, the larger fraction is the one with the larger numerator.

51) $\dfrac{1}{20} > \dfrac{1}{100}$ With like numerators, the larger fraction is the one with the smaller denominator.

53) $\dfrac{5}{12} > \dfrac{5}{20}$ With like numerators, the larger fraction is the one with the smaller denominator.

55) $\dfrac{12}{11} < \dfrac{13}{11}$ With like denominators, the larger fraction is the one with the larger numerator.

57)

59)

61) a) Connie sees that the first object has four parts, while the second has three, and argues that there is not a way to put them together.

b)

63) Since there are 12 equal-sized parts, the denominator is 12.
Since there are eight shaded parts, the numerator is 8.

Fraction $= \dfrac{8}{12}$

65) a) $14 + 11 = 25$
There were a total of 25 people at the party.

b) Since there were 14 men at the party, $\dfrac{14}{25}$ of the party-goers were men.

c) Since there were 11 women at the party, $\dfrac{11}{25}$ of the party-goers were women.

67) a) The pizza is cut into ten equal pieces.
The denominator is 10.
Seven pieces were eaten.
The numerator is 7.
$\dfrac{7}{10}$ of the pizza was eaten.

b) $10 - 7 = 3$
Three pieces of pizza were not eaten.
The numerator is 3.
$\dfrac{3}{10}$ of the pizza was not eaten.

69) $12 + 4 + 15 = 31$
There were a total of 31 fish in the tank.

a) Since there are 12 angelfish, $\dfrac{12}{31}$ of the fish are angelfish.

b) Since there are 15 discus fish, $\dfrac{15}{31}$ of the fish are discus fish.

c) $12 + 15 = 27$; $\dfrac{27}{31}$ of the fish are either angelfish or discus fish.

71) a) $\dfrac{8}{10}$ of the main tank's fuel remains.

b) $\dfrac{7}{10}$ of the auxiliary tank's fuel remains.

c) $8 + 7 = 15$, so $\dfrac{15}{10}$ of a single tank remains.

73) There are 13 shapes.

a) $\dfrac{7}{13}$ of the shapes are red.

b) $\dfrac{6}{13}$ of the shapes are square.

c) $\dfrac{4}{13}$ of the shapes are both red and square.

d) $\dfrac{9}{13}$ of the shapes are red or square.

2.2 Multiplying Fractions

GUIDED PRACTICE

1) a) Since the pictures for $\frac{4}{6}$ and $\frac{2}{3}$ are drawn differently but represent the same quantity, they are equivalent fractions.

   b) Of the fractions $\frac{4}{6}$ and $\frac{2}{3}$, $\frac{2}{3}$ is in simplest form.

2) Picture
   The greatest common factor of 6 and 9 is 3.
   Combine every three pieces into a single piece.

   Factoring
   Grouping every three pieces together is equivalent to factoring a 3 out of the numerator and denominator.

   $$\frac{6}{9} = \frac{2}{3}$$

3) $4 = 2 \cdot 2$
   $10 = 2 \cdot 5$

   $$\frac{4}{10} = \frac{2 \cdot 2}{10}$$
   $$= \frac{2 \cdot 2}{2 \cdot 5}$$
   $$= \frac{\cancel{2} \cdot 2}{\cancel{2} \cdot 5}$$
   $$= \frac{2}{5}$$

   $$\frac{4}{10} = \frac{2}{5}$$

4) $$\frac{18}{36} = \frac{2 \cdot 3 \cdot 3}{2 \cdot 2 \cdot 3 \cdot 3}$$
   $$= \frac{\cancel{2} \cdot \cancel{3} \cdot \cancel{3}}{\cancel{2} \cdot 2 \cdot \cancel{3} \cdot \cancel{3}}$$
   $$= \frac{1}{2}$$

   $$\frac{18}{36} = \frac{1}{2}$$

   18   36   (Scratch Paper)
   ╱╲   ╱╲
   2   9   6   6
   ╱ ╱╲   ╱╲ ╱╲
   2   3 3   2 3 2 3

5) $$\frac{240}{16} = \frac{2 \cdot 2 \cdot 2 \cdot 2 \cdot 3 \cdot 5}{2 \cdot 2 \cdot 2 \cdot 2}$$
   $$= \frac{\cancel{2} \cdot \cancel{2} \cdot \cancel{2} \cdot \cancel{2} \cdot 3 \cdot 5}{\cancel{2} \cdot \cancel{2} \cdot \cancel{2} \cdot \cancel{2}}$$
   $$= \frac{15}{1}$$
   $$= 15$$

   $$\frac{240}{16} = 240 \div 16$$
   $$= 15$$

6) Refer to the example in Focus on Estimating Fractions.

7) Refer to the example in Focus on Estimating Fractions.

8) Refer to the example in Focus on Estimating Fractions.

9) $$\frac{12}{5} \cdot \frac{10}{9} = \frac{12 \cdot 10}{5 \cdot 9}$$
   $$= \frac{2 \cdot 2 \cdot 3 \cdot 2 \cdot 5}{5 \cdot 3 \cdot 3}$$
   $$= \frac{2 \cdot 2 \cdot \cancel{3} \cdot 2 \cdot \cancel{5}}{\cancel{5} \cdot \cancel{3} \cdot 3}$$
   $$= \frac{8}{3}$$

10) $$\frac{14}{9} \cdot \frac{3}{35} = \frac{14 \cdot 3}{9 \cdot 35}$$
    $$= \frac{2 \cdot \cancel{7} \cdot \cancel{3}}{\cancel{3} \cdot 3 \cdot 5 \cdot \cancel{7}}$$
    $$= \frac{2}{15}$$

11) $$6 \cdot \frac{5}{12} = \frac{6}{1} \cdot \frac{5}{12}$$
    $$= \frac{6 \cdot 5}{1 \cdot 12}$$
    $$= \frac{\cancel{6} \cdot 5}{1 \cdot 2 \cdot \cancel{6}}$$
    $$= \frac{5}{2}$$

12) $$\frac{3}{5} \cdot 5,000 = \frac{3}{5} \cdot \frac{5,000}{1}$$
    $$= \frac{3 \cdot 5,000}{5 \cdot 1}$$
    $$= \frac{3 \cdot \cancel{5} \cdot 1,000}{\cancel{5} \cdot 1}$$
    $$= 3,000$$
    They will receive $3,000.

13) The missing multiplier is 7 because $5 \cdot 7 = 35$.

$$\frac{3}{5} \cdot \frac{7}{7} = \frac{21}{35}$$

$\frac{3}{5}$ and $\frac{21}{35}$ are equivalent fractions.

14) The missing multiplier is 3 because $7 \cdot 3 = 21$.

$$\frac{3}{7} \cdot \frac{3}{3} = \frac{9}{21}$$

$\frac{3}{7}$ and $\frac{9}{21}$ are equivalent fractions.

15) Since 12 inches = 1 foot, the two unit fractions

are $\frac{12 \text{ in.}}{1 \text{ ft}} = 1$ and $\frac{1 \text{ ft}}{12 \text{ in.}} = 1$.

16) 1 pint = 2 cups

$$\frac{5 \text{ pints}}{1} \cdot \left(\frac{2 \text{ cups}}{1 \text{ pint}} \right) = 10 \text{ cups}$$

5 pints = 10 cups

17) 1 mile = 5,280 feet

$$\frac{4 \text{ miles}}{1} \cdot \left(\frac{5,280 \text{ feet}}{1 \text{ mile}} \right) = 21,120 \text{ feet}$$

4 miles = 21,120 feet

18) $28 \text{ cups} = \dfrac{28 \text{ cups}}{1}$

$$= \frac{28 \text{ cups}}{1} \cdot \left(\frac{1 \text{ pint}}{2 \text{ cups}} \right)$$

$$= \frac{28 \text{ pints}}{2}$$

$$= 14 \text{ pints}$$

19) Convert 6 pounds into ounces.

$$6 \text{ lb} = \frac{6 \text{ lb}}{1}$$

$$= \frac{6 \text{ lb}}{1} \cdot \left(\frac{16 \text{ oz}}{1 \text{ lb}} \right)$$

$$= 96 \text{ oz}$$

No, 6 pounds of chocolate will not be enough to complete the recipe.

20) $\left(\dfrac{2}{3} \right)^4 = \dfrac{2}{3} \cdot \dfrac{2}{3} \cdot \dfrac{2}{3} \cdot \dfrac{2}{3}$

$$= \frac{2 \cdot 2 \cdot 2 \cdot 2}{3 \cdot 3 \cdot 3 \cdot 3}$$

$$= \frac{4 \cdot 2 \cdot 2}{9 \cdot 3 \cdot 3}$$

$$= \frac{8 \cdot 2}{27 \cdot 3}$$

$$= \frac{16}{81}$$

21) $\dfrac{2}{3} \cdot \left(\dfrac{9}{8} \cdot \dfrac{4}{3} \right)^3 = \dfrac{2}{3} \cdot \left(\dfrac{3 \cdot 3 \cdot 4}{2 \cdot 4 \cdot 3} \right)^3$

$$= \frac{2}{3} \cdot \left(\frac{3}{2} \right)^3$$

$$= \frac{2}{3} \cdot \left(\frac{3}{2} \cdot \frac{3}{2} \cdot \frac{3}{2} \right)$$

$$= \frac{2 \cdot 3 \cdot 3 \cdot 3}{3 \cdot 2 \cdot 2 \cdot 2}$$

$$= \frac{9}{4}$$

22) Area of the yard $= 30 \text{ ft} \cdot 60 \text{ ft}$

$$= 1,800 \text{ ft}^2$$

Area of the garden = "one ninth" of "the backyard"

$$= \frac{1}{9} \cdot \frac{1,800 \text{ ft}^2}{1}$$

$$= \frac{1,800 \text{ ft}^2}{9}$$

$$= \frac{9 \cdot 200 \text{ ft}^2}{9}$$

$$= 200 \text{ ft}^2$$

The area of the garden is 200 ft^2.

CONCEPT CHECKS AND PRACTICE EXERCISES

A1) Answers may vary. Example: Fractions in their simplest form are easy to understand. It is easier to visualize 1 part out of 2 than 9 parts out of 18.

A2) a) Since the pictures for $\frac{2}{6}$ and $\frac{1}{3}$ represent the same quantity, they are equivalent.

   b) $\frac{1}{3}$ is in simplest form.

A3) a) Since the pictures for $\frac{3}{9}$ and $\frac{1}{3}$ represent the same quantity, they are equivalent.

   b) $\frac{1}{3}$ is in simplest form.

A4) $\frac{3}{6} = \frac{1 \cdot \cancel{3}}{2 \cdot \cancel{3}} = \frac{1}{2}$

A5) $\frac{2}{6} = \frac{1 \cdot \cancel{2}}{\cancel{2} \cdot 3} = \frac{1}{3}$

A6) $\frac{4}{6} = \frac{\cancel{2} \cdot 2}{\cancel{2} \cdot 3}$

   $= \frac{2}{3}$

A7) $\frac{6}{8} = \frac{\cancel{2} \cdot 3}{\cancel{2} \cdot 4}$

   $= \frac{3}{4}$

A8) $\frac{12}{30} = \frac{\cancel{2} \cdot 2 \cdot \cancel{3}}{\cancel{2} \cdot \cancel{3} \cdot 5}$

   $= \frac{2}{5}$

A9) $\frac{16}{40} = \frac{\cancel{2} \cdot \cancel{2} \cdot \cancel{2} \cdot 2}{\cancel{2} \cdot \cancel{2} \cdot \cancel{2} \cdot 5}$

   $= \frac{2}{5}$

A10) $\frac{60}{45} = \frac{2 \cdot 2 \cdot \cancel{3} \cdot \cancel{5}}{\cancel{3} \cdot 3 \cdot \cancel{5}}$

   $= \frac{4}{3}$

A11) $\frac{50}{35} = \frac{2 \cdot \cancel{5} \cdot 5}{\cancel{5} \cdot 7}$

   $= \frac{10}{7}$

A12) $\frac{96}{16} = \frac{\cancel{2} \cdot \cancel{2} \cdot \cancel{2} \cdot \cancel{2} \cdot 2 \cdot 3}{\cancel{2} \cdot \cancel{2} \cdot \cancel{2} \cdot \cancel{2}}$

   $= 6$

A13) $\frac{84}{12} = \frac{\cancel{2} \cdot \cancel{2} \cdot \cancel{3} \cdot 7}{\cancel{2} \cdot \cancel{2} \cdot \cancel{3}}$

   $= 7$

A14) $\frac{15}{120} = \frac{\cancel{3} \cdot \cancel{5}}{2 \cdot 2 \cdot 2 \cdot \cancel{3} \cdot \cancel{5}}$

   $= \frac{1}{8}$

   The fees were $\frac{1}{8}$ of the sales price.

A15) $\frac{10}{120+10} = \frac{10}{130}$

   $= \frac{\cancel{10}}{\cancel{10} \cdot 13}$

   $= \frac{1}{13}$

   The shipping represented $\frac{1}{13}$ of the total.

FOCUS ON ESTIMATING FRACTIONS PRACTICE EXERCISES

1) $\frac{13}{30} \approx \frac{15}{30} = \frac{1}{2}$
   15 is half of 30. Since 13 is a little less than 15, shade a little less than one-half of the shape.

2) $\frac{9}{40} \approx \frac{10}{40} = \frac{1}{4}$
   10 is one-quarter of 40. Since 9 is a little less than 10, shade a little less than one-quarter of the shape.

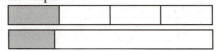

3) $\frac{20}{35} \approx \frac{20}{40} = \frac{1}{2}$
   20 is one-half of 40. Since 35 is a little less than 40, shade a little more than one-half of the shape.

4) $\dfrac{60}{94} \approx \dfrac{60}{90} = \dfrac{2}{3}$

60 is two-thirds of 90. Since 94 is a little more than 90, shade a little less than two-thirds of the shape.

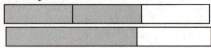

5) $\dfrac{60}{83} \approx \dfrac{60}{80} = \dfrac{3}{4}$

60 is three-fourths of 80. Since 83 is a little more than 80, shade a little less than than three-fourths of the shape.

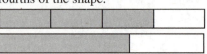

6) $\dfrac{18}{20} \approx \dfrac{20}{20} = 1$

18 is a little less than 20. If the numerator were 20, the fraction would represent a whole. Since 18 is a little less than 20, shade a little less than a whole.

B1)

B2)

B3) $\dfrac{3}{4} \cdot \dfrac{5}{7} = \dfrac{3 \cdot 5}{4 \cdot 7}$

$= \dfrac{15}{28}$

B4) $\dfrac{2}{5} \cdot \dfrac{9}{11} = \dfrac{2 \cdot 9}{5 \cdot 11}$

$= \dfrac{18}{55}$

B5) $7 \cdot \dfrac{1}{84} = \dfrac{7}{1} \cdot \dfrac{1}{84}$

$= \dfrac{7 \cdot 1}{1 \cdot 84}$

$= \dfrac{\cancel{7} \cdot 1}{1 \cdot \cancel{7} \cdot 12}$

$= \dfrac{1}{12}$

B6) $\dfrac{5}{6} \cdot 24 = \dfrac{5}{6} \cdot \dfrac{24}{1}$

$= \dfrac{5 \cdot 24}{6 \cdot 1}$

$= \dfrac{5 \cdot 4 \cdot \cancel{6}}{\cancel{6} \cdot 1}$

$= \dfrac{20}{1}$

$= 20$

B7) $\dfrac{5}{3} \cdot 12 = \dfrac{5}{3} \cdot \dfrac{12}{1}$

$= \dfrac{5 \cdot 12}{3 \cdot 1}$

$= \dfrac{5 \cdot \cancel{3} \cdot 4}{\cancel{3} \cdot 1}$

$= \dfrac{20}{1}$

$= 20$

B8) $3 \cdot \dfrac{5}{9} = \dfrac{3}{1} \cdot \dfrac{5}{9}$

$= \dfrac{3 \cdot 5}{1 \cdot 9}$

$= \dfrac{\cancel{3} \cdot 5}{1 \cdot \cancel{3} \cdot 3}$

$= \dfrac{5}{3}$

B9) $\dfrac{6}{35} \cdot \dfrac{21}{10} = \dfrac{6 \cdot 21}{35 \cdot 10}$

$= \dfrac{2 \cdot 3 \cdot 3 \cdot 7}{5 \cdot 7 \cdot 2 \cdot 5}$

$= \dfrac{\cancel{2} \cdot 3 \cdot 3 \cdot \cancel{7}}{5 \cdot \cancel{7} \cdot \cancel{2} \cdot 5}$

$= \dfrac{9}{25}$

B10) $\dfrac{3}{5} \cdot \dfrac{10}{17} = \dfrac{3 \cdot 10}{5 \cdot 17}$

$= \dfrac{3 \cdot 2 \cdot 5}{5 \cdot 17}$

$= \dfrac{3 \cdot 2 \cdot \cancel{5}}{\cancel{5} \cdot 17}$

$= \dfrac{6}{17}$

B11) Answers may vary.
Step 1: Multiply the numerators.
Step 2: Multiply the denominators.
Step 3: Simplify, if possible.

B12) $\frac{1}{5} \cdot 40,000 = \frac{1}{5} \cdot \frac{40,000}{1}$

$= \frac{1 \cdot 40,000}{5 \cdot 1}$

$= \frac{1 \cdot \cancel{5} \cdot 8,000}{\cancel{5} \cdot 1}$

$= 8,000$

Each family member will receive $8,000.

B13) $\frac{1}{15} \cdot 60,000 = \frac{1}{15} \cdot \frac{60,000}{1}$

$= \frac{1 \cdot 60,000}{15 \cdot 1}$

$= \frac{1 \cdot \cancel{15} \cdot 4,000}{\cancel{15} \cdot 1}$

$= 4,000$

He invested $4,000 in the retirement account.

C1) Answers may vary. Example: $\frac{3}{3}, \frac{7}{7}, \frac{10}{10}$.

C2) Answers may vary. Example: The number of shaded parts is the same as the number of parts in the whole.

C3) Answers may vary. Example: Any number divided by itself is one.

C4) $\frac{3}{7} \cdot \frac{5}{5} = \frac{3}{7}$ is true. $\frac{5}{5} = 1$. Any number multiplied by one is the same number.

C5) $7 \cdot \frac{9}{5} = 7$ is false. In order to end up the same value after multiplying, the multiplier must equal 1. $\frac{9}{5} \neq 1$.

C6) The missing multiplier is 3 because $4 \cdot 3 = 12$.

$\frac{1}{4} \cdot \frac{3}{3} = \frac{3}{12}$

C7) The missing multiplier is 5 because $3 \cdot 5 = 15$.

$\frac{1}{3} \cdot \frac{5}{5} = \frac{5}{15}$

C8) The missing multiplier is 4 because $4 \cdot 4 = 16$.

$\frac{3}{4} \cdot \frac{4}{4} = \frac{12}{16}$

C9) The missing multiplier is 7 because $3 \cdot 7 = 21$.

$\frac{5}{3} \cdot \frac{7}{7} = \frac{35}{21}$

C10) The missing multiplier is 5 because $9 \cdot 3 = 27$.

$\frac{7}{9} \cdot \frac{3}{3} = \frac{21}{27}$

C11) The missing multiplier is 2 because $12 \cdot 2 = 24$.

$\frac{11}{12} \cdot \frac{2}{2} = \frac{22}{24}$

C12) The missing multiplier is 4 because $21 \cdot 4 = 84$.

$\frac{16}{21} \cdot \frac{4}{4} = \frac{64}{84}$

C13) The missing multiplier is 2 because $23 \cdot 2 = 46$.

$\frac{5}{23} \cdot \frac{2}{2} = \frac{10}{46}$

C14) The missing multiplier is 3 because $13 \cdot 3 = 39$.

$\frac{10}{13} \cdot \frac{3}{3} = \frac{30}{39}$

C15) The missing multiplier is 4 because $16 \cdot 4 = 64$.

$\frac{15}{16} \cdot \frac{4}{4} = \frac{60}{64}$

C16) a) $\frac{1}{6} \cdot \frac{1}{2} = \frac{1}{12}$

Each person's piece will be $\frac{1}{12}$ of a whole pan.

b) Based on part (a), we know that $\frac{1}{2}$ is equivalent to $\frac{6}{12}$.

C17) a) $\frac{1}{4} \cdot \frac{2}{3} = \frac{1 \cdot 2}{4 \cdot 3}$

$= \frac{1 \cdot \cancel{2}}{\cancel{2} \cdot 2 \cdot 3}$

$= \frac{1}{6}$

Each piece will be $\frac{1}{6}$ of a whole pie.

b) Based on part (a), we know that $\frac{2}{3}$ is equivalent to $\frac{4}{6}$.

D1) Since 1 hour = 60 minutes, the two unit fractions are $\frac{1\text{ hr}}{60\text{ min}} = 1$ and $\frac{60\text{ min}}{1\text{ hr}} = 1$.

D2) Since 4 quarts = 1 gallon, the two unit fractions are $\frac{4\text{ qt}}{1\text{ gal}} = 1$ and $\frac{1\text{ gal}}{4\text{ qt}} = 1$.

D3) $\dfrac{8 \text{ qt}}{1} \cdot \left(\dfrac{2 \text{ pt}}{1 \text{ qt}} \right) = 16 \text{ pints}$

D4) $\dfrac{4,000 \text{ lb}}{1} \cdot \left(\dfrac{1 \text{ ton}}{2,000 \text{ lb}} \right) = 2 \text{ tons}$

D5) 1 foot = 12 inches

$\dfrac{5 \text{ ft}}{1} \cdot \left(\dfrac{12 \text{ in.}}{1 \text{ ft}} \right) = 60 \text{ in.}$

5 feet = 60 inches

D6) 1 yard = 3 feet

$\dfrac{4 \text{ yd}}{1} \cdot \left(\dfrac{3 \text{ ft}}{1 \text{ yd}} \right) = 12 \text{ ft}$

4 yards = 12 feet

D7) 2 cups = 1 pint

$\dfrac{8 \text{ cups}}{1} \cdot \left(\dfrac{1 \text{ pt}}{2 \text{ cups}} \right) = 4 \text{ pt}$

8 cups = 4 pints

D8) 2 pints = 1 quart

$\dfrac{12 \text{ pt}}{1} \cdot \left(\dfrac{1 \text{ qt}}{2 \text{ pt}} \right) = 6 \text{ qt}$

12 pints = 6 quarts

D9) 1 yard = 3 feet

$\dfrac{7 \text{ yd}}{1} \cdot \left(\dfrac{3 \text{ ft}}{1 \text{ yd}} \right) = 21 \text{ ft}$

There are 21 feet in 7 yards.

D10) 1 week = 7 days

$\dfrac{9 \text{ weeks}}{1} \cdot \left(\dfrac{7 \text{ days}}{1 \text{ week}} \right) = 63 \text{ days}$

There are 63 days in 9 weeks.

D11) 16 ounces = 1 pound

$\dfrac{192 \text{ oz.}}{1} \cdot \left(\dfrac{1 \text{ lb}}{16 \text{ oz.}} \right) = 12 \text{ lb}$

192 ounces is equal to 12 pounds.

D12) 24 hours = 1 day

$\dfrac{336 \text{ hr}}{1} \cdot \left(\dfrac{1 \text{ day}}{24 \text{ hr}} \right) = 14 \text{ days}$

336 hours is equal to 14 days.

D13) Convert 64 ounces into cups.

$64 \text{ oz} = \dfrac{64 \text{ oz}}{1}$

$= \dfrac{64 \text{ oz}}{1} \cdot \left(\dfrac{1 \text{ cup}}{8 \text{ oz}} \right)$

$= 8 \text{ cups}$

Yes, 64 ounces of soda will completely fill a 6-cup container.

D14) Convert 3 pints into cups.

$3 \text{ pt} = \dfrac{3 \text{ pt}}{1}$

$= \dfrac{3 \text{ pt}}{1} \cdot \left(\dfrac{2 \text{ cups}}{1 \text{ pt}} \right)$

$= 6 \text{ cups}$

No, 3 pints of milk will not be enough to complete the recipe.

SECTION 2.2   EXERCISES

For 1–14, refer to Concept Checks and Practice Exercises.

15) simplest form

17) equivalent

19) factor

21) $\dfrac{6}{9} = \dfrac{2 \cdot 3}{3 \cdot 3}$

$= \dfrac{2}{3}$

23) $\dfrac{25}{35} = \dfrac{5 \cdot 5}{5 \cdot 7}$

$= \dfrac{5}{7}$

25) $\dfrac{36}{54} = \dfrac{2 \cdot 2 \cdot 3 \cdot 3}{2 \cdot 3 \cdot 3 \cdot 3}$

$= \dfrac{2}{3}$

27) $\dfrac{10}{210} = \dfrac{1 \cdot 10}{10 \cdot 21}$

$= \dfrac{1}{21}$

29) $\dfrac{99}{117} = \dfrac{\cancel{3}\cdot\cancel{3}\cdot 11}{\cancel{3}\cdot\cancel{3}\cdot 13}$

$= \dfrac{11}{13}$

31) $\dfrac{42}{85} = \dfrac{2\cdot 3\cdot 7}{5\cdot 17}$

$= \dfrac{42}{85}$

33) Answers may vary. Example:
Step 1: Factor the numerator completely.
Step 2: Factor the denominator completely
Step 3: Divide out all common factors.

35) $\dfrac{1}{5}\cdot\dfrac{3}{4} = \dfrac{1\cdot 3}{5\cdot 4}$

$= \dfrac{3}{20}$

37) $\dfrac{5}{18}\cdot\dfrac{9}{2} = \dfrac{5\cdot 9}{18\cdot 2}$

$= \dfrac{5\cdot\cancel{9}}{2\cdot\cancel{9}\cdot 2}$

$= \dfrac{5}{4}$

39) $\dfrac{6}{5}\cdot\dfrac{9}{12} = \dfrac{6\cdot 9}{5\cdot 12}$

$= \dfrac{\cancel{2}\cdot\cancel{3}\cdot 3\cdot 3}{5\cdot\cancel{2}\cdot 2\cdot\cancel{3}}$

$= \dfrac{9}{10}$

41) $\dfrac{73}{18}\cdot\dfrac{9}{146} = \dfrac{73\cdot 9}{18\cdot 146}$

$= \dfrac{\cancel{73}\cdot\cancel{9}}{2\cdot\cancel{9}\cdot 2\cdot\cancel{73}}$

$= \dfrac{1}{4}$

43) $\dfrac{9}{4}\cdot\dfrac{10}{15} = \dfrac{9\cdot 10}{4\cdot 15}$

$= \dfrac{\cancel{3}\cdot 3\cdot\cancel{2}\cdot\cancel{5}}{\cancel{2}\cdot 2\cdot\cancel{3}\cdot\cancel{5}}$

$= \dfrac{3}{2}$

45) $8\cdot\dfrac{1}{4} = \dfrac{8\cdot 1}{1\cdot 4}$

$= \dfrac{2\cdot\cancel{4}}{1\cdot\cancel{4}}$

$= \dfrac{2}{1} = 2$

47) $\dfrac{3}{16}\cdot\dfrac{4}{2}\cdot\dfrac{2}{9} = \dfrac{3\cdot 4\cdot 2}{16\cdot 2\cdot 9}$

$= \dfrac{\cancel{3}\cdot\cancel{2}\cdot\cancel{2}\cdot\cancel{2}}{\cancel{2}\cdot\cancel{2}\cdot\cancel{2}\cdot 2\cdot 2\cdot\cancel{3}\cdot 3}$

$= \dfrac{1}{12}$

49) $\dfrac{14}{25}\cdot\dfrac{15}{21}\cdot\dfrac{3}{4} = \dfrac{14\cdot 15\cdot 3}{25\cdot 21\cdot 4}$

$= \dfrac{\cancel{2}\cdot\cancel{7}\cdot\cancel{3}\cdot\cancel{5}\cdot 3}{\cancel{5}\cdot 5\cdot\cancel{3}\cdot\cancel{7}\cdot\cancel{2}\cdot 2}$

$= \dfrac{3}{10}$

51) Answers may vary. Example:
Step 1: Multiply the numerators.
Step 2: Multiply the denominators.
Step 3: Simplify, if possible.

53) $\left(\dfrac{2}{7}\right)^3 = \dfrac{2}{7}\cdot\dfrac{2}{7}\cdot\dfrac{2}{7}$

55) $\left(\dfrac{6}{7}\right)^2 = \dfrac{6}{7}\cdot\dfrac{6}{7}$

$= \dfrac{36}{49}$

57) $\left(\dfrac{1}{3}\right)^3\cdot\dfrac{9}{20} = \left(\dfrac{1}{3}\cdot\dfrac{1}{3}\cdot\dfrac{1}{3}\right)\cdot\dfrac{9}{20}$

$= \dfrac{1}{27}\cdot\dfrac{9}{20}$

$= \dfrac{1\cdot\cancel{9}}{3\cdot\cancel{9}\cdot 20}$

$= \dfrac{1}{60}$

59) $\left(\dfrac{5}{6}\cdot\dfrac{3}{20}\right)^2 = \left(\dfrac{\cancel{5}\cdot\cancel{3}}{2\cdot\cancel{3}\cdot 4\cdot\cancel{5}}\right)^2$

$= \left(\dfrac{1}{8}\right)^2$

$= \dfrac{1}{8}\cdot\dfrac{1}{8}$

$= \dfrac{1}{64}$

61) $\dfrac{9}{12}\cdot\left(\dfrac{15}{27}\cdot3^3\right)=\dfrac{9}{12}\cdot\left(\dfrac{15}{27}\cdot3\cdot3\cdot3\right)$

$\qquad\qquad\quad=\dfrac{9}{12}\cdot\left(\dfrac{15}{\cancel{27}}\cdot\cancel{27}\right)$

$\qquad\qquad\quad=\dfrac{9}{12}\cdot\left(\dfrac{15}{1}\right)$

$\qquad\qquad\quad=\dfrac{\cancel{3}\cdot3\cdot3\cdot5}{2\cdot2\cdot\cancel{3}}$

$\qquad\qquad\quad=\dfrac{45}{4}$

63) The missing multiplier is 9 because $9\cdot9=81$.

$\qquad\dfrac{2}{9}\cdot\dfrac{9}{9}=\dfrac{18}{81}$

65) The missing multiplier is 4 because $16\cdot4=64$.

$\qquad\dfrac{14}{16}\cdot\dfrac{4}{4}=\dfrac{56}{64}$

67) The missing multiplier is 4 because $12\cdot4=48$.

$\qquad\dfrac{11}{12}\cdot\dfrac{4}{4}=\dfrac{44}{48}$

69) The missing multiplier is 3 because $16\cdot3=48$.

$\qquad\dfrac{7}{16}\cdot\dfrac{3}{3}=\dfrac{21}{48}$

71) 1 foot $=12$ inches

$\qquad\dfrac{10\ \cancel{ft}}{1}\cdot\left(\dfrac{12\ \text{in.}}{1\ \cancel{ft}}\right)=120$ in.

10 feet $=120$ inches

73) 7 days $=1$ week

$\qquad\dfrac{98\ \cancel{days}}{1}\cdot\left(\dfrac{1\ \text{week}}{7\ \cancel{days}}\right)=14$ weeks

98 days $=14$ weeks

75) 1 pound $=16$ ounces

$\qquad\dfrac{3\ \cancel{lb}}{1}\cdot\left(\dfrac{16\ oz}{1\ \cancel{lb}}\right)=48$ oz

3 pounds $=48$ ounces

77) 1 cup $=8$ ounces

$\qquad\dfrac{15\ \cancel{cups}}{1}\cdot\left(\dfrac{8\ oz}{1\ \cancel{cup}}\right)=120$ oz

There are 120 ounces in 15 cups.

79) 16 ounces $=1$ pound

$\qquad\dfrac{320\ \cancel{oz}}{1}\cdot\left(\dfrac{1\ lb}{16\ \cancel{oz}}\right)=20$ lb

320 ounces is equal to 20 pounds.

81) $\dfrac{75}{675+75}=\dfrac{75}{750}$

$\qquad\qquad\ =\dfrac{\cancel{75}}{10\cdot\cancel{75}}$

$\qquad\qquad\ =\dfrac{1}{10}$

The wireless router was $\dfrac{1}{10}$ of the total cost.

83) a) $\dfrac{1}{3}\cdot216=\dfrac{1}{3}\cdot\dfrac{216}{1}$

$\qquad\qquad\ =\dfrac{1\cdot\cancel{3}\cdot72}{\cancel{3}}$

$\qquad\qquad\ =72$

72 students prefer rap music.

b) $\dfrac{1}{6}\cdot216=\dfrac{1}{6}\cdot\dfrac{216}{1}$

$\qquad\qquad\ =\dfrac{1\cdot\cancel{6}\cdot36}{\cancel{6}}$

$\qquad\qquad\ =36$

36 students prefer country music.

c) $\dfrac{1}{2}\cdot216=\dfrac{1}{2}\cdot\dfrac{216}{1}$

$\qquad\qquad\ =\dfrac{1\cdot\cancel{2}\cdot108}{\cancel{2}}$

$\qquad\qquad\ =108$

108 students prefer rock music.

85) a) $\dfrac{1}{3}\cdot231=\dfrac{1}{3}\cdot\dfrac{231}{1}$

$\qquad\qquad\ =\dfrac{1\cdot\cancel{3}\cdot77}{\cancel{3}}$

$\qquad\qquad\ =77$

You saved \$77 on your purchase.

b) $231-77=154$

You spent \$154 on your purchase.

87) $A=l\cdot w$

$\qquad=2\text{ yd}\cdot\dfrac{1}{2}\text{ yd}$

$\qquad=1\text{ yd}^2$

The area of the planter box is 1 square yard.

89) a) Housing: $\frac{2}{5} \cdot 2{,}000 = 800$

Transportation: $\frac{1}{5} \cdot 2{,}000 = 400$

Credit Card: $\frac{1}{10} \cdot 2{,}000 = 200$

$800 + 400 + 200 = 1{,}400$

He spent a total of $1,400 on these expenses.

b) $2{,}000 - 1{,}400 = 600$

John had $600 left for other expenses.

2.3   Dividing Fractions

GUIDED PRACTICE

1) $\frac{5}{7} \cdot \frac{7}{5} = 1$

The reciprocal of $\frac{5}{7}$ is $\frac{7}{5}$.

2) $\frac{65}{23}$ is written as a fraction.

Interchange the numerator and denominator.

The reciprocal of $\frac{65}{23}$ is $\frac{23}{65}$.

3) 96 can be written as the fraction $\frac{96}{1}$.

Interchange the numerator and denominator.

The reciprocal of 96 is $\frac{1}{96}$.

4) $\frac{4}{7} \div \frac{16}{35} = \frac{4}{7} \cdot \frac{35}{16}$

$= \frac{\cancel{4} \cdot 5 \cdot \cancel{7}}{\cancel{7} \cdot \cancel{4} \cdot 4}$

$= \frac{5}{4}$

5) $\frac{3}{8} \div \frac{5}{2} = \frac{3}{8} \cdot \frac{2}{5}$

$= \frac{3 \cdot \cancel{2}}{\cancel{2} \cdot 2 \cdot 2 \cdot 5}$

$= \frac{3}{20}$

6) $4 \div \frac{3}{5} = \frac{4}{1} \div \frac{3}{5}$

$= \frac{4}{1} \cdot \frac{5}{3}$

$= \frac{4 \cdot 5}{1 \cdot 3}$

$= \frac{20}{3}$

7) This question asks us to find how many $\frac{3}{4}$'s fit into 3. Use division to answer this question.

$3 \div \frac{3}{4} = \frac{3}{1} \div \frac{3}{4}$

$= \frac{3}{1} \cdot \frac{4}{3}$

$= \frac{\cancel{3} \cdot 4}{1 \cdot \cancel{3}}$

$= 4$

It will take the child four years to grow three inches.

8) $\left(\frac{1}{3} \div \frac{3}{2}\right)^2 \cdot 7 = \left(\frac{1}{3} \cdot \frac{2}{3}\right)^2 \cdot 7$

$= \left(\frac{2}{9}\right)^2 \cdot 7$

$= \frac{4}{81} \cdot \frac{7}{1}$

$= \frac{28}{81}$

9) To find how many $\frac{1}{16}$'s are in seven cakes, divide 7 by $\frac{1}{16}$.

$7 \div \frac{1}{16} = \frac{7}{1} \cdot \frac{16}{1}$

$= \frac{7 \cdot 16}{1 \cdot 1}$

$= 112$

The seven cakes have a total of 112 pieces.

CONCEPT CHECKS AND PRACTICE EXERCISES

A1) Any number times its reciprocal is equal to one.

A2) a) $\frac{6}{7} \cdot \frac{7}{6} = \frac{\cancel{6} \cdot \cancel{7}}{\cancel{7} \cdot \cancel{6}} = 1$

b) $\frac{54}{12} \cdot \frac{12}{54} = \frac{\cancel{54} \cdot \cancel{12}}{\cancel{12} \cdot \cancel{54}} = 1$

c) $\frac{123}{321} \cdot \frac{321}{123} = 1$

d) $\frac{9}{10} \cdot \frac{10}{9} = 1$

e) $\frac{164}{73} \cdot \frac{73}{164} = 1$

f) $\frac{a}{b} \cdot \frac{b}{a} = 1$

A3) The reciprocal of $\frac{9}{7}$ is $\frac{7}{9}$.

A4) The reciprocal of $\frac{5}{2}$ is $\frac{2}{5}$.

A5) The reciprocal of $\frac{3}{13}$ is $\frac{13}{3}$.

A6) The reciprocal of $\frac{21}{52}$ is $\frac{52}{21}$.

A7) The reciprocal of $\frac{1}{3}$ is $\frac{3}{1}$ or 3.

A8) The reciprocal of $\frac{1}{9}$ is $\frac{9}{1}$ or 9.

A9) $22 = \frac{22}{1}$; the reciprocal of 22 is $\frac{1}{22}$.

A10) $16 = \frac{16}{1}$; the reciprocal of 16 is $\frac{1}{16}$.

B1) a) $8 \div 2 = 4$

$$8 \cdot \frac{1}{2} = \frac{8 \cdot 1}{1 \cdot 2} = \frac{\cancel{2} \cdot 4 \cdot 1}{1 \cdot \cancel{2}} = 4$$

   b) $12 \div 4 = 3$

$$12 \cdot \frac{1}{4} = \frac{12 \cdot 1}{1 \cdot 4} = \frac{3 \cdot \cancel{4} \cdot 1}{1 \cdot \cancel{4}} = 3$$

   c) $36 \div 9 = 4$

$$36 \cdot \frac{1}{9} = \frac{36 \cdot 1}{1 \cdot 9} = \frac{4 \cdot \cancel{9} \cdot 1}{1 \cdot \cancel{9}} = 4$$

B2) Based on the three exercises above, division by a number and multiplication by the number's reciprocal are equivalent.

B3) If you "flip" a fraction, you are finding its reciprocal.

B4) Answers may vary. Example:
Step 1: Rewrite the first fraction. "Skip."
Step 2: Write the reciprocal of the second fraction. "Flip."
Step 3: Multiply. Change the division sign to a multiplication sign.
Step 4: Simplify. Perform the multiplication and reduce.

B5) $$\frac{1}{2} \div \frac{3}{4} = \frac{1}{2} \cdot \frac{4}{3}$$
$$= \frac{1 \cdot \cancel{2} \cdot 2}{\cancel{2} \cdot 3}$$
$$= \frac{2}{3}$$

B6) $$\frac{3}{7} \div \frac{3}{5} = \frac{3}{7} \cdot \frac{5}{3}$$
$$= \frac{\cancel{3} \cdot 5}{7 \cdot \cancel{3}}$$
$$= \frac{5}{7}$$

B7) $$\frac{12}{16} \div \frac{9}{4} = \frac{12}{16} \cdot \frac{4}{9}$$
$$= \frac{\cancel{3} \cdot \cancel{4} \cdot \cancel{4}}{\cancel{4} \cdot \cancel{4} \cdot \cancel{3} \cdot 3}$$
$$= \frac{1}{3}$$

B8) $$\frac{15}{8} \div \frac{35}{6} = \frac{15}{8} \cdot \frac{6}{35}$$
$$= \frac{3 \cdot \cancel{5} \cdot \cancel{2} \cdot 3}{\cancel{2} \cdot 2 \cdot 2 \cdot \cancel{5} \cdot 7}$$
$$= \frac{9}{28}$$

B9) $$\frac{36}{11} \div 18 = \frac{36}{11} \div \frac{18}{1}$$
$$= \frac{36}{11} \cdot \frac{1}{18}$$
$$= \frac{2 \cdot \cancel{18} \cdot 1}{11 \cdot \cancel{18}}$$
$$= \frac{2}{11}$$

B10) $$\frac{5}{18} \div 15 = \frac{5}{18} \div \frac{15}{1}$$
$$= \frac{5}{18} \cdot \frac{1}{15}$$
$$= \frac{\cancel{5} \cdot 1}{18 \cdot 3 \cdot \cancel{5}}$$
$$= \frac{1}{54}$$

B11) $$5 \div \frac{1}{2} = \frac{5}{1} \div \frac{1}{2}$$
$$= \frac{5}{1} \cdot \frac{2}{1}$$
$$= 10$$

B12) $8 \div \dfrac{2}{5} = \dfrac{8}{1} \div \dfrac{2}{5}$

$= \dfrac{8}{1} \cdot \dfrac{5}{2}$

$= \dfrac{\cancel{2} \cdot 4 \cdot 5}{1 \cdot \cancel{2}}$

$= 20$

B13) $5 \div \dfrac{1}{8} = \dfrac{5}{1} \div \dfrac{1}{8}$

$= \dfrac{5}{1} \cdot \dfrac{8}{1}$

$= 40$

It will take the tree 40 years to grow five inches in diameter.

B14) $2 \div \dfrac{1}{3} = \dfrac{2}{1} \div \dfrac{1}{3}$

$= \dfrac{2}{1} \cdot \dfrac{3}{1}$

$= 6$

Yes, she can paint all five kitchen chairs.

SECTION 2.3   EXERCISES

For 1–6, refer to Concept Checks and Practice Exercises.

7) dividend

9) reciprocal

11) $\dfrac{1}{5} \div \dfrac{3}{10} = \dfrac{1}{5} \cdot \dfrac{10}{3}$

$= \dfrac{1 \cdot 2 \cdot \cancel{5}}{\cancel{5} \cdot 3}$

$= \dfrac{2}{3}$

13) $\dfrac{3}{2} \div \dfrac{9}{5} = \dfrac{3}{2} \cdot \dfrac{5}{9}$

$= \dfrac{\cancel{3} \cdot 5}{2 \cdot \cancel{3} \cdot 3}$

$= \dfrac{5}{6}$

15) $\dfrac{20}{9} \div \dfrac{50}{45} = \dfrac{20}{9} \cdot \dfrac{45}{50}$

$= \dfrac{2 \cdot \cancel{10} \cdot \cancel{5} \cdot \cancel{9}}{\cancel{9} \cdot \cancel{5} \cdot \cancel{10}}$

$= 2$

17) $\dfrac{7}{9} \div \dfrac{0}{3}$; undefined

19) $\dfrac{0}{5} \div 3 = 0$

21) $\dfrac{7}{12} \div \dfrac{12}{7} = \dfrac{7}{12} \cdot \dfrac{7}{12}$

$= \dfrac{7 \cdot 7}{12 \cdot 12}$

$= \dfrac{49}{144}$

23) $20 \div \dfrac{5}{8} = \dfrac{20}{1} \div \dfrac{5}{8}$

$= \dfrac{20}{1} \cdot \dfrac{8}{5}$

$= \dfrac{4 \cdot \cancel{5} \cdot 8}{1 \cdot \cancel{5}}$

$= 32$

25) $\dfrac{25}{64} \div \dfrac{5}{32} = \dfrac{25}{64} \cdot \dfrac{32}{5}$

$= \dfrac{\cancel{5} \cdot 5 \cdot \cancel{32}}{2 \cdot \cancel{32} \cdot \cancel{5}}$

$= \dfrac{5}{2}$

27) $\dfrac{12}{7} \div \dfrac{24}{35} = \dfrac{12}{7} \cdot \dfrac{35}{24}$

$= \dfrac{\cancel{12} \cdot 5 \cdot \cancel{7}}{\cancel{7} \cdot 2 \cdot \cancel{12}}$

$= \dfrac{5}{2}$

29) $\dfrac{1{,}000}{49} \cdot \dfrac{7}{100} = \dfrac{10 \cdot \cancel{100} \cdot \cancel{7}}{\cancel{7} \cdot 7 \cdot \cancel{100}}$

$= \dfrac{10}{7}$

31) $\dfrac{5}{3} \div \dfrac{3}{5} \cdot \dfrac{27}{10} = 1 \cdot \dfrac{27}{10}$

$= \dfrac{27}{10}$

33) $\left(\dfrac{5}{9} \cdot \dfrac{3}{10} \right)^2 = \left(\dfrac{\cancel{5} \cdot \cancel{3}}{\cancel{3} \cdot 3 \cdot 2 \cdot \cancel{5}} \right)^2$

$= \left(\dfrac{1}{6} \right)^2$

$= \dfrac{1}{36}$

35) $\left(\dfrac{1}{4} \div \dfrac{2}{3}\right)^2 = \left(\dfrac{1}{4} \cdot \dfrac{3}{2}\right)^2$

$= \left(\dfrac{3}{8}\right)^2$

$= \dfrac{9}{64}$

37) $\left(\dfrac{7}{9}\right)^2 \div 49 = \dfrac{49}{81} \div 49$

$= \dfrac{49}{81} \cdot \dfrac{1}{49}$

$= \dfrac{\cancel{49} \cdot 1}{81 \cdot \cancel{49}}$

$= \dfrac{1}{81}$

39) $\dfrac{4}{9} \div 2^2 = \dfrac{4}{9} \div 4$

$= \dfrac{4}{9} \cdot \dfrac{1}{4}$

$= \dfrac{\cancel{4} \cdot 1}{9 \cdot \cancel{4}}$

$= \dfrac{1}{9}$

41) $\dfrac{3}{5} \div \left(\dfrac{3}{5}\right)^2 \cdot 1 = \dfrac{3}{5} \div \dfrac{9}{25} \cdot 1$

$= \dfrac{3}{5} \cdot \dfrac{25}{9} \cdot 1$

$= \dfrac{5}{3} \cdot 1$

$= \dfrac{5}{3}$

43) $\dfrac{0}{3} \cdot \dfrac{7}{5} = 0$

45) $\left(4 \div \dfrac{1}{2}\right)^2 \div 8 = \left(4 \cdot 2\right)^2 \div 8$

$= (8)^2 \div 8$

$= 64 \div 8$

$= 8$

47) Dividing $\dfrac{8}{3}$ by $\dfrac{0}{1}$ is equivalent to dividing $\dfrac{8}{3}$ by 0, which is undefined.

49) a) The student multiplied instead of divided.

b) $\dfrac{3}{4} \div \dfrac{2}{9} = \dfrac{3}{4} \cdot \dfrac{9}{2}$

$= \dfrac{3 \cdot 9}{4 \cdot 2}$

$= \dfrac{27}{8}$

51) a) The student multiplied the 3 in the numerator and the denominator, instead of in the numerator only.

b) $3 \cdot \dfrac{3}{4} = \dfrac{3}{1} \cdot \dfrac{3}{4}$

$= \dfrac{3 \cdot 3}{1 \cdot 4}$

$= \dfrac{9}{4}$

53) Answers may vary.
B) and E) have the same answer.
C), D), and F) have the same answer.

55) $21 \div \dfrac{3}{8} = \dfrac{21}{1} \div \dfrac{3}{8}$

$= \dfrac{21}{1} \cdot \dfrac{8}{3}$

$= \dfrac{\cancel{3} \cdot 7 \cdot 8}{1 \cdot \cancel{3}}$

$= 56$

56 patients can be treated.

57) $13 \div \dfrac{1}{8} = 13 \cdot 8$

$= 104$

Yes, there will be enough pie for 98 people.

59) $\dfrac{1}{8} \div \dfrac{1}{32} = \dfrac{1}{8} \cdot \dfrac{32}{1}$

$= \dfrac{1 \cdot 4 \cdot \cancel{8}}{\cancel{8} \cdot 1}$

$= 4$

Four layers must be applied to make the polyurethane $\dfrac{1}{8}$ of an inch thick.

61) $24 \cdot \dfrac{3}{8} = \dfrac{24}{1} \cdot \dfrac{3}{8}$

$= \dfrac{3 \cdot \cancel{8} \cdot 3}{1 \cdot \cancel{8}}$

$= 9$

She will use a rod of length 9 inches.

63) $48 \div \dfrac{24}{10} = \dfrac{48}{1} \div \dfrac{24}{10}$

$\phantom{48 \div \dfrac{24}{10}} = \dfrac{48}{1} \cdot \dfrac{10}{24}$

$\phantom{48 \div \dfrac{24}{10}} = \dfrac{2 \cdot \cancel{24} \cdot 10}{1 \cdot \cancel{24}}$

$\phantom{48 \div \dfrac{24}{10}} = 20$

The mileage is 20 miles per gallon.

65) $24 \cdot \dfrac{7}{8} = \dfrac{24}{1} \cdot \dfrac{7}{8}$

$\phantom{24 \cdot \dfrac{7}{8}} = \dfrac{3 \cdot \cancel{8} \cdot 7}{1 \cdot \cancel{8}}$

$\phantom{24 \cdot \dfrac{7}{8}} = 21$

21 inches of wire is needed.

67) $3 \div \dfrac{3}{8} = \dfrac{3}{1} \div \dfrac{3}{8}$

$\phantom{3 \div \dfrac{3}{8}} = \dfrac{3}{1} \cdot \dfrac{8}{3}$

$\phantom{3 \div \dfrac{3}{8}} = \dfrac{\cancel{3} \cdot 8}{1 \cdot \cancel{3}}$

$\phantom{3 \div \dfrac{3}{8}} = 8$

Geoff must run 8 laps around the track.

2.4   Adding and Subtracting Fractions

GUIDED PRACTICE

1) $\dfrac{1}{6} + \dfrac{7}{6} = \dfrac{1+7}{6}$

$\phantom{\dfrac{1}{6} + \dfrac{7}{6}} = \dfrac{8}{6}$

$\phantom{\dfrac{1}{6} + \dfrac{7}{6}} = \dfrac{4 \cdot \cancel{2}}{3 \cdot \cancel{2}}$

$\phantom{\dfrac{1}{6} + \dfrac{7}{6}} = \dfrac{4}{3}$

2) $\dfrac{9}{12} - \dfrac{5}{12} = \dfrac{9-5}{12}$

$\phantom{\dfrac{9}{12} - \dfrac{5}{12}} = \dfrac{4}{12}$

$\phantom{\dfrac{9}{12} - \dfrac{5}{12}} = \dfrac{1 \cdot \cancel{4}}{3 \cdot \cancel{4}}$

$\phantom{\dfrac{9}{12} - \dfrac{5}{12}} = \dfrac{1}{3}$

3) List multiples of the larger denominator, 10, until the smaller denominator, 8, divides a multiple evenly.

$10 \cdot 1 = 10$ Keep going
$10 \cdot 2 = 20$ Keep going
$10 \cdot 3 = 30$ Keep going
$10 \cdot 4 = 40$ Stop
LCD = 40

4) $6 \cdot 1 = 6$ Keep going
$6 \cdot 2 = 12$ Stop, 4 divides 12 evenly.
LCD = 12

$\dfrac{3}{4} \cdot \left(\dfrac{\ }{\ } \right) = \dfrac{\ }{12}$

The missing multiplier is 3 because $4 \cdot 3 = 12$.

$\dfrac{3}{4} \cdot \dfrac{3}{3} = \dfrac{9}{12}$

$\dfrac{1}{6} \cdot \left(\dfrac{\ }{\ } \right) = \dfrac{\ }{12}$

The missing multiplier is 2 because $6 \cdot 2 = 12$.

$\dfrac{1}{6} \cdot \dfrac{2}{2} = \dfrac{2}{12}$

5) 7, 14, 21, 28, 35
5 divides 35 evenly.
35 is the LCD.

$\dfrac{4}{7} \cdot \dfrac{5}{5} = \dfrac{20}{35}$

$\dfrac{2}{5} \cdot \dfrac{7}{7} = \dfrac{14}{35}$

6) 12, 24, 36
9 divides 36 evenly.
36 is the LCD.

$\dfrac{1}{12} \cdot \dfrac{3}{3} = \dfrac{3}{36}$

$\dfrac{11}{9} \cdot \dfrac{4}{4} = \dfrac{44}{36}$

7) 5, 10, 15, 20
LCD = 20

$\dfrac{1}{4} \cdot \dfrac{5}{5} = \dfrac{5}{20}$

$\dfrac{3}{5} \cdot \dfrac{4}{4} = \dfrac{12}{20}$

$\dfrac{5}{20} + \dfrac{12}{20} = \dfrac{17}{20}$

$\dfrac{17}{20}$ is in simplest form.

8) 15, 30
    LCD = 30

$$\frac{7}{15} - \frac{1}{6} = \frac{7}{15} \cdot \frac{2}{2} - \frac{1}{6} \cdot \frac{5}{5}$$
$$= \frac{14}{30} - \frac{5}{30}$$
$$= \frac{9}{30}$$
$$= \frac{\cancel{3} \cdot 3}{\cancel{3} \cdot 10}$$
$$= \frac{3}{10}$$

9) 12, 24, 36
    LCD = 36

$$\frac{4}{9} - \frac{1}{12} = \frac{4}{9} \cdot \frac{4}{4} - \frac{1}{12} \cdot \frac{3}{3}$$
$$= \frac{16}{36} - \frac{3}{36}$$
$$= \frac{13}{36}$$

$\frac{13}{36}$ is in simplest form.

10) A whole number has a denominator of 1. So, the LCD is the denominator of the other fraction, 4.

$$6 + \frac{1}{4} = \frac{6}{1} + \frac{1}{4}$$
$$= \frac{6}{1} \cdot \frac{4}{4} + \frac{1}{4}$$
$$= \frac{24}{4} + \frac{1}{4}$$
$$= \frac{25}{4}$$

11) $18 = 2 \cdot 3 \cdot 3$
     $27 = 3 \cdot 3 \cdot 3$

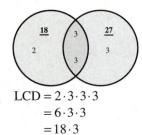

LCD $= 2 \cdot 3 \cdot 3 \cdot 3$
    $= 6 \cdot 3 \cdot 3$
    $= 18 \cdot 3$
    $= 54$

12) $14 = 2 \cdot 7$
     $10 = 2 \cdot 5$

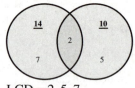

LCD $= 2 \cdot 5 \cdot 7$
    $= 10 \cdot 7$
    $= 70$
14 is missing a 5.
18 is missing a 7.

$$\frac{3}{14} \cdot \frac{5}{5} = \frac{15}{70}$$

$$\frac{1}{10} \cdot \frac{7}{7} = \frac{7}{70}$$

13) $12 = 2 \cdot 2 \cdot 3$
     $54 = 2 \cdot 3 \cdot 3 \cdot 3$

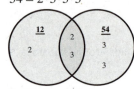

LCD $= 2 \cdot 2 \cdot 3 \cdot 3 \cdot 3$
    $= 4 \cdot 3 \cdot 3 \cdot 3$
    $= 12 \cdot 3 \cdot 3$
    $= 36 \cdot 3$
    $= 108$
12 is missing $3 \cdot 3 = 9$.
54 is missing 2.

$$\frac{13}{12} \cdot \frac{9}{9} = \frac{117}{108}$$

$$\frac{17}{54} \cdot \frac{2}{2} = \frac{34}{108}$$

14) $6 = 2 \cdot 3$
$50 = 2 \cdot 5 \cdot 5$

$LCD = 2 \cdot 3 \cdot 5 \cdot 5$
$= 6 \cdot 25$
$= 150$

6 is missing 25.
50 is missing 3.

$$\frac{1}{6} + \frac{3}{50} = \frac{1}{6} \cdot \frac{25}{25} + \frac{3}{50} \cdot \frac{3}{3}$$
$$= \frac{25}{150} + \frac{9}{150}$$
$$= \frac{34}{150}$$
$$= \frac{\cancel{2} \cdot 17}{\cancel{2} \cdot 75}$$
$$= \frac{17}{75}$$

15) $24 = 2 \cdot 2 \cdot 2 \cdot 3$
$18 = 2 \cdot 3 \cdot 3$

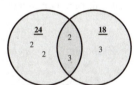

$LCD = 2 \cdot 2 \cdot 2 \cdot 3 \cdot 3$
$= 4 \cdot 6 \cdot 3$
$= 24 \cdot 3$
$= 72$

24 is missing 3.
18 is missing 4.

$$\frac{17}{24} - \frac{3}{18} = \frac{17}{24} \cdot \frac{3}{3} - \frac{3}{18} \cdot \frac{4}{4}$$
$$= \frac{51}{72} - \frac{12}{72}$$
$$= \frac{39}{72}$$
$$= \frac{\cancel{3} \cdot 13}{\cancel{3} \cdot 24}$$
$$= \frac{13}{24}$$

16) The LCD for 4^{ths} and 3^{rds} is 12.
The LCD of that demonimator, 12^{ths}, and 8^{ths}
is 24.

Build like fractions with a denominator of 24.

$$\frac{3}{4} - \frac{1}{3} + \frac{1}{8} = \frac{3}{4} \cdot \frac{6}{6} - \frac{1}{3} \cdot \frac{8}{8} + \frac{1}{8} \cdot \frac{3}{3}$$
$$= \frac{18}{24} - \frac{8}{24} + \frac{3}{24}$$
$$= \frac{10}{24} + \frac{3}{24}$$
$$= \frac{13}{24}$$

17) $$\frac{3}{4} - \frac{5}{8} = \frac{3}{4} \cdot \frac{2}{2} - \frac{5}{8}$$
$$= \frac{6}{8} - \frac{5}{8}$$
$$= \frac{1}{8}$$

The nail will stick out $\frac{1}{8}$ inch on the other side
of the board.

CONCEPT CHECKS AND PRACTICE EXERCISES

A1) Fractions are like fractions if they have the same
denominator.

A2) Answers may vary.
Example: $\frac{1}{6}, \frac{5}{6}$; $\frac{2}{7}, \frac{4}{7}$; $\frac{3}{10}, \frac{7}{10}$

A3) The numerator is 3.
The denominator is 10.

A4) a) 10 oranges $+$ 4 oranges $=$ 14 oranges

b) 1 fifth $+$ 3 fifths $=$ 4 fifths

c) 12 gumballs $-$ 8 gumballs $=$ 4 gumballs

d) 8 tenths $-$ 3 tenths $=$ 5 tenths

A5) $$\frac{1}{3} + \frac{1}{3} = \frac{1+1}{3}$$
$$= \frac{2}{3}$$

A6) $$\frac{3}{11} + \frac{4}{11} = \frac{3+4}{11}$$
$$= \frac{7}{11}$$

A7) $\dfrac{2}{5} - \dfrac{1}{5} = \dfrac{2-1}{5}$

$= \dfrac{1}{5}$

A8) $\dfrac{6}{7} - \dfrac{4}{7} = \dfrac{6-4}{7}$

$= \dfrac{2}{7}$

A9) $\dfrac{5}{8} + \dfrac{1}{8} = \dfrac{5+1}{8}$

$= \dfrac{6}{8}$

$= \dfrac{\cancel{2} \cdot 3}{\cancel{2} \cdot 4}$

$= \dfrac{3}{4}$

A10) $\dfrac{15}{24} + \dfrac{3}{24} = \dfrac{15+3}{24}$

$= \dfrac{18}{24}$

$= \dfrac{3 \cdot \cancel{6}}{4 \cdot \cancel{6}}$

$= \dfrac{3}{4}$

A11) $\dfrac{15}{26} - \dfrac{2}{26} = \dfrac{15-2}{26}$

$= \dfrac{13}{26}$

$= \dfrac{1 \cdot \cancel{13}}{2 \cdot \cancel{13}}$

$= \dfrac{1}{2}$

A12) $\dfrac{17}{22} - \dfrac{6}{22} = \dfrac{17-6}{22}$

$= \dfrac{11}{22}$

$= \dfrac{1 \cdot \cancel{11}}{2 \cdot \cancel{11}}$

$= \dfrac{1}{2}$

A13)

$\dfrac{4}{5} - \dfrac{2}{5} = \dfrac{2}{5}$

A14)

$\dfrac{3}{4} - \dfrac{1}{4} = \dfrac{2}{4} = \dfrac{1}{2}$

A15)

$\dfrac{5}{6} + \dfrac{1}{6} = \dfrac{6}{6} = 1$

A16)

$\dfrac{3}{4} + \dfrac{1}{4} = \dfrac{4}{4} = 1$

A17)

$\dfrac{1}{2} + \dfrac{1}{2} = \dfrac{2}{2} = 1$

A18)

$\dfrac{1}{3} + \dfrac{2}{3} = \dfrac{3}{3} = 1$

B1) Answers may vary. Example: The pieces were not the same size.

B2) Answers may vary. Example: The pieces were the same size.

B3) a) $\dfrac{4}{4} = 1$

b) $\dfrac{4}{4}$ and 1 are equal.

c) $\dfrac{2}{5} \cdot 1 = \dfrac{2}{5}$ True

$\dfrac{2}{5} \cdot \dfrac{4}{4} = \dfrac{8}{20}$ True

d) Answers may vary. Example: Both fractions were multiplied by one, which gives the beginning value.

e) Answers may vary. Example: In order not to change the value of the fraction, it can be multiplied only by 1.

B4) 2 divides 6 evenly.
6 is the LCD.

$$\frac{1}{2} \cdot \frac{3}{3} = \frac{3}{6}$$

$$\frac{1}{6} \cdot \frac{1}{1} = \frac{1}{6}$$

B5) 5 divides 15 evenly.
15 is the LCD.

$$\frac{2}{5} \cdot \frac{3}{3} = \frac{6}{15}$$

$$\frac{4}{15} \cdot \frac{1}{1} = \frac{4}{15}$$

B6) 10, 20, 30, 40
8 divides 40 evenly.
40 is the LCD.

$$\frac{3}{8} \cdot \frac{5}{5} = \frac{15}{40}$$

$$\frac{7}{10} \cdot \frac{4}{4} = \frac{28}{40}$$

B7) 8, 16, 24
3 divides 24 evenly.
24 is the LCD.

$$\frac{3}{8} \cdot \frac{3}{3} = \frac{9}{24}$$

$$\frac{2}{3} \cdot \frac{8}{8} = \frac{16}{24}$$

B8) 5 divides 20 evenly.
20 is the LCD.

$$\frac{3}{20} = \frac{3}{20}$$

$$\frac{3}{5} \cdot \frac{4}{4} = \frac{12}{20}$$

B9) 3 divides 6 evenly.
6 is the LCD.

$$\frac{1}{6} = \frac{1}{6}$$

$$\frac{2}{3} \cdot \frac{2}{2} = \frac{4}{6}$$

B10) 12, 24, 36
9 divides 36 evenly.
36 is the LCD.

$$\frac{4}{9} \cdot \frac{4}{4} = \frac{16}{36}$$

$$\frac{5}{12} \cdot \frac{3}{3} = \frac{15}{36}$$

B11) 9, 18, 27, 36
4 divides 36 evenly.
36 is the LCD.

$$\frac{5}{9} \cdot \frac{4}{4} = \frac{20}{36}$$

$$\frac{3}{4} \cdot \frac{9}{9} = \frac{27}{36}$$

B12) 25, 50, 75
3 divides 75 evenly.
24 is the LCD.

$$\frac{17}{25} \cdot \frac{3}{3} = \frac{51}{75}$$

$$\frac{1}{3} \cdot \frac{25}{25} = \frac{25}{75}$$

B13) 12, 24
8 divides 24 evenly.
24 is the LCD.

$$\frac{5}{8} \cdot \frac{3}{3} = \frac{15}{24}$$

$$\frac{3}{12} \cdot \frac{2}{2} = \frac{6}{24}$$

B14) 26, 52
4 divides 52 evenly.
52 is the LCD.

$$\frac{11}{26} \cdot \frac{2}{2} = \frac{22}{52}$$

$$\frac{1}{4} \cdot \frac{13}{13} = \frac{13}{52}$$

B15) 14, 28, 42
6 divides 42 evenly.
42 is the LCD.

$$\frac{1}{6} \cdot \frac{7}{7} = \frac{7}{42}$$

$$\frac{3}{14} \cdot \frac{3}{3} = \frac{9}{42}$$

C1) a) $\frac{1}{2} =$

$\frac{1}{3} =$

$\frac{2}{5} =$

b) No, the picture for $\frac{2}{5}$ does not look like it is the same area as the shaded areas for $\frac{1}{3}$ and $\frac{1}{2}$ combined.

c) No, it is not possible that $\frac{1}{3} + \frac{1}{2} = \frac{2}{5}$.

d) Answers may vary. Example: Your classmate added the numerators and added the denominators.

C2) Answers may vary. We must build like fractions to add two fractions with different denominators in order to be sure we are adding the same size pieces.

C3) LCD $= 4$

$$\frac{1}{2} + \frac{1}{4} = \frac{1}{2} \cdot \frac{2}{2} + \frac{1}{4}$$
$$= \frac{2}{4} + \frac{1}{4}$$
$$= \frac{3}{4}$$

C4) LCD $= 9$

$$\frac{2}{3} - \frac{2}{9} = \frac{2}{3} \cdot \frac{3}{3} - \frac{2}{9}$$
$$= \frac{6}{9} - \frac{2}{9}$$
$$= \frac{4}{9}$$

C5) LCD $= 30$

$$\frac{3}{5} - \frac{1}{6} = \frac{3}{5} \cdot \frac{6}{6} - \frac{1}{6} \cdot \frac{5}{5}$$
$$= \frac{18}{30} - \frac{5}{30}$$
$$= \frac{13}{30}$$

C6) LCD $= 20$

$$\frac{3}{4} + \frac{1}{5} = \frac{3}{4} \cdot \frac{5}{5} + \frac{1}{5} \cdot \frac{4}{4}$$
$$= \frac{15}{20} + \frac{4}{20}$$
$$= \frac{19}{20}$$

C7) LCD $= 30$

$$\frac{3}{10} + \frac{2}{15} = \frac{3}{10} \cdot \frac{3}{3} + \frac{2}{15} \cdot \frac{2}{2}$$
$$= \frac{9}{30} + \frac{4}{30}$$
$$= \frac{13}{30}$$

C8) LCD $= 18$

$$\frac{2}{9} + \frac{1}{6} = \frac{2}{9} \cdot \frac{2}{2} + \frac{1}{6} \cdot \frac{3}{3}$$
$$= \frac{4}{18} + \frac{3}{18}$$
$$= \frac{7}{18}$$

C9) LCD $= 24$

$$\frac{3}{8} - \frac{1}{6} = \frac{3}{8} \cdot \frac{3}{3} - \frac{1}{6} \cdot \frac{4}{4}$$
$$= \frac{9}{24} - \frac{4}{24}$$
$$= \frac{5}{24}$$

C10) LCD $= 12$

$$\frac{3}{4} - \frac{1}{6} = \frac{3}{4} \cdot \frac{3}{3} - \frac{1}{6} \cdot \frac{2}{2}$$
$$= \frac{9}{12} - \frac{2}{12}$$
$$= \frac{7}{12}$$

C11) Answers may vary. Example:
Step 1: Determine the LCD.
Step 2: Create the framework to build like fractions.
Step 3: Multiply the numerators and denominators by the missing multipliers.

C12) Answers may vary. Example:
Step 1: Add or subtract the numerators.
Step 2: Keep the denominators the same.
Step 3: Simplify, if possible.

D1) Answers may vary.

D2) Answers may vary.

D3) Answers may vary.

D4) $13 = 1 \cdot 13$
      $11 = 1 \cdot 11$

      $LCD = 11 \cdot 13$
      $\quad\quad = 143$
      13 is missing 11.
      11 is missing 13.

      $\dfrac{7}{13} \cdot \dfrac{11}{11} = \dfrac{77}{143}$

      $\dfrac{5}{11} \cdot \dfrac{13}{13} = \dfrac{65}{143}$

D5) $10 = 2 \cdot 5$
      $16 = 2 \cdot 2 \cdot 2 \cdot 2$

      $LCD = 2 \cdot 2 \cdot 2 \cdot 2 \cdot 5$
      $\quad\quad = 80$
      10 is missing $2 \cdot 2 \cdot 2 = 8$.
      16 is missing 5.

      $\dfrac{3}{10} \cdot \dfrac{8}{8} = \dfrac{24}{80}$

      $\dfrac{5}{16} \cdot \dfrac{5}{5} = \dfrac{25}{80}$

D6) $14 = 2 \cdot 7$
      $6 = 2 \cdot 3$

      $LCD = 2 \cdot 3 \cdot 7$
      $\quad\quad = 42$
      14 is missing 3.
      6 is missing 7.

      $\dfrac{3}{14} \cdot \dfrac{3}{3} = \dfrac{9}{42}$

      $\dfrac{1}{6} \cdot \dfrac{7}{7} = \dfrac{7}{42}$

D7) $8 = 2 \cdot 2 \cdot 2$
      $22 = 2 \cdot 11$

      $LCD = 2 \cdot 2 \cdot 2 \cdot 11$
      $\quad\quad = 88$
      8 is missing 11.
      22 is missing $2 \cdot 2 = 4$.

      $\dfrac{7}{8} \cdot \dfrac{11}{11} = \dfrac{77}{88}$

      $\dfrac{5}{22} \cdot \dfrac{4}{4} = \dfrac{20}{88}$

D8) $20 = 2 \cdot 2 \cdot 5$
      $35 = 5 \cdot 7$

      $LCD = 2 \cdot 2 \cdot 5 \cdot 7$
      $\quad\quad = 140$
      20 is missing 7.
      35 is missing $2 \cdot 2 = 4$.

      $\dfrac{3}{20} \cdot \dfrac{7}{7} = \dfrac{21}{140}$

      $\dfrac{6}{35} \cdot \dfrac{4}{4} = \dfrac{24}{140}$

D9) $21 = 3 \cdot 7$
      $49 = 7 \cdot 7$

      $LCD = 3 \cdot 7 \cdot 7$
      $\quad\quad = 147$
      21 is missing 7.
      49 is missing 3.

      $\dfrac{1}{21} \cdot \dfrac{7}{7} = \dfrac{7}{147}$

      $\dfrac{2}{49} \cdot \dfrac{3}{3} = \dfrac{6}{147}$

D10) $18 = 2 \cdot 3 \cdot 3$
      $42 = 2 \cdot 3 \cdot 7$

      $LCD = 2 \cdot 3 \cdot 3 \cdot 7$
      $\quad\quad = 126$
      18 is missing 7.
      42 is missing 3.

      $\dfrac{5}{18} \cdot \dfrac{7}{7} = \dfrac{35}{126}$

      $\dfrac{5}{42} \cdot \dfrac{3}{3} = \dfrac{15}{126}$

D11) $31 = 1 \cdot 31$
      $19 = 1 \cdot 19$

      $LCD = 19 \cdot 31$
      $\quad\quad = 589$
      31 is missing 19.
      19 is missing 31.

      $\dfrac{1}{31} \cdot \dfrac{19}{19} = \dfrac{19}{589}$

      $\dfrac{1}{19} \cdot \dfrac{31}{31} = \dfrac{31}{589}$

E1)   Answers may vary. As fractions become more complicated, LCDs will be difficult to find using the listing multiples technique.

E2) $20 = 2 \cdot 2 \cdot 5$
$24 = 2 \cdot 2 \cdot 2 \cdot 3$

$\text{LCD} = 2 \cdot 2 \cdot 5 \cdot 2 \cdot 3$
$= 20 \cdot 6$
$= 120$

20 is missing $2 \cdot 3 = 6$.
24 is missing 5.

$$\frac{7}{20} - \frac{1}{24} = \frac{7}{20} \cdot \frac{6}{6} - \frac{1}{24} \cdot \frac{5}{5}$$
$$= \frac{42}{120} - \frac{5}{120}$$
$$= \frac{37}{120}$$

E3) $20 = 2 \cdot 2 \cdot 5$
$12 = 2 \cdot 2 \cdot 3$

$\text{LCD} = 2 \cdot 2 \cdot 5 \cdot 3$
$= 20 \cdot 3$
$= 60$

20 is missing 3.
12 is missing 5.

$$\frac{11}{20} - \frac{5}{12} = \frac{11}{20} \cdot \frac{3}{3} - \frac{5}{12} \cdot \frac{5}{5}$$
$$= \frac{33}{60} - \frac{25}{60}$$
$$= \frac{8}{60}$$
$$= \frac{2 \cdot \cancel{4}}{\cancel{4} \cdot 15}$$
$$= \frac{2}{15}$$

E4) $21 = 3 \cdot 7$
$28 = 2 \cdot 2 \cdot 7$

$\text{LCD} = 3 \cdot 7 \cdot 2 \cdot 2$
$= 21 \cdot 4$
$= 84$

21 is missing $2 \cdot 2 = 4$.
28 is missing 3.

$$\frac{5}{21} + \frac{9}{28} = \frac{5}{21} \cdot \frac{4}{4} + \frac{9}{28} \cdot \frac{3}{3}$$
$$= \frac{20}{84} + \frac{27}{84}$$
$$= \frac{47}{84}$$

E5) $55 = 5 \cdot 11$
$22 = 2 \cdot 11$

$\text{LCD} = 5 \cdot 11 \cdot 2$
$= 55 \cdot 2$
$= 110$

55 is missing 2.
22 is missing 5.

$$\frac{4}{55} + \frac{7}{22} = \frac{4}{55} \cdot \frac{2}{2} + \frac{7}{22} \cdot \frac{5}{5}$$
$$= \frac{8}{110} + \frac{35}{110}$$
$$= \frac{43}{110}$$

E6) $8 = 2 \cdot 2 \cdot 2$
$18 = 2 \cdot 3 \cdot 3$

$\text{LCD} = 2 \cdot 2 \cdot 2 \cdot 3 \cdot 3$
$= 8 \cdot 9$
$= 72$

8 is missing $3 \cdot 3 = 9$.
18 is missing $2 \cdot 2 = 4$.

$$\frac{1}{8} + \frac{7}{18} = \frac{1}{8} \cdot \frac{9}{9} + \frac{7}{18} \cdot \frac{4}{4}$$
$$= \frac{9}{72} + \frac{28}{72}$$
$$= \frac{37}{72}$$

E7) $12 = 2 \cdot 2 \cdot 3$
$26 = 2 \cdot 13$

$\text{LCD} = 2 \cdot 2 \cdot 3 \cdot 13$
$= 12 \cdot 13$
$= 156$

12 is missing 13.
26 is missing $2 \cdot 3 = 6$.

$$\frac{7}{12} + \frac{5}{26} = \frac{7}{12} \cdot \frac{13}{13} + \frac{5}{26} \cdot \frac{6}{6}$$
$$= \frac{91}{156} + \frac{30}{156}$$
$$= \frac{121}{156}$$

E8) $15 = 3 \cdot 5$
$12 = 2 \cdot 2 \cdot 3$

$LCD = 3 \cdot 5 \cdot 2 \cdot 2$
$= 15 \cdot 4$
$= 60$

15 is missing $2 \cdot 2 = 4.$
12 is missing 5.

$\dfrac{8}{15} - \dfrac{5}{12} = \dfrac{8}{15} \cdot \dfrac{4}{4} - \dfrac{5}{12} \cdot \dfrac{5}{5}$
$= \dfrac{32}{60} - \dfrac{25}{60}$
$= \dfrac{7}{60}$

E9) $24 = 2 \cdot 2 \cdot 2 \cdot 3$
$16 = 2 \cdot 2 \cdot 2 \cdot 2$

$LCD = 2 \cdot 2 \cdot 2 \cdot 3 \cdot 2$
$= 24 \cdot 2$
$= 48$

24 is missing 2.
16 is missing 3.

$\dfrac{11}{24} - \dfrac{5}{16} = \dfrac{11}{24} \cdot \dfrac{2}{2} - \dfrac{5}{16} \cdot \dfrac{3}{3}$
$= \dfrac{22}{48} - \dfrac{15}{48}$
$= \dfrac{7}{48}$

SECTION 2.4   EXERCISES

For 1–15, refer to Concept Checks and Practice Exercises.

17) building like fractions

19) missing multiplier

21) $\dfrac{3}{7} + \dfrac{5}{5} = \dfrac{3+5}{7}$
$= \dfrac{8}{7}$

23) $\dfrac{9}{10} - \dfrac{3}{10} = \dfrac{9-3}{10}$
$= \dfrac{6}{10}$
$= \dfrac{\cancel{2} \cdot 3}{\cancel{2} \cdot 5}$
$= \dfrac{3}{5}$

25) $\dfrac{8}{15} + \dfrac{4}{15} = \dfrac{8+4}{15}$
$= \dfrac{12}{15}$
$= \dfrac{\cancel{3} \cdot 4}{\cancel{3} \cdot 5}$
$= \dfrac{4}{5}$

27) $\dfrac{13}{14} - \dfrac{5}{14} = \dfrac{13-5}{14}$
$= \dfrac{8}{14}$
$= \dfrac{4 \cdot \cancel{2}}{7 \cdot \cancel{2}}$
$= \dfrac{4}{7}$

29)
$\dfrac{1}{3} + \dfrac{1}{3} = \dfrac{2}{3}$

31)
$\dfrac{7}{8} - \dfrac{6}{8} = \dfrac{1}{8}$

33)
$\dfrac{2}{2} - \dfrac{1}{2} = \dfrac{1}{2}$

35) 5, 10
2 divides 10 evenly.
10 is the LCD.

37) 9, 18
6 divides 18 evenly.
18 is the LCD.

39) $15 = 3 \cdot 5$
$21 = 3 \cdot 7$

$LCD = 3 \cdot 5 \cdot 7$
$= 15 \cdot 7$
$= 105$

41) $7 = 7$
$12 = 2 \cdot 2 \cdot 3$

$LCD = 2 \cdot 2 \cdot 3 \cdot 7$
$= 4 \cdot 3 \cdot 7$
$= 12 \cdot 7$
$= 84$

43) LCD = 12

$$\frac{3}{4} + \frac{5}{12} = \frac{3}{4} \cdot \frac{3}{3} + \frac{5}{12}$$

$$= \frac{9}{12} + \frac{5}{12}$$

$$= \frac{14}{12}$$

$$= \frac{\cancel{2} \cdot 7}{\cancel{2} \cdot 6}$$

$$= \frac{7}{6}$$

45) LCD = 30

$$\frac{7}{10} - \frac{1}{15} = \frac{7}{10} \cdot \frac{3}{3} - \frac{1}{15} \cdot \frac{2}{2}$$

$$= \frac{21}{30} - \frac{2}{30}$$

$$= \frac{19}{30}$$

47) LCD = 45

$$\frac{8}{15} - \frac{2}{9} = \frac{8}{15} \cdot \frac{3}{3} - \frac{2}{9} \cdot \frac{5}{5}$$

$$= \frac{24}{45} - \frac{10}{45}$$

$$= \frac{14}{45}$$

49) LCD = 42

$$\frac{13}{14} - \frac{4}{21} = \frac{13}{14} \cdot \frac{3}{3} - \frac{4}{21} \cdot \frac{2}{2}$$

$$= \frac{39}{42} - \frac{8}{42}$$

$$= \frac{31}{42}$$

51) LCD = 60

$$\frac{17}{20} - \frac{1}{6} = \frac{17}{20} \cdot \frac{3}{3} - \frac{1}{6} \cdot \frac{10}{10}$$

$$= \frac{51}{60} - \frac{10}{60}$$

$$= \frac{41}{60}$$

53) LCD = 42

$$\frac{5}{21} + \frac{3}{14} = \frac{5}{21} \cdot \frac{2}{2} + \frac{3}{14} \cdot \frac{3}{3}$$

$$= \frac{10}{42} + \frac{9}{42}$$

$$= \frac{19}{42}$$

55) LCD = 54

$$\frac{1}{18} + \frac{16}{27} = \frac{1}{18} \cdot \frac{3}{3} + \frac{16}{27} \cdot \frac{2}{2}$$

$$= \frac{3}{54} + \frac{32}{54}$$

$$= \frac{35}{54}$$

57) LCD = 60

$$\frac{8}{15} - \frac{5}{12} = \frac{8}{15} \cdot \frac{4}{4} - \frac{5}{12} \cdot \frac{5}{5}$$

$$= \frac{32}{60} - \frac{25}{60}$$

$$= \frac{7}{60}$$

59) LCD = 144

$$\frac{3}{16} + \frac{7}{18} = \frac{3}{16} \cdot \frac{9}{9} + \frac{7}{18} \cdot \frac{8}{8}$$

$$= \frac{27}{144} + \frac{56}{144}$$

$$= \frac{83}{144}$$

61) LCD = 120

$$\frac{7}{20} - \frac{7}{24} = \frac{7}{20} \cdot \frac{6}{6} - \frac{7}{24} \cdot \frac{5}{5}$$

$$= \frac{42}{120} - \frac{35}{120}$$

$$= \frac{7}{120}$$

63) LCD = 12

$$\frac{3}{4} + \frac{1}{3} - \frac{1}{2} = \frac{3}{4} \cdot \frac{3}{3} + \frac{1}{3} \cdot \frac{4}{4} - \frac{1}{2} \cdot \frac{6}{6}$$

$$= \frac{9}{12} + \frac{4}{12} - \frac{6}{12}$$

$$= \frac{13}{12} - \frac{6}{12}$$

$$= \frac{7}{12}$$

65) LCD = 8

$$\frac{3}{4} - \frac{1}{8} - \frac{1}{2} = \frac{3}{4} \cdot \frac{2}{2} - \frac{1}{8} - \frac{1}{2} \cdot \frac{4}{4}$$

$$= \frac{6}{8} - \frac{1}{8} - \frac{4}{8}$$

$$= \frac{5}{8} - \frac{4}{8}$$

$$= \frac{1}{8}$$

67) $LCD = 15$

$$9 - \frac{1}{3} + \frac{1}{5} = \frac{9}{1} \cdot \frac{15}{15} - \frac{1}{3} \cdot \frac{5}{5} + \frac{1}{5} \cdot \frac{3}{3}$$
$$= \frac{135}{15} - \frac{5}{15} + \frac{3}{15}$$
$$= \frac{130}{15} + \frac{3}{15}$$
$$= \frac{133}{15}$$

69) $LCD = 24$

$$\frac{1}{6} + \frac{1}{8} + \frac{1}{12} = \frac{1}{6} \cdot \frac{4}{4} + \frac{1}{8} \cdot \frac{3}{3} + \frac{1}{12} \cdot \frac{2}{2}$$
$$= \frac{4}{24} + \frac{3}{24} + \frac{2}{24}$$
$$= \frac{7}{24} + \frac{2}{24}$$
$$= \frac{9}{24}$$
$$= \frac{\cancel{3} \cdot 3}{\cancel{3} \cdot 8}$$
$$= \frac{3}{8}$$

71) $LCD = 8$

a) $\frac{3}{4} \cdot \frac{2}{2} = \frac{6}{8}$ and $\frac{5}{8} = \frac{5}{8}$

b) $\frac{5}{8} < \frac{6}{8}$, so $\frac{5}{8} < \frac{3}{4}$

73) $LCD = 16$

a) $\frac{13}{16} = \frac{13}{16}$ and $\frac{3}{4} \cdot \frac{4}{4} = \frac{12}{16}$

b) $\frac{12}{16} < \frac{13}{16}$, so $\frac{3}{4} < \frac{13}{16}$

75) $LCD = 30$

a) $\frac{5}{6} \cdot \frac{5}{5} = \frac{25}{30}$ and $\frac{4}{5} \cdot \frac{6}{6} = \frac{24}{30}$

b) $\frac{24}{30} < \frac{25}{30}$, so $\frac{4}{5} < \frac{5}{6}$

77) $LCD = 30$

a) $\frac{2}{3} \cdot \frac{10}{10} = \frac{20}{30}$ and $\frac{7}{10} \cdot \frac{3}{3} = \frac{21}{30}$

b) $\frac{20}{30} < \frac{21}{30}$, so $\frac{2}{3} < \frac{7}{10}$

79) $LCD = 32$

$\frac{11}{32} = \frac{11}{32}$ and $\frac{5}{8} \cdot \frac{4}{4} = \frac{20}{32}$ and $\frac{1}{2} \cdot \frac{16}{16} = \frac{16}{32}$

$\frac{11}{32} < \frac{16}{32} < \frac{20}{32}$, so $\frac{11}{32} < \frac{1}{2} < \frac{5}{8}$

The sizes in order from smallest to largest are $\frac{11}{32}$ inch, $\frac{1}{2}$ inch, and $\frac{5}{8}$ inch.

81) $LCD = 32$

$\frac{11}{32} = \frac{11}{32}$ and $\frac{3}{8} \cdot \frac{4}{4} = \frac{12}{32}$ and $\frac{5}{16} \cdot \frac{2}{2} = \frac{10}{32}$

$\frac{12}{32} > \frac{11}{32} > \frac{10}{32}$, so $\frac{3}{8} > \frac{11}{32} > \frac{5}{16}$

The buttons in order from largest to smallest are $\frac{3}{8}$ inch, $\frac{11}{32}$ inch, and $\frac{5}{16}$ inch.

83) Answers may vary. The denominator tells the type of fraction being added. The type of fraction does not change.

85) $LCD = 12$

$$\frac{2}{3} + \frac{1}{4} = \frac{2}{3} \cdot \frac{4}{4} + \frac{1}{4} \cdot \frac{3}{3}$$
$$= \frac{8}{12} + \frac{3}{12}$$
$$= \frac{11}{12}$$

They purchased $\frac{11}{12}$ pound of grapes.

87) $LCD = 12$

$$\frac{3}{4} + \frac{5}{12} = \frac{3}{4} \cdot \frac{3}{3} + \frac{5}{12}$$
$$= \frac{9}{12} + \frac{5}{12}$$
$$= \frac{14}{12}$$
$$= \frac{\cancel{2} \cdot 7}{\cancel{2} \cdot 6}$$
$$= \frac{7}{6}$$

The bolts needed to be $\frac{7}{6}$ or $1\frac{1}{6}$ inches long.

89) $$\frac{84}{100} \cdot 50 = \frac{84}{100} \cdot \frac{50}{1}$$
$$= \frac{21 \cdot \cancel{4} \cdot 2 \cdot \cancel{25}}{\cancel{4} \cdot \cancel{25} \cdot 1}$$
$$= 42$$

You answered 42 questions correctly.

91) LCD = 35

a) $\frac{3}{7} \cdot \frac{5}{5} = \frac{15}{35}$ and $\frac{1}{5} \cdot \frac{7}{7} = \frac{7}{35}$

$\frac{15}{35} > \frac{7}{35}$, so $\frac{3}{7} > \frac{1}{5}$

Rosie has saved more for the house.

b) $\frac{3}{7} + \frac{1}{5} = \frac{3}{7} \cdot \frac{5}{5} + \frac{1}{5} \cdot \frac{7}{7}$

$= \frac{15}{35} + \frac{7}{35}$

$= \frac{22}{35}$

Together, they have $\frac{22}{35}$ of the down payment.

c) $1 - \frac{22}{35} = \frac{35}{35} - \frac{22}{35}$

$= \frac{13}{35}$

They still need to save $\frac{13}{35}$ of the down payment.

93) $\frac{2}{16} + \frac{3}{16} + \frac{3}{16} = \frac{5}{16} + \frac{3}{16}$

$= \frac{8}{16}$

$= \frac{1 \cdot \cancel{8}}{2 \cdot \cancel{8}}$

$= \frac{1}{2}$

The orthodontist has made $\frac{1}{2}$ turn in the last three months.

95) Quarter: $\frac{1}{4}$ dollar

Dime: $\frac{1}{10}$ dollar

Penny: $\frac{1}{100}$ dollar

$\frac{1}{4} + \frac{3}{10} + \frac{4}{100} = \frac{1}{4} \cdot \frac{25}{25} + \frac{3}{10} \cdot \frac{10}{10} + \frac{4}{100}$

$= \frac{25}{100} + \frac{30}{100} + \frac{4}{100}$

$= \frac{59}{100}$

One quarter, three dimes, and four pennies is $\frac{59}{100}$ of a dollar.

97) LCD = 10

$\frac{2}{5} + \frac{1}{2} + \frac{1}{10} = \frac{2}{5} \cdot \frac{2}{2} + \frac{1}{2} \cdot \frac{5}{5} + \frac{1}{10}$

$= \frac{4}{10} + \frac{5}{10} + \frac{1}{10}$

$= \frac{10}{10}$

$= 1$

1 kilometer of fencing is needed.

99) $\frac{4-2}{24} = \frac{2}{24}$

$= \frac{1 \cdot \cancel{2}}{\cancel{2} \cdot 12}$

$= \frac{1}{12}$

Steve's TV time exceeded his homework time by $\frac{1}{12}$ of the day.

2.5   Fractions and the Order of Operations

GUIDED PRACTICE

1) $\frac{5}{8} \div \frac{3}{9-5} = \frac{5}{8} \div \frac{3}{4}$

$= \frac{5}{8} \cdot \frac{4}{3}$

$= \frac{5 \cdot \cancel{4}}{2 \cdot \cancel{4} \cdot 3}$

$= \frac{5}{6}$

2) $\left(\frac{3}{5} + \frac{1}{10}\right) - \frac{7}{15} = \left(\frac{6}{10} + \frac{1}{10}\right) - \frac{7}{15}$

$= \left(\frac{7}{10}\right) - \frac{7}{15}$

$= \frac{21}{30} - \frac{14}{30}$

$= \frac{7}{30}$

3) $\frac{13}{20} - \frac{3}{4} \cdot \frac{9-5}{5} = \frac{13}{20} - \frac{3}{4} \cdot \frac{4}{5}$

$= \frac{13}{20} - \frac{3 \cdot \cancel{4}}{\cancel{4} \cdot 5}$

$= \frac{13}{20} - \frac{3}{5} \cdot \frac{4}{4}$

$= \frac{13}{20} - \frac{12}{20}$

$= \frac{1}{20}$

4) $\left(\dfrac{11}{6}\cdot\dfrac{1}{5}+\dfrac{17}{30}\right)\cdot\dfrac{6}{7}=\left(\dfrac{11}{30}+\dfrac{17}{30}\right)\cdot\dfrac{6}{7}$

$\qquad\qquad\qquad\quad=\left(\dfrac{28}{30}\right)\cdot\dfrac{6}{7}$

$\qquad\qquad\qquad\quad=\dfrac{2\cdot2\cdot\not7\cdot\not6}{5\cdot\not6\cdot\not7}$

$\qquad\qquad\qquad\quad=\dfrac{4}{5}$

5) $\dfrac{3}{5}+\dfrac{3}{10}=\dfrac{6}{10}+\dfrac{3}{10}=\dfrac{9}{10}$

Alex has set up $\dfrac{9}{10}$ of the chairs.

$1-\dfrac{9}{10}=\dfrac{10}{10}-\dfrac{9}{10}=\dfrac{1}{10}$

Alex will have to set up $\dfrac{1}{10}$ of the chairs after lunch.

$300\cdot\dfrac{1}{10}=\dfrac{\not{10}\cdot30\cdot1}{1\cdot\not{10}}=30$

Alex will have to set up 30 chairs after lunch.

6) $\qquad n-\dfrac{2}{3}=\dfrac{1}{2}$

$\quad n-\dfrac{2}{3}+\dfrac{2}{3}=\dfrac{1}{2}+\dfrac{2}{3}$

$\qquad\qquad\quad n=\dfrac{3}{6}+\dfrac{4}{6}$

$\qquad\qquad\quad n=\dfrac{7}{6}$

7) $\qquad \dfrac{3}{2}\cdot n=\dfrac{4}{5}$

$\quad \dfrac{2}{3}\cdot\dfrac{3}{2}\cdot n=\dfrac{2}{3}\cdot\dfrac{4}{5}$

$\qquad\qquad\quad n=\dfrac{8}{15}$

CONCEPT CHECKS AND PRACTICE EXERCISES

A1) Answers may vary. Example: There is a violation of the order of operations. The subtraction needs to be performed first, since it is to the left of the addition.

A2) Answers may vary. Example: There is a computation error. $\left(\dfrac{3}{5}\right)^2=\dfrac{9}{25}$, not $\dfrac{6}{10}$.

A3) Answers may vary. Example: Since the numerator and denominator represent groupings, these should be simplified first.

A4) $\dfrac{19}{20}-\left(\dfrac{1}{2}+\dfrac{2}{5}\right)=\dfrac{19}{20}-\left(\dfrac{5}{10}+\dfrac{4}{10}\right)$

$\qquad\qquad\qquad\quad=\dfrac{19}{20}-\dfrac{9}{10}$

$\qquad\qquad\qquad\quad=\dfrac{19}{20}-\dfrac{18}{20}$

$\qquad\qquad\qquad\quad=\dfrac{1}{20}$

A5) $\dfrac{7}{21}+\left(\dfrac{5}{7}-\dfrac{1}{3}\right)=\dfrac{7}{21}+\left(\dfrac{15}{21}-\dfrac{7}{21}\right)$

$\qquad\qquad\qquad\quad=\dfrac{7}{21}+\dfrac{8}{21}$

$\qquad\qquad\qquad\quad=\dfrac{15}{21}$

$\qquad\qquad\qquad\quad=\dfrac{\not3\cdot5}{\not3\cdot7}$

$\qquad\qquad\qquad\quad=\dfrac{5}{7}$

A6) $\dfrac{2}{3}\cdot\left(\dfrac{3}{4}\right)^2=\dfrac{2}{3}\cdot\dfrac{9}{16}$

$\qquad\qquad\quad=\dfrac{\not2\cdot\not3\cdot3}{\not3\cdot\not2\cdot8}$

$\qquad\qquad\quad=\dfrac{3}{8}$

A7) $\left(\dfrac{1}{5}\right)^2\div\dfrac{3}{5}=\dfrac{1}{25}\div\dfrac{3}{5}$

$\qquad\qquad\quad=\dfrac{1}{25}\cdot\dfrac{5}{3}$

$\qquad\qquad\quad=\dfrac{1\cdot\not5}{\not5\cdot5\cdot3}$

$\qquad\qquad\quad=\dfrac{1}{15}$

A8) $\dfrac{2}{3}-\dfrac{5}{12}+\dfrac{3}{4}=\left(\dfrac{8}{12}-\dfrac{5}{12}\right)+\dfrac{3}{4}$

$\qquad\qquad\qquad\quad=\dfrac{3}{12}+\dfrac{3}{4}$

$\qquad\qquad\qquad\quad=\dfrac{1}{4}+\dfrac{3}{4}$

$\qquad\qquad\qquad\quad=\dfrac{4}{4}$

$\qquad\qquad\qquad\quad=1$

A9) $\dfrac{7}{5}-\dfrac{3}{10}+\dfrac{1}{5}=\left(\dfrac{14}{10}-\dfrac{3}{10}\right)+\dfrac{1}{5}$

$\qquad\qquad\qquad\quad=\dfrac{11}{10}+\dfrac{1}{5}$

$\qquad\qquad\qquad\quad=\dfrac{11}{10}+\dfrac{2}{10}$

$\qquad\qquad\qquad\quad=\dfrac{13}{10}$

A10)
$$\frac{4}{5} \div \frac{8}{15} \cdot \frac{2}{9} = \left(\frac{4}{5} \cdot \frac{15}{8}\right) \cdot \frac{2}{9}$$
$$= \left(\frac{\cancel{4} \cdot 3 \cdot \cancel{5}}{\cancel{5} \cdot 2 \cdot \cancel{4}}\right) \cdot \frac{2}{9}$$
$$= \frac{3}{2} \cdot \frac{2}{9}$$
$$= \frac{\cancel{3} \cdot \cancel{2}}{\cancel{2} \cdot \cancel{3} \cdot 3}$$
$$= \frac{1}{3}$$

A11)
$$\frac{4}{5} \div \frac{1}{3} \cdot \frac{25}{6} = \left(\frac{4}{5} \cdot \frac{3}{1}\right) \cdot \frac{25}{6}$$
$$= \frac{12}{5} \cdot \frac{25}{6}$$
$$= \frac{2 \cdot \cancel{6} \cdot \cancel{5} \cdot 5}{\cancel{5} \cdot \cancel{6}}$$
$$= 10$$

A12)
$$\frac{13}{21} - \frac{5}{7} \cdot \left(\frac{9-7}{3}\right) = \frac{13}{21} - \frac{5}{7} \cdot \left(\frac{2}{3}\right)$$
$$= \frac{13}{21} - \frac{10}{21}$$
$$= \frac{3}{21}$$
$$= \frac{1 \cdot \cancel{3}}{\cancel{3} \cdot 7}$$
$$= \frac{1}{7}$$

A13)
$$\frac{7}{24} - \frac{1}{5} \div \left(\frac{6}{2+3}\right) = \frac{7}{24} - \frac{1}{5} \div \left(\frac{6}{5}\right)$$
$$= \frac{7}{24} - \frac{1}{5} \cdot \frac{5}{6}$$
$$= \frac{7}{24} - \frac{1 \cdot \cancel{5}}{\cancel{5} \cdot 6}$$
$$= \frac{7}{24} - \frac{1}{6}$$
$$= \frac{7}{24} - \frac{4}{24}$$
$$= \frac{3}{24}$$
$$= \frac{1 \cdot \cancel{3}}{\cancel{3} \cdot 8}$$
$$= \frac{1}{8}$$

A14)
$$\frac{25}{16} - \left(\frac{3}{4}\right)^2 + \frac{5}{8} = \frac{25}{16} - \frac{9}{16} + \frac{5}{8}$$
$$= \frac{16}{16} + \frac{5}{8}$$
$$= 1 + \frac{5}{8}$$
$$= \frac{8}{8} + \frac{5}{8}$$
$$= \frac{13}{8}$$

A15)
$$\frac{9}{25} + \left(\frac{4}{5}\right)^2 + \frac{1}{10} = \frac{9}{25} + \frac{16}{25} + \frac{1}{10}$$
$$= \frac{25}{25} + \frac{1}{10}$$
$$= 1 + \frac{1}{10}$$
$$= \frac{10}{10} + \frac{1}{10}$$
$$= \frac{11}{10}$$

FOCUS ON ALGEBRA PRACTICE EXERCISES

1)
$$n + \frac{1}{4} = \frac{3}{4}$$
$$n + \frac{1}{4} - \frac{1}{4} = \frac{3}{4} - \frac{1}{4}$$
$$n = \frac{2}{4}$$
$$n = \frac{1}{2}$$

2)
$$n + \frac{5}{8} = \frac{9}{8}$$
$$n + \frac{5}{8} - \frac{5}{8} = \frac{9}{8} - \frac{5}{8}$$
$$n = \frac{4}{8}$$
$$n = \frac{1}{2}$$

3)
$$n - \frac{1}{2} = \frac{3}{2}$$
$$n - \frac{1}{2} + \frac{1}{2} = \frac{3}{2} + \frac{1}{2}$$
$$n = \frac{4}{2}$$
$$n = 2$$

4)
$$n - \frac{1}{10} = \frac{4}{10}$$
$$n - \frac{1}{10} + \frac{1}{10} = \frac{4}{10} + \frac{1}{10}$$
$$n = \frac{5}{10}$$
$$n = \frac{1}{2}$$

5)
$$\frac{1}{5} \cdot n = \frac{1}{8}$$
$$\frac{5}{1} \cdot \frac{1}{5} \cdot n = \frac{5}{1} \cdot \frac{1}{8}$$
$$n = \frac{5}{8}$$

6)
$$\frac{1}{4} \cdot n = \frac{1}{9}$$
$$\frac{4}{1} \cdot \frac{1}{4} \cdot n = \frac{4}{1} \cdot \frac{1}{9}$$
$$n = \frac{4}{9}$$

7)
$$\frac{5}{6} \cdot n = 2$$
$$\frac{6}{5} \cdot \frac{5}{6} \cdot n = \frac{6}{5} \cdot \frac{2}{1}$$
$$n = \frac{12}{5}$$

8)
$$\frac{7}{3} \cdot n = 6$$
$$\frac{3}{7} \cdot \frac{7}{3} \cdot n = \frac{3}{7} \cdot \frac{6}{1}$$
$$n = \frac{18}{7}$$

9)
$$n + \frac{1}{3} = \frac{11}{18}$$
$$n + \frac{1}{3} - \frac{1}{3} = \frac{11}{18} - \frac{1}{3}$$
$$n = \frac{11}{18} - \frac{6}{18}$$
$$n = \frac{5}{18}$$

10)
$$n + \frac{1}{6} = \frac{3}{12}$$
$$n + \frac{1}{6} - \frac{1}{6} = \frac{3}{12} - \frac{1}{6}$$
$$n = \frac{3}{12} - \frac{2}{12}$$
$$n = \frac{1}{12}$$

11)
$$n - \frac{1}{3} = \frac{5}{6}$$
$$n - \frac{1}{3} + \frac{1}{3} = \frac{5}{6} + \frac{1}{3}$$
$$n = \frac{5}{6} + \frac{2}{6}$$
$$n = \frac{7}{6}$$

12)
$$n - \frac{1}{5} = \frac{3}{10}$$
$$n - \frac{1}{5} + \frac{1}{5} = \frac{3}{10} + \frac{1}{5}$$
$$n = \frac{3}{10} + \frac{2}{10}$$
$$n = \frac{5}{10}$$
$$n = \frac{1}{2}$$

13)
$$n + \frac{5}{6} = \frac{6}{5}$$
$$n + \frac{5}{6} - \frac{5}{6} = \frac{6}{5} - \frac{5}{6}$$
$$n = \frac{36}{30} - \frac{25}{30}$$
$$n = \frac{11}{30}$$

14)
$$n + \frac{8}{9} = \frac{3}{2}$$
$$n + \frac{8}{9} - \frac{8}{9} = \frac{3}{2} - \frac{8}{9}$$
$$n = \frac{27}{18} - \frac{16}{18}$$
$$n = \frac{11}{18}$$

15)
$$\frac{4}{5} \cdot n = \frac{3}{20}$$
$$\frac{5}{4} \cdot \frac{4}{5} \cdot n = \frac{5}{4} \cdot \frac{3}{20}$$
$$n = \frac{3}{16}$$

16)
$$\frac{9}{5} \cdot n = \frac{18}{7}$$
$$\frac{5}{9} \cdot \frac{9}{5} \cdot n = \frac{5}{9} \cdot \frac{18}{7}$$
$$n = \frac{10}{7}$$

SECTION 2.5   EXERCISES

For 1–5, refer to Concept Checks and Practice Exercises.

7) $\frac{1}{5} - \frac{1}{6} + \frac{1}{8} = \left(\frac{6}{30} - \frac{5}{30}\right) + \frac{1}{8}$

$\qquad = \frac{1}{30} + \frac{1}{8}$

$\qquad = \frac{4}{120} + \frac{15}{120}$

$\qquad = \frac{19}{120}$

9) $\frac{5}{12} \cdot \frac{9}{20} \div \frac{3}{10} = \left(\frac{\cancel{5} \cdot \cancel{3} \cdot 3}{\cancel{3} \cdot 4 \cdot 4 \cdot \cancel{5}}\right) \div \frac{3}{10}$

$\qquad = \frac{3}{16} \div \frac{3}{10}$

$\qquad = \frac{3}{16} \cdot \frac{10}{3}$

$\qquad = \frac{\cancel{3} \cdot \cancel{2} \cdot 5}{\cancel{2} \cdot 8 \cdot \cancel{3}}$

$\qquad = \frac{5}{8}$

11) $\frac{27}{32} - \frac{3}{8} \cdot \frac{1}{4} = \frac{27}{32} - \frac{3}{32}$

$\qquad = \frac{24}{32}$

$\qquad = \frac{3 \cdot \cancel{8}}{4 \cdot \cancel{8}}$

$\qquad = \frac{3}{4}$

13) $\frac{1}{2} \div \frac{1}{3} \cdot \frac{1}{4} = \left(\frac{1}{2} \cdot \frac{3}{1}\right) \cdot \frac{1}{4}$

$\qquad = \frac{3}{2} \cdot \frac{1}{4}$

$\qquad = \frac{3}{8}$

15) $\left(\frac{7}{8} - \frac{3}{8}\right) \cdot \frac{1}{4} = \frac{4}{8} \cdot \frac{1}{4}$

$\qquad = \frac{\cancel{4} \cdot 1}{8 \cdot \cancel{4}}$

$\qquad = \frac{1}{8}$

17) $\left(\frac{3}{7}\right)^2 \cdot \frac{14}{9} = \frac{9}{49} \cdot \frac{14}{9}$

$\qquad = \frac{\cancel{9} \cdot 2 \cdot \cancel{7}}{\cancel{7} \cdot 7 \cdot \cancel{9}}$

$\qquad = \frac{2}{7}$

19) $\frac{2}{3} - \left(\frac{2}{3}\right)^2 + \frac{1}{9} = \frac{2}{3} - \frac{4}{9} + \frac{1}{9}$

$\qquad = \frac{6}{9} - \frac{4}{9} + \frac{1}{9}$

$\qquad = \frac{2}{9} + \frac{1}{9}$

$\qquad = \frac{3}{9}$

$\qquad = \frac{1 \cdot \cancel{3}}{\cancel{3} \cdot 3}$

$\qquad = \frac{1}{3}$

21) $\frac{5}{12} \cdot \left(\frac{9}{20} \div \frac{3}{10}\right) = \frac{5}{12} \cdot \left(\frac{9}{20} \cdot \frac{10}{3}\right)$

$\qquad = \frac{5}{12} \cdot \left(\frac{\cancel{3} \cdot 3 \cdot \cancel{10}}{2 \cdot \cancel{10} \cdot \cancel{3}}\right)$

$\qquad = \frac{5}{12} \cdot \frac{3}{2}$

$\qquad = \frac{5 \cdot \cancel{3}}{\cancel{3} \cdot 4 \cdot 2}$

$\qquad = \frac{5}{8}$

23) $\left(\frac{4}{5}\right)^2 - \frac{3}{5} = \frac{16}{25} - \frac{3}{5}$

$\qquad = \frac{16}{25} - \frac{15}{25}$

$\qquad = \frac{1}{25}$

25) $\left(\frac{9}{2} - 4\right)^2 = \left(\frac{9}{2} - \frac{8}{2}\right)^2$

$\qquad = \left(\frac{1}{2}\right)^2$

$\qquad = \frac{1}{4}$

27) $\left(\frac{5}{8} - \frac{1}{4}\right) \cdot \frac{16}{21} = \left(\frac{5}{8} - \frac{2}{8}\right) \cdot \frac{16}{21}$

$\qquad = \frac{3}{8} \cdot \frac{16}{21}$

$\qquad = \frac{\cancel{3} \cdot 2 \cdot \cancel{8}}{\cancel{8} \cdot \cancel{3} \cdot 7}$

$\qquad = \frac{2}{7}$

29) $\left(\dfrac{1}{3}+\dfrac{1}{6}\right)\cdot\left(\dfrac{3}{4}-\dfrac{1}{2}\right)=\left(\dfrac{2}{6}+\dfrac{1}{6}\right)\cdot\left(\dfrac{3}{4}-\dfrac{2}{4}\right)$

$\qquad=\dfrac{3}{6}\cdot\dfrac{1}{4}$

$\qquad=\dfrac{\cancel{3}\cdot1}{2\cdot\cancel{3}\cdot4}$

$\qquad=\dfrac{1}{8}$

31) $\dfrac{9}{2}-\dfrac{6-1}{6}\div\dfrac{5}{9}=\dfrac{9}{2}-\dfrac{5}{6}\div\dfrac{5}{9}$

$\qquad=\dfrac{9}{2}-\dfrac{5}{6}\cdot\dfrac{9}{5}$

$\qquad=\dfrac{9}{2}-\dfrac{3}{2}$

$\qquad=\dfrac{6}{2}$

$\qquad=3$

33) $\left(\dfrac{3}{16-9}-\dfrac{3}{14}\right)+\dfrac{2}{5}=\left(\dfrac{3}{7}-\dfrac{3}{14}\right)+\dfrac{2}{5}$

$\qquad=\left(\dfrac{6}{14}-\dfrac{3}{14}\right)+\dfrac{2}{5}$

$\qquad=\dfrac{3}{14}+\dfrac{2}{5}$

$\qquad=\dfrac{15}{70}+\dfrac{28}{70}$

$\qquad=\dfrac{43}{70}$

35) $4-4\cdot\dfrac{1}{3}=4-\dfrac{4}{1}\cdot\dfrac{1}{3}$

$\qquad=4-\dfrac{4}{3}$

$\qquad=\dfrac{12}{3}-\dfrac{4}{3}$

$\qquad=\dfrac{8}{3}$

She has $\dfrac{8}{3}$ spools of thread left.

37) $n-\dfrac{1}{8}=\dfrac{5}{8}$

$n-\dfrac{1}{8}+\dfrac{1}{8}=\dfrac{5}{8}+\dfrac{1}{8}$

$n=\dfrac{6}{8}$

$n=\dfrac{3}{4}$

39) $n+\dfrac{1}{3}=\dfrac{1}{2}$

$n+\dfrac{1}{3}-\dfrac{1}{3}=\dfrac{1}{2}-\dfrac{1}{3}$

$n=\dfrac{3}{6}-\dfrac{2}{6}$

$n=\dfrac{1}{6}$

41) $\dfrac{4}{3}\cdot n=\dfrac{2}{3}$

$\dfrac{3}{4}\cdot\dfrac{4}{3}\cdot n=\dfrac{3}{4}\cdot\dfrac{2}{3}$

$n=\dfrac{1}{2}$

43) $n+\dfrac{7}{9}=\dfrac{8}{5}$

$n+\dfrac{7}{9}-\dfrac{7}{9}=\dfrac{8}{5}-\dfrac{7}{9}$

$n=\dfrac{72}{45}-\dfrac{35}{45}$

$n=\dfrac{37}{45}$

45) a) The equation expresses the fact that Victor began with some oranges, used $\dfrac{7}{2}$, and ended up with $\dfrac{5}{2}$.

b) $n-\dfrac{7}{2}=\dfrac{5}{2}$

$n-\dfrac{7}{2}+\dfrac{7}{2}=\dfrac{5}{2}+\dfrac{7}{2}$

$n=\dfrac{12}{2}$

$n=6$

Victor started with six oranges.

47) a) $P=2\cdot l+2\cdot w$

$\qquad=2\cdot80+2\cdot60$

$\qquad=160+120$

$\qquad=280$ feet

The perimeter is 280 feet.

b) $280-20-15=260-15$

$\qquad\qquad\qquad=245$

Antwan must buy 245 feet of fencing.

c) $\dfrac{20+15}{280}=\dfrac{35}{280}$

$\qquad=\dfrac{1\cdot\cancel{35}}{8\cdot\cancel{35}}$

$\qquad=\dfrac{1}{8}$

$\dfrac{1}{8}$ of the project is already completed.

2.6   Mixed Numbers

GUIDED PRACTICE

1) Since there are 3 whole objects and $\frac{1}{4}$ of another, the mixed number $3\frac{1}{4}$ represents this picture.

2)

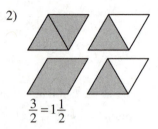

$$\frac{3}{2} = 1\frac{1}{2}$$

3) $\frac{3}{2} \rightarrow 2\overline{)3}$

Quotient = 1
Remainder = 1
Denominator = 2

$$\frac{3}{2} = 1\frac{1}{2}$$

4) $17 \div 6 = 2$ Remainder 5

$$\frac{17}{6} = 2\frac{5}{6}$$

5) $15 \div 8 = 1$ Remainder 7

$$\frac{15}{8} = 1\frac{7}{8}$$

6) $84 \div 7 = 12$ Remainder 0

$$\frac{84}{7} = 12$$

7)

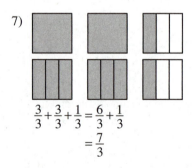

$$\frac{3}{3} + \frac{3}{3} + \frac{1}{3} = \frac{6}{3} + \frac{1}{3}$$
$$= \frac{7}{3}$$

8) a) Multiply the denominator and the whole number. There are $2 \cdot 3 = 6$ thirds in 2.

   b) There is 1 third in the fractional part of $2\frac{1}{3}$.

c) Add the results. There are $6 + 1 = 7$ thirds.

$$2\frac{1}{3} = \frac{7}{3}$$

9) a) Multiply the denominator and the whole number. There are $8 \cdot 5 = 40$ eighths in 5.

   b) There are 7 eighths in the fractional part of $5\frac{7}{8}$.

   c) Add the results. There are $40 + 7 = 47$ eighths.

$$5\frac{7}{8} = \frac{47}{8}$$

10) $4 \cdot 3 + 1 = 13$

$$3\frac{1}{4} = \frac{13}{4}$$

11) $9 \cdot 5 + 7 = 52$

$$5\frac{7}{9} = \frac{52}{9}$$

12) Since $\frac{9}{15} > \frac{1}{2}$, $5\frac{9}{15}$ rounds up to 6.

13) Since $\frac{3}{7} < \frac{1}{2}$, $12\frac{3}{7}$ rounds down to 12.

14) $7\frac{1}{7} \cdot 6\frac{3}{5} \approx 7 \cdot 7$
$$\approx 49$$

$$7\frac{1}{7} \cdot 6\frac{3}{5} = \frac{50}{7} \cdot \frac{33}{5}$$
$$= \frac{\cancel{5} \cdot 10 \cdot 33}{7 \cdot \cancel{5}}$$
$$= \frac{330}{7}$$
$$= 47\frac{1}{7}$$

15) $16\frac{1}{4} \div 8\frac{1}{3} \approx 16 \div 8$
$$\approx 2$$

$$16\frac{1}{4} \div 8\frac{1}{3} = \frac{65}{4} \div \frac{25}{3}$$
$$= \frac{65}{4} \cdot \frac{3}{25}$$
$$= \frac{\cancel{5} \cdot 13 \cdot 3}{4 \cdot \cancel{5} \cdot 5}$$
$$= \frac{39}{20}$$
$$= 1\frac{19}{20}$$

16) $3\frac{1}{8}+4\frac{3}{4}\approx 3+5$
$$\approx 8$$

$3\frac{1}{8}+4\frac{3}{4}=3\frac{1}{8}+4\frac{6}{8}$
$$=3+4+\frac{7}{8}$$
$$=7\frac{7}{8}$$

17) $7\frac{5}{6}+5\frac{2}{3}\approx 8+6$
$$\approx 14$$

$7\frac{5}{6}+5\frac{2}{3}=7\frac{5}{6}+5\frac{4}{6}$
$$=7+5+\frac{9}{6}$$
$$=7+5+1\frac{1}{2}$$
$$=13\frac{1}{2}$$

18) $5\frac{5}{8}-2\frac{1}{16}\approx 6-2$
$$\approx 4$$

$5\frac{5}{8}-2\frac{1}{16}=5\frac{10}{16}-2\frac{1}{16}$
$$=3\frac{9}{16}$$

19) a) We cannot subtract the fractions, because $\frac{5}{6}>\frac{1}{6}$. To subtract, we will need to borrow a 1 from the whole number 7.

b) $7\frac{1}{6}=6+1+\frac{1}{6}$
$$=6+\frac{6}{6}+\frac{1}{6}$$
$$=6\frac{7}{6}$$

c) $7\frac{1}{6}-3\frac{5}{6}=6\frac{7}{6}-3\frac{5}{6}$
$$=3\frac{2}{6}$$
$$=3\frac{1}{3}$$

20) We cannot subtract $\frac{1}{5}-\frac{2}{5}$. We need to borrow a 1 from the 3. Write the borrowed 1 as $\frac{5}{5}$ so we have common denominators.
$3\frac{1}{5}-1\frac{2}{5}=2+\frac{5}{5}+\frac{1}{5}-1\frac{2}{5}$
$$=2\frac{6}{5}-1\frac{2}{5}$$
$$=1\frac{4}{5}$$

21) $5\frac{5}{8}-2\frac{7}{8}\approx 6-3$
$$\approx 3$$

$5\frac{5}{8}-2\frac{7}{8}=4+\frac{8}{8}+\frac{5}{8}-2\frac{7}{8}$
$$=4\frac{13}{8}-2\frac{7}{8}$$
$$=2\frac{6}{8}$$
$$=2\frac{3}{4}$$

22) $9\frac{1}{4}-4\frac{1}{3}\approx 9-4$
$$\approx 5$$

$9\frac{1}{4}-4\frac{1}{3}=9\frac{3}{12}-4\frac{4}{12}$
$$=8+\frac{12}{12}+\frac{3}{12}-4\frac{4}{12}$$
$$=8\frac{15}{12}-4\frac{4}{12}$$
$$=4\frac{11}{12}$$

23) $4-2\frac{1}{5}\approx 4-2$
$$\approx 2$$

$4-2\frac{1}{5}=3+\frac{5}{5}-2\frac{1}{5}$
$$=1\frac{4}{5}$$

24) $\left(3\frac{1}{6}-1\frac{1}{3}\right)^2 \div \left(\frac{1}{2}\right) = \left(\frac{19}{6}-\frac{8}{6}\right)^2 \cdot \left(\frac{2}{1}\right)$

$$= \left(\frac{11}{6}\right)^2 \cdot \left(\frac{2}{1}\right)$$

$$= \frac{121}{36} \cdot \left(\frac{2}{1}\right)$$

$$= \frac{121 \cdot \cancel{2}}{\cancel{2} \cdot 18 \cdot 1}$$

$$= \frac{121}{18}$$

$$= 6\frac{13}{18}$$

21)   Total Packaged $= 5 \cdot \frac{3}{4} + 1 \cdot \left(1\frac{1}{2}\right) + 2 \cdot 1$

$$= 5 \cdot \frac{3}{4} + 1 \cdot \frac{3}{2} + 2 \cdot 1$$

$$= \frac{15}{4} + \frac{3}{2} + 2$$

$$= \frac{15}{4} + \frac{6}{4} + \frac{8}{4}$$

$$= \frac{29}{4}$$

$$= 7\frac{1}{4}$$

Coffee Left $= 10 - 7\frac{1}{4}$

$$= 9\frac{4}{4} - 7\frac{1}{4}$$

$$= 2\frac{3}{4}$$

There will be $2\frac{3}{4}$ pounds of coffee left over.

CONCEPT CHECKS AND PRACTICE EXERCISES

A1)   Answers may vary. Example: The quotient is the whole number, the remainder is the numerator, and the divisor is the denominator.

A2)   a) $\frac{15}{8}$

      b) $1\frac{7}{8}$

A3)   a) $\frac{8}{5}$

      b) $1\frac{3}{5}$

A4)   a) $\frac{19}{8}$

      b) $2\frac{3}{8}$

A5)   a) $\frac{32}{9}$

      b) $3\frac{5}{9}$

A6) $1\frac{1}{2}; \frac{3}{2}$

A7) $2\frac{3}{4}; \frac{11}{4}$

A8) $12 \div 5 = 2$ Remainder 2

$$\frac{12}{5} = 2\frac{2}{5}$$

A9) $17 \div 3 = 5$ Remainder 2

$$\frac{17}{3} = 5\frac{2}{3}$$

A10) $42 \div 10 = 4$ Remainder 2

$$\frac{42}{10} = 4\frac{2}{10} = 4\frac{1}{5}$$

A11) $39 \div 9 = 4$ Remainder 3

$$\frac{39}{9} = 4\frac{3}{9} = 4\frac{1}{3}$$

A12) $48 \div 8 = 6$ Remainder 0

$$\frac{48}{8} = 6$$

A13) $26 \div 2 = 13$ Remainder 0

$$\frac{26}{2} = 13$$

B1)   Answers may vary. Example:

      a) Multiplying the whole number by the denominator gives the numerator of a fraction equal to the whole number with the same denominator as the fraction.

      b) The improper fraction can be formed by adding the "like fractions."

B2)   a) Multiply the denominator and the whole number. There are $8 \cdot 6 = 48$ eighths in 6.

      b) There are 7 eighths in $\frac{7}{8}$.

      c) Add the results. There are $48 + 7 = 55$ eighths.

$$6\frac{7}{8} = \frac{55}{8}$$

B3) a) Multiply the denominator and the whole
         number. There are $3 \cdot 9 = 27$ thirds in 9.

     b) There are 2 thirds in $\frac{2}{3}$.

     c) Add the results. There are $27 + 2 = 29$
         thirds.
         $$9\frac{2}{3} = \frac{29}{3}$$

B4) $7 \cdot 1 + 6 = 13$
     $$1\frac{6}{7} = \frac{13}{7}$$

B5) $5 \cdot 3 + 2 = 17$
     $$3\frac{2}{5} = \frac{17}{5}$$

B6) $2 \cdot 2 + 1 = 5$
     $$2\frac{1}{2} = \frac{5}{2}$$

B7) $5 \cdot 6 + 3 = 33$
     $$6\frac{3}{5} = \frac{33}{5}$$

B8) $3 \cdot 10 + 1 = 31$
     $$10\frac{1}{3} = \frac{31}{3}$$

B9) $10 \cdot 3 + 1 = 31$
     $$3\frac{1}{10} = \frac{31}{10}$$

B10) $9 \cdot 9 + 1 = 82$
      $$9\frac{1}{9} = \frac{82}{9}$$

B11) $8 \cdot 7 + 5 = 61$
      $$7\frac{5}{8} = \frac{61}{8}$$

FOCUS ON ESTIMATION EXERCISES

1) $5\frac{3}{5} \approx 6$ since $\frac{3}{5} > \frac{1}{2}$

2) $3\frac{2}{4} \approx 4$ since $\frac{2}{4} = \frac{1}{2}$

3) $7\frac{4}{9} \approx 7$ since $\frac{4}{9} < \frac{1}{2}$

4) $17\frac{9}{20} \approx 17$ since $\frac{9}{20} < \frac{1}{2}$

5) $12\frac{3}{4} \approx 13$ since $\frac{3}{4} > \frac{1}{2}$

6) $1\frac{5}{8} \approx 2$ since $\frac{5}{8} > \frac{1}{2}$

7) $6\frac{9}{20} \approx 6$ since $\frac{9}{20} < \frac{1}{2}$

8) $43\frac{2}{3} \approx 44$ since $\frac{2}{3} > \frac{1}{2}$

9) $7\frac{58}{100} \approx 8$ since $\frac{58}{100} > \frac{1}{2}$

10) $1\frac{1}{1000} \approx 1$ since $\frac{1}{1000} < \frac{1}{2}$

11) $12\frac{16}{33} \approx 12$ since $\frac{16}{33} < \frac{1}{2}$

12) $13\frac{32}{63} \approx 14$ since $\frac{32}{63} > \frac{1}{2}$

C1) Answers may vary. Example: Determine if the
     fraction is less than or more than $\frac{1}{2}$. If it is less,
     the mixed number is approximately equal to the
     whole number. If it is greater than or equal to
     $\frac{1}{2}$, the mixed number is approximately equal to
     the next larger whole number.

C2) Answers may vary. Example: Estimating allows
     us to determine whether or not the answer is
     reasonable.

C3) $3\frac{3}{7} \cdot 4\frac{1}{8} \approx 3 \cdot 4$
     $$\approx 12$$
     $$3\frac{3}{7} \cdot 4\frac{1}{8} = \frac{24}{7} \cdot \frac{33}{8}$$
     $$= \frac{3 \cdot \cancel{8} \cdot 33}{7 \cdot \cancel{8}}$$
     $$= \frac{99}{7}$$
     $$= 14\frac{1}{7}$$

C4) $\quad 1\frac{1}{2} \cdot 2\frac{2}{3} \approx 2 \cdot 3$

$\qquad\qquad \approx 6$

$\quad 1\frac{1}{2} \cdot 2\frac{2}{3} = \frac{3}{2} \cdot \frac{8}{3}$

$\qquad\qquad = \frac{\cancel{3} \cdot \cancel{2} \cdot 4}{\cancel{2} \cdot \cancel{3}}$

$\qquad\qquad = 4$

C5) $\quad 9\frac{1}{3} \cdot 7\frac{3}{4} \approx 9 \cdot 8$

$\qquad\qquad \approx 72$

$\quad 9\frac{1}{3} \cdot 7\frac{3}{4} = \frac{28}{3} \cdot \frac{31}{4}$

$\qquad\qquad = \frac{\cancel{4} \cdot 7 \cdot 31}{3 \cdot \cancel{4}}$

$\qquad\qquad = \frac{217}{3}$

$\qquad\qquad = 72\frac{1}{3}$

C6) $\quad 5\frac{1}{7} \cdot 6\frac{1}{3} \approx 5 \cdot 6$

$\qquad\qquad \approx 30$

$\quad 5\frac{1}{7} \cdot 6\frac{1}{3} = \frac{36}{7} \cdot \frac{19}{3}$

$\qquad\qquad = \frac{\cancel{3} \cdot 12 \cdot 19}{7 \cdot \cancel{3}}$

$\qquad\qquad = \frac{228}{7}$

$\qquad\qquad = 32\frac{4}{7}$

C7) $\quad 2\frac{1}{4} \cdot \frac{1}{2} \approx 2 \cdot 1$

$\qquad\qquad \approx 2$

$\quad 2\frac{1}{4} \cdot \frac{1}{2} = \frac{9}{4} \cdot \frac{1}{2}$

$\qquad\qquad = \frac{9}{8}$

$\qquad\qquad = 1\frac{1}{8}$

You will need $1\frac{1}{8}$ cups of flour.

C8) $\quad \frac{3}{4} \cdot 1\frac{1}{2} \approx 1 \cdot 2$

$\qquad\qquad \approx 2$

$\quad \frac{3}{4} \cdot 1\frac{1}{2} = \frac{3}{4} \cdot \frac{3}{2}$

$\qquad\qquad = \frac{9}{8}$

$\qquad\qquad = 1\frac{1}{8}$

You will need $1\frac{1}{8}$ pounds of chicken.

C9) $\quad 5\frac{5}{8} \div 1\frac{2}{7} \approx 6 \div 1$

$\qquad\qquad \approx 6$

$\quad 5\frac{5}{8} \div 1\frac{2}{7} = \frac{45}{8} \div \frac{9}{7}$

$\qquad\qquad = \frac{45}{8} \cdot \frac{7}{9}$

$\qquad\qquad = \frac{5 \cdot \cancel{9} \cdot 7}{8 \cdot \cancel{9}}$

$\qquad\qquad = \frac{35}{8}$

$\qquad\qquad = 4\frac{3}{8}$

C10) $\quad 3\frac{7}{9} \div 3\frac{1}{5} \approx 4 \div 3$

$\qquad\qquad \approx 1\frac{1}{3}$

$\quad 3\frac{7}{9} \div 3\frac{1}{5} = \frac{34}{9} \div \frac{16}{5}$

$\qquad\qquad = \frac{34}{9} \div \frac{5}{16}$

$\qquad\qquad = \frac{\cancel{2} \cdot 17 \cdot 5}{9 \cdot \cancel{2} \cdot 8}$

$\qquad\qquad = \frac{85}{72}$

$\qquad\qquad = 1\frac{13}{72}$

C11) $\quad \frac{7}{8} \div 2\frac{2}{3} \approx 1 \div 3$

$\qquad\qquad = \frac{1}{3}$

$\quad \frac{7}{8} \div 2\frac{2}{3} = \frac{7}{8} \div \frac{8}{3}$

$\qquad\qquad = \frac{7}{8} \cdot \frac{3}{8}$

$\qquad\qquad = \frac{21}{64}$

C12) $\quad \frac{3}{4} \div 2\frac{2}{3} \approx 1 \div 3$

$\qquad\qquad = \frac{1}{3}$

$\quad \frac{3}{4} \div 2\frac{2}{3} = \frac{3}{4} \div \frac{8}{3}$

$\qquad\qquad = \frac{3}{4} \cdot \frac{3}{8}$

$\qquad\qquad = \frac{9}{32}$

D1) Answers may vary. Example: It is a good idea to add the fractional portions first in case we need to carry a whole number.

D2) $3\frac{3}{7}+4\frac{1}{8}=3\frac{24}{56}+4\frac{7}{56}$

$=3+4+\frac{31}{56}$

$=7\frac{31}{56}$

D3) $1\frac{1}{2}+2\frac{1}{3}=1\frac{3}{6}+2\frac{2}{6}$

$=1+2+\frac{5}{6}$

$=3\frac{5}{6}$

D4) $9\frac{1}{3}+7\frac{3}{5}=9\frac{5}{15}+7\frac{9}{15}$

$=9+7+\frac{14}{15}$

$=16\frac{14}{15}$

D5) $10\frac{3}{5}+6\frac{3}{10}=10\frac{6}{10}+6\frac{3}{10}$

$=10+6+\frac{9}{10}$

$=16\frac{9}{10}$

D6) $2\frac{1}{4}+1\frac{1}{8}=2\frac{2}{8}+1\frac{1}{8}$

$=2+1+\frac{3}{8}$

$=3\frac{3}{8}$

The weight on the right weighs $3\frac{3}{8}$ pounds.

D7) $5\frac{1}{3}+3\frac{1}{12}=5\frac{4}{12}+3\frac{1}{12}$

$=5+3+\frac{5}{12}$

$=8\frac{5}{12}$

The weight on the right weighs $8\frac{5}{12}$ pounds.

D8) $4\frac{8}{9}+3\frac{1}{6}=4\frac{16}{18}+3\frac{3}{18}$

$=4+3+\frac{19}{18}$

$=4+3+1\frac{1}{18}$

$=8\frac{1}{18}$

D9) $9\frac{7}{9}+10\frac{5}{9}=9+10+\frac{12}{9}$

$=9+10+1\frac{3}{9}$

$=20\frac{1}{3}$

D10) $13\frac{13}{14}+6\frac{6}{7}=13\frac{13}{14}+6\frac{12}{14}$

$=13+6+\frac{25}{14}$

$=13+6+1\frac{11}{14}$

$=20\frac{11}{14}$

D11) $98\frac{3}{5}+1\frac{8}{10}=98\frac{6}{10}+1\frac{8}{10}$

$=98+1+\frac{14}{10}$

$=98+1+1\frac{4}{10}$

$=100\frac{2}{5}$

E1) $5\frac{1}{6}=4+1+\frac{1}{6}$

$=4+1\frac{1}{6}$

$=4+\frac{7}{6}$

$=4\frac{7}{6}$

E2) $8\frac{2}{7}=7+1+\frac{2}{7}$

$=7+1\frac{2}{7}$

$=7+\frac{9}{7}$

$=7\frac{9}{7}$

E3) Answers may vary. You are borrowing "one" from the whole number. In order to add 1 to the fraction, it must be in fraction form with the same denominator as the fraction to which 1 is being added. One in fraction form has the same numerator as denominator. Thus, the numerator of the mixed number is increased by the value of the denominator.

E4) $7\frac{1}{5}-3\frac{1}{10}=7\frac{2}{10}-3\frac{1}{10}$

$=4\frac{1}{10}$

E5) $12\frac{5}{8}-3\frac{3}{16}=12\frac{10}{16}-3\frac{3}{16}$

$\qquad\qquad=9\frac{7}{16}$

E6) $4\frac{2}{3}-3\frac{3}{5}=4\frac{10}{15}-3\frac{9}{15}$

$\qquad\qquad=1\frac{1}{15}$

E7) $10\frac{7}{12}-6\frac{1}{4}=10\frac{7}{12}-6\frac{3}{12}$

$\qquad\qquad=4\frac{4}{12}$

$\qquad\qquad=4\frac{1}{3}$

E8) $9\frac{1}{9}-3\frac{2}{3}=9\frac{1}{9}-3\frac{6}{9}$

$\qquad\qquad=8+\frac{9}{9}+\frac{1}{9}-3\frac{6}{9}$

$\qquad\qquad=8\frac{10}{9}-3\frac{6}{9}$

$\qquad\qquad=5\frac{4}{9}$

E9) $13\frac{2}{7}-\frac{10}{21}=13\frac{6}{21}-\frac{10}{21}$

$\qquad\qquad=12+\frac{21}{21}+\frac{6}{21}-\frac{10}{21}$

$\qquad\qquad=12\frac{27}{21}-\frac{10}{21}$

$\qquad\qquad=12\frac{17}{21}$

E10) $23\frac{5}{6}-10\frac{9}{10}=23\frac{25}{30}-10\frac{27}{30}$

$\qquad\qquad=22+\frac{30}{30}+\frac{25}{30}-10\frac{27}{30}$

$\qquad\qquad=22\frac{55}{30}-10\frac{27}{30}$

$\qquad\qquad=12\frac{28}{30}$

$\qquad\qquad=12\frac{14}{15}$

E11) $20\frac{1}{5}-\frac{2}{3}=20\frac{3}{15}-\frac{10}{15}$

$\qquad\qquad=19+\frac{15}{15}+\frac{3}{15}-\frac{10}{15}$

$\qquad\qquad=19\frac{18}{15}-\frac{10}{15}$

$\qquad\qquad=19\frac{8}{15}$

E12) $8-3\frac{1}{3}=7+\frac{3}{3}-3\frac{1}{3}$

$\qquad\qquad=4\frac{2}{3}$

E13) $7-\frac{1}{6}=6+\frac{6}{6}-\frac{1}{6}$

$\qquad\qquad=6\frac{5}{6}$

E14) $6\frac{1}{6}-4=2\frac{1}{6}$

E15) $8\frac{4}{5}-5=3\frac{4}{5}$

SECTION 2.6 EXERCISES

For 1–17, refer to Concept Checks and Practice Exercises.

19) proper fraction

21) mixed number

23) a) $\frac{15}{4}$

b) $3\frac{3}{4}$

25) a) $\frac{22}{7}$

b) $3\frac{1}{7}$

27) a) $\frac{35}{8}$

b) $4\frac{3}{8}$

29) a) $\frac{8}{6}$

b) $1\frac{2}{6}$

31) $8\cdot3+1=25$

$3\frac{1}{8}=\frac{25}{8}$

33) $16\div3=5\text{ Remainder }1$

$\frac{16}{3}=5\frac{1}{3}$

35) $3\cdot13+1=40$

$13\frac{1}{3}=\frac{40}{3}$

37) $102\div7=14\text{ Remainder }4$

$\frac{102}{7}=14\frac{4}{7}$

39) $41 \div 14 = 2$ Remainder 13

$$\frac{41}{14} = 2\frac{13}{14}$$

41) a) $2\frac{3}{4} \cdot 3\frac{2}{3} \approx 3 \cdot 4$

$$\approx 12$$

b) $2\frac{3}{4} \cdot 3\frac{2}{3} = \frac{11}{4} \cdot \frac{11}{3}$

$$= \frac{121}{12}$$

$$= 10\frac{1}{12}$$

43) a) $2\frac{2}{3} \div 4\frac{1}{3} \approx 3 \div 4$

$$\approx \frac{3}{4}$$

b) $2\frac{2}{3} \div 4\frac{1}{3} = \frac{8}{3} \div \frac{13}{3}$

$$= \frac{8}{3} \cdot \frac{3}{13}$$

$$= \frac{8 \cdot \cancel{3}}{\cancel{3} \cdot 13}$$

$$= \frac{8}{13}$$

45) a) $10\frac{3}{5} \div 2\frac{4}{5} \approx 11 \div 3$

$$\approx \frac{11}{3} = 3\frac{2}{3}$$

b) $10\frac{3}{5} \div 2\frac{4}{5} = \frac{53}{5} \div \frac{14}{5}$

$$= \frac{53}{5} \cdot \frac{5}{14}$$

$$= \frac{53 \cdot \cancel{5}}{\cancel{5} \cdot 14}$$

$$= \frac{53}{14}$$

$$= 3\frac{11}{14}$$

47) a) $5\frac{2}{3} \cdot 2\frac{1}{7} \approx 6 \cdot 2$

$$\approx 12$$

b) $5\frac{2}{3} \cdot 2\frac{1}{7} = \frac{17}{3} \cdot \frac{15}{7}$

$$= \frac{17 \cdot \cancel{3} \cdot 5}{\cancel{3} \cdot 7}$$

$$= \frac{85}{7}$$

$$= 12\frac{1}{7}$$

49) a) $21\frac{5}{9} + 4\frac{1}{9} \approx 22 + 4$

$$\approx 26$$

b) $21\frac{5}{9} + 4\frac{1}{9} = 21 + 4 + \frac{6}{9}$

$$= 21 + 4 + \frac{2}{3}$$

$$= 25\frac{2}{3}$$

51) a) $62\frac{9}{11} - 52\frac{3}{11} \approx 63 - 52$

$$\approx 11$$

b) $62\frac{9}{11} - 52\frac{3}{11} = 10\frac{6}{11}$

53) a) $50\frac{1}{10} - 25\frac{2}{5} \approx 50 - 25$

$$\approx 25$$

b) $50\frac{1}{10} - 25\frac{2}{5} = 50\frac{1}{10} - 25\frac{4}{10}$

$$= 49 + 1\frac{1}{10} - 25\frac{4}{10}$$

$$= 49\frac{11}{10} - 25\frac{4}{10}$$

$$= 24\frac{7}{10}$$

55) a) $10 - 5\frac{1}{10} \approx 10 - 5$

$$\approx 5$$

b) $10 - 5\frac{1}{10} = 9\frac{10}{10} - 5\frac{1}{10}$

$$= 4\frac{9}{10}$$

57) a) $5\frac{7}{20} - 3\frac{1}{5} \approx 5 - 3$

$$\approx 2$$

b) $5\frac{7}{20} - 3\frac{1}{5} = 5\frac{7}{20} - 3\frac{4}{20}$

$$= 2\frac{3}{20}$$

The weight on the left weighs $2\frac{3}{20}$ pounds.

59) a) $8\frac{7}{10} + 4\frac{3}{5} \approx 9 + 5$

$$\approx 14$$

59) b) $8\frac{7}{10}+4\frac{3}{5}=8\frac{7}{10}+4\frac{6}{10}$

$=8+4+\frac{13}{10}$

$=8+4+1\frac{3}{10}$

$=13\frac{3}{10}$

61) a) $53-26\frac{3}{5}\approx 53-27$

≈ 26

b) $53-26\frac{3}{5}=52\frac{5}{5}-26\frac{3}{5}$

$=26\frac{2}{5}$

63) a) $7\frac{2}{3}\div 4\approx 8\div 4$

≈ 2

b) $7\frac{2}{3}\div 4=7\frac{2}{3}\cdot\frac{1}{4}$

$=\frac{23}{3}\cdot\frac{1}{4}$

$=\frac{23}{12}=1\frac{11}{12}$

65) a) $100\frac{15}{16}+25\frac{3}{4}\approx 101+26$

≈ 127

b) $100\frac{15}{16}+25\frac{3}{4}=100\frac{15}{16}+25\frac{12}{16}$

$=100+25+\frac{27}{16}$

$=100+25+1\frac{11}{16}$

$=126\frac{11}{16}$

67) a) $1\frac{2}{3}\cdot 3\frac{4}{5}\approx 2\cdot 4$

≈ 8

b) $1\frac{2}{3}\cdot 3\frac{4}{5}=\frac{5}{3}\cdot\frac{19}{5}$

$=\frac{\cancel{5}\cdot 19}{3\cdot\cancel{5}}$

$=\frac{19}{3}=6\frac{1}{3}$

69) a) $7\frac{2}{5}-6\frac{13}{30}\approx 7-6$

≈ 1

b) $7\frac{2}{5}-6\frac{13}{30}=7\frac{12}{30}-6\frac{13}{30}$

$=6\frac{42}{30}-6\frac{13}{30}$

$=\frac{29}{30}$

71) a) $2\frac{3}{4}\cdot 3\frac{2}{3}\approx 3\cdot 4$

≈ 12

b) $2\frac{3}{4}\cdot 3\frac{2}{3}=\frac{11}{4}\cdot\frac{11}{3}$

$=\frac{121}{12}=10\frac{1}{12}$

73) a) $9\frac{1}{4}\div 5\approx 9\div 5$

$\approx\frac{9}{5}=1\frac{4}{5}$

b) $9\frac{1}{4}\div 5=\frac{37}{4}\div\frac{5}{1}$

$=\frac{37}{4}\cdot\frac{1}{5}$

$=\frac{37}{20}$

$=1\frac{17}{20}$

75) a) $17\frac{3}{4}+21\frac{1}{2}+12\frac{1}{2}=17\frac{3}{4}+21\frac{2}{4}+12\frac{2}{4}$

$=17+21+12+\frac{7}{4}$

$=17+21+12+1\frac{3}{4}$

$=51\frac{3}{4}$

Vladimir skied a total distance of $51\frac{3}{4}$ kilometers.

b) $51\frac{3}{4}\div 3=51\frac{3}{4}\cdot\frac{1}{3}$

$=\frac{207}{4}\cdot\frac{1}{3}$

$=\frac{207}{12}$

$=17\frac{3}{12}$

$=17\frac{1}{4}$

The average daily distance was $17\frac{1}{4}$ kilometers.

77) $\left(4\frac{1}{8}\right) - \left(2\frac{3}{8}\right) + \left(\frac{3}{4}\right) = \frac{33}{8} - \frac{19}{8} + \frac{3}{4}$

$= \frac{14}{8} + \frac{6}{8}$

$= \frac{20}{8}$

$= 2\frac{4}{8}$

$= 2\frac{1}{2}$

79) $\left(4\frac{1}{5}\right) \div \left(1\frac{2}{5}\right) \cdot \left(3\frac{1}{2}\right) = \left(\frac{21}{5} \div \frac{7}{5}\right) \cdot \frac{7}{2}$

$= \left(\frac{21}{5} \cdot \frac{5}{7}\right) \cdot \frac{7}{2}$

$= \frac{3}{1} \cdot \frac{7}{2}$

$= \frac{21}{2}$

$= 10\frac{1}{2}$

81) $\left(2\frac{1}{3}\right) + \left(1\frac{1}{3}\right) \cdot \left(8\frac{1}{4}\right) = \frac{7}{3} + \left(\frac{4}{3} \cdot \frac{33}{4}\right)$

$= \frac{7}{3} + 11$

$= 2\frac{1}{3} + 11$

$= 13\frac{1}{3}$

83) $16\frac{3}{4} \div \left(1\frac{1}{8} - \frac{3}{8}\right) = \frac{67}{4} \div \left(\frac{9}{8} - \frac{3}{8}\right)$

$= \frac{67}{4} \div \frac{6}{8}$

$= \frac{67}{4} \cdot \frac{4}{3}$

$= \frac{67}{3}$

$= 22\frac{1}{3}$

85) $\left(1\frac{1}{8}\right) \cdot \left(3\frac{1}{4}\right) - \left(\frac{7}{8}\right) = \left(\frac{9}{8} \cdot \frac{13}{4}\right) - \left(\frac{7}{8}\right)$

$= \frac{117}{32} - \frac{7}{8}$

$= \frac{117}{32} - \frac{28}{32}$

$= \frac{89}{32} = 2\frac{25}{32}$

87) $\frac{14}{23} \div \left(2 - \left(\frac{2}{3}\right)^2\right) = \frac{14}{23} \div \left(2 - \frac{4}{9}\right)$

$= \frac{14}{23} \div \left(\frac{18}{9} - \frac{4}{9}\right)$

$= \frac{14}{23} \div \frac{14}{9}$

$= \frac{14}{23} \cdot \frac{9}{14}$

$= \frac{9}{23}$

89) $3 - 1\frac{1}{8} - \frac{3}{4} = \frac{24}{8} - \frac{9}{8} - \frac{6}{8}$

$= \frac{9}{8}$

$= 1\frac{1}{8}$

$1\frac{1}{8}$ pounds of coffee were left over.

91) $2\frac{1}{4} - \left(4 - 2\frac{1}{4}\right) = 2\frac{1}{4} - \left(3\frac{4}{4} - 2\frac{1}{4}\right)$

$= 2\frac{1}{4} - 1\frac{3}{4}$

$= 1\frac{5}{4} - 1\frac{3}{4}$

$= \frac{2}{4}$

$= \frac{1}{2}$

Patrick took $\frac{1}{2}$ cup of flour from JR.

93) $1{,}000 \cdot \left(1 - \frac{1}{4} - \frac{1}{5}\right)$

$= 1{,}000 \cdot \left(\frac{20}{20} - \frac{5}{20} - \frac{4}{20}\right)$

$= 1{,}000 \cdot \frac{11}{20}$

$= \frac{1{,}000}{1} \cdot \frac{11}{20}$

$= 550$

$550 was left to purchase a savings bond.

CHAPTER 2   REVIEW EXERCISES

1) Since there are 15 equal-sized parts, the denominator is 15.
Since there are a total of five shaded parts, the numerator is 5.

Fraction $= \frac{5}{15}$

2) Since there are five equal parts in each object, the denominator is 5.
Since there are a total of 14 shaded parts, the numerator is 14.

Fraction $= \dfrac{14}{5}$

3) Since the denominator is 5, we draw an object with five equal parts.
Since the numerator is 2, we shade two of those parts.

4) Since the denominator is 3, we draw objects with three equal parts.
The numerator is 7, so we must shade seven parts.
A total of three whole objects must be drawn to shade seven parts.

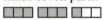

5) a)

  b) $\dfrac{6}{9}$

6) a)

  b) $\dfrac{2}{4}$

7) $\dfrac{5}{8} > \dfrac{5}{9}$

With like numerators, the larger fraction is the one with the smaller denominator.

8) $\dfrac{5}{12} < \dfrac{7}{12}$

With like denominators, the larger fraction is the one with the larger numerator.

9) $\dfrac{6}{6+7} = \dfrac{6}{13}$

$\dfrac{6}{13}$ of Adalina's work is at Snicker's.

10) $\dfrac{8}{7+8} = \dfrac{8}{15}$

$\dfrac{8}{15}$ of the players are on the Flowers.

11) a) B and C are equivalent.

  b) Of those, C is in simplest form.

12) $\dfrac{6}{8} = \dfrac{\cancel{2}\cdot 3}{\cancel{2}\cdot 4} = \dfrac{3}{4}$

13) $\dfrac{15}{20} = \dfrac{3\cdot\cancel{5}}{4\cdot\cancel{5}}$

$= \dfrac{3}{4}$

14) $\dfrac{21}{49} = \dfrac{3\cdot\cancel{7}}{\cancel{7}\cdot 7}$

$= \dfrac{3}{7}$

15) $\dfrac{7}{2}\cdot\dfrac{3}{2} = \dfrac{7\cdot 3}{2\cdot 2}$

$= \dfrac{21}{4}$

16) $\dfrac{1}{3}\cdot\dfrac{9}{5} = \dfrac{1\cdot\cancel{3}\cdot 3}{\cancel{3}\cdot 5}$

$= \dfrac{3}{5}$

17) $\dfrac{4}{7}\cdot\dfrac{21}{5} = \dfrac{4\cdot 3\cdot\cancel{7}}{\cancel{7}\cdot 5}$

$= \dfrac{12}{5}$

18) $\dfrac{7}{10}\cdot\dfrac{15}{14} = \dfrac{\cancel{7}\cdot 3\cdot\cancel{5}}{2\cdot\cancel{5}\cdot 2\cdot\cancel{7}}$

$= \dfrac{3}{4}$

19) $\dfrac{3}{6}\cdot\dfrac{18}{7}\cdot\dfrac{14}{3} = \dfrac{\cancel{3}\cdot 3\cdot\cancel{6}\cdot 2\cdot\cancel{7}}{\cancel{6}\cdot\cancel{7}\cdot\cancel{3}}$

$= 6$

20) $\left(\dfrac{5}{3}\right)^{2} = \dfrac{5}{3}\cdot\dfrac{5}{3}$

$= \dfrac{25}{9}$

21) $\dfrac{200-40}{200} = \dfrac{160}{200}$

$= \dfrac{4 \cdot \cancel{40}}{5 \cdot \cancel{40}}$

$= \dfrac{4}{5}$

Antonio paid $\dfrac{4}{5}$ of the original cost.

22) $240 - \dfrac{1}{4} \cdot 240 = 240 - \dfrac{1}{4} \cdot \dfrac{240}{1}$

$= 240 - \dfrac{1}{4} \cdot \dfrac{240}{1}$

$= 240 - \dfrac{\cancel{4} \cdot 60}{\cancel{4}}$

$= 240 - 60$

$= 180$

Falicia paid $180.

23) 1 foot $= 12$ inches

$\dfrac{15 \ \cancel{ft}}{1} \cdot \left(\dfrac{12 \text{ in.}}{1 \ \cancel{ft}} \right) = 180 \text{ in.}$

15 feet $= 180$ inches

24) 1 cup $= 8$ ounces

$\dfrac{11 \ \cancel{cups}}{1} \cdot \left(\dfrac{8 \text{ oz}}{1 \ \cancel{cup}} \right) = 88 \text{ oz}$

11 cups $= 88$ ounces

25) The reciprocal of $\dfrac{5}{9}$ is $\dfrac{9}{5}$.

26) The reciprocal of $\dfrac{1}{45}$ is $\dfrac{45}{1} = 45$.

27) The reciprocal of $3 = \dfrac{3}{1}$ is $\dfrac{1}{3}$.

28) $\dfrac{5}{6} \div \dfrac{2}{3} = \dfrac{5}{6} \cdot \dfrac{3}{2}$

$= \dfrac{5 \cdot \cancel{3}}{2 \cdot \cancel{3} \cdot 2}$

$= \dfrac{5}{4}$

29) $\dfrac{6}{7} \div \dfrac{3}{14} = \dfrac{6}{7} \cdot \dfrac{14}{3}$

$= \dfrac{2 \cdot \cancel{3} \cdot 2 \cdot \cancel{7}}{\cancel{7} \cdot \cancel{3}}$

$= 4$

30) $\dfrac{5}{6} \div \dfrac{6}{5} = \dfrac{5}{6} \cdot \dfrac{5}{6}$

$= \dfrac{5 \cdot 5}{6 \cdot 6}$

$= \dfrac{25}{36}$

31) $\dfrac{14}{5} \div \dfrac{7}{10} = \dfrac{14}{5} \cdot \dfrac{10}{7}$

$= \dfrac{2 \cdot \cancel{7} \cdot 2 \cdot \cancel{5}}{\cancel{5} \cdot \cancel{7}}$

$= 4$

32) $\dfrac{3}{7} \div \dfrac{3}{7} = 1$

33) $\dfrac{0}{2} \div \dfrac{8}{5} = 0$

34) $\left(\dfrac{3}{5} \div \dfrac{6}{5} \right)^2 = \left(\dfrac{3}{5} \cdot \dfrac{5}{6} \right)^2$

$= \left(\dfrac{\cancel{3} \cdot \cancel{5}}{\cancel{6} \cdot 2 \cdot \cancel{3}} \right)^2$

$= \left(\dfrac{1}{2} \right)^2$

$= \dfrac{1}{4}$

35) $\left(\dfrac{3}{2} \right)^2 \div \dfrac{15}{8} = \dfrac{9}{4} \div \dfrac{15}{8}$

$= \dfrac{9}{4} \cdot \dfrac{8}{15}$

$= \dfrac{\cancel{3} \cdot 3 \cdot 2 \cdot \cancel{4}}{\cancel{4} \cdot \cancel{3} \cdot 5}$

$= \dfrac{6}{5}$

36) $4 \div \dfrac{8}{10} = \dfrac{4}{1} \div \dfrac{8}{10}$

$= \dfrac{4}{1} \cdot \dfrac{10}{8}$

$= \dfrac{\cancel{4} \cdot \cancel{2} \cdot 5}{\cancel{2} \cdot \cancel{4}}$

$= 5$

Five pieces can be made.

37) $14 \div \dfrac{4}{5} = \dfrac{14}{1} \div \dfrac{4}{5}$

$= \dfrac{14}{1} \cdot \dfrac{5}{4}$

$= \dfrac{\cancel{2} \cdot 7 \cdot 5}{\cancel{2} \cdot 2}$

$= \dfrac{35}{2}$

$= 17\dfrac{1}{2}$

$17\dfrac{1}{2}$ bags of jelly beans will equal 14 pounds.

38) $\dfrac{3}{8} \div \dfrac{1}{64} = \dfrac{3}{8} \cdot \dfrac{64}{1}$

$= \dfrac{3 \cdot \cancel{8} \cdot 8}{\cancel{8} \cdot 1}$

$= 24$

The go-cart got 24 miles per gallon.

39)

$\dfrac{11}{12} - \dfrac{6}{12} = \dfrac{5}{12}$

40)

$\dfrac{3}{10} + \dfrac{1}{10} = \dfrac{4}{10}$

41) a) 12, 24
       8 divides 24 evenly.
       24 is the LCD.

   b) $\dfrac{5}{12} \cdot \dfrac{2}{2} = \dfrac{10}{24}$

       $\dfrac{1}{8} \cdot \dfrac{3}{3} = \dfrac{3}{24}$

42) a) 8, 16, 24
       6 divides 24 evenly.
       24 is the LCD.

   b) $\dfrac{7}{8} \cdot \dfrac{3}{3} = \dfrac{21}{24}$

       $\dfrac{5}{6} \cdot \dfrac{4}{4} = \dfrac{20}{24}$

43) $\dfrac{7}{9} + \dfrac{1}{9} = \dfrac{8}{9}$

44) $\dfrac{6}{7} - \dfrac{2}{7} = \dfrac{4}{7}$

45) $\dfrac{7}{18} + \dfrac{2}{9} = \dfrac{7}{18} + \dfrac{4}{18}$

$= \dfrac{11}{18}$

46) $\dfrac{3}{5} - \dfrac{1}{3} = \dfrac{9}{15} - \dfrac{5}{15}$

$= \dfrac{4}{15}$

47) $\dfrac{3}{14} - \dfrac{2}{21} = \dfrac{9}{42} - \dfrac{4}{42}$

$= \dfrac{5}{42}$

48) $\dfrac{7}{10} + \dfrac{3}{25} = \dfrac{35}{50} + \dfrac{6}{50}$

$= \dfrac{41}{50}$

49) $\dfrac{4}{5} - \dfrac{3}{5} + \dfrac{3}{10} = \dfrac{8}{10} - \dfrac{6}{10} + \dfrac{3}{10}$

$= \dfrac{5}{10}$

$= \dfrac{1}{2}$

50) $\dfrac{5}{8} + \dfrac{3}{4} - \dfrac{1}{8} = \dfrac{5}{8} + \dfrac{6}{8} - \dfrac{1}{8}$

$= \dfrac{10}{8}$

$= \dfrac{\cancel{2} \cdot 5}{\cancel{2} \cdot 4}$

$= \dfrac{5}{4}$

51) LCD = 16

$\dfrac{7}{16} = \dfrac{7}{16}$ and $\dfrac{1}{4} \cdot \dfrac{4}{4} = \dfrac{4}{16}$ and $\dfrac{3}{8} \cdot \dfrac{2}{2} = \dfrac{6}{16}$

$\dfrac{7}{16} > \dfrac{6}{16} > \dfrac{4}{16}$, so $\dfrac{7}{16} > \dfrac{3}{8} > \dfrac{1}{4}$

52) LCD = 30

$\dfrac{4}{5} \cdot \dfrac{6}{6} = \dfrac{24}{30}$ and $\dfrac{5}{6} \cdot \dfrac{5}{5} = \dfrac{25}{30}$ and $\dfrac{9}{15} \cdot \dfrac{2}{2} = \dfrac{18}{30}$

$\dfrac{18}{30} < \dfrac{24}{30} < \dfrac{25}{30}$, so $\dfrac{9}{15} < \dfrac{4}{5} < \dfrac{5}{6}$

53) $1\dfrac{1}{2} - \dfrac{3}{8} = \dfrac{3}{2} - \dfrac{3}{8}$

$= \dfrac{12}{8} - \dfrac{3}{8}$

$= \dfrac{9}{8}$

$\dfrac{9}{8}$ gallon of water must be added.

54) $\dfrac{7}{8} - \dfrac{3}{4} \cdot \dfrac{1}{2} = \dfrac{7}{8} - \dfrac{3}{8}$

$= \dfrac{4}{8}$

$= \dfrac{1 \cdot \cancel{4}}{2 \cdot \cancel{4}}$

$= \dfrac{1}{2}$

55) $\dfrac{1}{7-3} + \dfrac{1}{5} = \dfrac{1}{4} + \dfrac{1}{5}$

$= \dfrac{5}{20} + \dfrac{4}{20}$

$= \dfrac{9}{20}$

56) $\dfrac{6}{7}\left(\dfrac{2}{3}\right)^2 = \dfrac{6}{7} \cdot \dfrac{4}{9}$

$= \dfrac{2 \cdot \cancel{3} \cdot 4}{7 \cdot \cancel{3} \cdot 3}$

$= \dfrac{8}{21}$

57) $\dfrac{7}{9} \div \dfrac{14}{3} \cdot \dfrac{2}{3} = \left(\dfrac{7}{9} \cdot \dfrac{3}{14}\right) \cdot \dfrac{2}{3}$

$= \left(\dfrac{\cancel{7} \cdot \cancel{3}}{\cancel{3} \cdot 3 \cdot 2 \cdot \cancel{7}}\right) \cdot \dfrac{2}{3}$

$= \dfrac{1}{6} \cdot \dfrac{2}{3}$

$= \dfrac{1 \cdot \cancel{2}}{\cancel{2} \cdot 3 \cdot 3}$

$= \dfrac{1}{9}$

58) $\dfrac{4}{5} + \dfrac{1}{7} - \dfrac{1}{3} = \dfrac{84}{105} + \dfrac{15}{105} - \dfrac{35}{105}$

$= \dfrac{64}{105}$

59) $\left(\dfrac{1}{3} + \dfrac{1}{2}\right)^2 = \left(\dfrac{2}{6} + \dfrac{3}{6}\right)^2$

$= \left(\dfrac{5}{6}\right)^2$

$= \dfrac{25}{36}$

60) $\dfrac{5-2}{6} + \dfrac{4}{5-2} = \dfrac{3}{6} + \dfrac{4}{3}$

$= \dfrac{3}{6} + \dfrac{8}{6}$

$= \dfrac{11}{6}$

61) $\left(\dfrac{7}{2} \div \dfrac{3}{7}\right) \div 49 = \left(\dfrac{7}{2} \cdot \dfrac{7}{3}\right) \div 49$

$= \dfrac{49}{6} \div 49$

$= \dfrac{49}{6} \cdot \dfrac{1}{49}$

$= \dfrac{\cancel{49} \cdot 1}{6 \cdot \cancel{49}}$

$= \dfrac{1}{6}$

62) $n - \dfrac{1}{2} = \dfrac{1}{3}$

$n - \dfrac{1}{2} + \dfrac{1}{2} = \dfrac{1}{3} + \dfrac{1}{2}$

$n = \dfrac{2}{6} + \dfrac{3}{6}$

$n = \dfrac{5}{6}$

63) $n + \dfrac{4}{5} = \dfrac{13}{15}$

$n + \dfrac{4}{5} - \dfrac{4}{5} = \dfrac{13}{15} - \dfrac{4}{5}$

$n = \dfrac{13}{15} - \dfrac{12}{15}$

$n = \dfrac{1}{15}$

64) $\dfrac{1}{3} \cdot n = \dfrac{5}{9}$

$\dfrac{3}{1} \cdot \dfrac{1}{3} \cdot n = \dfrac{3}{1} \cdot \dfrac{5}{9}$

$n = \dfrac{5}{3}$

65) $\dfrac{3}{5} \cdot n = \dfrac{7}{10}$

$\dfrac{5}{3} \cdot \dfrac{3}{5} \cdot n = \dfrac{5}{3} \cdot \dfrac{7}{10}$

$n = \dfrac{7}{6}$

66) $50 \cdot 7 \cdot \left(1 - \dfrac{4}{7}\right) = 50 \cdot 7 \cdot \left(\dfrac{7}{7} - \dfrac{4}{7}\right)$

$= 50 \cdot 7 \cdot \dfrac{3}{7}$

$= \dfrac{50 \cdot \cancel{7} \cdot 3}{\cancel{7}}$

$= 150$

He must still inventory 150 shelves.

67) $5\left(\dfrac{3}{4}+\dfrac{1}{8}\right)=5\left(\dfrac{6}{8}+\dfrac{1}{8}\right)$

$\quad=\dfrac{5}{1}\cdot\dfrac{7}{8}$

$\quad=\dfrac{35}{8}=4\dfrac{3}{8}$

$4\dfrac{3}{8}$ cups of milk are needed.

68) a) $\dfrac{19}{5}$

b) $3\dfrac{4}{5}$

69) a) $\dfrac{7}{4}$

b) $1\dfrac{3}{4}$

70) $6\cdot7+1=43$

$7\dfrac{1}{6}=\dfrac{43}{6}$

71) $53\div8=6$ Remainder 5

$\dfrac{53}{8}=6\dfrac{5}{8}$

72) $13\dfrac{3}{5}-5\dfrac{1}{2}\approx14-6$

$\qquad\approx8$

73) $5\dfrac{1}{4}\cdot8\dfrac{7}{9}\approx5\cdot9$

$\qquad\approx45$

74) $4\dfrac{5}{7}+1\dfrac{1}{14}=4\dfrac{10}{14}+1\dfrac{1}{14}$

$\qquad=4+1+\dfrac{11}{14}$

$\qquad=5\dfrac{11}{14}$

75) $3\dfrac{1}{2}\div5\dfrac{3}{7}=\dfrac{7}{2}\div\dfrac{38}{7}$

$\qquad=\dfrac{7}{2}\cdot\dfrac{7}{38}$

$\qquad=\dfrac{49}{76}$

76) $9\dfrac{1}{8}-2\dfrac{5}{8}=8+1\dfrac{1}{8}-2\dfrac{5}{8}$

$\qquad=8\dfrac{9}{8}-2\dfrac{5}{8}$

$\qquad=6\dfrac{4}{8}$

$\qquad=6\dfrac{1}{2}$

77) $4\dfrac{2}{3}\cdot2\dfrac{1}{7}=\dfrac{14}{3}\cdot\dfrac{15}{7}$

$\qquad=\dfrac{2\cdot\cancel{7}\cdot\cancel{3}\cdot5}{\cancel{3}\cdot\cancel{7}}$

$\qquad=10$

78) $3\dfrac{2}{3}+4\dfrac{7}{9}=3\dfrac{6}{9}+4\dfrac{7}{9}$

$\qquad=3+4+\dfrac{13}{9}$

$\qquad=3+4+1\dfrac{4}{9}$

$\qquad=8\dfrac{4}{9}$

79) $8-2\dfrac{5}{6}=7\dfrac{6}{6}-2\dfrac{5}{6}$

$\qquad=5\dfrac{1}{6}$

80) $3\dfrac{3}{7}\div\dfrac{5}{14}=\dfrac{24}{7}\div\dfrac{5}{14}$

$\qquad=\dfrac{24}{7}\cdot\dfrac{14}{5}$

$\qquad=\dfrac{24\cdot2\cdot\cancel{7}}{\cancel{7}\cdot5}$

$\qquad=\dfrac{48}{5}$

$\qquad=9\dfrac{3}{5}$

81) $3\dfrac{3}{4}\div3=\dfrac{15}{4}\div\dfrac{3}{1}$

$\qquad=\dfrac{15}{4}\cdot\dfrac{1}{3}$

$\qquad=\dfrac{\cancel{3}\cdot5\cdot1}{4\cdot\cancel{3}}$

$\qquad=\dfrac{5}{4}$

$\qquad=1\dfrac{1}{4}$

82) $100\dfrac{1}{5}-10\dfrac{1}{3}=100\dfrac{3}{15}-10\dfrac{5}{15}$

$\qquad=99+1\dfrac{3}{15}-10\dfrac{5}{15}$

$\qquad=99\dfrac{18}{15}-10\dfrac{5}{15}$

$\qquad=89\dfrac{13}{15}$

83) $\left(3\dfrac{1}{4}\right)^2=\left(\dfrac{13}{4}\right)^2$

$\qquad=\dfrac{169}{16}$

$\qquad=10\dfrac{9}{16}$

CHAPTER 2   TEST

1) Since the denominator is 6, we draw an object with six equal parts.
Since the numerator is 5, we shade five of those parts.

2) a) A and C are equivalent.

b) Of those, C is in simplest form.

3) $\frac{5}{6} - \frac{2}{6} = \frac{3}{6}$

4) $\frac{3}{8} + \frac{2}{8} = \frac{5}{8}$

5) $\frac{1}{5} \cdot \frac{1}{3} = \frac{1}{15}$

6) $\frac{5}{6} - \frac{2}{6} = \frac{3}{6}$
$= \frac{1 \cdot \cancel{3}}{2 \cdot \cancel{3}}$
$= \frac{1}{2}$

7) $\frac{5}{8} \div \frac{3}{4} = \frac{5}{8} \cdot \frac{4}{3}$
$= \frac{5 \cdot \cancel{4}}{2 \cdot \cancel{4} \cdot 3}$
$= \frac{5}{6}$

8) $\frac{7}{12} + \frac{1}{3} = \frac{7}{12} + \frac{4}{12}$
$= \frac{11}{12}$

9) $\frac{7}{12} + \frac{1}{14} = \frac{49}{84} + \frac{6}{84}$
$= \frac{55}{84}$

10) $\frac{3}{14} \div \frac{2}{7} = \frac{3}{14} \cdot \frac{7}{2}$
$= \frac{3 \cdot \cancel{7}}{2 \cdot \cancel{7} \cdot 2}$
$= \frac{3}{4}$

11) $\frac{7}{20} - \frac{1}{16} = \frac{28}{80} - \frac{5}{80}$
$= \frac{23}{80}$

12) $\frac{5}{6} \cdot 3 = \frac{5}{6} \cdot \frac{3}{1}$
$= \frac{5 \cdot \cancel{3}}{2 \cdot \cancel{3} \cdot 1}$
$= \frac{5}{2}$

13) $\frac{2}{7} - \frac{1}{14} + \frac{3}{7} = \frac{4}{14} - \frac{1}{14} + \frac{6}{14}$
$= \frac{9}{14}$

14) $\frac{8}{5} + \frac{1}{3} \cdot \frac{2}{5} = \frac{8}{5} + \frac{2}{15}$
$= \frac{24}{15} + \frac{2}{15}$
$= \frac{26}{15}$

15) $\frac{3}{8} - \frac{1}{14} = \frac{21}{56} - \frac{4}{56}$
$= \frac{17}{56}$

16) $\frac{8}{5} \cdot \frac{1}{2} \cdot \frac{10}{4} = \frac{\cancel{2} \cdot \cancel{4} \cdot 1 \cdot 2 \cdot \cancel{5}}{\cancel{5} \cdot \cancel{2} \cdot \cancel{4}}$
$= 2$

17) $\left(\frac{5}{4}\right)^2 - \frac{7}{16} = \frac{25}{16} - \frac{7}{16}$
$= \frac{18}{16}$
$= \frac{9}{8}$

18) $\frac{4}{3} \cdot \frac{9}{24} + \frac{3}{2} = \frac{\cancel{4} \cdot \cancel{3} \cdot \cancel{3}}{\cancel{3} \cdot 2 \cdot \cancel{3} \cdot \cancel{4}} + \frac{3}{2}$
$= \frac{1}{2} + \frac{3}{2}$
$= \frac{4}{2}$
$= 2$

19)   LCD = 24

$$\frac{3}{4} \cdot \frac{6}{6} = \frac{18}{24} \text{ and } \frac{15}{24} = \frac{15}{24} \text{ and } \frac{7}{12} \cdot \frac{2}{2} = \frac{14}{24}$$

$$\frac{18}{24} > \frac{15}{24} > \frac{14}{24}, \text{ so } \frac{3}{4} > \frac{15}{24} > \frac{7}{12}$$

20) $\dfrac{60-36}{60} = \dfrac{24}{60}$

$$= \frac{2 \cdot \cancel{12}}{5 \cdot \cancel{12}}$$

$$= \frac{2}{5}$$

$\frac{2}{5}$ of the guests have not replied.

21) $99\left(1 - \dfrac{2}{3}\right) = 99\left(\dfrac{3}{3} - \dfrac{2}{3}\right)$

$$= \frac{99}{1} \cdot \frac{1}{3}$$

$$= \frac{\cancel{3} \cdot 33}{1 \cdot \cancel{3}}$$

$$= 33$$

The judge must still read 33 documents.

22) $7,000\left(1 - \dfrac{2}{1000}\right) = 7,000\left(\dfrac{1000}{1000} - \dfrac{2}{1000}\right)$

$$= \frac{7000}{1} \cdot \frac{998}{1000}$$

$$= \frac{7 \cdot \cancel{1000} \cdot 998}{1 \cdot \cancel{1000}}$$

$$= 6,986$$

6,986 people do not have malaria.

23) $11 \div 4 = 2 \text{ remainder } 3$

$$\frac{11}{4} = 2\frac{3}{4}$$

24) $7 \cdot 3 + 1 = 22$

$$3\frac{1}{7} = \frac{22}{7}$$

25) $6 \cdot 3 + 2 = 20$

$$3\frac{2}{6} = \frac{20}{6}$$

26) $33 \div 5 = 6 \text{ Remainder } 3$

$$\frac{33}{5} = 6\frac{3}{5}$$

27) $\dfrac{3}{4} - \dfrac{1}{4} \cdot \dfrac{1}{4} = \dfrac{3}{4} - \dfrac{1}{16}$

$$= \frac{12}{16} - \frac{1}{16}$$

$$= \frac{11}{16}$$

28) $3\dfrac{1}{8} + 1\dfrac{1}{2} = 3\dfrac{1}{8} + 1\dfrac{4}{8}$

$$= 3 + 1 + \frac{5}{8}$$

$$= 4\frac{5}{8}$$

29) $\left(\dfrac{1}{2} + \dfrac{1}{4}\right) \cdot \dfrac{10}{9} = \left(\dfrac{2}{4} + \dfrac{1}{4}\right) \cdot \dfrac{10}{9}$

$$= \frac{3}{4} \cdot \frac{10}{9}$$

$$= \frac{\cancel{3} \cdot \cancel{2} \cdot 5}{\cancel{2} \cdot 2 \cdot \cancel{3} \cdot 3}$$

$$= \frac{5}{6}$$

30) $4\dfrac{1}{2} \cdot 3\dfrac{1}{5} = \dfrac{9}{2} \cdot \dfrac{16}{5}$

$$= \frac{9 \cdot \cancel{2} \cdot 8}{\cancel{2} \cdot 5}$$

$$= \frac{72}{5}$$

$$= 14\frac{2}{5}$$

31) $5\dfrac{3}{4} - 3\dfrac{5}{8} = 5\dfrac{6}{8} - 3\dfrac{5}{8}$

$$= 2\frac{1}{8}$$

32) $\dfrac{8-3}{3} \cdot \dfrac{1}{12+3} = \dfrac{5}{3} \cdot \dfrac{1}{15}$

$$= \frac{\cancel{5} \cdot 1}{3 \cdot 3 \cdot \cancel{5}}$$

$$= \frac{1}{9}$$

33) $6\dfrac{1}{4} - 2\dfrac{7}{8} = 6\dfrac{2}{8} - 2\dfrac{7}{8}$

$$= 5 + 1\frac{2}{8} - 2\frac{7}{8}$$

$$= 5\frac{10}{8} - 2\frac{7}{8}$$

$$= 3\frac{3}{8}$$

34) $2\dfrac{4}{6} \div 5\dfrac{1}{3} = \dfrac{16}{6} \div \dfrac{16}{3}$

$\qquad\qquad = \dfrac{16}{6} \cdot \dfrac{3}{16}$

$\qquad\qquad = \dfrac{\cancel{16} \cdot \cancel{3}}{\cancel{3} \cdot 2 \cdot \cancel{16}}$

$\qquad\qquad = \dfrac{1}{2}$

35) $2 \div \left(\dfrac{3}{4} - \dfrac{1}{2} \right) = 2 \div \left(\dfrac{3}{4} - \dfrac{2}{4} \right)$

$\qquad\qquad\quad = 2 \div \dfrac{1}{4}$

$\qquad\qquad\quad = 2 \cdot 4$

$\qquad\qquad\quad = 8$

36) $\dfrac{1}{2} \div \dfrac{2}{3} \cdot \dfrac{3}{4} = \left(\dfrac{1}{2} \cdot \dfrac{3}{2} \right) \cdot \dfrac{3}{4}$

$\qquad\qquad\quad = \dfrac{3}{4} \cdot \dfrac{3}{4}$

$\qquad\qquad\quad = \dfrac{9}{16}$

37) $150 \div \dfrac{1}{4} = 150 \cdot 4$

$\qquad\qquad = 600$

Mila must get 600 points.

Chapter 3   Decimals

3.1   Understanding Decimal Numbers

GUIDED PRACTICE

1) The digit 5 is in the third place to the right of the decimal.
Tenths, hundredths, thousandths
5 is in the thousandths place.

2) The digit 1 is in the fifth place to the right of the decimal.
Tenths, hundredths, thousandths,
ten thousandths, hundred thousandths
1 is in the hundred thousandths place.

3) $8.19 = 8 + \dfrac{1}{10} + \dfrac{9}{100}$

4) $300.40012 = 300 + \dfrac{4}{10} + \dfrac{1}{10,000} + \dfrac{2}{100,000}$

5) Write the words for 43 and the place value of the 3.
0.43 is forty-three hundredths.

6) Write the words for 987 and the place value of the 7.
0.00987 is nine hundred eighty-seven hundred thousandths.

7) We must write the whole number portion and the decimal portion.
thirteen and ninety-four thousandths

8) We must write the whole number portion and the decimal portion.
five hundred three and one thousand six ten thousandths

9) $6.839 =$ six and eight hundred thirty-nine thousandths
$6.839 = 6\dfrac{839}{1,000}$

10) $320.0004 =$ three hundred twenty and four ten thousandths
$320.0004 = 320\dfrac{4}{10,000}$

11) $\dfrac{23}{100} =$ twenty-three hundredths
$\dfrac{23}{100} = 0.23$

12) $74\dfrac{103}{100,000} =$ seventy-four and one hundred three hundred thousandths
$74\dfrac{103}{100,000} = 74.00103$

13) Since 1.35 is closer to 1.346, it is the best approximation. Round the number up.

14) Since 3.4 is closer to 3.44, it is the best approximation. Round the number down.

15) It's a tie. Neither 9.94 nor 9.93 is closer to 9.935. We always round up when a tie occurs. Round the number up to 9.94.

16) The digit 2 is in the thousandths place.
Round up to 6.873 or down to 6.872.
The digit to the right is 6, so we round up.
$6.8726 \approx 6.873$ is the best approximation.

17) The digit 7 is in the hundredths place.
Round up to 89.08 or down to 89.07.
The digit to the right is 2, so we round down.
$89.07239 \approx 89.07$ is the best approximation.

18) The digit 5 is in the tenths place.
Round up to 18.6 or down to 18.5.
The digit to the right is 2, so we round down.
$18.523 \approx 18.5$ miles per hour is the best approximation.

19) $0.073 = 0.073$
$0.2 = 0.200$
Since both numbers are thousandths, we can easily compare them. $200 > 73$, so $0.2 > 0.073$.

20) $0.4 = 0.400$
$0.039 = 0.039$
$0.38 = 0.380$
Since the numbers are all thousandths, we can easily compare them. $39 < 380 < 400$, so $0.039 < 0.38 < 0.4$.

113

CONCEPT CHECKS AND PRACTICE EXERCISES

A1) Answers may vary. Example: In our number system, each place value is one tenth of the place value to its left. One tenth of the ones place is $1 \cdot \frac{1}{10} = \frac{1}{10}$, so the place to the right of the ones place is the tenths place.

A2) Because of the symmetry of place values, "thousands" and "thousandths" are 3 places to the left and right, respectively, of the digit in the ones place.

A3) $4.56 = 4 + \frac{5}{10} + \frac{6}{100}$

A4)

ten thousands	thousands	hundreds	tens	ones	decimal point	tenths	hundredths	thousandths	ten thousandths
10,000	1,000	100	10	1	.	$\frac{1}{10}$	$\frac{1}{100}$	$\frac{1}{1,000}$	$\frac{1}{10,000}$

A5) The digit 4 is in the third place to the right of the decimal.
Tenths, hundredths, thousandths
4 is in the thousandths place.

A6) The digit 4 is in the second place to the right of the decimal.
Tenths, hundredths
4 is in the hundredths place.

A7) The digit 4 is in the first place to the right of the decimal.
Tenths
4 is in the tenths place.

A8) The digit 4 is in the fourth place to the right of the decimal.
Tenths, hundredths, thousandths, ten thousandths
4 is in the ten thousandths place.

A9) The digit 4 is in the fourth place to the right of the decimal.
Tenths, hundredths, thousandths, ten thousandths
4 is in the ten thousandths place.

A10) The digit 4 is in the fifth place to the right of the decimal.
Tenths, hundredths, thousandths, ten thousandths, hundred thousandths
4 is in the hundred thousandths place.

A11) The digit 4 is in the first place to the right of the decimal.
Tenths
"4" is in the tenths place.

A12) The digit 4 is in the third place to the right of the decimal.
Tenths, hundredths, thousandths
"4" is in the thousandths place.

A13) $0.8 = \frac{8}{10}$

A14) $0.9 = \frac{9}{10}$

A15) $0.768 = \frac{7}{10} + \frac{6}{100} + \frac{8}{1,000}$

A16) $0.475 = \frac{4}{10} + \frac{7}{100} + \frac{5}{1,000}$

A17) $0.0087 = \frac{8}{1,000} + \frac{7}{10,000}$

A18) $0.0567 = \frac{5}{100} + \frac{6}{1,000} + \frac{7}{10,000}$

A19) $4.909 = 4 + \frac{9}{10} + \frac{9}{1,000}$

A20) $7.076 = 7 + \frac{7}{100} + \frac{6}{1,000}$

B1) 1) 0.08   →   d) eight hundredths
2) 8,000   →   a) eight thousand
3) 0.008   →   b) eight thousandths
4) 800   →   c) eight hundred

B2) The word "and" is used to separate the whole portion from the decimal portion.

B3) 1.5 = one and five tenths
$1\frac{5}{10}$ = one and five tenths
They are the same.

B4) 0.8 = eight tenths

B5)   0.9 = nine tenths

B6)   0.0087 = eighty-seven ten thousandths

B7)   0.0567 = five hundred sixty-seven ten thousandths

B8)   1.53 = one and fifty-three hundredths

B9)   5.49 = five and forty-nine hundredths

B10)   60.02 = sixty and two hundredths

B11)   55.05 = fifty-five and five hundredths

B12)   5.834 = five and eight hundred thirty-four thousandths

B13)   4.892 = four and eight hundred ninety-two thousandths

B14)   5.9006 = five and nine thousand six ten thousandths

B15)   6.90003 = six and ninety thousand three hundred thousandths

C1)   310.67 = three hundred ten and sixty-seven hundredths

C2)   Answers may vary. Example: The word "and" places the decimal point incorrectly.

C3) $3\dfrac{4}{10,000} = 3.0004$; the 4 is in the ten thousandths place.

C4) $4.09 = 4\dfrac{9}{100}$

C5) $3.12 = 3\dfrac{12}{100}$

C6) $54.4 = 54\dfrac{4}{10}$

C7) $23.2 = 23\dfrac{2}{10}$

C8) $3,000.003 = 3,000\dfrac{3}{1,000}$

C9) $4,000.004 = 4,000\dfrac{4}{1,000}$

C10) $230.00043 = 230\dfrac{43}{100,000}$

C11) $837.0439 = 837\dfrac{439}{10,000}$

C12) $\dfrac{3}{10} = 0.3$

C13) $\dfrac{7}{10} = 0.7$

C14) $\dfrac{37}{100} = 0.37$

C15) $\dfrac{73}{100} = 0.73$

C16) $5\dfrac{1}{100} = 5.01$

C17) $3\dfrac{13}{1000} = 3.013$

C18) $36\dfrac{27}{10,000} = 36.0027$

C19) $25\dfrac{29}{100,000} = 25.00029$

D1)   Answers may vary. Example: 310.143 rounded to the hundreds place is 300. Rounded to the hundredths place, 310.143 is 310.14.

D2)   Answers may vary. Example: The digit in the place value directly to the right of the given place value indicates whether the number is less than half way to the next larger number. Knowing this, it's possible to state which number it is approximately equal to.

D3)   Answers may vary. If the 0 was not included in 240, the number would be 24, which is not a good approximation of 241.

D4)   The digit 7 is in the hundredths place. Round up to 5.68 or down to 5.67. Since the digit to the right is 8, round up. $5.678 \approx 5.68$ is the better approximation.

D5)   The digit 5 is in the tenths place. Round up to 2.6 or down to 2.5. Since the digit to the right is 6, round up. $2.567 \approx 2.6$ is the better approximation.

D6)   The digit 2 is in the ones place.
Round up to 3 or down to 2.
Since the digit to the right is 4, round down.
$2.456 \approx 2$ is the better approximation.

D7)   The digit 3 is in the tenths place.
Round up to 0.4 or down to 0.3.
Since the digit to the right is 5, round up.
$0.3567 \approx 0.4$ is the better approximation.

D8)   The digit 8 is in the hundredths place.
Round up to 8.99 or down to 8.98.
Since the digit to the right is 7, round up.
$8.987 \approx 8.99$

D9)   The digit 3 is in the tenths place.
Round up to 37.4 or down to 37.3.
Since the digit to the right is 6, round up.
$37.369 \approx 37.4$

D10) The digit 9 is in the thousandths place.
Round up to 100.01 or down to 100.009.
Since the digit to the right is 0, round down.
$100.00907 \approx 100.009$

D11) The digit 6 is in the ten thousandths place.
Round up to 782.1287 or down to 782.1286.
Since the digit to the right is 1, round down.
$782.12861 \approx 782.1286$

D12) The digit 2 is in the thousands place.
Round up to 13,000 or down to 12,000.
Since the digit to the right is 6, round up.
$12,679.6728 \approx 13,000$

D13) The digit 7 is in the hundreds place.
Round up to 54,800 or down to 54,700.
Since the digit to the right is 8, round up.
$54,789.782 \approx 54,800$

D14) The digit 0 is in the ones place.
Round up to 11 or down to 10.
Since the digit to the right is 9, round up.
$10.978 \approx 11$

D15) The digit 0 is in the ones place.
Round up to 21 or down to 20.
Since the digit to the right is 4, round down.
$20.432 \approx 20$

D16) The digit 9 is in the ones place.
Round up to 120 or down to 119.
Since the digit to the right is 5, round up.
$119.57 \approx 120$
The speed is approximately 120 miles per hour.

D17) The digit 2 is in the tens place.
Round up to 130 or down to 120.
Since the digit to the right is 7, round up.
$127.53 \approx 130$
The speed is approximately 130 miles per hour.

E1) a) $0.003 = \dfrac{3}{1,000}$

$0.04 = \dfrac{4}{100}$

$0.2 = \dfrac{2}{10}$

b) $0.003 = \dfrac{3}{1,000}$

$0.04 = \dfrac{4}{100} \cdot \dfrac{10}{10} = \dfrac{40}{1,000}$

$0.2 = \dfrac{2}{10} \cdot \dfrac{100}{100} = \dfrac{200}{1,000}$

c) $200 > 40 > 3$, so $0.2 > 0.04 > 0.003$

E2) a) $\dfrac{2}{10} \cdot \dfrac{10}{10} = \dfrac{20}{100}$

$\dfrac{5}{100} = \dfrac{5}{100}$

$\dfrac{20}{100} > \dfrac{5}{100}$, so $\dfrac{2}{10} > \dfrac{5}{100}$.

b) $0.2 = 0.20$
$0.05 = 0.05$
$0.20 > 0.05$, so $0.2 > 0.05$.

c)   When comparing quantities, building like
fractions is equivalent to writing each decimal
number to the same place value.

E3) $0.4 = 0.40$
$0.06 = 0.06$
$0.40 > 0.06$, so $0.4 > 0.06$.

E4) $0.01 = 0.010$
$0.004 = 0.004$
$0.010 > 0.004$, so $0.01 > 0.004$.

E5) $0.07 = 0.07$
$0.6 = 0.60$
$0.07 < 0.60$, so $0.07 < 0.6$

E6) $0.024 = 0.024$
$0.15 = 0.150$
$0.024 < 0.150$, so $0.024 < 0.15$

E7) $0.04 = \dfrac{4}{100} \cdot \dfrac{10}{10} = \dfrac{40}{1,000}$

$0.042 = \dfrac{42}{1,000}$

$0.12 = \dfrac{12}{100} \cdot \dfrac{10}{10} = \dfrac{120}{1,000}$

$120 > 42 > 40$, so $0.12 > 0.042 > 0.04$.

E8) $0.087 = \dfrac{87}{1,000}$

$0.53 = \dfrac{53}{100} \cdot \dfrac{10}{10} = \dfrac{530}{1,000}$

$0.056 = \dfrac{56}{1,000}$

$530 > 87 > 56$, so $0.53 > 0.087 > 0.056$.

E9) $0.76 = \dfrac{76}{100}$

$0.7 = \dfrac{7}{10} \cdot \dfrac{10}{10} = \dfrac{70}{100}$

$0.73 = \dfrac{73}{100}$

$70 < 73 < 76$, so $0.7 < 0.73 < 0.76$.

The Mobile Metro, with a thickness of 0.7 inches, is the thinnest laptop.

E10) $0.428 = \dfrac{428}{1,000} \cdot \dfrac{10}{10} = \dfrac{4,280}{10,000}$

$0.3733 = \dfrac{3,733}{10,000}$

$0.275 = \dfrac{275}{1,000} \cdot \dfrac{10}{10} = \dfrac{2,750}{10,000}$

$4,280 > 3,733 > 2,750$, so

$0.428 > 0.3733 > 0.275$.

The Motorola L6, with a thickness of 0.428 inches, has the greatest thickness.

SECTION 3.1   EXERCISES

For 1–15, refer to Concept Checks and Practice Exercises.

17) like decimals

19) digits

21) The digit 4 is in the third place to the right of the decimal.
Tenths, hundredths, thousandths
4 is in the thousandths place.

23) The digit 4 is in the second place to the right of the decimal.
Tenths, hundredths
4 is in the hundredths place.

25) The digit 4 is in the first place to the right of the decimal.
Tenths
4 is in the tenths place.

27) The digit 4 is in the first place to the left of the decimal.
Ones
4 is in the ones place.

29) $1,004.0123 = 1,000 + 4 + \dfrac{1}{100} + \dfrac{2}{1,000} + \dfrac{3}{10,000}$

31) $0.03204 = \dfrac{3}{100} + \dfrac{2}{1,000} + \dfrac{4}{100,000}$

33) $0.65 =$ sixty-five hundredths

35) $203.129 =$ two hundred three and one hundred twenty-nine thousandths

37) $9.0002 =$ nine and two ten thousandths

39) $12,000.0012 =$ twelve thousand and twelve ten thousandths

41) 15.67 is fifteen and 67/100

43) 102.45 is one hundred two and 45/100

45) 504.98 is five hundred four and 98/100

47) $0.12 = \dfrac{12}{100}$

49) $1.2 = 1\dfrac{2}{10}$

51) $12.009 = 12\dfrac{9}{1,000}$

53) $102.908 = 102\dfrac{908}{1,000}$

55) $12,902.12902 = 12,902\dfrac{12,902}{100,000}$

or $12,902\dfrac{6451}{50,000}$

57) $0.1004 = \dfrac{1004}{10,000}$ or $\dfrac{251}{2500}$

59) $\dfrac{3}{10} = 0.3$

61) $2\dfrac{5}{10} = 2.5$

63) $14\dfrac{15}{100} = 14.15$

65) $3\dfrac{7}{100} = 3.07$

67) $54\dfrac{5}{1000} = 54.005$

69) $10,000\dfrac{10}{1,000} = 10,000.010 = 10,000.01$

71) $\dfrac{1}{100,000} = 0.00001$

73) $165\dfrac{1,674}{10,000} = 165.1674$

75) The digit 4 is in the tenths place.
Round up to 163.5 or down to 163.4.
Since the digit to the right is 5, round up.
$163.45 \approx 163.5$

77) The digit 6 is in the tens place.
Round up to 170 or down to 160.
Since the digit to the right is 3, round down.
$163.45 \approx 160$

79) The digit 7 is in the hundredths place.
Round up to 236.98 or down to 236.97.
Since the digit to the right is 9, round up.
$236.979 \approx 236.98$

81) The digit 2 is in the hundreds place.
Round up to 300 or down to 200.
Since the digit to the right is 3, round down.
$236.979 \approx 200$

83) The digit 6 is in the thousandths place.
Round up to 872.827 or down to 872.826.
Since the digit to the right is 0, round down.
$872.8260 \approx 872.826$

85) The digit 7 is in the thousandths place.
Round up to 7,327.828 or down to 7,327.827.
Since the digit to the right is 7, round up.
$7,327.8277 \approx 7,327.828$

87) The digit 5 is in the tenths place.
Round up to 864.6 or down to 864.5.
Since the digit to the right is 4, round down.
$864.54 \approx 864.5$
The average temperature on Venus is
approximately 864.5 degrees Fahrenheit.

89) To round to the nearest cent, we must round to
the hundredths place.
The digit 8 is in the hundredths place.
Round up to 345.69 or down to 345.68.
Since the digit to the right is 7, round up.
$345.6872 \approx 345.69$
The bank will charge $345.69.

91) $0.45 = \dfrac{45}{100} \cdot \dfrac{10}{10} = \dfrac{450}{1,000}$

$0.103 = \dfrac{103}{1,000}$

$0.08 = \dfrac{8}{100} \cdot \dfrac{10}{10} = \dfrac{80}{1,000}$

$0.7 = \dfrac{7}{10} \cdot \dfrac{100}{100} = \dfrac{700}{1,000}$

$700 > 450 > 103 > 80$,
so $0.7 > 0.45 > 0.103 > 0.08$.

93) $1.38 = 1\dfrac{38}{100} \cdot \dfrac{10}{10} = 1\dfrac{380}{1,000}$

$1.83 = 1\dfrac{83}{100} \cdot \dfrac{10}{10} = 1\dfrac{830}{1,000}$

$1.083 = 1\dfrac{83}{1,000}$

$1.308 = 1\dfrac{308}{1,000}$

$1.8 = 1\dfrac{8}{10} \cdot \dfrac{100}{100} = 1\dfrac{800}{1,000}$

$830 > 800 > 380 > 308 > 83$,
so $1.83 > 1.8 > 1.38 > 1.308 > 1.083$.

95) $149.335 = 149\dfrac{335}{1,000}$

$148.295 = 148\dfrac{295}{1,000}$

$149.601 = 149\dfrac{601}{1,000}$

$149 > 148$ and $601 > 335$, so
$149.601 > 149.335 > 148.295$.
The speeds from greatest to least are Hamilton 1970 (149.601 mph), Harvick 2007 (149.335 mph), and Gordon 1997 (148.295 mph).

97) $2.2 = 2\dfrac{2}{10}$

$5.5 = 5\dfrac{5}{10}$

$2 = 2\dfrac{0}{10}$

$2 < 5$ and $0 < 2$, so $2 < 2.2 < 5.5$
The items organized from softest to hardest are gypsum, fingernail, and glass.

99) The digit 9 is in the hundredths place.
Round up to 87.50 or down to 87.49.
Since the digit to the right is 3, round down.
$87.493 \approx 87.49$
Since $87.49 < 87.50$, Brittany will not get a B+ on her final exam.

101) To round to the nearest pound, we must round to the ones place.
The digit "4" is in the ones place.
Round up to 155 or down to 154.
Since the digit to the right is 7, round up.
$154.798 \approx 155$
The wrestler's weight is approximately 155 pounds.

3.2   Adding and Subtracting Decimal Numbers

GUIDED PRACTICE

1)
$$\begin{array}{r} \overset{1}{}3.879 \\ +\,5.710 \\ \hline 9.589 \end{array}$$

2)
$$\begin{array}{r} 0.002378 \\ +\,3.392000 \\ \hline 3.394378 \end{array}$$

3)
$$\begin{array}{r} \overset{5}{6}.\overset{9}{\cancel{0}}\,{}^{1}0\,2\,3\,7\,8 \\ -\,3.\,3\,9\,2\,0\,0\,0 \\ \hline 2.\,6\,1\,0\,3\,7\,8 \end{array}$$

4)
$$\begin{array}{r} \overset{3}{\cancel{4}}\,\overset{9}{\cancel{0}}.\overset{9}{\cancel{0}}\,{}^{1}0 \\ -\,7.\,3\,2 \\ \hline 3\,2.\,6\,8 \end{array}$$

5) A rise represents addition.
A decrease represents subtraction.
A fall represents subtraction.
new value
$= 13{,}562.33 + 129.43 - 39.87 - 67.13$
$= 13{,}691.76 - 39.87 - 67.13$
$= 13{,}651.89 - 67.13$
$= 13{,}584.76$
The new value is 13,584.76.

6) The greatest increase occurred from 2006 to 2007. The increase is equal to the difference in values: $(47.5 - 41.2 = 6.3)$. In that year, the account increased by \$6.3 thousand or \$6,300 in value.

The greatest decrease occurred from 2008 to 2009. The decrease is equal to the difference in values: $(36.3 - 19.5 = 16.8)$. In that year, the account decreased by \$16.8 thousand or \$16,800 in value.

CONCEPT CHECKS AND PRACTICE EXERCISES

A1) When adding decimal numbers, line up the decimal points to make sure that the correct digits are added together.

A2) Answers may vary. Example: Both represent tenths. Adding the numbers is similar to adding the numerators of the corresponding fractions.

A3) $0.3 = \dfrac{3}{10}$

$0.30 = \dfrac{30}{100} = \dfrac{3 \cdot \cancel{10}}{10 \cdot \cancel{10}} = \dfrac{3}{10}$

A4) Answers may vary. Example: Adding zeros on the far right of a decimal is similar to the process of building equivalent fractions.

A5)   1.3
     + 2.5

       3.8

A6)   2.5
     + 3.3

       5.8

A7)   14.91
     + 2.07

      16.98

A8)   3.08
     + 2.91

       5.99

A9)   4.0
     + 3.4

       7.4

A10)   8.00
     + 9.21

      17.21

A11) $\overset{0}{\cancel{1}}{}^{1}2.\,5\;6$
     −   3. 2 4
     ----
          9. 3 2

A12) $\overset{0}{\cancel{1}}{}^{1}1.\,9\;3$
     −   4. 7 1
     ----
          7. 2 2

A13) $9.\overset{0}{\cancel{1}}{}^{1}2$
     − 5.0 4

       4.0 8

A14) $1\,4.\overset{2}{\cancel{3}}{}^{1}6$
     −   3. 1 7
     ----
       1 1. 1 9

A15) $8.\overset{8}{\cancel{9}}{}^{1}1\overset{1}{\cancel{2}}{}^{1}0$
     − 4. 1 9 1 2
     ----
       4. 7 2 0 8

A16) $\overset{6}{\cancel{7}}{}^{1}5.\,7\;8\;\overset{8}{\cancel{9}}{}^{1}0$
     − 6. 0 7 8 9
     ----
       6 9.7 1 0 1

A17) $\overset{3}{\cancel{4}}.\,\overset{9}{\cancel{0}}{}^{1}0$
     − 0. 1 4
     ----
       3. 8 6

A18) $\overset{6}{\cancel{7}}.\,\overset{9}{\cancel{0}}{}^{1}0$
     − 0. 8 7
     ----
       6. 1 3

A19) $\overset{1}{1}1.6$
     + 1.6

      13.2

The total number of knives and lighters
intercepted is 13.2 million, or 13,200,000.

A20)   13.7
     − 13.2

       0.5

0.5 million (or 500,000) of the intercepted items
were neither knives or lighters.

SECTION 3.2   EXERCISES

For 1–3, refer to Concept Checks and Practice
Exercises.

5) sum

7) difference

9)   7.4
    + 4.3

     11.7

11) $\overset{1\;\;1}{6}7.12$
    + 2.93

     70.05

13) $\overset{1}{2}.90$
    + 5.42

      8.32

15)
$$\begin{array}{r} \overset{1\ \ 1}{7.982} \\ +\ 2.730 \\ \hline 10.712 \end{array}$$

17)
$$\begin{array}{r} \overset{1}{61.57} \\ +\ 17.34 \\ \hline 78.91 \end{array}$$
The total bill was \$78.91.

19)
$$\begin{array}{r} \overset{1\ \ 1}{45.80} \\ +\ 39.55 \\ \hline 85.35 \end{array}$$
They raised a total of \$85.35.

21)
$$\begin{array}{r} 6.4 \\ -\ 2.3 \\ \hline 4.1 \end{array}$$

23)
$$\begin{array}{r} \overset{0\ \ ^{14}}{1\,5.^{1}6\,5} \\ -\ 6.8\,2 \\ \hline 8.8\,3 \end{array}$$

25)
$$\begin{array}{r} 1\,\overset{8}{9}.\overset{^{1}1}{2}\,0 \\ -\ 5.7\,5 \\ \hline 1\,3.4\,5 \end{array}$$

27)
$$\begin{array}{r} 8\,\overset{2}{3}.\overset{^{17}8}{}\,\overset{9}{9}\,0 \\ -\ 1.9\,8\,3 \\ \hline 8\,1.8\,1\,7 \end{array}$$

29)
$$\begin{array}{r} 9\,\overset{4}{5}.\overset{^{12}3}{}\,4 \\ -\ 2\,3.8\,5 \\ \hline 7\,1.4\,9 \end{array}$$
The new grocery bill is \$71.49.

31)
$$\begin{array}{r} \overset{0}{1}\,\overset{9}{0}\,\overset{9}{0}.\overset{9}{0}\,0 \\ -\ \ 8\,6.2\,4 \\ \hline 1\,3.7\,6 \end{array}$$
Aaron should receive \$13.76 in change.

33)
$$\begin{array}{r} 4\,2\,3.\overset{1}{1}\,7\,8\,4 \\ +\ \ 4.6\,4\,0\,0 \\ \hline 4\,2\,7.8\,1\,8\,4 \end{array}$$

35)
$$\begin{array}{r} 9\,2\,\overset{3}{4}.\overset{9}{0}\,\overset{9}{0}\,0 \\ -\ \ 0.1\,2\,3 \\ \hline 9\,2\,3.8\,7\,7 \end{array}$$

37)
$$\begin{array}{r} \overset{1}{1}\,2\,8.0\,0 \\ +\ \ 8.8\,3 \\ \hline 1\,3\,6.8\,3 \end{array}$$

39)
$$\begin{array}{r} 3\,\overset{0}{1}\,\overset{^{14}5}{}.\overset{9}{0}\,0 \\ -\ \ 7.4\,5 \\ \hline 3\,0\,7.5\,5 \end{array}$$

41)
$$\begin{array}{r} 5\,4.9\,2\,3\,7 \\ -\ 3.9\,0\,1\,0 \\ \hline 5\,1.0\,2\,2\,7 \end{array}$$

43)
$$\begin{array}{r} \overset{1\ 1\ \ 1}{78.99} \\ +\ 43.87 \\ \hline 122.86 \end{array}$$

45)
$$\begin{array}{r} \overset{6}{7}.^{1}2\,7 \\ -\ 2.3\,0 \\ \hline 4.9\,7 \end{array}$$
The weight of the package is 4.97 pounds.

47) a)
$$\begin{array}{r} 3.15 \\ -\ 2.10 \\ \hline 1.05 \end{array}$$
It is 1.05 miles from Bob's Boats to the intersection.

b)
$$\begin{array}{r} 5.\overset{1}{2}\,0 \\ -\ 1.0\,5 \\ \hline 4.1\,5 \end{array}$$
It is 4.15 miles from Anita's Grocery to the intersection.

49)
$$\begin{array}{r} \overset{1}{2}\,\overset{^{12}3}{}.\overset{^{16}7}{}\,\overset{^{15}6}{}\,\overset{9}{0}\,0 \\ -\ \ 4.9\,7\,2\,4 \\ \hline 1\,8.7\,8\,7\,6 \end{array}$$

51)
$$\begin{array}{r} \overset{1\ 1\ \ 1\ 1}{45.972} \\ +\ 54.028 \\ \hline 100.000 \end{array}$$

53) a) $325.00 \approx \overset{1}{3}30$
          $212.78 \approx 210$
          $37.87 \approx 40$
          $16.21 \approx 20$
          $39.99 \approx 40$
          $35.00 \approx + 40$
          $\overline{680}$

The monthly bills are approximately $680.

b) 698.43
       -680.00
       $\overline{18.43}$

It appears that you will be able to pay extra on the credit card.

c) $325 + 212.78 + 37.87 + 39.99 + 35$
     $= 537.78 + 37.87 + 16.21 + 39.99 + 35$
     $= 575.65 + 16.21 + 39.99 + 35$
     $= 591.86 + 39.99 + 35$
     $= 631.85 + 35$
     $= 666.85$

The actual total of all the bills is $666.85.

d) $6\,9\,\overset{7}{\cancel{8}}.\overset{13}{\cancel{4}}\,{}^{1}3$
      $-\,6\,6\,6\,.\,8\,5$
      $\overline{3\,1\,.\,5\,8}$

$31.58 will be left in the account.

55) 6-Feb: $438.45 - $74.32 = $364.13

7-Feb: $364.13 - $275.00 = $89.13

7-Feb: $89.13 - $25.00 - $3.00 = $61.13

15-Feb: $61.13 - $49.95 = $11.18

15-Feb: $11.18 + $256.98 = $268.16

57) $17.2 - (9.4 + 6.9) = 17.2 - 16.3$
     $ = 0.9$

There is 0.9 million square miles less area in North and South America combined than in Asia.

59) $1\,\overset{6}{\cancel{7}}.{}^{1}2$
      $-1\,1\,.\,6$
      $\overline{5\,.\,6}$

Africa would need to be increased by more than 5.6 million square miles for it to be the largest continent.

3.3   Multiplying Decimal Numbers

GUIDED PRACTICE

1) 3 decimal places $\rightarrow$ 0.003
    2 decimal places $\rightarrow$ $\times\ \ 0.02$
    $3 + 2 = 5$ decimal places $\rightarrow$ $\overline{0.00006}$

2) 1 decimal place $\rightarrow$ 0.7
    3 decimal places $\rightarrow$ $\times 0.009$
    $1 + 3 = 4$ decimal places $\rightarrow$ $\overline{0.0063}$

3) 0 decimal places $\rightarrow$ 5
    7 decimal places $\rightarrow$ $\times 0.0000003$
    $0 + 7 = 7$ decimal places $\rightarrow$ $\overline{0.0000015}$

4) 2 decimal places $\rightarrow$ $\overset{2\ \,2}{50.67}$
    2 decimal places $\rightarrow$ $\times\ \ 0.03$
    $2 + 2 = 4$ decimal places $\rightarrow$ $\overline{1.5201}$

5) 5 decimal places $\rightarrow$ 0.00345
    0 decimal places $\rightarrow$ $\times\ \ \ \ \ \ 27$
     $\overline{2415}$
     6900
    $5 + 0 = 5$ decimal places $\rightarrow$ $\overline{0.09315}$

6) 0 decimal places $\rightarrow$ 1500
    5 decimal places $\rightarrow$ $\times 0.00073$
     $\overline{4500}$
     $1\,05000$
    $0 + 5 = 5$ decimal places $\rightarrow$ $\overline{1.09500}$

7) Multiplying by 0.01 will make the product smaller than 501. Move the decimal point two places to the left.
    $501 \times 0.01 = 5.01$

8) Even though 3.01 ends in .01, it is not a power of 10. Look at 0.001 to see how to move the decimal point. Multiplying by 0.001 will make the product smaller than 3.01. Move the decimal point three places to the left.
    $3.01 \times 0.001 = 0.00301$

9) Multiplying by 100,000 will make the product larger than 23.2. Move the decimal point five places to the right.
    $23.2 \times 100,000 = 2,320,000$

10) Multiplying by 1,000,000 will make the product larger than 3.76. Move the decimal point six places to the right.
$3.76 \times 1,000,000 = 3,760,000$

11) $0.00373 \times 0.065 \approx 0.004 \times 0.07$
$$= 0.00028$$
1) After rounding, multiply the non-zero digits.
$4 \cdot 7 = 28$
2) The first digit came from the thousandths place. Move the decimal point 3 places to the left. $28 \rightarrow 0.028$
3) The second digit came from the hundredths place. Move the decimal point 2 places to the left. $0.028 \rightarrow 0.00028$

12) $8.953 \times 0.076 \approx 9 \times 0.08$
$$= 0.72$$
1) After rounding, multiply the non-zero digits.
$9 \cdot 8 = 72$
2) The first digit came from the ones place. Move the decimal point 0 places. $72 \rightarrow 72$
3) The second digit came from the hundredths place. Move the decimal point 2 places to the left. $72 \rightarrow 0.72$

13) $4,723 \times 0.0523 \approx 5,000 \times 0.05$
$$= 250$$
1) After rounding, multiply the non-zero digits.
$5 \cdot 5 = 25$
2) The first digit came from the thousands place. Move the decimal point 3 places to the right. $25 \rightarrow 25,000$
3) The second digit came from the hundredths place. Move the decimal point 2 places to the left. $25,000 \rightarrow 250$

14) The radius is half the diameter.
$$\text{Radius} = \text{Diameter} \div 2$$
$$= 4.8 \div 2$$
$$= 2.4$$
The radius is 2.4 cm.

15) The diameter is twice the radius.
$$\text{Diameter} = \text{Radius} \times 2$$
$$= 3.75 \times 2$$
$$= 7.5$$
The diameter is 7.5 in.

16) $C = 2\pi r$
$$= 2(3.14)(20 \text{ meters})$$
$$= 6.28(20 \text{ meters})$$
$$= 125.6 \text{ meters}$$

17) $r = 8 \div 2 = 4$ centimeters
$A = \pi r^2$
$$= (3.14)(4 \text{ centimeters})^2$$
$$= 3.14(16 \text{ square centimeters})$$
$$= 50.24 \text{ square centimeters}$$

CONCEPT CHECKS AND PRACTICE EXERCISES

A1) a) $0.3 = \dfrac{3}{10}$

$0.5 = \dfrac{5}{10}$

b) $\dfrac{3}{10} \cdot \dfrac{5}{10} = \dfrac{15}{100}$

c) The number of decimal places in the decimal form indicates the number of zeros in the denominator of the fractional form.

d) $0.3 \times 0.5 = 0.15$

e) The total number of decimal places in 0.3 and 0.5 is equal to the number of decimal places in the product 0.3×0.5.

A2)
$$\begin{array}{r} 3 \text{ decimal places} \rightarrow \quad 0.003 \\ 2 \text{ decimal places} \rightarrow \quad \times \; 0.04 \\ \hline 3+2 = 5 \text{ decimal places} \rightarrow \quad 0.00012 \end{array}$$

A3)
$$\begin{array}{r} 4 \text{ decimal places} \rightarrow \quad 0.0004 \\ 2 \text{ decimal places} \rightarrow \quad \times \; 0.04 \\ \hline 4+2 = 6 \text{ decimal places} \rightarrow \quad 0.000016 \end{array}$$

A4)
$$\begin{array}{r} 3 \text{ decimal places} \rightarrow \quad 0.006 \\ 5 \text{ decimal places} \rightarrow \quad \times \; 0.00008 \\ \hline 3+5 = 8 \text{ decimal places} \rightarrow \quad 0.00000048 \end{array}$$

A5)
$$\begin{array}{r} 1 \text{ decimal place} \rightarrow \quad 0.1 \\ 3 \text{ decimal places} \rightarrow \quad \times \; 0.003 \\ \hline 1+3 = 4 \text{ decimal places} \rightarrow \quad 0.0003 \end{array}$$

A6)
$$\begin{array}{r} 0 \text{ decimal places} \rightarrow \quad 9 \\ 4 \text{ decimal places} \rightarrow \quad \times \; 0.0004 \\ \hline 0+4 = 4 \text{ decimal places} \rightarrow \quad 0.0036 \end{array}$$

A7)
$$\begin{array}{r} 0 \text{ decimal places} \rightarrow \quad 5 \\ 4 \text{ decimal places} \rightarrow \quad \times \; 0.0006 \\ \hline 0+4 = 4 \text{ decimal places} \rightarrow \quad 0.0030 \end{array}$$

A8)   1 decimal place →   8.0
      1 decimal place →   × 9.0
      1+1 = 2 decimal places →   72.00

A9)   1 decimal place →   8.0
      1 decimal place →   × 2.0
      1+1 = 2 decimal places →   16.00

A10) 0 decimal places →   80
      3 decimal places →   × 0.004
      0+3 = 3 decimal places →   0.320
      The thickness of 80 sheets of paper is 0.32 inch.

A11) 0 decimal places →   40
      1 decimal place →   ×   0.3
      0+1 = 1 decimal place →   12.0
      The height of 40 stacked tiles is 12 inches.

B1)   Answers may vary. Example: Add the number
      of decimal places in the factors. The answer will
      have the same number of decimal places.

B2)   Answers may vary. Example: The zero does not
      add value to the number. It is simply a
      placeholder.

B3)   3 decimal places →   0.5$\overset{5\,2}{7}$3
      2 decimal places →   × 0.07
      3+2 = 5 decimal places →   0.04011

B4)   4 decimal places →   0.0$\overset{1\,1}{6}$89
      2 decimal places →   ×   0.02
      4+2 = 6 decimal places →   0.001378

B5)   2 decimal places →   3.65
      1 decimal place →   × 8.2
                                         730
                                      29 200
      2+1 = 3 decimal places →   29.930

B6)   2 decimal places →   4.93
      1 decimal place →   × 2.4
                                       1 972
                                       9 860
      2+1 = 3 decimal places →   11.832

B7)   3 decimal places →   45.965
      4 decimal places →   × 0.0032
                                         91930
                                      1378950
      3+4 = 7 decimal places →   0.1470880

B8)   3 decimal places →   85.403
      3 decimal places →   × 0.023
                                       256209
                                     1 708060
      3+3 = 6 decimal places →   1.964269

B9)   0 decimal places →   7600
      2 decimal places →   × 1.06
                                       456 00
                                       000 00
                                      7600 00
      0+2 = 2 decimal places →   8056.00

B10) 0 decimal places →   3800
      2 decimal places →   × 2.12
                                        76 00
                                       380 00
                                      7600 00
      0+2 = 2 decimal places →   8056.00

B11) 0 decimal places →   700
      3 decimal places →   × 0.375
                                       3 500
                                      49 000
                                     210 000
      0+3 = 3 decimal places →   262.500
      In 2004, a person could deduct $262.50 on her
      tax return if she drove 700 miles for business.

B12) 0 decimal places →   900
      3 decimal places →   × 0.505
                                       4 500
                                       0 000
                                     450 000
      0+3 = 3 decimal places →   454.500
      In 2008, a person could deduct $454.50 on his
      tax return if he drove 900 miles for business.

C1)   a) Answers may vary. Example: 42.2
      b) Answers may vary. Example: 4.22
      c) 4.22 is less than 42.2. If the decimal point is
          moved to the left in a number, the result is
          smaller than the original number.

C2) Moving the decimal point three places to the right in 45.9 is the same as multiplying 45.9 by 1,000.

C3) Multiplying by 100 will make the product larger. Move the decimal point two places to the right.
$13.6 \times 100 = 1,360$

C4) Multiplying by 1,000 will make the product larger. Move the decimal point three places to the right.
$3.71 \times 1,000 = 3,710$

C5) Even though 10.001 ends in .001, it is not a power of 10. Look at 0.001 to see how to move the decimal point. Multiplying by 0.001 will make the product smaller. Move the decimal point three places to the left.
$0.001 \times 10.001 = 10.001 \times 0.001 = 0.010001$

C6) 101 is not a power of 10. Look at 0.1 to see how to move the decimal point. Multiplying by 0.1 will make the product smaller. Move the decimal point one place to the left.
$0.1 \times 101 = 101 \times 0.1 = 10.1$

C7) Multiplying by 1,000,000 will make the product larger. Move the decimal point six places to the right.
$0.00098 \times 1,000,000 = 980$

C8) Multiplying by 0.00001 will make the product smaller. Move the decimal point five places to the left.
$854,982 \times 0.00001 = 8.54982$

C9) Multiplying by 0.0000001 will make the product smaller. Move the decimal point seven places to the left.
$987 \times 0.0000001 = 0.0000987$

C10) Multiplying by 100,000 will make the product larger. Move the decimal point five places to the right.
$87 \times 100,000 = 8,700,000$

D1) Move the decimal point four places to the left.

D2) Move the decimal point three places to the right.

D3) Don't move the decimal point.

D4) $0.423 \times 0.622 \approx 0.4 \times 0.6$
$= 0.24$

D5) $0.842 \times 0.21 \approx 0.8 \times 0.2$
$= 0.16$

D6) $36.5 \times 0.74 \approx 40 \times 0.7$
$= 28$

D7) $85.4 \times 0.39 \approx 90 \times 0.4$
$= 36$

D8) $26,984 \times 0.0034 \approx 30,000 \times 0.003$
$= 90$

D9) $54,987 \times 0.00074 \approx 50,000 \times 0.0007$
$= 35$

D10) $0.00034 \times 0.56347 \approx 0.0003 \times 0.6$
$= 0.00018$

D11) $0.67843 \times 0.00387 \approx 0.7 \times 0.004$
$= 0.0028$

D12) $1,260.67 \times 8 \approx 1,000 \times 8$
$= 8,000$
The cost of eight credits is approximately $8,000.

D13) $34.33 \times 8 \approx 30 \times 8$
$= 240$
The cost of eight credits is approximately $240.

E1) Circumference is the perimeter of a circle.

E2) b) 18.86 square cm, or c) $78.5\,\text{ft}^2$ could be used to describe the area of a circle since they are in square units.

E3) If you know a diameter of a circle, you can find the radius by dividing the diameter by 2.

E4) $A = \pi r^2$
$= (3.14)(3 \text{ kilometers})^2$
$= 3.14(9 \text{ square kilometers})$
$= 28.26$ square kilometers

E5) $A = \pi r^2$
$= (3.14)(2 \text{ miles})^2$
$= 3.14(4 \text{ square miles})$
$= 12.56$ square miles

E6) $r = 42 \div 2 = 21$ inches

$C = 2\pi r$
  $= 2(3.14)(21 \text{ inches})$
  $= 6.28(21 \text{ inches})$
  $= 131.88$ inches

E7) $r = 58 \div 2 = 29$ centimeters

$C = 2\pi r$
  $= 2(3.14)(29 \text{ centimeters})$
  $= 6.28(29 \text{ centimeters})$
  $= 182.12$ centimeters

E8) $r = 14 \div 2 = 7$ inches

$A = \pi r^2$
  $= (3.14)(7 \text{ inches})^2$
  $= 3.14(49 \text{ square inches})$
  $= 153.86$ square inches

E9) $r = 28 \div 2 = 14$ meters

$A = \pi r^2$
  $= (3.14)(14 \text{ meters})^2$
  $= 3.14(196 \text{ square meters})$
  $= 615.44$ square meters

E10) $C = 2\pi r$
  $= 2(3.14)(3.8 \text{ millimeters})$
  $= 6.28(3.8 \text{ millimeters})$
  $= 23.864$ millimeters

E11) $C = 2\pi r$
  $= 2(3.14)(2.5 \text{ inches})$
  $= 6.28(2.5 \text{ inches})$
  $= 15.7$ inches

SECTION 3.3   EXERCISES

For 1–15, refer to Concept Checks and Practice Exercises.

17) Pi or π

19) product

21) circumference

23)   3 decimal places $\rightarrow$ 0.005
3 decimal places $\rightarrow$ $\times$ 0.005
$3 + 3 = 6$ decimal places $\rightarrow$ 0.000025

25)   1 decimal place $\rightarrow$ 0.9
4 decimal places $\rightarrow$ $\times \, 0.0007$
$1 + 4 = 5$ decimal places $\rightarrow$ 0.00063

27)   2 decimal places $\rightarrow$ $\overset{1}{2}.31$
1 decimal place $\rightarrow$ $\times \, 0.4$
$2 + 1 = 3$ decimal places $\rightarrow$ 0.924

29)   1 decimal place $\rightarrow$ 4.2
1 decimal place $\rightarrow$ $\times \, 1.7$
    2 94
    4 20
$1 + 1 = 2$ decimal places $\rightarrow$ 7.14

31)   1 decimal place $\rightarrow$ 212.6
1 decimal place $\rightarrow$ $\times \;\; 5.3$
    63 78
    1063 00
$1 + 1 = 2$ decimal places $\rightarrow$ 1126.78

33)   3 decimal places $\rightarrow$ 0.946
3 decimal places $\rightarrow$ $\times \, 0.051$
    946
    47300
$3 + 3 = 6$ decimal places $\rightarrow$ 0.048246

35)   2 decimal places $\rightarrow$ $\overset{1\;\;1}{4}.75$
0 decimal places $\rightarrow$ $\times \;\; 2$
$2 + 0 = 2$ decimal places $\rightarrow$ 9.50

The price of two hotdogs at the baseball stadium in Los Angeles is $9.50.

37)   1 decimal place $\rightarrow$ $\overset{1}{3}.5$
0 decimal places $\rightarrow$ $\times \, 32$
    7 0
    105 0
$1 + 0 = 1$ decimal place $\rightarrow$ 112.0

It will cost $112 to park in Honolulu for 3.5 days.

39) Multiplying by 0.1 will make the product smaller. Move the decimal point one place to the left.
$43.2 \times 0.1 = 4.32$

41) Multiplying by 100 will make the product larger. Move the decimal point two places to the right.
$12.65 \times 100 = 1,265$

43) Multiplying by 0.001 will make the product smaller. Move the decimal point three places to the left.
$0.00078 \times 0.001 = 0.00000078$

45) Multiplying by 100 will make the product larger. Move the decimal point two places to the right.
$0.021 \times 100 = 2.1$

47) a) $125 \times 3.2 \approx 100 \times 3$

   b) $100 \times 3 = 300$

   c)
   $$\begin{array}{r} 0 \text{ decimal places} \rightarrow \quad 125 \\ 1 \text{ decimal place} \rightarrow \quad \times\ 3.2 \\ \hline 25\ 0 \\ 375\ 0 \\ \hline 0+1 = 1 \text{ decimal place} \rightarrow \quad 400.0 \end{array}$$
   $400 \approx 400$

49) a) $43.2 \times 0.006 \approx 40 \times 0.006$

   b) $40 \times 0.006 = 0.24$

   c)
   $$\begin{array}{r} 1 \text{ decimal place} \rightarrow \quad \overset{1\ 1}{43.2} \\ 3 \text{ decimal places} \rightarrow \quad \times\ 0.006 \\ \hline 1+3 = 4 \text{ decimal places} \rightarrow \quad 0.2592 \end{array}$$
   $0.2592 \approx 0.26$

51) a) $2.0008 \times 0.0067 \approx 2 \times 0.007$

   b) $2 \times 0.007 = 0.014$

   c)
   $$\begin{array}{r} 4 \text{ decimal places} \rightarrow \quad 2.0008 \\ 4 \text{ decimal places} \rightarrow \quad \times\ 0.0067 \\ \hline 140056 \\ 1200480 \\ \hline 4+4 = 8 \text{ decimal places} \rightarrow \quad 0.01340536 \end{array}$$
   $0.01340536 \approx 0.013$

53) a) $879.9 \times 0.0452 \approx 900 \times 0.05$

   b) $900 \times 0.05 = 45$

c)
$$\begin{array}{r} 1 \text{ decimal place} \rightarrow \quad 879.9 \\ 4 \text{ decimal places} \rightarrow \quad \times\ 0.0452 \\ \hline 17598 \\ 4\ 39950 \\ 35\ 19600 \\ \hline 1+4 = 5 \text{ decimal places} \rightarrow \quad 39.77148 \end{array}$$
$39.77148 \approx 40$

55) $C = 2\pi r$
$= 2(3.14)(4 \text{ kilometers})$
$= 6.28(4 \text{ kilometers})$
$= 25.12 \text{ kilometers}$
$\approx 25.1 \text{ kilometers}$

57) $r = 3 \cdot 0.5 = 1.5 \text{ yards}$
$A = \pi r^2$
$= (3.14)(1.5 \text{ yards})^2$
$= 3.14(2.25 \text{ square yards})$
$= 7.065 \text{ square yards}$
$\approx 7.1 \text{ square yards}$

59) $C = 2\pi r$
$= 2(3.14)(4.5 \text{ inches})$
$= 6.28(4.5 \text{ inches})$
$= 28.26 \text{ inches}$
$\approx 28.3 \text{ inches}$
The circumference of this decorative plate is approximately 28.3 inches.

61) $r = 60 \cdot 0.5 = 30 \text{ centimeters}$
$A = \pi r^2$
$= (3.14)(30 \text{ centimeters})^2$
$= 3.14(900 \text{ square centimeters})$
$= 2,826 \text{ square centimeters}$
The area of the lid is approximately 2,826 square centimeters.

63)
$$\begin{array}{r} 3 \text{ decimal places} \rightarrow \quad 3.549 \\ 2 \text{ decimal places} \rightarrow \quad \times\ 8.54 \\ \hline 14196 \\ 1\ 77450 \\ 28\ 39200 \\ \hline 3+2 = 5 \text{ decimal places} \rightarrow \quad 30.30846 \end{array}$$
$30.30846 \approx 30.31$
The total bill is $30.31.

65) 3 decimal places $\rightarrow$ 0.713
     1 decimal place $\rightarrow$ $\times$ 71.1
                                 ─────────
                                       713
                                      7130
                                    49 9100
                                 ─────────
     $3+1=4$ decimal places $\rightarrow$ 50.6943
     $50.6943 \approx 50.69$
     The charge for the gas was $50.69.

67) a) 2 decimal places $\rightarrow$ 14.99
        1 decimal place $\rightarrow$ $\times$ 3.5
                               ─────────
                                     7 495
                                    44 970
                               ─────────
        $2+1=3$ decimal places $\rightarrow$ 52.465
        $52.465 \approx 52.47$
        The per-hour charge is $52.47.

    b) $29.99 + 52.47 = 82.46$
        The entire bill is $82.46.

    c) $82.46 < $89.99, so it would not be cheaper
        to rent a truck for a daily rate of $89.99.

3.4 Dividing Decimal Numbers

GUIDED PRACTICE

1) The divisor, 0.91, has two decimal places.
    Move both decimal points two places to the right.

    $9\,1\overline{)3\,4\,8\,6.}$

2) The divisor, 1.2, has one decimal place.
    Move both decimal points one place to the right.

    $1\,2\overline{)3\,2.1}$

3) To move the decimal point three places to the
    right in 12.3, we must write two placeholder
    zeros.

    $1\,2\overline{)1\,2\,3\,0\,0.}$

4)
    $0.4\overline{)2\,2} \rightarrow$
    $4\overline{)2\,2\,0}$ quotient $5\,5$
    $-\,2\,0$
    $\;\;\;\;2\,0$
    $-\,2\,0$
    $\;\;\;\;\;\;0$

5) To round to the tenths place, continue dividing
    until the hundredths place.

    $3\overline{)1\,4.\,0\,0}$ quotient $\#.\#\#$

    To divide to the hundredths place, annex two
    zeros.

6)
    $0.05\overline{)0.006} \rightarrow 5\overline{)0.6\,0}$ quotient $0.1\,2$
    $-\,5$
    $\;\;\;1\,0$
    $-\,1\,0$
    $\;\;\;\;\;\;0$

    $0.006 \div 0.05 \approx 0.1$

7)
    $0.6\overline{)7} \rightarrow 6\overline{)7\,0.0\,0\,0}$ quotient $1\,1.6\,6\,6$
    $-\,6$
    $\;\;1\,0$
    $-\,6$
    $\;\;\;\;4\,0$
    $-\,3\,6$
    $\;\;\;\;\;\;4\,0$
    $-\,3\,6$
    $\;\;\;\;\;\;\;\;4\,0$

    $7 \div 0.6 \approx 11.67$

8) $\dfrac{3}{4} \rightarrow 4\overline{)3.0\,0}$ quotient $0.7\,5$
    $-\,2\,8$
    $\;\;\;2\,0$
    $-\,2\,0$
    $\;\;\;\;\;0$

    $\dfrac{3}{4} = 0.75$

9) $\dfrac{1}{6} \rightarrow 6\overline{)1.0\,0\,0}$ quotient $0.1\,6\,6$
    $-\,6$
    $\;\;\;4\,0$
    $-\,3\,6$
    $\;\;\;\;\;4\,0$

    $\dfrac{1}{6} = 0.1\overline{6}$

10) $\dfrac{3}{7} \rightarrow 7\overline{)3.0\,0\,0}$ quotient $0.4\,2\,8$
    $-\,2\,8$
    $\;\;\;2\,0$
    $-\,1\,4$
    $\;\;\;\;\;6\,0$

    $\dfrac{3}{7} \approx 0.43$

11)
$$\frac{1}{5} \rightarrow$$

$$\begin{array}{r} 0.\ 2 \\ 5\overline{)1.\ 0} \\ -1\ 0 \\ \hline 0 \end{array}$$

$$4\frac{1}{5} = 4.2$$

12) $9.32 \approx 9$
$0.0425 \approx 0.04$

$$0.04\overline{)9} \rightarrow \quad 4\overline{)9\ 0\ 0}$$

$$\begin{array}{r}
2\ 2\ 5 \\
4\overline{)9\ 0\ 0} \\
-8 \\
\hline
1\ 0 \\
-8 \\
\hline
2\ 0 \\
-2\ 0 \\
\hline
0
\end{array}$$

$9.32 \div 0.0425 \approx 225$

13) $0.0784 \approx 0.08$
$8.726 \approx 9$

$$\begin{array}{r}
0.\ 0\ 0\ 8\ 8 \\
9\overline{)0.\ 0\ 8\ 0\ 0} \\
-7\ 2 \\
\hline
8\ 0 \\
-7\ 2 \\
\hline
8
\end{array}$$

$0.0784 \div 8.726 \approx 0.009$

14) $0.004643 \approx 0.005$
$0.0835 \approx 0.08$

$$0.08\overline{)0.005} \rightarrow$$

$$\begin{array}{r}
0.\ 0\ 6\ 2\ 5 \\
8\overline{)0.\ 5\ 0\ 0\ 0} \\
-4\ 8 \\
\hline
2\ 0 \\
-1\ 6 \\
\hline
4\ 0
\end{array}$$

$0.004643 \div 0.0835 \approx 0.063$

15)
$$16\overline{)280} \rightarrow \quad 20\overline{)300}$$
$$\begin{array}{r} 15 \\ 20\overline{)300} \end{array}$$

$2.8 \div 0.16 \approx 15$

$$\begin{array}{r}
1\ 7.\ 5 \\
16\overline{)2\ 8\ 0.\ 0} \\
-1\ 6 \\
\hline
1\ 2\ 0 \\
-1\ 1\ 2 \\
\hline
8\ 0
\end{array}$$

$2.8 \div 0.16 = 17.5$

16)
$$52\overline{)863.4} \rightarrow \quad 50\overline{)900}$$
$$\begin{array}{r} 18 \\ 50\overline{)900} \end{array}$$

$8.634 \div 0.52 \approx 18$

$$\begin{array}{r}
1\ 6.\ 6\ 0 \\
52\overline{)8\ 6\ 3.\ 4\ 0} \\
-5\ 2 \\
\hline
3\ 4\ 3 \\
-3\ 1\ 2 \\
\hline
3\ 1\ 4 \\
-3\ 1\ 2 \\
\hline
2\ 0
\end{array}$$

$8.634 \div 0.52 \approx 16.6$

17)
$$77\overline{)3.51} \rightarrow \quad 80\overline{)4.00}$$
$$\begin{array}{r} 0.05 \\ 80\overline{)4.00} \end{array}$$

$3.51 \div 77 \approx 0.05$

$$\begin{array}{r}
0.\ 0\ 4\ 5\ 5 \\
77\overline{)3.\ 5\ 1\ 0\ 0} \\
-3\ 0\ 8 \\
\hline
4\ 3\ 0 \\
-3\ 8\ 5 \\
\hline
4\ 5\ 0 \\
-3\ 8\ 5 \\
\hline
6\ 5
\end{array}$$

$3.51 \div 77 \approx 0.046$

CONCEPT CHECKS AND PRACTICE EXERCISES

A1) a) The dividend is 1.2.

   b) The divisor is 0.06.

A2) If you move the decimal point three places in the divisor, you should move the decimal point three places in the dividend.

A3) The divisor, 0.4, has one decimal place.
Move both decimal points one place to the right.

$$4\overline{)1\ 2\ 5.}$$

A4) The divisor, 0.5, has one decimal place.
Move both decimal points one place to the right.

$$5\overline{)1\ 3\ 2.}$$

A5) The divisor, 2, has zero decimal places.
Do not move the decimal point.

$$2\overline{)7\ 1.\ 4\ 5}$$

A6) The divisor, 3, has zero decimal places.
Do not move the decimal point.

$$3\overline{)1\;0.\;9\;7}$$

A7) To move the decimal point three places to the
right in 3.1, we must write two placeholder
zeros.

$$2\overline{)3\;1\;0\;0.}$$

A8) To move the decimal point three places to the
right in 4.3, we must write two placeholder
zeros.

$$2\overline{)4\;3\;0\;0.}$$

A9) To move the decimal point three places to the
right in 3.56, we must write one placeholder
zero.

$$1332\overline{)3\;5\;6\;0.}$$

A10) To move the decimal point three places to the
right in 4.26, we must write one placeholder zero.

$$2672\overline{)4\;2\;6\;0.}$$

B1) You must find the hundredths place value to
round to the tenths place value.

B2) a) Answers may vary. Example: To annex a
zero means to write a zero at the end of a
number.

b) Zeros can be annexed only to the right of the
decimal point because this does not change
the value of the number. It is similar to
writing equivalent fractions.

B3)

$$0.04\overline{)2.3} \;\rightarrow\; 4\overline{)230.0}$$

$$
\begin{array}{r}
57.5 \\
4\overline{)230.0} \\
-20\;\;\;\;\; \\
\hline
30\;\;\;\; \\
-28\;\;\;\; \\
\hline
2\,0\;\; \\
-2\,0\;\; \\
\hline
0
\end{array}
$$

$$2.3 \div 0.04 = 57.5$$

B4)

$$0.008\overline{)5.3} \;\rightarrow\;
\begin{array}{r}
662.5 \\
8\overline{)5300.0} \\
-48\;\;\;\;\;\; \\
\hline
50\;\;\;\; \\
-48\;\;\;\; \\
\hline
20\;\;\; \\
-16\;\;\; \\
\hline
4\,0\;\; \\
-4\,0\;\; \\
\hline
0
\end{array}
$$

$$5.3 \div 0.008 = 662.5$$

B5)

$$0.003\overline{)0.00045} \;\rightarrow\;
\begin{array}{r}
0.15 \\
3\overline{)0.45} \\
-3\;\;\; \\
\hline
15 \\
-15 \\
\hline
0
\end{array}
$$

$$0.00045 \div 0.003 = 0.15$$

B6)

$$0.06\overline{)0.00073} \;\rightarrow\;
\begin{array}{r}
0.0121 \\
6\overline{)0.0730} \\
-6\;\;\;\;\; \\
\hline
13\;\; \\
-12\;\; \\
\hline
10
\end{array}
$$

$$0.00073 \div 0.06 \approx 0.012$$

B7)

$$0.02\overline{)12} \;\rightarrow\;
\begin{array}{r}
600 \\
2\overline{)1200} \\
-12\;\; \\
\hline
00
\end{array}
$$

$$12 \div 0.02 = 600$$

B8)

$$0.005\overline{)14} \;\rightarrow\;
\begin{array}{r}
2,800 \\
5\overline{)14,000} \\
-10\;\;\;\;\; \\
\hline
4\,0\;\;\; \\
-4\,0\;\;\; \\
\hline
00
\end{array}
$$

$$14 \div 0.005 = 2,800$$

B9)

$$0.9\overline{)17.683} \;\rightarrow\;
\begin{array}{r}
19.6477 \\
9\overline{)176.8300} \\
-9\;\;\;\;\;\;\;\; \\
\hline
86\;\;\;\;\; \\
-81\;\;\;\;\; \\
\hline
5\,8\;\;\; \\
-5\,4\;\;\; \\
\hline
43\;\; \\
-36\;\; \\
\hline
70\; \\
-63\; \\
\hline
70
\end{array}
$$

$$17.683 \div 0.9 \approx 19.648$$

B10)

$$0.08\overline{)42.876} \rightarrow \quad 8\overline{)4287.60} \quad \begin{array}{r} 535.95 \end{array}$$

$$\begin{array}{r} -40 \\ \hline 28 \\ -24 \\ \hline 47 \\ -40 \\ \hline 7\,6 \\ -7\,2 \\ \hline 40 \\ -40 \\ \hline 0 \end{array}$$

$$42.876 \div 0.08 = 535.95$$

B11)

$$0.4\overline{)33.75} \rightarrow \quad 4\overline{)337.500} \quad \begin{array}{r} 84.375 \end{array}$$

$$\begin{array}{r} -32 \\ \hline 17 \\ -16 \\ \hline 15 \\ -12 \\ \hline 30 \\ -28 \\ \hline 20 \\ -20 \\ \hline 0 \end{array}$$

84 DVD cases can fit on the top shelf.

B12)

$$6\overline{)43.20} \rightarrow \quad 6\overline{)43.20} \quad \begin{array}{r} 7.20 \end{array}$$

$$\begin{array}{r} -42 \\ \hline 1\,2 \\ -1\,2 \\ \hline 00 \end{array}$$

Each paid $7.20.

C1) a) $\dfrac{1}{2} = 0.5$

b) $\dfrac{1}{4} = 0.25$

c) $\dfrac{1}{5} = 0.2$

d) $\dfrac{1}{10} = 0.1$

C2) a) $1\dfrac{1}{4} = 1.25$

b) $2\dfrac{1}{4} = 2.25$

c) $3\dfrac{1}{4} = 3.25$

d) $4\dfrac{1}{4} = 4.25$

C3) Answers may vary. Example: The whole number of the mixed number becomes the whole number portion of the decimal.

C4) $\dfrac{2}{5} \rightarrow \quad 5\overline{)2.0} \quad \begin{array}{r} 0.4 \end{array}$

$$\begin{array}{r} -2\,0 \\ \hline 0 \end{array}$$

$$\dfrac{2}{5} = 0.4$$

C5) $\dfrac{7}{10} \rightarrow \quad 10\overline{)7.00} \quad \begin{array}{r} 0.7 \end{array}$

$$\begin{array}{r} -70 \\ \hline 00 \end{array}$$

$$\dfrac{7}{10} = 0.7$$

C6) $\dfrac{5}{8} \rightarrow \quad 8\overline{)5.000} \quad \begin{array}{r} 0.625 \end{array}$

$$\begin{array}{r} -4\,8 \\ \hline 20 \\ -16 \\ \hline 40 \\ -40 \\ \hline 0 \end{array}$$

$$\dfrac{5}{8} = 0.625$$

C7) $\dfrac{1}{8} \rightarrow \quad 8\overline{)1.000} \quad \begin{array}{r} 0.125 \end{array}$

$$\begin{array}{r} -8 \\ \hline 20 \\ -16 \\ \hline 40 \\ -40 \\ \hline 0 \end{array}$$

$$\dfrac{1}{8} = 0.125$$

C8) $\dfrac{1}{3} \rightarrow \quad 3\overline{)1.000} \quad \begin{array}{r} 0.333 \end{array}$

$$\begin{array}{r} -9 \\ \hline 10 \\ -9 \\ \hline 10 \end{array}$$

$$\dfrac{1}{3} = 0.\overline{3}$$

C9) $\dfrac{5}{6} \rightarrow$ $6)\overline{5.000}$
$\dfrac{-4\,8}{20}$
$\dfrac{-18}{20}$

$\dfrac{5}{6} = 0.8\overline{3}$

C10) $7\dfrac{3}{5} \rightarrow$ $5)\overline{3.0}$
$\dfrac{-30}{0}$

$7\dfrac{3}{5} = 7.6$

C11) $8\dfrac{3}{10} \rightarrow$ $10)\overline{3.0}$
$\dfrac{-30}{0}$

$8\dfrac{3}{10} = 8.3$

C12) $\dfrac{3}{8} \rightarrow$ $8)\overline{3.000}$
$\dfrac{-24}{60}$
$\dfrac{-56}{40}$
$\dfrac{-40}{0}$

$\dfrac{3}{8} = 0.375 \approx 0.38$

C13) $\dfrac{2}{3} \rightarrow$ $3)\overline{2.000}$
$\dfrac{-18}{20}$
$\dfrac{-18}{20}$

$\dfrac{2}{3} = 0.\overline{6} \approx 0.67$

C14) $1\dfrac{5}{12} \rightarrow$ $12)\overline{5.000}$
$\dfrac{-48}{20}$
$\dfrac{-12}{80}$

$1\dfrac{5}{12} \approx 1.42$

C15) $3\dfrac{7}{8} \rightarrow$ $8)\overline{7.000}$
$\dfrac{-6\,4}{60}$
$\dfrac{-56}{40}$
$\dfrac{-40}{0}$

$3\dfrac{7}{8} = 3.875 \approx 3.88$

D1) Answers may vary.

D2) Answers may vary.

D3) Answers may vary. Example: The rest of the answers have the decimal point in an incorrect position.

D4) $87.56 \approx 90$
     $6.32 \approx 6$
     $6)\overline{90}$ (quotient 15)
     $\dfrac{-6}{30}$
     $\dfrac{-30}{0}$
     $87.56 \div 6.32 \approx 15$

D5) $45.765 \approx 50$
     $2.35 \approx 2$
     $2)\overline{50}$ (quotient 25)
     $\dfrac{-4}{10}$
     $\dfrac{-10}{0}$
     $45.765 \div 2.35 \approx 25$

D6) $0.00657 \approx 0.007$
     $23.87 \approx 20$
     $20)\overline{0.00700}$ (quotient 0.00035)
     $\dfrac{-60}{100}$
     $\dfrac{-100}{0}$
     $0.00657 \div 23.87 \approx 0.00035$

D7) $0.0252 \approx 0.03$

$11.96 \approx 10$

$$\begin{array}{r} 0.003 \\ 10\overline{)0.030} \\ -30 \\ \hline 0 \end{array}$$

$0.0252 \div 11.96 \approx 0.003$

D8) $0.04845 \approx 0.05$

$0.0185 \approx 0.02$

$$0.02\overline{)0.05} \rightarrow \begin{array}{r} 2.5 \\ 2\overline{)5.0} \\ -4 \\ \hline 10 \\ -10 \\ \hline 0 \end{array}$$

$0.04845 \div 0.0185 \approx 2.5$

D9) $0.00857 \approx 0.009$

$0.0642 \approx 0.06$

$$0.06\overline{)0.009} \rightarrow \begin{array}{r} 0.15 \\ 6\overline{)0.90} \\ -6 \\ \hline 30 \\ -30 \\ \hline 0 \end{array}$$

$0.00857 \div 0.0642 \approx 0.15$

D10) $739.28 \approx 700$

$43.6 \approx 40$

$$\begin{array}{r} 17.5 \\ 40\overline{)700.0} \\ -40 \\ \hline 300 \\ -280 \\ \hline 20\ 0 \\ -20\ 0 \\ \hline 0 \end{array}$$

$739.28 \div 43.6 \approx 17.5$

D11) $867.623 \approx 900$

$23.7 \approx 20$

$$\begin{array}{r} 45 \\ 20\overline{)900} \\ -80 \\ \hline 100 \\ -100 \\ \hline 0 \end{array}$$

$867.623 \div 23.7 \approx 45$

D12) $221.7 \approx 200$

$72.3 \approx 70$

$$\begin{array}{r} 2.8 \\ 70\overline{)200.0} \\ -140 \\ \hline 60\ 0 \\ -56\ 0 \\ \hline 4\ 0 \end{array}$$

It will take you approximately 3 hours to complete your trip.

D13) $226.32 \approx 200$

$11.67 \approx 10$

$$\begin{array}{r} 20 \\ 10\overline{)200} \\ -20 \\ \hline 00 \end{array}$$

$226.32 \div 11.67 \approx 20$

You must work approximately 20 hours to make the car payment.

E1) To round an answer to the hundredths place, continue dividing to the thousandths place.

E2) $$3.8\overline{)42.0} \rightarrow 38\overline{)420} \rightarrow \begin{array}{r} 10 \\ 40\overline{)400} \end{array}$$

$42.0 \div 3.8 \approx 10$

E3) $$0.77\overline{)3.70} \rightarrow 77\overline{)370} \rightarrow \begin{array}{r} 5 \\ 80\overline{)400} \end{array}$$

$3.70 \div 0.77 \approx 5$

E4) $$57\overline{)6.3} \rightarrow \begin{array}{r} 0.1 \\ 60\overline{)6.0} \end{array}$$

$6.3 \div 57 \approx 0.1$

E5) $$28\overline{)2.5} \rightarrow \begin{array}{r} 0.1 \\ 30\overline{)3.0} \end{array}$$

$2.5 \div 28 \approx 0.1$

E6) $$1.4\overline{)0.045} \rightarrow \begin{array}{r} 0.032 \\ 14\overline{)0.450} \\ -42 \\ \hline 30 \\ -28 \\ \hline 2 \end{array}$$

$0.045 \div 1.4 \approx 0.03$

E7)
$$\begin{array}{r} 0.050 \\ 0.63\overline{)0.032} \rightarrow \quad 63\overline{)3.200} \\ -3\,15 \\ \hline 50 \\ -0 \\ \hline 50 \end{array}$$

$0.032 \div 0.63 \approx 0.05$

E8)
$$\begin{array}{r} 16.088 \\ 0.34\overline{)5.47} \rightarrow \quad 34\overline{)547.000} \\ -34 \\ \hline 207 \\ -204 \\ \hline 3\,0 \\ -0 \\ \hline 3\,00 \\ -2\,72 \\ \hline 280 \\ -272 \\ \hline 8 \end{array}$$

$5.47 \div 0.34 \approx 16.09$

E9)
$$\begin{array}{r} 13.680 \\ 0.72\overline{)9.85} \rightarrow \quad 72\overline{)985.000} \\ -72 \\ \hline 265 \\ -216 \\ \hline 49\,0 \\ -432 \\ \hline 580 \\ -576 \\ \hline 40 \end{array}$$

$9.85 \div 0.72 \approx 13.68$

E10)
$$\begin{array}{r} 130 \\ 0.055\overline{)7.15} \rightarrow \quad 55\overline{)7150} \\ -55 \\ \hline 165 \\ -165 \\ \hline 0 \end{array}$$

$7.15 \div 0.055 = 130$

E11)
$$\begin{array}{r} 11 \\ 0.63\overline{)6.93} \rightarrow \quad 63\overline{)693} \\ -63 \\ \hline 63 \\ -63 \\ \hline 0 \end{array}$$

$6.93 \div 0.63 = 11$

E12)
$$\begin{array}{r} 11.56 \\ 0.042\overline{)0.4858} \rightarrow \quad 42\overline{)485.80} \\ -42 \\ \hline 65 \\ -42 \\ \hline 23\,8 \\ -21\,0 \\ \hline 2\,80 \end{array}$$

$0.4858 \div 0.042 \approx 11.6$

E13)
$$\begin{array}{r} 15.61 \\ 0.45\overline{)7.026} \rightarrow \quad 45\overline{)702.60} \\ -45 \\ \hline 252 \\ -225 \\ \hline 27\,6 \\ -27\,0 \\ \hline 60 \end{array}$$

$7.026 \div 0.45 \approx 15.6$

E14)
$$\begin{array}{r} 6.74 \\ 64\overline{)431.70} \\ -384 \\ \hline 47\,7 \\ -44\,8 \\ \hline 2\,90 \\ -2\,56 \\ \hline 34 \end{array}$$

It will take about 6.7 hours to get to the Grand Canyon.

E15)
$$\begin{array}{r} 46.8 \\ 320\overline{)15,000.0} \\ -12\,80 \\ \hline 2\,200 \\ -1\,920 \\ \hline 280\,0 \\ -256\,0 \\ \hline 24\,0 \end{array}$$

It will take about 47 months to save $15,000.

SECTION 3.4   EXERCISES

For 1–15, refer to Concept Checks and Practice Exercises.

17) divisor

19) $6 \div 3 = 2$
   6 is the dividend.
   3 is the divisor.
   2 is the quotient.

$$5\overline{)35}^{\;7}$$

   5 is the divisor.
   35 is the dividend.
   7 is the quotient.

   $\dfrac{16}{8} = 2$

   16 is the dividend.
   8 is the divisor.
   2 is the quotient.

21)   The divisor, 2.1, has one decimal place.
   Move both decimal points one place to the right.

   $21\overline{)\;3\;4.\;5\;}$

23)   The divisor, 0.012, has three decimal places.
   Move both decimal points three places to the
   right.

   $12\overline{)\;3.\;4\;}$

25) $$6\overline{)2.34}^{\;0.39}$$
   $\quad -1\,8$
   $\quad \overline{\quad 54}$
   $\quad -54$
   $\quad \overline{\quad\;\; 0}$
   $2.34 \div 6 = 0.39$

27) $0.2\overline{)2.19} \rightarrow$ $2\overline{)21.90}^{\;10.95}$
   $\qquad\qquad\qquad\quad -2$
   $\qquad\qquad\qquad\quad \overline{\quad 1}$
   $\qquad\qquad\qquad\quad -0$
   $\qquad\qquad\qquad\quad \overline{\quad 19}$
   $\qquad\qquad\qquad\quad -18$
   $\qquad\qquad\qquad\quad \overline{\quad 10}$
   $\qquad\qquad\qquad\quad -10$
   $\qquad\qquad\qquad\quad \overline{\quad\; 0}$
   $2.19 \div 0.2 = 10.95$

29) $0.03\overline{)0.00763} \rightarrow$ $3\overline{)0.7630}^{\;0.2543}$
   $\qquad\qquad\qquad\qquad -6$
   $\qquad\qquad\qquad\qquad \overline{\quad 16}$
   $\qquad\qquad\qquad\qquad -15$
   $\qquad\qquad\qquad\qquad \overline{\quad 13}$
   $\qquad\qquad\qquad\qquad -12$
   $\qquad\qquad\qquad\qquad \overline{\quad 10}$
   $0.00763 \div 0.03 \approx 0.254$

31) $0.06\overline{)5} \rightarrow$ $6\overline{)500.0000}^{\;83.3333}$
   $\qquad\qquad\qquad -48$
   $\qquad\qquad\qquad \overline{\quad 20}$
   $\qquad\qquad\qquad -18$
   $\qquad\qquad\qquad \overline{\quad 2\,0}$
   $\qquad\qquad\qquad -1\,8$
   $\qquad\qquad\qquad \overline{\quad 20\text{ etc.}}$
   $5 \div 0.06 \approx 83.333$

33) $0.004\overline{)20.005} \rightarrow$ $4\overline{)20,005.00}^{\;5,001.25}$
   $\qquad\qquad\qquad\qquad -20$
   $\qquad\qquad\qquad\qquad \overline{\quad 005}$
   $\qquad\qquad\qquad\qquad -4$
   $\qquad\qquad\qquad\qquad \overline{\quad 1\,0}$
   $\qquad\qquad\qquad\qquad -8$
   $\qquad\qquad\qquad\qquad \overline{\quad 20}$
   $\qquad\qquad\qquad\qquad -20$
   $\qquad\qquad\qquad\qquad \overline{\quad\; 0}$
   $20.005 \div 0.004 = 5,001.25$

35) $0.06\overline{)13.2} \rightarrow$ $6\overline{)1320}^{\;220}$
   $\qquad\qquad\qquad -12$
   $\qquad\qquad\qquad \overline{\quad 12}$
   $\qquad\qquad\qquad -12$
   $\qquad\qquad\qquad \overline{\quad\; 0}$
   $13.2 \div 0.06 = 220$

37) $$5\overline{)112.230}^{\;22.446}$$
   $\quad -10$
   $\quad \overline{\quad 12}$
   $\quad -10$
   $\quad \overline{\quad 2\,2}$
   $\quad -2\,0$
   $\quad \overline{\quad 23}$
   $\quad -20$
   $\quad \overline{\quad 30}$
   Each person will need to pay \$22.45.

39) $\dfrac{6}{5} = 1\dfrac{1}{5} = 1.2$

41) $\frac{5}{9} = 0.\overline{5} \approx 0.56$

43) $4\frac{1}{4} = 4.25$

45) $\frac{7}{16} = 0.4375 \approx 0.44$

47) $1\frac{4}{25} = 1.16$

49) $3\frac{2}{7} \approx 3.29$

51)
$$1.8\overline{)4.16} \to \begin{array}{r} 2.311 \\ 18\overline{)41.600} \\ -36 \\ \hline 5\,6 \\ -5\,4 \\ \hline 20 \\ -18 \\ \hline 20 \end{array}$$

$4.16 \div 1.8 \approx 2.31$

53)
$$1.7\overline{)84} \to \begin{array}{r} 49.411 \\ 17\overline{)840.000} \\ -68 \\ \hline 160 \\ -153 \\ \hline 7\,0 \\ -6\,8 \\ \hline 20 \\ -17 \\ \hline 30 \end{array}$$

$84 \div 1.7 \approx 49.41$

55)
$$0.8\overline{)32.973} \to \begin{array}{r} 41.216 \\ 8\overline{)329.730} \\ -32 \\ \hline 9 \\ -8 \\ \hline 1\,7 \\ -1\,6 \\ \hline 13 \\ -8 \\ \hline 50 \end{array}$$

$32.973 \div 0.8 \approx 41.22$

57)
$$0.027\overline{)0.0034} \to \begin{array}{r} 0.125 \\ 27\overline{)3.400} \\ -27 \\ \hline 70 \\ -54 \\ \hline 160 \\ -135 \\ \hline 25 \end{array}$$

$0.0034 \div 0.027 \approx 0.13$

59)
$$0.0024\overline{)0.2168} \to \begin{array}{r} 90.333 \\ 24\overline{)2168.000} \\ -216 \\ \hline 8 \\ -0 \\ \hline 8\,0 \\ -7\,2 \\ \hline 80 \\ -72 \\ \hline 80 \end{array}$$

$0.2168 \div 0.0024 \approx 90.33$

61)
$$8.3\overline{)25.87} \to \begin{array}{r} 3.116 \\ 83\overline{)258.700} \\ -249 \\ \hline 9\,7 \\ -8\,3 \\ \hline 1\,40 \\ -83 \\ \hline 570 \end{array}$$

$25.87 \div 8.3 \approx 3.12$

63)
$$12\overline{)3855.00} \begin{array}{r} 321.25 \\ \\ -36 \\ \hline 25 \\ -24 \\ \hline 15 \\ -12 \\ \hline 3\,0 \\ -2\,4 \\ \hline 60 \\ -60 \\ \hline 0 \end{array}$$

Each payment will be $321.25.

65)
$$3.1\overline{)39.6} \to \begin{array}{r} 12.77 \\ 31\overline{)396.00} \\ -31 \\ \hline 86 \\ -62 \\ \hline 24\,0 \\ -217 \\ \hline 230 \end{array}$$

$39.6 \div 3.1 \approx 12.8$

67)
$$0.63\overline{)12.8} \rightarrow 63\overline{)1280.00}$$
$$\begin{array}{r} 20.31 \\ \underline{-126} \\ 20 \\ \underline{-0} \\ 20\,0 \\ \underline{-18\,9} \\ 1\,10 \end{array}$$

$12.8 \div 0.63 \approx 20.3$

69)
$$0.24\overline{)315} \rightarrow 24\overline{)31,500.0}$$
$$\begin{array}{r} 1,312.5 \\ \underline{-24} \\ 7\,5 \\ \underline{-7\,2} \\ 30 \\ \underline{-24} \\ 60 \\ \underline{-48} \\ 120 \\ \underline{-120} \\ 0 \end{array}$$

$315 \div 0.24 = 1,312.5$

71)
$$0.89\overline{)756} \rightarrow 89\overline{)75,600.00}$$
$$\begin{array}{r} 849.43 \\ \underline{-71\,2} \\ 4\,40 \\ \underline{-3\,56} \\ 840 \\ \underline{-801} \\ 39\,0 \\ \underline{-35\,6} \\ 3\,40 \end{array}$$

$756 \div 0.89 \approx 849.4$

73)
$$0.019\overline{)33.68} \rightarrow 19\overline{)33,680.000}$$
$$\begin{array}{r} 1,772.631 \\ \underline{-19} \\ 14\,6 \\ \underline{-133} \\ 138 \\ \underline{-133} \\ 50 \\ \underline{-38} \\ 120 \\ \underline{-114} \\ 60 \\ \underline{-57} \\ 30 \end{array}$$

$33.68 \div 0.019 \approx 1,772.63$

75)
$$0.089\overline{)0.0162} \rightarrow 89\overline{)16.200}$$
$$\begin{array}{r} 0.182 \\ \underline{-89} \\ 730 \\ \underline{-712} \\ 180 \end{array}$$

$0.0162 \div 0.089 \approx 0.18$

77)
$$1.3\overline{)45.5} \rightarrow 13\overline{)455}$$
$$\begin{array}{r} 35 \\ \underline{-39} \\ 65 \\ \underline{-65} \\ 0 \end{array}$$

The quotient should be 36. The package is missing one iPod.

79) $8 \times 12 = 96$
8 feet is equivalent to 96 inches.

$$1.4\overline{)96} \rightarrow 14\overline{)960}$$
$$\begin{array}{r} 68 \\ \underline{-84} \\ 120 \\ \underline{-112} \\ 8 \end{array}$$

The greatest number of complete parts that can be made is 68.

CHAPTER 3   REVIEW EXERCISES

1) $1.2 =$ one and two tenths

$1\frac{2}{10} =$ one and two tenths

The numbers are equal; both represent the same quantity.

2) The missing place values are tenths, hundredths and thousandths.

3) Finding the best approximation for a number to a certain place value is called rounding the number.

4) When writing a decimal number in words, the word *and* separates the whole number and decimal portions.

5) The digit 3 is in the fourth place to the left of the decimal.
Ones, tens, hundreds, thousands
3 is in the thousands place.

6) The digit 3 is in the third place to the right of the decimal.
Tenths, hundredths, thousandths
3 is in the thousandths place.

7) The digit 3 is in the fifth place to the right of the decimal.
Tenths, hundredths, thousandths, ten thousandths, hundred thousandths
3 is in the hundred thousandths place.

8) The digit 3 is in the fourth place to the right of the decimal.
Tenths, hundredths, thousandths, ten thousandths
3 is in the ten thousandths place.

9) $76.34 =$ seventy-six and thirty-four hundredths

10) $2.057 =$ two and fifty-seven thousandths

11) $12.6003 =$ twelve and six thousand three ten thousandths

12) $5.07002 =$ five and seven thousand two hundred thousandths

13) $0.035 = \dfrac{35}{1,000}$

14) $0.45 = \dfrac{45}{100}$

15) $\dfrac{14}{1,000} = 0.014$

16) $\dfrac{8}{10,000} = 0.0008$

17) The digit 8 is in the tenths place.
Round up to 67.9 or down to 67.8.
Since the digit to the right is 4, round down.
$67.841 \approx 67.8$

18) The digit 7 is in the hundredths place.
Round up to 578.88 or down to 578.87.
Since the digit to the right is 6, round up.
$578.8762 \approx 578.88$

19) The digit 7 is in the thousandths place.
Round up to 3.568 or down to 3.567.
Since the digit to the right is 2, round down.
$3.56721 \approx 3.567$

20) The digit 3 is in the hundreds place.
Round up to 1,400 or down to 1,300.
Since the digit to the right is 5, round up.
$1,356.57 \approx 1,400$

21) $0.0435 = \dfrac{435}{10,000}$

$0.14 = \dfrac{14}{100} \cdot \dfrac{100}{100} = \dfrac{1,400}{10,000}$

$0.0436 = \dfrac{436}{10,000}$

$435 < 436 < 1400,$
so $0.0435 < 0.0436 < 0.14$

22) $0.567 = \dfrac{567}{1,000} \cdot \dfrac{10}{10} = \dfrac{5,670}{10,000}$

$0.0987 = \dfrac{987}{10,000}$

$7.003 = 7\dfrac{3}{1,000} \cdot \dfrac{10}{10} = 7\dfrac{30}{10,000}$

$987 < 5,670$ and $0 < 7,$
so $0.0987 < 0.567 < 7.003$

23) When subtracting decimal numbers, the decimal points should be aligned to make sure like place values get subtracted.

24) Answers may vary. Example: It's the same as writing equivalent fractions.

25) $\begin{array}{r} 12.56 \\ -\ 9.45 \\ \hline 3.11 \end{array}$

26) $\begin{array}{r} 768.9 \\ -\ 34.5 \\ \hline 734.4 \end{array}$

27) $\begin{array}{r} \overset{1\ \ 1}{26.890} \\ +\ 1\,2.57\,1 \\ \hline 39.46\,1 \end{array}$

28) $\begin{array}{r} \overset{1}{123.40} \\ +\ 5.89 \\ \hline 129.29 \end{array}$

29) $\begin{array}{r} \overset{1}{19.72} \\ +\ 9.00 \\ \hline 28.72 \end{array}$

30) $\overset{1}{3}45.9$
       $+\ 15.0$
       _____
       360.9

31) $2\ \overset{3}{\cancel{4}}.\ \overset{9}{\cancel{0}}\ ^{1}0$
       $-1\ 1.\ 3\ 1$
       _____
       $1\ 2.\ 6\ 9$

32) $\overset{4}{\cancel{5}}.\ \overset{9}{\cancel{0}}\ ^{1}0$
       $-\ 2.\ 8\ 9$
       _____
       $2.\ 1\ 1$

33) a) $\overset{0}{\cancel{1}}.^{1}3$
          $-\ 0.\ 5$
          _____
          $0.\ 8$

    The distance from McDonald's to Wendy's is 0.8 miles.

    b) $\overset{1}{0}.6$
          $+\ 0.5$
          _____
          $1.\ 1$

    The distance from the Holiday Inn to McDonald's is 1.1 miles.

34) a) $\overset{1\ 1\ 1}{3}45.65$
          $+\ 57.80$
          _____
          403.45

    He spent $403.45.

    b) $1\ 4\ 5\ \overset{5}{\cancel{6}}.^{1}2\ 5$
          $-4\ 0\ 3.4\ 5$
          _____
          $1\ 0\ 5\ 2.8\ 0$

    The new balance in his account is $1,052.80.

35) When multiplying decimal numbers, the sum of the number of decimal places in the factors must match the number of decimal places in the **product**.

36) Moving the decimal point three places to the right in 4.57 is equivalent to multiplying 4.57 by **1,000**.

37) If you know the radius of a circle, you can find the diameter of the circle by multiplying the radius by two.

38)   3 decimal places →   0.006
       2 decimal places →   × 0.03
                                 _____
       3 + 2 = 5 decimal places → 0.00018

39)   2 decimal places →   0.09
       4 decimal places →   × 0.0002
                                 _____
       2 + 4 = 6 decimal places → 0.000018

40)   0 decimal places →   8
       2 decimal places →   × 0.04
                                 _____
       0 + 2 = 2 decimal places →   0.32

41)   0 decimal places →   7
       3 decimal places →   × 0.005
                                 _____
       0 + 3 = 3 decimal places → 0.035

42) Multiplying by 0.01 will make the product smaller. Move the decimal point two places to the left.
    $623 \times 0.01 = 6.23$

43) Multiplying by 0.0001 will make the product smaller. Move the decimal point four places to the left.
    $70 \times 0.0001 = 0.007$

44) Multiplying by 1000 will make the product larger. Move the decimal point three places to the right.
    $1,000 \times 0.041 = 41$

45) Multiplying by 10,000 will make the product larger. Move the decimal point four places to the right.
    $10,000 \times 0.0087 = 87$

46)   3 decimal places →   0.008
       0 decimal places →   ×   60
                                 _____
       3 + 0 = 3 decimal places → 0.480

    The height of 60 pieces of construction paper is 0.48 inch.

47)   2 decimal places →   0.15
       0 decimal places →   ×   35
                                 _____
                                     75
                                    450
                                 _____
       2 + 0 = 2 decimal places →   5.25

    It will cost the teacher $5.25.

48) a) $0.568 \times 0.0324 \approx 0.6 \times 0.03$
$$= 0.018$$

   b) 3 decimal places $\rightarrow$ 0.568
   4 decimal places $\rightarrow$ $\times\,0.0324$

                                         2272
                                    11360
                               170400

   $3 + 4 = 7$ decimal places $\rightarrow$ 0.0184032

49) a) $0.684 \times 0.00026 \approx 0.7 \times 0.0003$
$$= 0.00021$$

   b) 3 decimal places $\rightarrow$ 0.684
   5 decimal places $\rightarrow$ $\times\,0.00026$

                                         4104
                                    13680

   $3 + 5 = 8$ decimal places $\rightarrow$ 0.00017784

50) a) $235 \times 0.0386 \approx 200 \times 0.04$
$$= 8$$

   b) 0 decimal places $\rightarrow$ 235
   4 decimal places $\rightarrow$ $\times\,0.0386$

                                         1410
                                    18800
                                    70500

   $0 + 4 = 4$ decimal places $\rightarrow$ 9.0710

51) a) $478 \times 0.06739 \approx 500 \times 0.07$
$$= 35$$

   b) 0 decimal places $\rightarrow$ 478
   5 decimal places $\rightarrow$ $\times\,0.06739$

                                         4302
                                    14340
                                334600
                              2868000

   $0 + 5 = 5$ decimal places $\rightarrow$ 32.21242

52) $C = 2\pi r$
$$= 2(3.14)(8 \text{ meters})$$
$$= 6.28(8 \text{ meters})$$
$$= 50.24 \text{ meters}$$

53) $r = 6 \div 2 = 3 \text{ millimeters}$

   $C = 2\pi r$
$$= 2(3.14)(3 \text{ millimeters})$$
$$= 6.28(3 \text{ millimeters})$$
$$= 18.84 \text{ millimeters}$$

54) $A = \pi r^2$
$$= (3.14)(3 \text{ feet})^2$$
$$= 3.14(9 \text{ square feet})$$
$$= 28.26 \text{ square feet}$$

55) $r = 10 \div 2 = 5 \text{ inches}$

   $A = \pi r^2$
$$= (3.14)(5 \text{ inches})^2$$
$$= 3.14(25 \text{ square inches})$$
$$= 78.5 \text{ square inches}$$

56) When setting up the division exercise $1.2 \div 0.15$ you should move the decimal point in the divisor and dividend **2** places to the **right**.

57) To round a quotient to the thousandths place, the digit in the **ten thousandths** place must be known.

58)
$$0.03\overline{)6} \rightarrow \quad 3\overline{)600} \quad \begin{array}{r} 200 \\ \hline \end{array}$$
$$\begin{array}{r} -6 \\ \hline 00 \end{array}$$

   $6 \div 0.03 = 200$

59)
$$0.0002\overline{)8} \rightarrow \quad 2\overline{)80,000} \quad \begin{array}{r} 40,000 \\ \hline \end{array}$$
$$\begin{array}{r} -8 \\ \hline 0\ 000 \end{array}$$

   $8 \div 0.0002 = 40,000$

60)
$$0.04\overline{)0.016} \rightarrow \quad 4\overline{)1.6} \quad \begin{array}{r} 0.4 \\ \hline \end{array}$$
$$\begin{array}{r} -1\ 6 \\ \hline 0 \end{array}$$

   $0.016 \div 0.04 = 0.4$

61)
$$0.005\overline{)0.07} \rightarrow \quad 5\overline{)70} \quad \begin{array}{r} 14 \\ \hline \end{array}$$
$$\begin{array}{r} -5 \\ \hline 20 \\ -20 \\ \hline 0 \end{array}$$

   $0.07 \div 0.005 = 14$

62)
$$0.014\overline{)0.623} \rightarrow$$

```
        44.5
   14)623.0
      −56
       63
      −56
        7 0
       −7 0
          0
```

$$0.623 \div 0.014 = 44.5$$

63)
$$0.012\overline{)0.576} \rightarrow$$

```
       48
   12)576
     −48
      96
     −96
       0
```

$$0.5765 \div 0.012 = 48$$

64)
$$0.71\overline{)18.91} \rightarrow$$

```
         26.633
   71)1891.000
     −142
      471
     −426
       450
      −426
        240
       −213
        270
```

$$18.91 \div 0.71 \approx 26.63$$

65)
$$0.58\overline{)25.87} \rightarrow$$

```
          44.603
   58)2587.000
     −232
      267
     −232
       350
      −348
        20
        −0
       200
```

$$25.87 \div 0.58 \approx 44.60$$

66)
$$70.8\overline{)235.3} \rightarrow$$

```
          3.32
   708)2353.00
      −2124
       229 0
      −212 4
        16 60
```

It will take you about 3.3 hours to get to Venice Beach.

67)
$$3\overline{)56.870}$$

```
        18.956
   3)56.870
     −3
      26
     −24
      2 8
     −2 7
       17
      −15
       20
```

Each person will need to pay $18.96.

68) $\dfrac{3}{5} = 0.6$

69) $\dfrac{5}{8} = 0.625$

70) $\dfrac{7}{6} = 1\dfrac{1}{6} = 1.1\overline{6}$

71) $\dfrac{11}{6} = 1\dfrac{5}{6} = 1.8\overline{3}$

72) $5\dfrac{1}{3} = 5.\overline{3}$

73) $7\dfrac{2}{3} = 7.\overline{6}$

CHAPTER 3   TEST

1) The missing place values are hundredths, thousandths and ten thousandths.

2) Answers may vary. Example: Count the total number of decimal places in the factors.

3) When setting up the division exercise $4.5 \div 0.678$ you should move the decimal point in the divisor and dividend **3** places to the **right**.

4) To round a quotient to the tenths place, you must know the digit in the **hundredths** place of the quotient.

5) $216.67 =$ Two hundred sixteen and 67/100

6) $0.16 = \dfrac{16}{100}$

7) $\dfrac{9}{1000} = 0.009$

8) The digit 2 is in the tenths place.
Round up to 145.3 or down to 145.2.
Since the digit to the right is 6, round up.
$145.261 \approx 145.3$

9) The digit 7 is in the hundredths place.
Round up to 5.88 or down to 5.87.
Since the digit to the right is 2, round down.
$5.8729 \approx 5.87$

10) The digit 7 is in the thousandths place.
Round up to 10.548 or down to 10.547.
Since the digit to the right is 8, round up.
$10.54781 \approx 10.548$

11) The digit 1 is in the thousands place.
Round up to 2,000 or down to 1,000.
Since the digit to the right is 3, round down.
$1,375.9872 \approx 1,000$

12) $0.0655 = \dfrac{655}{10,000}$

$0.101 = \dfrac{101}{1,000} \cdot \dfrac{10}{10} = \dfrac{1,010}{10,000}$

$0.0499 = \dfrac{499}{10,000}$

$499 < 655 < 1,010$,
so $0.0499 < 0.0655 < 0.101$

13)
$$\begin{array}{r} \overset{1\;1}{33.56} \\ + 10.45 \\ \hline 44.01 \end{array}$$

14)
$$\begin{array}{r} 7\,\overset{5}{\cancel{6}}.{}^1 0 \\ - 5\,4 . 6 \\ \hline 2\,1 . 4 \end{array}$$

15)
$$\begin{array}{r} \overset{1\;1}{26}89.000 \\ + \;\;1\,9.371 \\ \hline 2708.371 \end{array}$$

16)
$$\begin{array}{r} 3\,4\,\overset{1}{\cancel{2}}.\,\overset{9}{\cancel{0}}\,\overset{9}{\cancel{0}}\,{}^1 0 \\ - \quad 0 . 0\;8\;9 \\ \hline 3\,4\,1 . 9\;1\;1 \end{array}$$

17) a)
$$\begin{array}{r} 1.5 \\ - 0.3 \\ \hline 1.2 \end{array}$$
The distance from Cracker Barrel to Taco Bell is 1.2 miles.

b)
$$\begin{array}{r} 3.4 \\ + 1.5 \\ \hline 4.9 \end{array}$$
The distance from American Inn to Taco Bell is 4.9 miles.

18) a)
$$\begin{array}{r} \overset{7}{\cancel{8}}\,\overset{{}^1 4}{\cancel{5}}\,\overset{9}{\cancel{0}}.\,\overset{9}{\cancel{0}}\,{}^1 0 \\ - 4\,5\;\;2 . 3\;3 \\ \hline 3\,9\,7 . 6\;7 \end{array}$$
$397.67 was left to apply towards the principal.

b)
$$\begin{array}{r} 8\,7,\,\overset{4}{\cancel{5}}\,\overset{{}^1 5}{\cancel{6}}\,\overset{9}{\cancel{0}}.\,\overset{9}{\cancel{0}}\,{}^1 0 \\ - \qquad 3\;9\,7 . 6\;7 \\ \hline 8\,7,\,1\,6\,2 . 3\;3 \end{array}$$
The new principal was $87,162.33.

19) a) $1.83 \times 0.55 \approx 2 \times 0.6$
$= 1.2$

b)
2 decimal places →	1.83
2 decimal places →	× 0.55

$$\begin{array}{r} 915 \\ 9150 \\ \hline \end{array}$$
$2 + 2 = 4$ decimal places → $\quad 1.0065$

20) a) $7.35 \times 0.089 \approx 7 \times 0.09$
$= 0.63$

b)
2 decimal places →	7.35
3 decimal places →	× 0.089

$$\begin{array}{r} 6615 \\ 58800 \\ \hline \end{array}$$
$2 + 3 = 5$ decimal places → $\quad 0.65415$

21) a) $25.56 \times 0.0578 \approx 30 \times 0.06$
$= 1.8$

b)
2 decimal places →	25.56
4 decimal places →	× 0.0578

$$\begin{array}{r} 20448 \\ 178920 \\ 1278000 \\ \hline \end{array}$$
$2 + 4 = 6$ decimal places → $\quad 1.477368$

22) 0 decimal places →   568
    3 decimal places → $\times\,0.505$

$$
\begin{array}{r}
568 \\
\times\,0.505 \\
\hline
2840 \\
0000 \\
284000 \\
\hline
286.840
\end{array}
$$

0 + 3 = 3 decimal places →   286.840

The person could deduct \$286.84.

23) $C = 2\pi r$

$\quad = 2(3.14)(7\text{ feet})$

$\quad = 6.28(7\text{ feet})$

$\quad = 43.96\text{ feet}$

24) $r = 16 \div 2 = 8\text{ inches}$

$A = \pi r^2$

$\quad = (3.14)(8\text{ inches})^2$

$\quad = 3.14(64\text{ square inches})$

$\quad = 200.96\text{ square inches}$

25)

$$
0.041\overline{)0.075} \rightarrow 41\overline{)75.0000}
$$

$$
\begin{array}{r}
1.8292 \\
\hline
75.0000 \\
-41 \\
\hline
340 \\
-328 \\
\hline
120 \\
-82 \\
\hline
380 \\
-369 \\
\hline
110
\end{array}
$$

$0.075 \div 0.041 \approx 1.829$

26)

$$
0.152\overline{)0.065} \rightarrow 152\overline{)65.0000}
$$

$$
\begin{array}{r}
0.4276 \\
\hline
65.0000 \\
-608 \\
\hline
420 \\
-304 \\
\hline
1160 \\
-1064 \\
\hline
960
\end{array}
$$

$0.065 \div 0.152 \approx 0.428$

27)

$$
75\overline{)1356.87}
$$

$$
\begin{array}{r}
18.09 \\
\hline
1356.87 \\
-75 \\
\hline
606 \\
-600 \\
\hline
6\,8 \\
-0 \\
\hline
6\,87
\end{array}
$$

Since the quotient is greater than 18, it will take you 19 weeks to pay for the laptop.

28)

$$
82\overline{)2320.00}
$$

$$
\begin{array}{r}
28.29 \\
\hline
2320.00 \\
-164 \\
\hline
680 \\
-656 \\
\hline
240 \\
-164 \\
\hline
760
\end{array}
$$

Kobe averaged about 28.3 points per game.

29) $\dfrac{3}{8} = 0.375$

30) $4\dfrac{2}{5} = 4.4$

31) $\dfrac{5}{6} = 0.8\overline{3}$

32) $\dfrac{9}{4} = 2\dfrac{1}{4} = 2.25$

Chapter 4   Ratios, Rates, and Proportions

4.1   Ratios and Rates

GUIDED PRACTICE

1) Use the word *to* to write the ratio as 5 to 7.

   Use a fraction to write the ratio as $\frac{5}{7}$.

   Use a colon to write the ratio as 5:7.

2) $10¢$ to $1¢ = \dfrac{10¢}{1¢} = \dfrac{10}{1}$

   Since the coins have the same units, cents, this fraction is a ratio. The common units, cents, divide out, giving the ratio $\dfrac{10}{1}$.

3) The ratio of 3 to 7 is $\dfrac{3}{7}$.

4) The ratio of $6:13$ is $\dfrac{6}{13}$.

5) Julian's hot dogs to Mikito's hot dogs

   $= \dfrac{\text{Julian's hot dogs}}{\text{Mikito's hot dogs}}$

   $= \dfrac{40 \text{ hot dogs}}{13 \text{ hot dogs}}$

   $= \dfrac{40}{13}$.

   The ratio of Julian's hot dog consumption to Mikito's is $\dfrac{40}{13}$.

6) State taxes to federal taxes $= \dfrac{\text{State Taxes}}{\text{Federal Taxes}}$

   $= \dfrac{\$23}{\$88}$

   $= \dfrac{23}{88}$.

   The ratio of state taxes to federal taxes is $\dfrac{23}{88}$.

7) ratio $= \dfrac{42 \text{ kilograms}}{30 \text{ kilograms}}$

   $= \dfrac{\overset{1}{\cancel{6}} \cdot 7}{\underset{}{\overset{1}{\cancel{6}} \cdot 5}}$

   $= \dfrac{7}{5}$

8) ratio $= \dfrac{0.13 \text{ liters}}{0.5 \text{ liters}}$

   $= \dfrac{0.13}{0.5} \cdot \dfrac{100}{100}$

   $= \dfrac{13}{50}$

9) $\dfrac{1\frac{1}{2}}{2\frac{1}{4}} = \dfrac{\frac{3}{2}}{\frac{9}{4}}$

   $= \dfrac{3}{2} \div \dfrac{9}{4}$

   $= \dfrac{3}{2} \cdot \dfrac{4}{9}$

   $= \dfrac{\overset{1}{\cancel{3}}}{\cancel{2}^{1}} \cdot \dfrac{\cancel{2}^{1} \cdot 2}{\overset{1}{\cancel{3}} \cdot 3}$

   $= \dfrac{2}{3}$

10) $2 \text{ days} = \dfrac{2 \text{ days}}{1} \cdot \dfrac{24 \text{ hours}}{1 \text{ day}}$

    $= 48 \text{ hours}$

    ratio $= \dfrac{14.4 \text{ hours}}{48 \text{ hours}}$

    $= \dfrac{14.4}{48} \cdot \dfrac{10}{10}$

    $= \dfrac{144}{480}$

    $= \dfrac{3 \cdot \cancel{48}^{1}}{10 \cdot \cancel{48}^{1}}$

    $= \dfrac{3}{10}$

11) rate $= \dfrac{5 \text{ miles}}{42 \text{ minutes}}$

12) $\text{rate} = \dfrac{\$200}{15 \text{ hours}}$

$= \dfrac{\cancel{5} \cdot \$40}{3 \cdot \cancel{5} \text{ hours}}$

$= \dfrac{\$40}{3 \text{ hours}}$

13) $\text{rate} = \dfrac{40 \text{ milligrams}}{120 \text{ pounds}}$

$= \dfrac{\overset{1}{\cancel{40}} \cdot 1 \text{ milligram}}{\underset{1}{\cancel{40}} \cdot 3 \text{ pounds}}$

$= \dfrac{1 \text{ milligram}}{3 \text{ pounds}}$

14) $\text{rate} = \dfrac{\$42}{5 \text{ hours}}$

$= \dfrac{\$8.40}{1 \text{ hour}}$

$= \$8.40 \text{ per hour}$

15) $\text{rate} = \dfrac{160 \text{ miles}}{4 \text{ gallons}}$

$\text{unit rate} = \dfrac{40 \text{ miles}}{1 \text{ gallon}}$

The unit rate is 40 miles per gallon.

16) $\text{rate} = \dfrac{\$138}{12 \text{ hours}}$

$\text{unit rate} = \dfrac{\$11.50}{1 \text{ hour}}$

The unit rate is $11.50 per hour.

17) $\text{rate} = \dfrac{20 \text{ milligrams}}{100 \text{ pounds}}$

$\text{unit rate} = \dfrac{0.2 \text{ milligram}}{1 \text{ pound}}$

The unit rate is 0.2 milligram per pound.

18) a) $\dfrac{20 \text{ oz}}{\$5.00} = 4 \text{ oz per dollar}$

b) $\dfrac{27 \text{ oz}}{\$5.40} = 5 \text{ oz per dollar}$

Purchasing the larger box will give you more cereal for every dollar. Therefore, the 27 oz box is the better buy.

19) a) $\dfrac{\$1.50}{6 \text{ rutabagas}} = \$0.25 \text{ per rutabaga}$

b) $\dfrac{\$2.60}{10 \text{ rutabagas}} = \$0.26 \text{ per rutabaga}$

Since the "per rutabaga" cost is less at Hector's stand, it is offering the better buy.

CONCEPT CHECKS AND PRACTICE EXERCISES

A1)   A ratio is a quotient that compares two quantities with the same units.

A2)   Answers may vary. Example: $\dfrac{30 \text{ miles}}{1 \text{ gallon}}$ is not a ratio because the units in the numerator and the denominator are not the same, nor can one unit be converted to the other.

A3)   The ratio of 4 to 6 is $\dfrac{4}{6} = \dfrac{\overset{1}{\cancel{2}} \cdot 2}{\underset{1}{\cancel{2}} \cdot 3} = \dfrac{2}{3}$

A4)   The ratio of 5 to 10 is $\dfrac{5}{10} = \dfrac{\overset{1}{\cancel{5}} \cdot 1}{\underset{1}{\cancel{5}} \cdot 2} = \dfrac{1}{2}$

A5)   The ratio of 33:15 is $\dfrac{33}{15} = \dfrac{\overset{1}{\cancel{3}} \cdot 11}{\underset{1}{\cancel{3}} \cdot 5} = \dfrac{11}{5}$

A6)   The ratio of 40:54 is $\dfrac{40}{54} = \dfrac{\overset{1}{\cancel{2}} \cdot 20}{\underset{1}{\cancel{2}} \cdot 27} = \dfrac{20}{27}$

A7) $\dfrac{3\frac{1}{2} \cancel{\text{ minutes}}}{4\frac{1}{4} \cancel{\text{ minutes}}} = \dfrac{\frac{7}{2}}{\frac{17}{4}}$

$= \dfrac{7}{2} \div \dfrac{17}{4}$

$= \dfrac{7}{2} \cdot \dfrac{4}{17}$

$= \dfrac{7}{\underset{1}{\cancel{2}}} \cdot \dfrac{\overset{1}{\cancel{2}} \cdot 2}{17}$

$= \dfrac{14}{17}$

A8) $\dfrac{1\frac{1}{6}\ \text{tons}}{2\frac{1}{3}\ \text{tons}}=\dfrac{\frac{7}{6}}{\frac{7}{3}}$

$=\dfrac{7}{6}\div\dfrac{7}{3}$

$=\dfrac{7}{6}\cdot\dfrac{3}{7}$

$=\dfrac{\overset{1}{\cancel{7}}}{\cancel{3}\cdot 2}\cdot\dfrac{\overset{1}{\cancel{3}}\cdot 1}{\overset{1}{\cancel{7}}}$

$=\dfrac{1}{2}$

A9) ratio $=\dfrac{1.2\ \text{centimeters}}{71\ \text{centimeters}}$

$=\dfrac{1.2}{71}\cdot\dfrac{10}{10}$

$=\dfrac{12}{710}$

$=\dfrac{6\cdot\overset{1}{\cancel{2}}}{355\cdot\cancel{2}}$

$=\dfrac{6}{355}$

A10) ratio $=\dfrac{3.6\ \text{inches}}{53\ \text{inches}}$

$=\dfrac{3.6}{53}\cdot\dfrac{10}{10}$

$=\dfrac{36}{530}$

$=\dfrac{18\cdot\overset{1}{\cancel{2}}}{265\cdot\cancel{2}}$

$=\dfrac{18}{265}$

A11) ratio $=\dfrac{2,000\ \text{rpm}}{600\ \text{rpm}}$

$=\dfrac{\overset{1}{\cancel{200}}\cdot 10}{\overset{1}{\cancel{200}}\cdot 3}$

$=\dfrac{10}{3}$

The ratio between the engine's rpm and the wheel's rpm is $\dfrac{10}{3}$.

A12) ratio $=\dfrac{3,506\ \text{feet}}{600\ \text{feet}}$

$=\dfrac{3,506}{750}$

$=\dfrac{\overset{1}{\cancel{2}}\cdot 1,753}{\cancel{2}\cdot 375}$

$=\dfrac{1,753}{375}$

A13) 1.5 hrs $=\dfrac{1.5\ \text{hrs}}{1}\cdot\dfrac{60\ \text{minutes}}{1\ \text{hr}}=90$ minutes

ratio $=\dfrac{45\ \text{minutes}}{90\ \text{minutes}}$

$=\dfrac{45}{90}$

$=\dfrac{1\cdot\overset{1}{\cancel{45}}}{2\cdot\cancel{45}}$

$=\dfrac{1}{2}$

A14) 3.5 hours $=\dfrac{3.5\ \text{hours}}{1}\cdot\dfrac{60\ \text{minutes}}{1\ \text{hour}}$

$=210$ minutes

ratio $=\dfrac{210\ \text{minutes}}{20\ \text{minutes}}$

$=\dfrac{210}{20}$

$=\dfrac{21\cdot\overset{1}{\cancel{10}}}{2\cdot\cancel{10}}$

$=\dfrac{21}{2}$

A15) 25 feet $=\dfrac{25\ \text{feet}}{1}\cdot\dfrac{12\ \text{inches}}{1\ \text{foot}}$

$=300$ inches

ratio $=\dfrac{\text{measurements at building site}}{\text{measurements on blueprint}}$

$=\dfrac{25\ \text{feet}}{1\ \text{inch}}$

$=\dfrac{300\ \text{inches}}{1\ \text{inch}}$

$=\dfrac{300}{1}$

The ratio of measurements at the building site to measurements on the blueprint is $\dfrac{300}{1}$.

A16) $\text{ratio} = \dfrac{\text{size of football field}}{\text{size of toy field}}$

$= \dfrac{100 \ \text{yards}}{1 \ \text{yard}}$

$= \dfrac{100}{1}$

The ratio of the size of a football field to the child's toy field is $\dfrac{100}{1}$.

B1)   Answers may vary. A ratio compares quantities with like units, while a rate compares quantities with unlike units.

B2)   Answers may vary. A unit rate is a rate whose denominator is always one unit.

B3)   a) $\text{rate} = \dfrac{315 \ \text{miles}}{15 \ \text{gallons}}$

$\text{unit rate} = \dfrac{21 \ \text{miles}}{1 \ \text{gallon}}$

The unit rate for Trevor is 21 miles per gallon.

$\text{rate} = \dfrac{138 \ \text{miles}}{6 \ \text{gallons}}$

$\text{unit rate} = \dfrac{23 \ \text{miles}}{1 \ \text{gallon}}$

The unit rate for Valerie is 23 miles per gallon.

b)   Valerie has the better mileage (23 mpg) than Trevor (21 mpg).

c)   Answers may vary. Example: Expressing fuel consumption as a unit rate makes comparison easier.

B4) $\text{rate} = \dfrac{7 \ \text{miles}}{42 \ \text{minutes}}$

$= \dfrac{\overset{1}{\cancel{7}} \cdot 1 \ \text{miles}}{\underset{}{\cancel{7}} \cdot 6 \ \text{minutes}}$

$= \dfrac{1 \ \text{mile}}{6 \ \text{minutes}}$

The rate is 1 mile per 6 minutes.

B5) $\text{rate} = \dfrac{48 \ \text{miles}}{15 \ \text{minutes}}$

$= \dfrac{\overset{1}{\cancel{3}} \cdot 16 \ \text{miles}}{\underset{}{\cancel{3}} \cdot 5 \ \text{minutes}}$

$= \dfrac{16 \ \text{miles}}{5 \ \text{minutes}}$

The rate is 16 miles per 5 minutes.

B6) $\text{rate} = \dfrac{\$64}{6 \ \text{hours}}$

$= \dfrac{\overset{1}{\cancel{2}} \cdot \$32}{\underset{}{\cancel{2}} \cdot 3 \ \text{hours}}$

$= \dfrac{\$32}{3 \ \text{hours}}$

The rate is $32 per 3 hours.

B7) $\text{rate} = \dfrac{\$46}{4 \ \text{hours}}$

$= \dfrac{\overset{1}{\cancel{2}} \cdot \$23}{\underset{}{\cancel{2}} \cdot 2 \ \text{hours}}$

$= \dfrac{\$23}{2 \ \text{hours}}$

The rate is $23 per 2 hours.

B8) $\text{rate} = \dfrac{\$250}{40 \ \text{hours}}$

$\text{unit rate} = \dfrac{\$6.25}{1 \ \text{hour}}$

The unit rate is $6.25 per hour.

B9) $\text{rate} = \dfrac{\$165}{30 \ \text{hours}}$

$\text{unit rate} = \dfrac{\$5.50}{1 \ \text{hour}}$

The unit rate is $5.50 per hour.

B10) $\text{rate} = \dfrac{\$3.90}{15 \ \text{ounces}}$

$\text{unit rate} = \dfrac{\$0.26}{1 \ \text{ounce}}$

The unit rate is $0.26 per ounce.

B11) $\text{rate} = \dfrac{\$4.80}{20 \ \text{ounces}}$

$\text{unit rate} = \dfrac{\$0.24}{1 \ \text{ounce}}$

The unit rate is $0.24 per ounce.

B12) $\text{rate} = \dfrac{750 \text{ milligrams}}{150 \text{ pounds}}$

$\text{unit rate} = \dfrac{5 \text{ milligrams}}{1 \text{ pound}}$

The unit rate is 5 milligrams per pound.

B13) $\text{rate} = \dfrac{400 \text{ milligrams}}{80 \text{ pounds}}$

$\text{unit rate} = \dfrac{5 \text{ milligrams}}{1 \text{ pound}}$

The unit rate is 5 milligrams per pound.

B14) $\text{rate} = \dfrac{318 \text{ miles}}{6 \text{ hours}}$

$\text{unit rate} = \dfrac{53 \text{ miles}}{1 \text{ hour}}$

The unit rate is 53 miles per hour.

B15) $\text{rate} = \dfrac{360 \text{ miles}}{8 \text{ hours}}$

$\text{unit rate} = \dfrac{45 \text{ miles}}{1 \text{ hour}}$

The unit rate is 45 miles per hour.

SECTION 4.1   EXERCISES

For 1–6, refer to Concept Checks and Practice Exercises.

7) unit rate

9) ratio

11) a) $7:35$ is a ratio.

  b) With no units indicated, the assumption is that the units are the same.

  c) The ratio of $7:35$ is $\dfrac{7}{35} = \dfrac{\cancel{7} \cdot 1}{\cancel{7} \cdot 5} = \dfrac{1}{5}$, or 1:5.

13) a) $8:14$ is a ratio.

  b) With no units indicated, the assumption is that the units are the same.

  c) The ratio of $8:14$ is $\dfrac{8}{14} = \dfrac{\cancel{2} \cdot 4}{\cancel{2} \cdot 7} = \dfrac{4}{7}$, or 4:7.

15) a) 14 girls to 27 students is a rate.

  b) The units are not the same.

  c) The rate of 14 girls to 27 students is already in simplest form.

17) a) 5 yards per dollar is a rate.

  b) The units are not the same.

  c) The rate of 5 yards per dollar is already in simplest form.

19) a) $135 to 3 credit hours is a rate.

  b) The units are not the same.

  c) The rate of $135 to 3 credit hours is
  $\dfrac{\$135}{3 \text{ credit hours}} = \dfrac{\$45}{1 \text{ credit hour}}$, or $45 per credit hour

21) a) $84 for 15 hours of work is a rate.

  b) The units are not the same.

  c) The rate of $84 for 15 hours of work is
  $\dfrac{\$84}{15 \text{ hours}} = \dfrac{\cancel{3}^1 \cdot \$28}{\cancel{3}^1 \cdot 5 \text{ hours}} = \dfrac{\$28}{5 \text{ hours}}$, or $28 per 5 hours.

23) a) $3\frac{1}{3}$ rpm to $6\frac{1}{6}$ rpm is a ratio.

  b) The units are the same.

  c) The ratio of $3\frac{1}{3}$ rpm to $6\frac{1}{6}$ rpm is

  $\dfrac{3\frac{1}{3} \text{ rpm}}{6\frac{1}{6} \text{ rpm}} = \dfrac{\frac{10}{3}}{\frac{37}{6}}$

  $= \dfrac{10}{3} \div \dfrac{37}{6}$

  $= \dfrac{10}{3} \cdot \dfrac{6}{37}$

  $= \dfrac{10}{\cancel{3}^1} \cdot \dfrac{\cancel{3}^1 \cdot 2}{37}$

  $= \dfrac{20}{37}$

  The ratio is 20:37.

25) $1 \text{ dollar} = \dfrac{1 \text{ dollar}}{1} \cdot \dfrac{100 \text{ cents}}{1 \text{ dollar}} = 100 \text{ cents}$

  The ratio of 6 cents per dollar is

  $\dfrac{6 \text{ cents}}{100 \text{ cents}} = \dfrac{3 \cdot \cancel{2}^1}{50 \cdot \cancel{2}^1} = \dfrac{3}{50}$, or 3 to 50.

27) $1 \text{ week} = \frac{1 \text{ week}}{1} \cdot \frac{7 \text{ days}}{1 \text{ week}} = 7 \text{ days}$

The rate of 3 cloudy days every week is

$\frac{3 \text{ days}}{7 \text{ days}} = \frac{3}{7}$, or 3 to 7.

29) Answers may vary. Example: 120 to 78 (measure of blood pressure); 1 cup oatmeal to 2 cups water (recipe for oatmeal cereal); 16:9 (the aspect ratio of an HDTV).

31) $\text{rate} = \frac{380 \text{ miles}}{6.5 \text{ hours}}$

$\text{unit rate} \approx \frac{58.46 \text{ miles}}{1 \text{ hour}} \approx \frac{58 \text{ miles}}{1 \text{ hour}}$

The unit rate is approximately 58 miles per hour.

33) $\text{rate} = \frac{508 \text{ students}}{21 \text{ classes}}$

$\text{unit rate} \approx \frac{24.19 \text{ students}}{1 \text{ class}} \approx \frac{24 \text{ students}}{1 \text{ class}}$

The unit rate is approximately 24 students per class.

35) $\text{rate} = \frac{\$8.50}{\$42.35}$

$\text{unit rate} \approx \frac{0.2007}{1} \approx \frac{0.20}{1}$

The unit rate is approximately $0.20 per dollar.

37) $\text{rate} = \frac{500 \text{ square feet}}{2.5 \text{ hours}}$

$\text{unit rate} = \frac{200 \text{ square feet}}{1 \text{ hour}}$

The unit rate is 200 square feet per hour.

39) $\text{rate} = \frac{120 \text{ millligrams}}{14 \text{ minutes}}$

$\text{unit rate} \approx \frac{8.57 \text{ milligrams}}{1 \text{ minute}} \approx \frac{8.6 \text{ milligrams}}{1 \text{ minute}}$

The unit rate is approximately 8.6 milligrams per minute.

41) $\text{rate} = \frac{440 \text{ people}}{11 \text{ buses}}$

$\text{unit rate} = \frac{40 \text{ people}}{1 \text{ bus}}$

The unit rate is 40 people per bus.

43) a) Factor the numerator and the denominator.

b) Divide out all common factors.

45) Divide the numerator by the denominator and use the units "miles per gallon" in the answer.

47) $\frac{\$35}{6 \text{ issues}} \approx \$5.83 \text{ per issue}$

$\frac{\$48}{12 \text{ issues}} = \$4.00 \text{ per issue}$

The 12-issue subscription costs less per issue and is the better buy.

49) $\frac{10 \text{ oz}}{\$2.12} \approx 4.72 \text{ oz per dollar}$

$\frac{15 \text{ oz}}{\$2.32} \approx 6.47 \text{ oz per dollar}$

$\frac{20 \text{ oz}}{\$4.89} \approx 4.09 \text{ oz per dollar}$

Compare the ounces per dollar.
a) The 15-ounce box is the best buy.

b) The 20-ounce box is the worst buy.

51) $\frac{60 \text{ miles}}{1.2 \text{ hours}} = \frac{50 \text{ miles}}{1 \text{ hour}} = 50 \text{ mph}$

$\frac{60 \text{ miles}}{2.5 \text{ gallons}} = \frac{24 \text{ miles}}{1 \text{ gallon}} = 24 \text{ mpg}$

$\frac{42 \text{ miles}}{2.1 \text{ hours}} = \frac{20 \text{ miles}}{1 \text{ hour}} = 20 \text{ mph}$

$\frac{42 \text{ miles}}{2.8 \text{ gallons}} = \frac{15 \text{ miles}}{1 \text{ gallon}} = 15 \text{ mpg}$

53) $\$35,880 - \$32,160 = \$3,720$

$\frac{\$3,720}{120 \text{ PDAs}} = \frac{\$31}{1 \text{ PDA}}$

The retailer's profit is $31 per PDA.

55) $\frac{24,000 \text{ teeth}}{10 \text{ years}} = \frac{2,400 \text{ teeth}}{1 \text{ year}}$

A tiger shark grows approximately 2,400 teeth per year.

57) $\$13,769$ per month $\approx \dfrac{\$10,000}{1 \text{ month}}$

$\$1,385$ per month $\approx \dfrac{\$1,000}{1 \text{ month}}$

$\dfrac{\dfrac{\$10,000}{1 \text{ month}}}{\dfrac{\$1,000}{1 \text{ month}}} = \dfrac{10,000}{1,000} = \dfrac{10}{1};$

The ratio of a general's salary to a private's salary is about 10,000 to 1,000 or 10 to 1.

59) a) Mark McGwire:

$\dfrac{583 \text{ home runs}}{6,187 \text{ attempts}} \approx \dfrac{0.094 \text{ home run}}{1 \text{ attempt}}$

Babe Ruth:

$\dfrac{714 \text{ home runs}}{8,399 \text{ attempts}} \approx \dfrac{0.085 \text{ home runs}}{1 \text{ attempt}}$

McGwire was more likely to hit a home run.

b) Mark McGwire:

$\dfrac{1,596 \text{ strike outs}}{6,187 \text{ attempts}} \approx \dfrac{0.258 \text{ strike outs}}{1 \text{ attempt}}$

Babe Ruth:

$\dfrac{1,330 \text{ strike outs}}{8,399 \text{ attempts}} \approx \dfrac{0.158 \text{ strike outs}}{1 \text{ attempt}}$

McGwire was more likely to strike out.

c) Answers may vary.

61) a) $\dfrac{240 \text{ parts}}{8 \text{ hours}} = \dfrac{30 \text{ parts}}{1 \text{ hour}}$

The worker made 30 parts per hour.

b) $\dfrac{1 \text{ hour}}{30 \text{ parts}} \cdot \dfrac{\$45}{1 \text{ hour}} = \dfrac{\$1.50}{1 \text{ part}}$

The worker earns $\$1.50$ per part.

63) Kareem Abdul-Jabbar scored the most points in a career.

65) Michael Jordan scored the most points per game over his career.

67) Answers may vary.

69) Answers may vary.

4.2   Writing and Solving Proportions

GUIDED PRACTICE

1) Write a reference fraction that compares the good parts to the total parts.

Reference fraction $= \dfrac{\text{good parts}}{\text{total parts}}$

Write the proportion.

$\dfrac{480 \text{ good parts}}{500 \text{ total parts}} = \dfrac{384 \text{ good parts}}{400 \text{ total parts}}$

2) Write a reference fraction that compares the price to the number of boxes.

Reference fraction $= \dfrac{\text{price}}{\text{number of boxes}}$

Write the proportion.

$\dfrac{\$1.50}{1 \text{ box}} = \dfrac{\$4.50}{3 \text{ boxes}}$

3) Write a reference fraction that compares the amount of medication to the weight of the bull.

Reference fraction $= \dfrac{\text{milligrams}}{\text{pounds}}$

Write the proportion.

$\dfrac{4,000 \text{ milligrams}}{1,500 \text{ pounds}} = \dfrac{x \text{ milligrams}}{1,200 \text{ pounds}}$

4) $15 \cdot 98 \overset{?}{=} 14 \cdot 105$

$1,470 = 1,470$

The statement is a proportion because the cross products are equal.

5) $175 \cdot 40 \overset{?}{=} 50 \cdot 140$

$7,000 = 7,000$

The statement is a proportion because the cross products are equal.

6) $360 \cdot 8 \overset{?}{=} 280 \cdot 10.1$

$2,880 \neq 2,828$

The statement isn't a proportion because the cross products aren't equal.

7) The missing multiplier is 2 because $6 \cdot 2 = 12$.

$$\frac{3}{6} \cdot \frac{?}{?} = \frac{n}{12}$$

$$\frac{3}{6} \cdot \frac{2}{2} = \frac{n}{12}$$

$$\frac{6}{12} = \frac{n}{12}$$

$$6 = n$$

8) $24 = 8 \cdot 3$

denominator $= 8 \cdot$ numerator

Use the same relationship in the fraction with the unknown.

denominator $= 8 \cdot$ numerator

$$n = 8 \cdot 13$$

$$n = 104$$

9) $$\frac{n}{28} = \frac{3}{14}$$

$$\frac{\overset{1}{\cancel{28}}}{1} \cdot \frac{n}{\underset{1}{\cancel{28}}} = \frac{2 \cdot \overset{1}{\cancel{14}}}{1} \cdot \frac{3}{\underset{1}{\cancel{14}}}$$

$$n = 6$$

Check: $$\frac{6}{28} \overset{?}{=} \frac{3}{14}$$

$$6 \cdot 14 \overset{?}{=} 3 \cdot 28$$

$$84 = 84$$

10) $$\frac{4}{n} = \frac{16}{7}$$

$$\frac{n}{4} = \frac{7}{16}$$

$$\frac{\overset{1}{\cancel{4}}}{1} \cdot \frac{n}{\underset{1}{\cancel{4}}} = \frac{\overset{1}{\cancel{4}}}{1} \cdot \frac{7}{\underset{1}{\cancel{4}} \cdot 4}$$

$$n = \frac{7}{4}$$

Check: $$\frac{4}{\frac{7}{4}} \overset{?}{=} \frac{16}{7}$$

$$4 \cdot 7 \overset{?}{=} 16 \cdot \frac{7}{4}$$

$$4 \cdot 7 \overset{?}{=} \frac{16}{1} \cdot \frac{7}{4}$$

$$4 \cdot 7 \overset{?}{=} \frac{\overset{1}{\cancel{4}} \cdot 4}{1} \cdot \frac{7}{\underset{1}{\cancel{4}}}$$

$$28 = 28$$

11) $$\frac{1.5}{4} = \frac{n}{12}$$

$$\frac{1.5}{4} \cdot \frac{12}{1} = \frac{n}{\underset{1}{\cancel{12}}} \cdot \frac{\overset{1}{\cancel{12}}}{1}$$

$$\frac{1.5}{\underset{1}{\cancel{4}}} \cdot \frac{3 \cdot \overset{1}{\cancel{4}}}{1} = n$$

$$4.5 = n$$

Check: $$\frac{1.5}{4} \overset{?}{=} \frac{4.5}{12}$$

$$1.5 \cdot 12 \overset{?}{=} 4 \cdot 4.5$$

$$18 = 18$$

CONCEPT CHECKS AND PRACTICE EXERCISES

A1) A proportion states that two ratios or rates are equal.

A2) Answers may vary. Example: A rate is a comparison of two quantities with unlike units. A proportion compares two rates (or ratios).

A3) Answers may vary. Example: The units in the numerators of the two rates are not the same. It could become a proportion by interchanging the numerator and denominator of one of the rates. For example: 14 pounds is to 4 bricks as 28 pounds is to 8 bricks.

A4) $$\frac{5 \text{ cups}}{3 \text{ pounds}} = \frac{15 \text{ cups}}{9 \text{ pounds}}$$

A5) $$\frac{8 \text{ liters}}{7 \text{ kilograms}} = \frac{32 \text{ liters}}{28 \text{ kilograms}}$$

A6) $$\frac{9 \text{ boards}}{10\text{-foot wall}} = \frac{18 \text{ boards}}{20\text{-foot wall}}$$

A7) $$\frac{3 \text{ pounds}}{8 \text{ vases}} = \frac{9 \text{ pounds}}{24 \text{ vases}}$$

A8) $$\frac{5 \text{ copies of pamphlet}}{45 \text{ pieces of paper}} = \frac{50 \text{ copies of pamphlet}}{450 \text{ pieces of paper}}$$

A9) $$\frac{1 \text{ packet}}{1 \text{ row}} = \frac{30 \text{ packets}}{30 \text{ rows}}$$

A10) $$\frac{\$8}{1 \text{ hour}} = \frac{\$n}{7.5 \text{ hours}}$$

A11) $\dfrac{\$18}{1\text{ hour}}=\dfrac{\$n}{16.4\text{ hours}}$

B1)   Answers may vary. Example: A statement is a proportion if the cross products are equal.

B2)   Answers may vary. Example:

$$\frac{3}{5}=\frac{5.1}{8.5}$$
$$(3)(8.5)=(5)(5.1)$$
$$25.5=25.5$$

B3) $\dfrac{7}{13}\overset{?}{=}\dfrac{2}{3}$

$13\cdot2\overset{?}{=}7\cdot3$

$26\ne21$

No, the statement is not a proportion.

B4) $\dfrac{14}{18}\overset{?}{=}\dfrac{35}{45}$

$18\cdot35\overset{?}{=}14\cdot45$

$630=630$

Yes, the statement is a proportion.

B5) $\dfrac{10}{18}\overset{?}{=}\dfrac{25}{45}$

$18\cdot25\overset{?}{=}10\cdot45$

$450=450$

Yes, the statement is a proportion.

B6) $\dfrac{42}{100}\overset{?}{=}\dfrac{22}{55}$

$100\cdot22\overset{?}{=}42\cdot55$

$2,200\ne2,310$

No, the statement is not a proportion.

B7) $\dfrac{1\frac{1}{5}}{8}\overset{?}{=}\dfrac{1\frac{1}{2}}{10}$

$8\cdot1\frac{1}{2}\overset{?}{=}1\frac{1}{5}\cdot10$

$8\cdot\dfrac{3}{2}\overset{?}{=}\dfrac{6}{5}\cdot10$

$\dfrac{\cancel{2}^{1}\cdot4}{1}\cdot\dfrac{3}{\cancel{2}^{1}\cdot1}\overset{?}{=}\dfrac{6}{\cancel{5}^{1}\cdot1}\cdot\dfrac{\cancel{5}^{1}\cdot2}{1}$

$12=12$

Yes, the statement is a proportion.

B8) $\dfrac{1\frac{1}{3}}{9}\overset{?}{=}\dfrac{1\frac{7}{9}}{12}$

$9\cdot1\frac{7}{9}\overset{?}{=}1\frac{1}{3}\cdot12$

$9\cdot\dfrac{16}{9}\overset{?}{=}\dfrac{4}{3}\cdot12$

$\dfrac{\cancel{9}^{1}\cdot1}{1}\cdot\dfrac{16}{\cancel{9}^{1}\cdot1}\overset{?}{=}\dfrac{4}{\cancel{3}^{1}\cdot1}\cdot\dfrac{\cancel{3}^{1}\cdot4}{1}$

$16=16$

Yes, the statement is a proportion.

B9) $\dfrac{3.1}{2}\overset{?}{=}\dfrac{9.4}{6}$

$2\cdot9.4\overset{?}{=}3.1\cdot6$

$18.8\ne18.6$

No, the statement is not a proportion.

B10) $\dfrac{2.5}{3}\overset{?}{=}\dfrac{15}{18}$

$3\cdot15\overset{?}{=}2.5\cdot18$

$45=45$

Yes, the statement is a proportion.

B11) a) Game 1: $\dfrac{24\text{ points}}{10\text{ minutes}}$

Game 2: $\dfrac{50\text{ points}}{25\text{ minutes}}$

b) $\dfrac{24\text{ points}}{10\text{ minutes}}\overset{?}{=}\dfrac{50\text{ points}}{25\text{ minutes}}$

$10\cdot50\overset{?}{=}24\cdot25$

$500\ne600$

No, the two rates are not proportional.

B12) a) Day 1: $\dfrac{5\text{ driveways plowed}}{2\text{ hours}}$

Day 2: $\dfrac{7\text{ driveways plowed}}{3\text{ hours}}$

b) $\dfrac{5\text{ driveways plowed}}{2\text{ hours}}\overset{?}{=}\dfrac{7\text{ driveways plowed}}{3\text{ hours}}$

$5\cdot3\overset{?}{=}2\cdot7$

$15\ne14$

No, the two rates are not proportional.

FOCUS ON PROPORTIONAL REASONING PRACTICE EXERCISES

1) The missing multiplier is 3 because $6 \cdot 3 = 18$.

$$\frac{5}{6} \cdot \frac{?}{?} = \frac{n}{18}$$
$$\frac{5}{6} \cdot \frac{3}{3} = \frac{n}{18}$$
$$\frac{15}{18} = \frac{n}{18}$$
$$15 = n$$

2) The missing multiplier is 6 because $4 \cdot 6 = 24$.

$$\frac{5}{4} \cdot \frac{?}{?} = \frac{n}{24}$$
$$\frac{5}{4} \cdot \frac{6}{6} = \frac{n}{24}$$
$$\frac{30}{24} = \frac{n}{24}$$
$$30 = n$$

3) The missing multiplier is 6 because $8 \cdot 6 = 48$.

$$\frac{n}{8} \cdot \frac{?}{?} = \frac{36}{48}$$
$$\frac{n}{8} \cdot \frac{6}{6} = \frac{36}{48}$$
$$\frac{6n}{48} = \frac{36}{48}$$
$$6n = 36$$
$$n = 6$$

4) The missing multiplier is 9 because $7 \cdot 9 = 63$.

$$\frac{n}{7} \cdot \frac{?}{?} = \frac{9}{63}$$
$$\frac{n}{7} \cdot \frac{9}{9} = \frac{9}{63}$$
$$\frac{9n}{63} = \frac{9}{63}$$
$$9n = 9$$
$$n = 1$$

5) The missing multiplier is 5 because $15 = 3 \cdot 5$.

$$\frac{15}{n} = \frac{3}{4} \cdot \frac{?}{?}$$
$$\frac{15}{n} = \frac{3}{4} \cdot \frac{5}{5}$$
$$\frac{15}{n} = \frac{15}{20}$$
$$n = 20$$

6) The missing multiplier is 7 because $6 \cdot 7 = 42$.

$$\frac{6}{7} \cdot \frac{?}{?} = \frac{42}{n}$$
$$\frac{6}{7} \cdot \frac{7}{7} = \frac{42}{n}$$
$$\frac{42}{49} = \frac{42}{n}$$
$$49 = n$$

7) The missing multiplier is 3 because $6 = 2 \cdot 3$.

$$\frac{6}{9} = \frac{2}{n} \cdot \frac{?}{?}$$
$$\frac{6}{9} = \frac{2}{n} \cdot \frac{3}{3}$$
$$\frac{6}{9} = \frac{6}{3n}$$
$$9 = 3n$$
$$3 = n$$

8) The missing multiplier is 5 because $10 = 2 \cdot 5$.

$$\frac{10}{45} = \frac{2}{n} \cdot \frac{?}{?}$$
$$\frac{10}{45} = \frac{2}{n} \cdot \frac{5}{5}$$
$$\frac{10}{45} = \frac{10}{5n}$$
$$45 = 5n$$
$$9 = n$$

9) The missing multiplier is 4 because $4 = 1 \cdot 4$.

$$\frac{n}{4} = \frac{8}{1} \cdot \frac{?}{?}$$
$$\frac{n}{4} = \frac{8}{1} \cdot \frac{4}{4}$$
$$\frac{n}{4} = \frac{32}{4}$$
$$n = 32$$

10) The missing multiplier is 2 because $2 = 1 \cdot 2$.

$$\frac{n}{2} = \frac{3}{1} \cdot \frac{?}{?}$$
$$\frac{n}{2} = \frac{3}{1} \cdot \frac{2}{2}$$
$$\frac{n}{2} = \frac{6}{2}$$
$$n = 6$$

11)
$$\frac{27}{9} = \frac{n}{12}$$
$$27 = 3 \cdot 9$$
$$\text{numerator} = 3 \cdot \text{denominator}$$
$$n = 3 \cdot 12$$
$$n = 36$$

12)
$$\frac{20}{4} = \frac{n}{11}$$
$$20 = 5 \cdot 4$$
$$\text{numerator} = 5 \cdot \text{denominator}$$
$$n = 5 \cdot 11$$
$$n = 55$$

13)
$$\frac{n}{15} = \frac{2}{6}$$
$$6 = 3 \cdot 2$$
denominator= 3 · numerator
$$15 = 3 \cdot n$$
$$5 = n$$
$$n = 5$$

14)
$$\frac{n}{20} = \frac{8}{32}$$
$$32 = 4 \cdot 8$$
denominator= 4 · numerator
$$20 = 4 \cdot n$$
$$5 = n$$
$$n = 5$$

15)
$$\frac{3}{12} = \frac{7}{n}$$
$$12 = 4 \cdot 3$$
denominator = 4 · numerator
$$n = 4 \cdot 7$$
$$n = 28$$

16)
$$\frac{6}{n} = \frac{2}{12}$$
$$12 = 6 \cdot 2$$
denominator = 6 · numerator
$$n = 6 \cdot 6$$
$$n = 36$$

17)
$$\frac{12}{n} = \frac{20}{5}$$
$$20 = 4 \cdot 5$$
numerator = 4 · denominator
$$12 = 4 \cdot n$$
$$3 = n$$
$$n = 3$$

18)
$$\frac{72}{8} = \frac{18}{n}$$
$$72 = 9 \cdot 8$$
numerator = 9 · denominator
$$18 = 9 \cdot n$$
$$2 = n$$
$$n = 2$$

19)
$$\frac{n}{\frac{1}{2}} = \frac{9}{3}$$
$$9 = 3 \cdot 3$$
numerator= 3 · denominator
$$n = 3 \cdot \frac{1}{2}$$
$$n = \frac{3}{2} = 1\frac{1}{2}$$

20)
$$\frac{24}{4} = \frac{n}{\frac{1}{4}}$$
$$24 = 6 \cdot 4$$
numerator= 6 · denominator
$$n = 6 \cdot \frac{1}{4}$$
$$n = \frac{6}{4} = \frac{3}{2} = 1\frac{1}{2}$$

21) 43 is greater than 32, but 15 is less than 23.

22) 15 is three times 5, but 38 is not three times 13.

23) 88 is very close to 90, but 41 is not very close to 63.

24) 100 is five times 20, but 36 is not five times 6.

25) 1 is less than half of 3, but 2 is exactly half of 4.

26) 2 is half of 4, but 7 is not half of 10.

27) 53 is not very close to 30, but 62 is close to 60.

28) 2 is larger than 1, but 500 is smaller than 1,000.

C1) Answers may vary. Example: A woman on a diet ate 3 cookies. According to the cookie's package, one cookie has 250 calories. How many cookies did the woman consume?

C2) $\dfrac{250 \text{ calories}}{1 \text{ cookie}} = \dfrac{x \text{ calories}}{3 \text{ cookies}}$

C3) Answers may vary. Example: The first ratio, $\dfrac{4}{5}$, is the ratio of cloudy days (numerator) to all days (denominator). The second ratio, $\dfrac{30}{C}$, has "all days" in the numerator.

C4)
$$\frac{3}{5} = \frac{n}{15}$$

$$\frac{15}{1} \cdot \frac{3}{5} = \frac{\cancel{15}^{1}}{1} \cdot \frac{n}{\cancel{15}^{1}}$$

$$\frac{3 \cdot \cancel{5}^{1}}{1} \cdot \frac{3}{\cancel{5}^{1}} = n$$

$$9 = n, \text{ or } n = 9$$

Check: $$\frac{3}{5} \overset{?}{=} \frac{9}{15}$$

$$5 \cdot 9 \overset{?}{=} 3 \cdot 15$$

$$45 = 45$$

C5)
$$\frac{7}{12} = \frac{n}{24}$$

$$\frac{24}{1} \cdot \frac{7}{12} = \frac{\cancel{24}^{1}}{1} \cdot \frac{n}{\cancel{24}^{1}}$$

$$\frac{2 \cdot \cancel{12}^{1}}{1} \cdot \frac{7}{\cancel{12}^{1}} = n$$

$$14 = n, \text{ or } n = 14$$

Check: $$\frac{7}{12} \overset{?}{=} \frac{14}{24}$$

$$7 \cdot 24 \overset{?}{=} 12 \cdot 14$$

$$168 = 168$$

C6)
$$\frac{2}{5} = \frac{7}{n}$$

$$\frac{5}{2} = \frac{n}{7}$$

$$\frac{7}{1} \cdot \frac{5}{2} = \frac{\cancel{7}^{1}}{1} \cdot \frac{n}{\cancel{7}^{1}}$$

$$\frac{35}{2} = 17\frac{1}{2} = n, \text{ or } n = 17\frac{1}{2}$$

Check: $$\frac{2}{5} \overset{?}{=} \frac{7}{17\frac{1}{2}}$$

$$2 \cdot 17\frac{1}{2} \overset{?}{=} 5 \cdot 7$$

$$\frac{\cancel{2}^{1}}{1} \cdot \frac{35}{\cancel{2}^{1}} \overset{?}{=} 5 \cdot 7$$

$$35 = 35$$

C7)
$$\frac{13}{11} = \frac{2}{n}$$

$$\frac{11}{13} = \frac{n}{2}$$

$$\frac{2}{1} \cdot \frac{11}{13} = \frac{\cancel{2}^{1}}{1} \cdot \frac{n}{\cancel{2}^{1}}$$

$$\frac{22}{13} = 1\frac{9}{13} = n, \text{ or } n = 1\frac{9}{13}$$

Check: $$\frac{13}{11} \overset{?}{=} \frac{2}{1\frac{9}{13}}$$

$$13 \cdot 1\frac{9}{13} \overset{?}{=} 11 \cdot 2$$

$$\frac{\cancel{13}^{1}}{1} \cdot \frac{22}{\cancel{13}^{1}} \overset{?}{=} 11 \cdot 2$$

$$22 = 22$$

C8)
$$\frac{5}{24} = \frac{n}{16}$$

$$\frac{16}{1} \cdot \frac{5}{24} = \frac{\cancel{16}^{1}}{1} \cdot \frac{n}{\cancel{16}^{1}}$$

$$\frac{2 \cdot \cancel{8}^{1}}{1} \cdot \frac{5}{3 \cdot \cancel{8}^{1}} = n$$

$$\frac{10}{3} = 3\frac{1}{3} = n, \text{ or } n = 3\frac{1}{3}$$

Check: $$\frac{5}{24} \overset{?}{=} \frac{3\frac{1}{3}}{16}$$

$$24 \cdot 3\frac{1}{3} \overset{?}{=} 5 \cdot 16$$

$$\frac{\cancel{3}^{1} \cdot 8}{1} \cdot \frac{10}{\cancel{3}^{1}} \overset{?}{=} 5 \cdot 16$$

$$80 = 80$$

C9) $\dfrac{3}{35}=\dfrac{n}{21}$

$$\dfrac{3\cdot\cancel{7}^{1}}{1}\cdot\dfrac{3}{5\cdot\cancel{7}^{1}}=\dfrac{\cancel{21}^{1}}{1}\cdot\dfrac{n}{\cancel{21}^{1}}$$

$$\dfrac{9}{5}=1\dfrac{4}{5}=n,\ \text{or}\ n=1\dfrac{4}{5}$$

Check: $\dfrac{3}{35}\overset{?}{=}\dfrac{1\dfrac{4}{5}}{21}$

$$35\cdot1\dfrac{4}{5}\overset{?}{=}3\cdot21$$

$$\dfrac{\cancel{35}^{1}\cdot7}{1}\cdot\dfrac{9}{\cancel{5}^{1}}\overset{?}{=}3\cdot21$$

$$63=63$$

C10) $\dfrac{1.2}{7}=\dfrac{n}{10}$

$$\dfrac{1.2}{7}\cdot\dfrac{10}{1}=\dfrac{n}{\cancel{10}^{1}}\cdot\dfrac{\cancel{10}^{1}}{1}$$

$$\dfrac{12}{7}=1\dfrac{5}{7}=n,\ \text{or}\ n=1\dfrac{5}{7}$$

Check: $\dfrac{1.2}{7}\overset{?}{=}\dfrac{1\dfrac{5}{7}}{10}$

$$7\cdot1\dfrac{5}{7}\overset{?}{=}1.2\cdot10$$

$$\dfrac{\cancel{7}^{1}}{1}\cdot\dfrac{12}{\cancel{7}^{1}}\overset{?}{=}1.2\cdot10$$

$$12=12$$

C11) $\dfrac{3.8}{3}=\dfrac{n}{5}$

$$\dfrac{3.8}{3}\cdot\dfrac{5}{1}=\dfrac{n}{\cancel{5}^{1}}\cdot\dfrac{\cancel{5}^{1}}{1}$$

$$\dfrac{19}{3}=6\dfrac{1}{3}=n,\ \text{or}\ n=6\dfrac{1}{3}$$

Check: $\dfrac{3.8}{3}\overset{?}{=}\dfrac{6\dfrac{1}{3}}{5}$

$$3\cdot6\dfrac{1}{3}\overset{?}{=}3.8\cdot5$$

$$\dfrac{\cancel{3}^{1}}{1}\cdot\dfrac{19}{\cancel{3}^{1}}\overset{?}{=}3.8\cdot5$$

$$19=19$$

C12) $\dfrac{2\ \text{quarts}}{3\ \text{lemons}}=\dfrac{n\ \text{quarts}}{24\ \text{lemons}}$

$$\dfrac{24}{1}\cdot\dfrac{2}{3}=\dfrac{\cancel{24}^{1}}{1}\cdot\dfrac{n}{\cancel{24}^{1}}$$

$$\dfrac{8\cdot\cancel{3}^{1}}{1}\cdot\dfrac{2}{\cancel{3}^{1}}=n$$

$$16=n,\ \text{or}\ n=16\ \text{quarts}$$

Check: $\dfrac{3}{2}=\dfrac{24}{16}$

$$3\cdot16\overset{?}{=}2\cdot24$$

$$48=48$$

Cassandra can make 16 quarts of lemonade.

C13) $\dfrac{20\ \text{min}}{16\ \text{mi}}=\dfrac{n\ \text{min}}{40\ \text{mi}}$

$$\dfrac{40}{1}\cdot\dfrac{20}{16}=\dfrac{\cancel{40}^{1}}{1}\cdot\dfrac{n}{\cancel{40}^{1}}$$

$$\dfrac{5\cdot\cancel{8}^{1}}{1}\cdot\dfrac{10\cdot\cancel{2}^{1}}{1\cdot\cancel{2}^{1}\cdot\cancel{8}^{1}}=n$$

Check: $\dfrac{16}{20}=\dfrac{40}{50}$

$$16\cdot50\overset{?}{=}20\cdot40$$

$$800=800$$

Her entire commute will take 50 minutes.

SECTION 4.2   EXERCISES

For 1–11, refer to Concept Checks and Practice Exercises.

13)   proportion

15) $\dfrac{5}{3}=\dfrac{n}{7}$

17) $\dfrac{8}{11}=\dfrac{0.06}{n}$

19) $\dfrac{53\ \text{opposed}}{100\ \text{people}}=\dfrac{n\ \text{opposed}}{4{,}300\ \text{people}}$, where n represents the number of people who oppose the proposition.

21) $\dfrac{\$0.261}{1\ \text{in}^2}=\dfrac{\$n}{766\ \text{in}^2}$, where n represents the purchase price.

23) $\dfrac{\$8.55}{1\ \text{hour}}=\dfrac{\$n}{30\ \text{hours}}$, where n represents Sansui's earnings.

25) $\dfrac{1\ \text{page}}{250\ \text{words}}=\dfrac{n\ \text{pages}}{3{,}000\ \text{words}}$, where n represents the number of pages in Jessica's report.

27) $\dfrac{1\ \text{batch}}{\frac{1}{4}\ \text{cup}}=\dfrac{n\ \text{batches}}{1\frac{1}{2}\ \text{cups}}$, where n represents the number of batches of chili made.

29) $\dfrac{15}{7}\overset{?}{=}\dfrac{60}{28}$

$7\cdot 60\overset{?}{=}15\cdot 28$

$420=420$

Yes, the statement is a proportion.

31) $\dfrac{15}{25}\overset{?}{=}\dfrac{21}{34}$

$25\cdot 21\overset{?}{=}15\cdot 34$

$525\neq 510$

No, the statement is not a proportion.

33) $\dfrac{2\frac{1}{3}}{3}\overset{?}{=}\dfrac{7}{15}$

$3\cdot 7\overset{?}{=}2\frac{1}{3}\cdot 15$

$3\cdot 7\overset{?}{=}\dfrac{7}{3}\cdot 15$

$3\cdot 7\overset{?}{=}\dfrac{7}{\cancel{3}^{1}\cdot 1}\cdot\dfrac{\cancel{3}^{1}\cdot 5}{1}$

$21\neq 35$

No, the statement is not a proportion.

35) $\dfrac{3.5}{\frac{1}{3}}\overset{?}{=}\dfrac{61}{6}$

$\dfrac{1}{3}\cdot 61\overset{?}{=}3.5\cdot 6$

$\dfrac{61}{3}\overset{?}{=}21$

$20\frac{1}{3}\neq 21$

No, the statement is not a proportion.

37) $\dfrac{82\ \text{miles}}{2\ \text{hours}}\overset{?}{=}\dfrac{123\ \text{miles}}{3\ \text{hours}}$

$2\cdot 123\overset{?}{=}82\cdot 3$

$246=246$

Yes, the statement is a proportion.

39) $\dfrac{38\ \text{people}}{2\ \text{buses}}\overset{?}{=}\dfrac{19\ \text{buses}}{1\ \text{person}}$

No. The units are not the same in the numerators or the denominators. The statement is not a proportion.

41) a) Rate 1: $\dfrac{30\ \text{mph}}{5°\ \text{F}}$

   Rate 2: $\dfrac{5\ \text{mph}}{33°\ \text{F}}$

 b) $\dfrac{30\ \text{mph}}{5°\ \text{F}}\overset{?}{=}\dfrac{5\ \text{mph}}{33°\ \text{F}}$

   $30\cdot 33\overset{?}{=}5\cdot 5$

   $990\neq 25$

   No, the two events are not proportional.

43) a) Ratio 1: $\dfrac{3\ \text{inches}}{\frac{1}{3}\ \text{inch}}$

   Ratio 2: $\dfrac{9\ \text{square inches}}{\frac{1}{9}\ \text{square inch}}$

 b) $\dfrac{3\ \text{inches}}{\frac{1}{3}\ \text{inch}}\overset{?}{=}\dfrac{9\ \text{square inches}}{\frac{1}{9}\ \text{square inch}}$

c) $\dfrac{1}{3} \cdot 9 \overset{?}{=} 3 \cdot \dfrac{1}{9}$

$\dfrac{1}{\cancel{3}^1 \cdot 1} \cdot \dfrac{\cancel{3}^1 \cdot 3}{1} \overset{?}{=} \dfrac{\cancel{3}^1 \cdot 1}{1} \cdot \dfrac{1}{\cancel{3}^1 \cdot 3}$

$3 \neq \dfrac{1}{3}$

No, the two ratios are not proportional.

45) $\dfrac{3 \text{ ounces}}{\$0.87} \overset{?}{=} \dfrac{8 \text{ ounces}}{\$2.07}$

$3 \cdot 2.07 \overset{?}{=} 0.87 \cdot 8$

$6.21 \neq 6.96$

No, the postage was not proportional to the weight of the letters.

47) a) Rate 1: $\dfrac{320 \text{ feet}}{9.6 \text{ inches}}$

Rate 2: $\dfrac{60 \text{ feet}}{1.8 \text{ inches}}$

b) $\dfrac{320 \text{ feet}}{9.6 \text{ inches}} \overset{?}{=} \dfrac{60 \text{ feet}}{1.8 \text{ inches}}$

$320 \cdot 1.8 \overset{?}{=} 9.6 \cdot 60$

$576 = 576$

Yes, Giuseppe's model is proportional to the actual ship.

49) $\dfrac{n}{5} = \dfrac{6}{15}$

$\dfrac{\cancel{5}^1}{1} \cdot \dfrac{n}{\cancel{5}^1} = \dfrac{5}{1} \cdot \dfrac{6}{15}$

$n = \dfrac{\cancel{5}^1}{1} \cdot \dfrac{\cancel{3}^1 \cdot 2}{\cancel{5}^1 \cdot \cancel{3}^1}$

$n = 2$

51) $\dfrac{1}{6} = \dfrac{2}{n}$

$\dfrac{6}{1} = \dfrac{n}{2}$

$\dfrac{2}{1} \cdot \dfrac{6}{1} = \dfrac{\cancel{2}^1}{1} \cdot \dfrac{n}{\cancel{2}^1}$

$12 = n, \text{ or } n = 12$

53) $\dfrac{n}{20} = \dfrac{3.2}{10}$

$\dfrac{\cancel{20}^1}{1} \cdot \dfrac{n}{\cancel{20}^1} = \dfrac{20}{1} \cdot \dfrac{3.2}{10}$

$n = \dfrac{\cancel{10}^1 \cdot 2}{1} \cdot \dfrac{3.2}{\cancel{10}^1 \cdot 1}$

$n = 6.4$

55) $\dfrac{450}{n} = \dfrac{5}{3}$

$\dfrac{n}{450} = \dfrac{3}{5}$

$\dfrac{\cancel{450}^1}{1} \cdot \dfrac{n}{\cancel{450}^1} = \dfrac{450}{1} \cdot \dfrac{3}{5}$

$n = \dfrac{\cancel{5}^1 \cdot 90}{1} \cdot \dfrac{3}{\cancel{5}^1 \cdot 1}$

$n = 270$

57) $\dfrac{11}{27} = \dfrac{n}{3}$

$\dfrac{11}{27} \cdot \dfrac{3}{1} = \dfrac{n}{\cancel{3}^1} \cdot \dfrac{\cancel{3}^1}{1}$

$\dfrac{11}{\cancel{3}^1 \cdot 9} \cdot \dfrac{\cancel{3}^1 \cdot 1}{1} = n$

$\dfrac{11}{9} = 1\dfrac{2}{9} = n, \text{ or } n = 1\dfrac{2}{9}$

59) $\dfrac{3.5}{7} = \dfrac{10.5}{n}$

$\dfrac{7}{3.5} = \dfrac{n}{10.5}$

$\dfrac{10.5}{1} \cdot \dfrac{7}{3.5} = \dfrac{\cancel{10.5}^1}{1} \cdot \dfrac{n}{\cancel{10.5}^1}$

$\dfrac{10.5}{1} \cdot \dfrac{\cancel{3.5}^1 \cdot 2}{\cancel{3.5}^1 \cdot 1}$

$21 = n, \text{ or } n = 21$

61) $\dfrac{\$n}{40\ \text{hours}} \overset{?}{=} \dfrac{\$100}{8\ \text{hours}}$

$\dfrac{\cancel{40}^{1}}{1} \cdot \dfrac{n}{\cancel{40}^{1}} = \dfrac{40}{1} \cdot \dfrac{100}{8}$

$n = \dfrac{\cancel{8}^{1} \cdot 5}{1} \cdot \dfrac{100}{\cancel{8}^{1} \cdot 1}$

$n = \$500$

63) $\dfrac{3\ \text{kilograms}}{60\ \text{pounds}} = \dfrac{5\ \text{kilograms}}{n\ \text{pounds}}$

$\dfrac{60\ \text{pounds}}{3\ \text{kilograms}} = \dfrac{n\ \text{pounds}}{5\ \text{kilograms}}$

$\dfrac{5}{1} \cdot \dfrac{60}{3} = \dfrac{\cancel{5}^{1}}{1} \cdot \dfrac{n}{\cancel{5}^{1}}$

$\dfrac{5}{1} \cdot \dfrac{20 \cdot \cancel{3}^{1}}{1 \cdot \cancel{3}^{1}} = n$

$100 = n, \text{ or } n = 100\ \text{pounds}$

65) $\dfrac{16\ \text{credit hours}}{\$1,000} = \dfrac{1\ \text{credit hour}}{n\ \text{dollars}}$

$\dfrac{\$1,000}{16\ \text{credit hours}} = \dfrac{n\ \text{dollars}}{1\ \text{credit hour}}$

$\dfrac{62.5 \cdot \cancel{16}^{1}}{\cancel{16}^{1} \cdot 1} = n$

$62.5 = n, \text{ or } n = \$62.50$

67) $\dfrac{2\ \text{meters}}{3\ \text{meters}} \overset{?}{=} \dfrac{n\ \text{meters}}{30\ \text{meters}}$

$\dfrac{30}{1} \cdot \dfrac{2}{3} = \dfrac{n}{\cancel{30}^{1}} \cdot \dfrac{\cancel{30}^{1}}{1}$

$\dfrac{\cancel{3}^{1} \cdot 10}{1} \cdot \dfrac{2}{\cancel{3}^{1} \cdot 1} = n$

$20 = n, \text{ or } n = 20\ \text{meters}$

The tree is 20 meters tall.

69) $\dfrac{850\ \text{miles}}{2\ \text{hours}} = \dfrac{1,275\ \text{miles}}{n\ \text{hours}}$

$\dfrac{2\ \text{hours}}{850\ \text{miles}} = \dfrac{n\ \text{hours}}{1,275\ \text{miles}}$

$\dfrac{1,275}{1} \cdot \dfrac{2}{850} = \dfrac{\cancel{1,275}^{1}}{1} \cdot \dfrac{n}{\cancel{1,275}^{1}}$

$\dfrac{\cancel{425}^{1} \cdot 3}{1} \cdot \dfrac{\cancel{2}^{1} \cdot 1}{\cancel{425}^{1} \cdot \cancel{2}^{1}} = n$

$3 = n, \text{ or } n = 3\ \text{hours}$

It will take the plane three hours.

4.3   Applications of Ratios, Rates, and Proportions

GUIDED PRACTICE

1) On the blueprint, the width of the cabin is 5 inches.

$\dfrac{5\ \text{ft}}{1\ \text{in.}} = \dfrac{n\ \text{ft}}{5\ \text{in.}}$

$\dfrac{5\ \text{ft}}{1\ \cancel{\text{in.}}} \cdot \dfrac{5\ \cancel{\text{in.}}}{1} = \dfrac{n\ \text{ft}}{5\ \cancel{\text{in.}}} \cdot \dfrac{5\ \cancel{\text{in.}}}{1}$

$25\ \text{ft} = n$

The width of the cabin is 25 feet.

2) Length of the Bathroom

$\dfrac{5\ \text{ft}}{1\ \text{in.}} = \dfrac{n\ \text{ft}}{3.5\ \text{in.}}$

$\dfrac{5\ \text{ft}}{1\ \cancel{\text{in.}}} \cdot \dfrac{3.5\ \cancel{\text{in.}}}{1} = \dfrac{n\ \text{ft}}{3.5\ \cancel{\text{in.}}} \cdot \dfrac{3.5\ \cancel{\text{in.}}}{1}$

$17.5\ \text{ft} = n$

Width of the Bathroom

$\dfrac{5\ \text{ft}}{1\ \text{in.}} = \dfrac{n\ \text{ft}}{2\ \text{in.}}$

$\dfrac{5\ \text{ft}}{1\ \cancel{\text{in.}}} \cdot \dfrac{2\ \cancel{\text{in.}}}{1} = \dfrac{n\ \text{ft}}{2\ \cancel{\text{in.}}} \cdot \dfrac{2\ \cancel{\text{in.}}}{1}$

$10\ \text{ft} = n$

Area of the Bathroom
$A = l \cdot w$
$\quad = 17.5\ \text{feet} \cdot 10\ \text{feet}$
$\quad = 175\ \text{square feet}$

The area of the bathroom is 175 ft^2.

3) The rate $= \dfrac{86 \text{ white jelly beans}}{1.3 \text{ lb}}$

$\dfrac{86 \text{ white jelly beans}}{1.3 \text{ lb}} = \dfrac{n}{100 \text{ lb}}$

$\dfrac{86 \text{ white jelly beans}}{1.3 \text{ lb}} \cdot \dfrac{100 \text{ lb}}{1} = \dfrac{n}{100 \text{ lb}} \cdot \dfrac{100 \text{ lb}}{1}$

$6{,}615 \text{ white jelly beans} \approx n$

There are approximately 6,615 white jelly beans in a 100-pound box.

4) A maximum of "$\dfrac{50 \text{ mg}}{1 \text{ kg}}$ a day in divided doses"

Divide by 2 to get $\dfrac{25 \text{ mg}}{1 \text{ kg}}$ for a single dose, twice daily.

$\text{Dose} = \dfrac{300 \text{ mg}}{5 \text{ kg}} = \dfrac{60 \text{ mg}}{1 \text{ kg}}$

$\dfrac{60 \text{ mg}}{1 \text{ kg}} > \dfrac{25 \text{ mg}}{1 \text{ kg}}$

Since the prescribed dose is more than the usual dose, do not administer the drug.

5) Prescribed dose is 125 mg, twice daily. Since the usual adult does is 300 mg, the dose is safe.

$\dfrac{1 \text{ mL}}{50 \text{ mg}} = \dfrac{n \text{ mL}}{125 \text{ mg}}$

$\dfrac{125 \text{ mg}}{1} \times \dfrac{1 \text{ mL}}{50 \text{ mg}} = \dfrac{n \text{ mL}}{125 \text{ mg}} \times \dfrac{125 \text{ mg}}{1}$

$\dfrac{125 \text{ mL}}{50} = n$

$2.5 \text{ mL} = n$

2.5 mL must be given to the patient.

CONCEPT CHECKS AND PRACTICE EXERCISES

A1) If a scale drawing shows a 2-inch line that represents an 8-foot wall, then the scale is 1 in. = 4 ft.

A2) 1 in. = 6 ft

$\dfrac{6 \text{ ft}}{1 \text{ in.}} = \dfrac{n \text{ ft}}{5 \text{ in.}}$

$\dfrac{6 \text{ ft}}{1 \text{ in.}} \cdot \dfrac{5 \text{ in.}}{1} = \dfrac{n \text{ ft}}{5 \text{ in.}} \cdot \dfrac{5 \text{ in.}}{1}$

$30 \text{ ft} = n$

A3) $\dfrac{5 \text{ ft}}{1 \text{ in.}} = \dfrac{n \text{ ft}}{5 \text{ in.}}$

$\dfrac{5 \text{ ft}}{1 \text{ in.}} \cdot \dfrac{5 \text{ in.}}{1} = \dfrac{n \text{ ft}}{5 \text{ in.}} \cdot \dfrac{5 \text{ in.}}{1}$

$25 \text{ ft} = n$

North to south, the width of the living room is 25 feet.

A4) $\dfrac{5 \text{ ft}}{1 \text{ in.}} = \dfrac{n \text{ ft}}{3.5 \text{ in.}}$

$\dfrac{5 \text{ ft}}{1 \text{ in.}} \cdot \dfrac{3.5 \text{ in.}}{1} = \dfrac{n \text{ ft}}{3.5 \text{ in.}} \cdot \dfrac{3.5 \text{ in.}}{1}$

$17.5 \text{ ft} = n$

West to east, the length of the living room is 17.5 feet.

A5) Area of the Living Room

$A = l \cdot w$

$= 17.5 \text{ ft} \cdot 25 \text{ ft}$

$= 437.5 \text{ ft}^2$

$\dfrac{437.5 \text{ ft}^2}{1} \cdot \dfrac{\$1.50}{1 \text{ ft}^2} = \$656.25$

It will cost $656.25 to carpet the living room.

A6) As given in Example 2, the area of the bedroom is 262.5 ft^2.

$\dfrac{262.5 \text{ ft}^2}{1} \cdot \dfrac{\$2.50}{1 \text{ ft}^2} = \$656.25$

It will cost $656.25 to install hardwood in the bedroom.

A7) $\dfrac{25 \text{ mi}}{1 \text{ in.}} = \dfrac{n \text{ mi}}{4.5 \text{ in.}}$

$\dfrac{25 \text{ mi}}{1 \text{ in.}} \cdot \dfrac{4.5 \text{ in.}}{1} = \dfrac{n \text{ mi}}{4.5 \text{ in.}} \cdot \dfrac{4.5 \text{ in.}}{1}$

$112.5 \text{ mi} = n$

You still have to travel 112.5 miles.

A8)
$$\frac{1}{87} = \frac{8 \text{ in.}}{n}$$
$$\frac{87}{1} = \frac{n}{8 \text{ in.}}$$
$$\frac{87}{1} \cdot \frac{8 \text{ in.}}{1} = \frac{n}{8 \text{ in.}} \cdot \frac{8 \text{ in.}}{1}$$
$$696 \text{ in.} = n$$
The actual locomotive is 696 inches long.

A9)
$$\frac{1.25 \text{ ft}}{1 \text{ in.}} = \frac{n \text{ ft}}{5.2 \text{ in.}}$$
$$\frac{1.25 \text{ ft}}{1 \text{ in.}} \cdot \frac{5.2 \text{ in.}}{1} = \frac{n \text{ ft}}{5.2 \text{ in.}} \cdot \frac{5.2 \text{ in.}}{1}$$
$$6.5 \text{ ft} = n$$
The real life person will be 6.5 feet tall.

A10) a) Answers may vary.

b) Answers may vary.

B1) a) The population is the students at the college.

b) The sample is the students in the dining hall.

B2) a) The population is the total 500 trees.

b) The sample is 20 trees.

B3) Answers may vary. Example: Sampling techniques do not give an exact amount.

B4)
$$\frac{120}{300} = \frac{n}{4,500}$$
$$\frac{120}{300} \cdot \frac{4,500}{1} = \frac{n}{4,500} \cdot \frac{4,500}{1}$$
$$1,800 = n$$
Approximately 1,800 students take night classes.

B5)
$$\frac{14}{20} = \frac{n}{1,600}$$
$$\frac{14}{20} \cdot \frac{1,600}{1} = \frac{n}{1,600} \cdot \frac{1,600}{1}$$
$$1,120 = n$$
Brian received approximately 1,120 pieces of junk mail.

B6)
$$\frac{40 \text{ fish}}{3 \text{ min}} = \frac{n \text{ fish}}{24 \text{ hr}}$$
$$\frac{40 \text{ fish}}{3 \text{ min}} \cdot \frac{60 \text{ min}}{1 \text{ hr}} \cdot \frac{24 \text{ hr}}{1} = \frac{n \text{ fish}}{24 \text{ hr}} \cdot \frac{24 \text{ hr}}{1}$$
$$19,200 = n$$
Approximately 19,200 fish swam upstream in 24 hours.

B7)
$$\frac{68 \text{ words}}{1 \text{ page}} = \frac{n \text{ words}}{1,540 \text{ pages}}$$
$$\frac{68 \text{ words}}{1 \text{ page}} \cdot \frac{1,540 \text{ pages}}{1} = \frac{n \text{ words}}{1,540 \text{ page}} \cdot \frac{1,540 \text{ pages}}{1}$$
$$104,720 = n$$
There are approximately 104,720 words in the dictionary.

C1)
$$\frac{60 \text{ mg}}{1 \text{ kg}} \div 3 = \frac{60 \text{ mg}}{1 \text{ kg}} \cdot \frac{1}{3} = \frac{20 \text{ mg}}{1 \text{ kg}}$$
The single dose for this medication is 20 mg/kg.

C2) The usual dose of RyPed 200 is 30–50 mg per kg per day in two divided doses.

C3) The usual dose of Terracin for adults is 250 mg per day or 300 mg given in divided doses at 8 to 12 hour intervals. The usual dose for children is 12–25 mg/kg, up to a maximum of 250 mg per single daily injection.

C4) a) $\dfrac{300 \text{ mg}}{12 \text{ kg}} = \dfrac{25 \text{ mg}}{1 \text{ kg}}$, or 25 mg per kg

b) Yes. Since $\dfrac{25 \text{ mg}}{1 \text{ kg}} < \dfrac{50 \text{ mg}}{1 \text{ kg}}$, this prescribed dose is safe.

C5) a) $\dfrac{375 \text{ mg}}{25 \text{ kg}} = \dfrac{15 \text{ mg}}{1 \text{ kg}}$, or 15 mg per kg

b) No. Although $\dfrac{15 \text{ mg}}{1 \text{ kg}} \leq \dfrac{25 \text{ mg}}{1 \text{ kg}}$, the dose is larger than the maximum dose for children of 250 mg. The prescribed dose is not safe.

D1) a) $\dfrac{50 \text{ mg}}{1 \text{ mL}} \cdot 2 \text{ mL} = 100 \text{ mg}$

b) $\dfrac{50 \text{ mg}}{1 \text{ mL}} \cdot 3 \text{ mL} = 150 \text{ mg}$

c) There are 125 mg of the drug in a 2 mL injection.

D2) a) The usual dose of RyPed 200 is 30–50 mg per kg per day in divided doses.

b) $\dfrac{350 \text{ mg}}{8 \text{ kg}} = \dfrac{43.75 \text{ mg}}{1 \text{ kg}}$

Yes. Since $\dfrac{43.75 \text{ mg}}{1 \text{ kg}} < \dfrac{50 \text{ mg}}{1 \text{ kg}}$, this prescribed dose is considered safe.

D2) c) The drug concentration of RyPed 200 is 200 mg per 5 mL.

d)
$$\frac{5 \text{ mL}}{200 \text{ mg}} = \frac{n \text{ mL}}{350 \text{ mg}}$$

$$\frac{5 \text{ mL}}{200 \text{ mg}} \cdot \frac{350 \text{ mg}}{1} = \frac{n \text{ mL}}{350 \text{ mg}} \cdot \frac{350 \text{ mg}}{1}$$

$$8.75 \text{ mL} = n$$

e)

D3) a) The usual dose of Terracin is 250 mg per day or 300 mg given in divided doses at 8- to 12-hour intervals. For children, the usual dose is 15–25 mg/kg up to a maximum of 250 mg.

b)
$$\frac{440 \text{ mg}}{30 \text{ kg}} \approx \frac{14.7 \text{ mg}}{1 \text{ kg}}$$

Yes. Since $\dfrac{14.7 \text{ mg}}{1 \text{ kg}} < \dfrac{25 \text{ mg}}{1 \text{ kg}}$, this prescribed dose is considered safe.

c) The drug concentration of Terracin is 50 mg per mL.

d)
$$\frac{1 \text{ mL}}{50 \text{ mg}} = \frac{n \text{ mL}}{440 \text{ mg}}$$

$$\frac{1 \text{ mL}}{50 \text{ mg}} \cdot \frac{440 \text{ mg}}{1} = \frac{n \text{ mL}}{440 \text{ mg}} \cdot \frac{440 \text{ mg}}{1}$$

$$8.8 \text{ mL} = n$$

e)

SECTION 4.3   EXERCISES

For 1–12, refer to Concept Checks and Practice Exercises.

13) prescribed dose

15) usual dose

17) drug concentration

19)
$$\frac{1.5 \text{ ft}}{1 \text{ in.}} = \frac{n \text{ ft}}{13 \text{ in.}}$$

$$\frac{1.5 \text{ ft}}{1 \text{ in.}} \cdot \frac{13 \text{ in.}}{1} = \frac{n \text{ ft}}{13 \text{ in.}} \cdot \frac{13 \text{ in.}}{1}$$

$$19.5 \text{ ft} = n$$

The overall length of the truck is about 19.5 feet.

21)
$$\frac{1.5 \text{ ft}}{1 \text{ in.}} = \frac{n \text{ ft}}{2 \text{ in.}}$$

$$\frac{1.5 \text{ ft}}{1 \text{ in.}} \cdot \frac{2 \text{ in.}}{1} = \frac{n \text{ ft}}{2 \text{ in.}} \cdot \frac{2 \text{ in.}}{1}$$

$$3 \text{ ft} = n$$

The diameter of the tires is about 3 feet.

23)
$$\frac{1.5 \text{ ft}}{1 \text{ in.}} = \frac{n \text{ ft}}{5 \text{ in.}}$$

$$\frac{1.5 \text{ ft}}{1 \text{ in.}} \cdot \frac{5 \text{ in.}}{1} = \frac{n \text{ ft}}{5 \text{ in.}} \cdot \frac{5 \text{ in.}}{1}$$

$$7.5 \text{ ft} = n$$

The length of the truck bed is about 7.5 feet.

25)
$$\frac{1.5 \text{ ft}}{1 \text{ in.}} = \frac{n \text{ ft}}{1.5 \text{ in.}}$$

$$\frac{1.5 \text{ ft}}{1 \text{ in.}} \cdot \frac{1.5 \text{ in.}}{1} = \frac{n \text{ ft}}{1.5 \text{ in.}} \cdot \frac{1.5 \text{ in.}}{1}$$

$$2.25 \text{ ft} = n$$

The bottom of the rear bumper is about 2.25 feet off the ground.

27) Length of the Kitchen
$$\frac{5 \text{ ft}}{1 \text{ in.}} = \frac{n \text{ ft}}{4 \text{ in.}}$$

$$\frac{5 \text{ ft}}{1 \text{ in.}} \cdot \frac{4 \text{ in.}}{1} = \frac{n \text{ ft}}{4 \text{ in.}} \cdot \frac{4 \text{ in.}}{1}$$

$$20 \text{ ft} = n$$

Width of the Kitchen
$$\frac{5 \text{ ft}}{1 \text{ in.}} = \frac{n \text{ ft}}{3 \text{ in.}}$$

$$\frac{5 \text{ ft}}{1 \text{ in.}} \cdot \frac{3 \text{ in.}}{1} = \frac{n \text{ ft}}{3 \text{ in.}} \cdot \frac{3 \text{ in.}}{1}$$

$$15 \text{ ft} = n$$

The kitchen is 20 feet long by 15 feet wide.

29) Area of the kitchen
$$A = l \cdot w$$
$$= 20 \text{ ft} \cdot 15 \text{ ft}$$
$$= 300 \text{ ft}^2$$

The kitchen will require about 300 square feet of flooring.

31)   On the diagram, six windows are 1 inch wide, while one window is 2 inches wide. The total area of the windows is therefore

$$A = w \cdot h$$

$$= 6 \cdot \left[4 \text{ ft} \cdot \left(1 \text{ in.} \cdot \frac{5 \text{ ft}}{1 \text{ in.}} \right) \right] + 4 \text{ ft} \cdot \left(2 \text{ in.} \cdot \frac{5 \text{ ft}}{1 \text{ in.}} \right)$$

$$= (6 \cdot 4 \text{ ft} \cdot 5 \text{ ft}) + (4 \text{ ft} \cdot 10 \text{ ft})$$

$$= 120 \text{ ft}^2 + 40 \text{ ft}^2$$

$$= 160 \text{ ft}^2$$

The total area of all the windows is about 160 ft^2.

33)
$$\frac{5 \text{ ft}}{1 \text{ in.}} = \frac{n \text{ ft}}{8 \text{ in.} + 10 \text{ in.}}$$

$$\frac{5 \text{ ft}}{1 \text{ in.}} = \frac{n \text{ ft}}{18 \text{ in.}}$$

$$\frac{5 \text{ ft}}{1 \text{ in.}} \cdot \frac{18 \text{ in.}}{1} = \frac{n \text{ ft}}{18 \text{ in.}} \cdot \frac{18 \text{ in.}}{1}$$

$$90 \text{ ft} = n$$

The amount of baseboard needed is about 90 feet.

35)
$$\frac{\text{length}}{\text{width}} = \frac{4 \text{ in.}}{2 \text{ in.}} = \frac{2}{1}$$

No. The ratio of length to width for the living room is 2 to 1 rather than the golden ratio of 1.6 to 1.

37)
$$\frac{10,000}{n} = \frac{250}{1,000,000}$$

$$\frac{n}{10,000} = \frac{1,000,000}{250}$$

$$\frac{10,000}{1} \cdot \frac{n}{10,000} = \frac{10,000}{1} \cdot \frac{1,000,000}{250}$$

$$n = 40,000,000$$

The total population of lake trout in Lake Superior is approximately 40,000,000.

39)
$$\frac{15}{10,000} = \frac{n}{500,000}$$

$$\frac{15}{10,000} \cdot \frac{500,000}{1} = \frac{n}{500,000} \cdot \frac{500,000}{1}$$

$$750 = n$$

About 750 nails are defective.

41)
$$\frac{165}{300} = \frac{n}{9,000,000}$$

$$\frac{9,000,000}{1} \cdot \frac{165}{300} = \frac{n}{9,000,000} \cdot \frac{9,000,000}{1}$$

$$4,950,000 = n$$

About 4,950,000 viewers learned important health information.

43) a) Answers may vary.

b) Answers may vary.

c) Answers may vary.

d) Answers may vary.

e) Answers may vary.

f) Answers may vary.

g) Answers may vary.

h) Answers may vary.

45) a) The usual dose of RyPed is 30–50 mg per kg per day in divided doses.

$$\frac{500 \text{ mg}}{25 \text{ kg}} = \frac{20 \text{ mg}}{\text{kg}}; \quad 20 \text{ mg} \cdot 3 = 60 \text{ mg}$$

Since 60 mg > 50 mg, the prescribed dose is not considered safe.

47) a) The usual adult dose of Maclor is 375 mg two times a day.
Since 300 mg < 375 mg, the prescribed dose is considered safe.

b) The drug concentration of Maclor is 375 mg per 5 mL.

$$\frac{5 \text{ mL}}{375 \text{ mg}} = \frac{n \text{ mL}}{300 \text{ mg}}$$

$$\frac{5 \text{ mL}}{375 \text{ mg}} \cdot \frac{300 \text{ mg}}{1} = \frac{n \text{ mL}}{300 \text{ mg}} \cdot \frac{300 \text{ mg}}{1}$$

$$4 \text{ mL} = n$$

4 mL should be administered.

c)

49) a) The usual adult dose of Terracin is 250 mg administered once every 24 hours or 300 mg given in divided doses at 8-12 hour intervals.
Since 100 mg · 3 = 300 mg, the prescribed dose is considered safe.

b) The drug concentration of Terracin is 50 mg per mL.

$$\frac{1 \text{ mL}}{50 \text{ mg}} = \frac{n \text{ mL}}{100 \text{ mg}}$$

$$\frac{1 \text{ mL}}{50 \text{ mg}} \cdot \frac{100 \text{ mg}}{1} = \frac{n \text{ mL}}{100 \text{ mg}} \cdot \frac{100 \text{ mg}}{1}$$

$$2 \text{ mL} = n$$

2 mL should be administered.

c)

51) a) The usual dose of RyPed is 33–50 mg per kg per day in divided doses.

$$\frac{300 \text{ mg}}{30 \text{ kg}} = \frac{10 \text{ mg}}{\text{kg}}; \quad 10 \text{ mg} \cdot 3 = 30 \text{ mg}$$

Since 30 mg ≤ 50 mg, the prescribed dose is considered safe.

b) The drug concentration of RyPed is 200 mg per 5 mL.

$$\frac{5 \text{ mL}}{200 \text{ mg}} = \frac{n \text{ mL}}{300 \text{ mg}}$$

$$\frac{5 \text{ mL}}{200 \text{ mg}} \cdot \frac{300 \text{ mg}}{1} = \frac{n \text{ mL}}{300 \text{ mg}} \cdot \frac{300 \text{ mg}}{1}$$

$$7.5 \text{ mL} = n$$

c) 7.5 mL should be administered.

CHAPTER 4   REVIEW EXERCISES

1) A ratio is a quotient that compares two quantities with like units.

2) The ratio of 21 miles to 15 miles is

$$\frac{21 \text{ miles}}{15 \text{ miles}} = \frac{3 \cdot 7}{3 \cdot 5} = \frac{7}{5}.$$

3) The ratio of 30 mph to 64 mph is

$$\frac{30 \text{ mph}}{64 \text{ mph}} = \frac{2 \cdot 15}{2 \cdot 32} = \frac{15}{32}.$$

4) The ratio of 1.4 cm to 50 cm is

$$\frac{1.4 \text{ cm}}{50 \text{ cm}} = \frac{1.4}{50} \cdot \frac{10}{10} = \frac{14}{500} = \frac{2 \cdot 7}{2 \cdot 250} = \frac{7}{250}.$$

5) The ratio of 3.5 inches to 40 inches is

$$\frac{3.5 \text{ inches}}{40 \text{ inches}} = \frac{3.5}{40} \cdot \frac{10}{10} = \frac{35}{400} = \frac{5 \cdot 7}{5 \cdot 80} = \frac{7}{80}.$$

6) A ratio is a comparison of items with the same units. A rate is a comparison of items with different units.

7) A unit rate has a denominator of 1 unit.

8) $\text{rate} = \dfrac{\$28}{3 \text{ hours}}$

$\text{unit rate} \approx \dfrac{\$9.33}{1 \text{ hour}}$

The unit rate is $9.33 per hour.

9) $\text{rate} = \dfrac{\$36}{5 \text{ hours}}$

$\text{unit rate} = \dfrac{\$7.20}{1 \text{ hour}}$

The unit rate is $7.20 per hour.

10) $\text{rate} = \dfrac{\$2.90}{20 \text{ ounces}}$

$\text{unit rate} \approx \dfrac{\$0.15}{1 \text{ ounce}}$

The unit rate is $0.15 per ounce.

11) $\text{rate} = \dfrac{\$3.80}{30 \text{ ounces}}$

$\text{unit rate} \approx \dfrac{\$0.13}{1 \text{ ounce}}$

The unit rate is $0.13 per ounce.

12) $\dfrac{12 \text{ oz}}{\$1.00} = 12$ oz per dollar

$\dfrac{68 \text{ oz}}{\$1.56} \approx 43.59$ oz per dollar

Purchasing the larger bottle will give you more soda for every dollar. Therefore, the 68 ounce bottle is the better buy.

13) $\dfrac{128 \text{ oz}}{\$2.55} \approx 50.20$ oz per dollar

$\dfrac{100 \text{ oz}}{\$2.10} \approx 47.62$ oz per dollar

Purchasing the larger bottle will give you more bleach for every dollar. Therefore, the 128 ounce bottle is the better buy.

14) $\dfrac{\$3.25}{128 \text{ oz}} \approx \$0.03$ per oz

$\dfrac{\$1.25}{58 \text{ oz}} \approx \$0.02$ per oz

savings $= \dfrac{\$0.03}{1 \text{ oz}} - \dfrac{\$0.02}{1 \text{ oz}} \approx \$0.01$ per oz

The savings gained from buying the smaller bottle is about \$0.01 per ounce.

15) $\dfrac{\$2.25}{10 \text{ oz}} \approx \$0.23$ per oz

$\dfrac{\$3.45}{24 \text{ oz}} \approx \$0.14$ per oz

savings $= \dfrac{\$0.23}{1 \text{ oz}} - \dfrac{\$0.14}{1 \text{ oz}} \approx \$0.09$ per oz

The savings gained from buying the larger jar is about \$0.09 per ounce.

16) A proportion states that two rates are equal.

17) The statement is not a proportion because the same units are not in the numerators. Similarly, the units in the denominators are not the same.

18) $\dfrac{15}{45} = \dfrac{1}{3}$

19) $\dfrac{11}{55} = \dfrac{22}{110}$

20) $\dfrac{4 \text{ days}}{60 \text{ meters}} = \dfrac{x \text{ days}}{100 \text{ meters}}$

21) $\dfrac{3 \text{ hours}}{5 \text{ pages}} = \dfrac{h \text{ hours}}{8 \text{ pages}}$

22) $\dfrac{6}{10} \overset{?}{=} \dfrac{2}{3}$

$6 \cdot 3 \overset{?}{=} 10 \cdot 2$

$18 \neq 20$

No. The statement is not a proportion.

23) $\dfrac{4}{11} \overset{?}{=} \dfrac{77}{28}$

$4 \cdot 28 \overset{?}{=} 11 \cdot 77$

$112 \neq 847$

No. The statement is not a proportion.

24) $\dfrac{3.5}{3} \overset{?}{=} \dfrac{14}{12}$

$3.5 \cdot 12 \overset{?}{=} 3 \cdot 14$

$42 = 42$

Yes. The statement is a proportion.

25) $\dfrac{1\frac{1}{2}}{9} \overset{?}{=} \dfrac{2\frac{1}{9}}{16}$

$1\frac{1}{2} \cdot 16 \overset{?}{=} 9 \cdot 2\frac{1}{9}$

$24 \neq 19$

No. The statement is not a proportion.

26) $\dfrac{2}{5} = \dfrac{n}{30}$

$\dfrac{30}{1} \cdot \dfrac{2}{5} = \dfrac{\cancel{30}}{1} \cdot \dfrac{n}{\cancel{30}}$

$\dfrac{6 \cdot \cancel{5}}{1} \cdot \dfrac{2}{\cancel{5}} = n$

$12 = n$

27) $\dfrac{7}{20} = \dfrac{n}{80}$

$\dfrac{80}{1} \cdot \dfrac{7}{20} = \dfrac{\cancel{80}}{1} \cdot \dfrac{n}{\cancel{80}}$

$\dfrac{4 \cdot \cancel{20}}{1} \cdot \dfrac{7}{\cancel{20}} = n$

$28 = n$

28) $\dfrac{3}{8} = \dfrac{7}{n}$

$\dfrac{8}{3} = \dfrac{n}{7}$

$\dfrac{7}{1} \cdot \dfrac{8}{3} = \dfrac{\cancel{7}}{1} \cdot \dfrac{n}{\cancel{7}}$

$\dfrac{56}{3} = 18\frac{2}{3} = n$

29) $\dfrac{9}{11} = \dfrac{2}{n}$

$\dfrac{11}{9} = \dfrac{n}{2}$

$\dfrac{2}{1} \cdot \dfrac{11}{9} = \dfrac{\cancel{2}}{1} \cdot \dfrac{n}{\cancel{2}}$

$\dfrac{22}{9} = 2\frac{4}{9} = n$

30) $\dfrac{\$n}{14\ \text{hours}} = \dfrac{\$85}{10\ \text{hours}}$

$\dfrac{\cancel{14}}{1} \cdot \dfrac{n}{\cancel{14}} = \dfrac{14}{1} \cdot \dfrac{85}{10}$

$n = \dfrac{\cancel{2} \cdot 7}{1} \cdot \dfrac{17 \cdot \cancel{5}}{\cancel{2} \cdot \cancel{5}}$

$n = 119$

She will earn $119.

31) $\dfrac{\$n}{4\ \text{hours}} = \dfrac{\$25}{3\ \text{hours}}$

$\dfrac{\cancel{4}}{1} \cdot \dfrac{n}{\cancel{4}} = \dfrac{4}{1} \cdot \dfrac{25}{3}$

$n = \dfrac{100}{3} = 33.33$

They will need to pay the babysitter approximately $33.33.

32) $\dfrac{25\ \text{miles}}{1\ \text{in.}} = \dfrac{n\ \text{miles}}{2.5\ \text{in.}}$

$\dfrac{25\ \text{miles}}{1\ \cancel{\text{in.}}} \cdot \dfrac{2.5\ \cancel{\text{in.}}}{1} = \dfrac{n\ \text{miles}}{\cancel{2.5\ \text{in.}}} \cdot \dfrac{\cancel{2.5\ \text{in.}}}{1}$

$62.5\ \text{miles} = n$

You still have to travel about 62.5 miles.

33) $\dfrac{10\ \text{miles}}{1\ \text{in.}} = \dfrac{n\ \text{miles}}{5.6\ \text{in.}}$

$\dfrac{10\ \text{miles}}{1\ \cancel{\text{in.}}} \cdot \dfrac{5.6\ \cancel{\text{in.}}}{1} = \dfrac{n\ \text{miles}}{\cancel{5.6\ \text{in.}}} \cdot \dfrac{\cancel{5.6\ \text{in.}}}{1}$

$56\ \text{miles} = n$

You still have to travel about 56 miles.

34) $\dfrac{4\ \text{in.}}{1\ \text{in.}} = \dfrac{n\ \text{in.}}{5\ \text{in.}}$

$\dfrac{4\ \text{in.}}{1\ \cancel{\text{in.}}} \cdot \dfrac{5\ \cancel{\text{in.}}}{1} = \dfrac{n\ \text{in.}}{\cancel{5\ \text{in.}}} \cdot \dfrac{\cancel{5\ \text{in.}}}{1}$

$20\ \text{in.} = n$

The length of the model plane is about 20 inches.

35) $\dfrac{4\ \text{in.}}{1\ \text{in.}} = \dfrac{n\ \text{in.}}{7\ \text{in.}}$

$\dfrac{4\ \text{in.}}{1\ \cancel{\text{in.}}} \cdot \dfrac{7\ \cancel{\text{in.}}}{1} = \dfrac{n\ \text{in.}}{\cancel{7\ \text{in.}}} \cdot \dfrac{\cancel{7\ \text{in.}}}{1}$

$28\ \text{in.} = n$

The wingspan of the model plane is about 28 inches.

36) $\dfrac{4\ \text{in.}}{1\ \text{in.}} = \dfrac{n\ \text{in.}}{3.5\ \text{in.}}$

$\dfrac{4\ \text{in.}}{1\ \cancel{\text{in.}}} \cdot \dfrac{3.5\ \cancel{\text{in.}}}{1} = \dfrac{n\ \text{in.}}{\cancel{3.5\ \text{in.}}} \cdot \dfrac{\cancel{3.5\ \text{in.}}}{1}$

$14\ \text{in.} = n$

The length of one wing from the center of the model plane is about 14 inches.

37) $\dfrac{4\ \text{in.}}{1\ \text{in.}} = \dfrac{n\ \text{in.}}{2\ \text{in.}}$

$\dfrac{4\ \text{in.}}{1\ \cancel{\text{in.}}} \cdot \dfrac{2\ \cancel{\text{in.}}}{1} = \dfrac{n\ \text{in.}}{\cancel{2\ \text{in.}}} \cdot \dfrac{\cancel{2\ \text{in.}}}{1}$

$8\ \text{in.} = n$

The width of the tail is about 8 inches.

38) $\dfrac{150}{400} = \dfrac{n}{3{,}000}$

$\dfrac{150}{400} \cdot \dfrac{3{,}000}{1} = \dfrac{n}{\cancel{3{,}000}} \cdot \dfrac{\cancel{3{,}000}}{1}$

$1{,}125 = n$

Approximately 1,125 students are taking a mathematics class.

39) $\dfrac{12}{26} = \dfrac{n}{1{,}500}$

$\dfrac{12}{26} \cdot \dfrac{1{,}500}{1} = \dfrac{n}{\cancel{1{,}500}} \cdot \dfrac{\cancel{1{,}500}}{1}$

$692 \approx n$

Brenda received approximately 692 pieces of junk mail.

40) $\dfrac{35\ \text{fish}}{2\ \text{min}} = \dfrac{n\ \text{fish}}{1\ \text{hr}}$

$\dfrac{35\ \text{fish}}{2\ \cancel{\text{min}}} \cdot \dfrac{60\ \cancel{\text{min}}}{1\ \cancel{\text{hr}}} \cdot \dfrac{1\ \cancel{\text{hr}}}{1} = \dfrac{n\ \text{fish}}{1\ \cancel{\text{hr}}} \cdot \dfrac{1\ \cancel{\text{hr}}}{1}$

$1{,}050 = n$

Approximately 1,050 fish go upstream in one hour.

41) $\dfrac{580\ \text{words}}{1\ \text{page}} = \dfrac{n\ \text{words}}{425\ \text{pages}}$

$\dfrac{580\ \text{words}}{1\ \cancel{\text{page}}} \cdot \dfrac{425\ \cancel{\text{pages}}}{1} = \dfrac{n\ \text{words}}{\cancel{425\ \text{pages}}} \cdot \dfrac{\cancel{425\ \text{pages}}}{1}$

$246{,}500 = n$

There were approximately 246,500 words.

42) a) The usual dose of Florazine is 75 to 400 mg per day. Since

$$\frac{750 \text{ mg}}{1 \text{ dose}} \cdot \frac{2 \text{ doses}}{1 \text{ day}} = \frac{1,500 \text{ mg}}{1 \text{ day}} > \frac{400 \text{ mg}}{1 \text{ day}},$$

the prescribed dose is not considered safe.

43) a) The usual dose of RyPed 200 is 33–50 mg per kg per day in divided doses.

$$\frac{250 \text{ mg}}{10 \text{ kg}} = \frac{25 \text{ mg}}{1 \text{ kg}};$$

$$\frac{\frac{25 \text{ mg}}{1 \text{ kg}}}{1 \text{ dose}} \cdot \frac{2 \text{ doses}}{1 \text{ day}} = \frac{50 \text{ mg}}{1 \text{ kg}} \text{ per day}$$

Since $\frac{50 \text{ mg}}{1 \text{ kg}}$ per day $\geq \frac{50 \text{ mg}}{1 \text{ kg}}$ per day, the prescribed dose is considered safe.

b) The drug concentration of RyPed 200 is 200 mg per 5 mL.

$$\frac{5 \text{ mL}}{200 \text{ mg}} = \frac{n \text{ mL}}{250 \text{ mg}}$$

$$\frac{5 \text{ mL}}{200 \text{ mg}} \cdot \frac{250 \text{ mg}}{1} = \frac{n \text{ mL}}{250 \text{ mg}} \cdot \frac{250 \text{ mg}}{1}$$

$$6.25 \text{ mL} = n$$

c) 6.25 mL should be administered.

44) a) The usual dose of Terracin for a child is 12–25 mg per kg up to a maximum of 250 mg per single daily injection.

$$\frac{70 \text{ mg}}{20 \text{ kg}} = \frac{3.5 \text{ mg}}{1 \text{ kg}} \text{ per dose; or } 3.5 \cdot 3 = 10.5$$

mg/kg per day
Since 10.5 mg < 25 mg, the prescribed dose is considered safe.

b) The drug concentration of Terracin is 50 mg per mL.

$$\frac{1 \text{ mL}}{50 \text{ mg}} = \frac{n \text{ mL}}{70 \text{ mg}}$$

$$\frac{1 \text{ mL}}{50 \text{ mg}} \cdot \frac{70 \text{ mg}}{1} = \frac{n \text{ mL}}{70 \text{ mg}} \cdot \frac{70 \text{ mg}}{1}$$

$$1.4 \text{ mL} = n$$

1.4 mL should be administered.

c)

CHAPTER 4   TEST

1) $\text{rate} = \dfrac{2 \text{ cups}}{3 \text{ loaves}}$

It is a rate because the units are different.

2) $\text{ratio} = \dfrac{3 \text{ miles}}{2 \text{ miles}} = \dfrac{3}{2}$

It is a ratio because the units are the same.

3) $\text{ratio} = \dfrac{\$1.24}{\$41.27} = \dfrac{1.24}{41.27}$

It is a ratio because the units are the same.

4) $\text{rate} = \dfrac{2 \text{ cartridges}}{1 \text{ box}}$

It is a rate because the units are different.

5) $\text{rate} = \dfrac{\$1.28}{8 \text{ ounces}}$

$\text{unit rate} = \dfrac{\$0.16}{1 \text{ ounce}}$

The unit rate is $0.16 per ounce.

6) $\text{rate} = \dfrac{144 \text{ miles}}{6 \text{ gallons}}$

$\text{unit rate} = \dfrac{24 \text{ miles}}{\text{gallon}}$

The unit rate is 24 miles per gallon.

7) $\text{rate} = \dfrac{145 \text{ miles}}{2.5 \text{ hours}}$

$\text{unit rate} = \dfrac{58 \text{ miles}}{\text{hour}}$

The unit rate is 58 miles per hour.

8) $\text{rate} = \dfrac{\$328}{40 \text{ hours}}$

$\text{unit rate} = \dfrac{\$8.20}{1 \text{ hour}}$

The unit rate is $8.20 per hour.

9) $\dfrac{\$3.85}{20 \text{ oz}} = \$0.1925 \text{ per ounce}$

$\dfrac{\$2.69}{15 \text{ oz}} \approx \$0.1793 \text{ per ounce}$

The smaller box costs less per ounce. Therefore, the 15-ounce box is the better buy at $2.69.

10) $\dfrac{\$1.64}{4 \text{ lb}} = \$0.41$ per lb

$\dfrac{\$1.99}{5 \text{ lb}} \approx \$0.40$ per pound

The larger package costs less per pound.
Therefore, the 5-pound package is the better
buy at $1.99.

11) $\dfrac{\$45}{2 \text{ hours}} = \dfrac{\$n}{3 \text{ hours}}$

$\dfrac{3}{1} \cdot \dfrac{45}{2} = \dfrac{\cancel{3}}{1} \cdot \dfrac{n}{\cancel{3}}$

$67.5 = n$
She earns $67.50.

12) a)   rate $= \dfrac{300 \text{ mg}}{80 \text{ kg}}$

unit rate $= \dfrac{3.75 \text{ mg}}{1 \text{ kg}}$

The unit rate is 3.75 mg per kg.

b) $\dfrac{3.75 \text{ mg}}{1 \text{ kg}} = \dfrac{n \text{ mg}}{50 \text{ kg}}$

$\dfrac{50}{1} \cdot \dfrac{3.75}{1} = \dfrac{\cancel{50}}{1} \cdot \dfrac{n}{\cancel{50}}$

$\dfrac{50 \cdot 3.75}{1} = n$

$187.5 = n$
The dose should be 187.5 mg.

13) $\dfrac{10}{14} \overset{?}{=} \dfrac{5}{7}$

$10 \cdot 7 \overset{?}{=} 14 \cdot 5$

$70 = 70$
Yes. The statement is a proportion.

14) $\dfrac{13.6 \text{ kg}}{8 \text{ m}} \overset{?}{=} \dfrac{15.3 \text{ kg}}{9 \text{ m}}$

$13.6 \cdot 9 \overset{?}{=} 8 \cdot 15.3$

$122.4 = 122.4$
Yes. The statement is a proportion.

15) $\dfrac{12}{47} \overset{?}{=} \dfrac{3}{11}$

$12 \cdot 11 \overset{?}{=} 47 \cdot 3$

$132 \neq 141$
No. The statement is not a proportion.

16) $\dfrac{42 \text{ feet}}{17 \text{ hours}} \overset{?}{=} \dfrac{33 \text{ feet}}{13 \text{ hours}}$

$42 \cdot 13 \overset{?}{=} 17 \cdot 33$

$546 \neq 561$
No. The statement is not a proportion.

17) $\dfrac{24}{20} = \dfrac{15}{n}$

$\dfrac{20}{24} = \dfrac{n}{15}$

$\dfrac{15}{1} \cdot \dfrac{20}{24} = \dfrac{\cancel{15}}{1} \cdot \dfrac{n}{\cancel{15}}$

$12.5 = n$

18) $\dfrac{70 \text{ m}}{16 \text{ sec}} = \dfrac{n \text{ m}}{24 \text{ sec}}$

$\dfrac{70 \text{ m}}{16 \text{ sec}} \cdot \dfrac{24 \text{ sec}}{1} = \dfrac{n \text{ m}}{24 \text{ sec}} \cdot \dfrac{24 \text{ sec}}{1}$

$105 \text{ m} = n$

19) $\dfrac{n \text{ gallons}}{1,400 \text{ ft}^2} = \dfrac{3 \text{ gallons}}{1,050 \text{ ft}^2}$

$\dfrac{n}{1,400} = \dfrac{3}{1,050}$

$\dfrac{\cancel{1,400}}{1} \cdot \dfrac{n}{\cancel{1,400}} = \dfrac{1400}{1} \cdot \dfrac{3}{1050}$

$n = 4$
He needs 4 gallons.

20) $\dfrac{n \text{ hours}}{1 \text{ report}} = \dfrac{3 \text{ hours}}{\frac{2}{3} \text{ report}}$

$\dfrac{n}{1} = \dfrac{3}{\frac{2}{3}}$

$n = 3 \cdot \dfrac{3}{2}$

$n = \dfrac{9}{2} = 4.5$

It will take 4.5 hours to finish the entire report.

21) For the family room:

$l = 8 \text{ gridlines} \cdot \dfrac{2 \text{ ft}}{1 \text{ gridline}} = 16 \text{ ft}$

$w = 5 \text{ gridlines} \cdot \dfrac{2 \text{ ft}}{1 \text{ gridline}} = 10 \text{ ft}$

$\begin{aligned} A_{family} &= l \cdot w \\ &= 16 \text{ ft} \cdot 10 \text{ ft} \\ &= 160 \text{ ft}^2 \end{aligned}$

For the living room:

$$l = 6 \text{ gridlines} \cdot \frac{2 \text{ ft}}{1 \text{ gridline}} = 12 \text{ ft}$$

$$w = 5 \text{ gridlines} \cdot \frac{2 \text{ ft}}{1 \text{ gridline}} = 10 \text{ ft}$$

$$\begin{aligned} A_{living} &= l \cdot w \\ &= 12 \text{ ft} \cdot 10 \text{ ft} \\ &= 120 \text{ ft}^2 \end{aligned}$$

The area of both rooms is
$160 \text{ ft}^2 + 120 \text{ ft}^2 = 280 \text{ ft}^2$

22)　$l = 7 \text{ ft} \cdot \dfrac{1 \text{ gridline}}{2 \text{ ft}} = 3\dfrac{1}{2} \text{ gridlines}$

$w = 3 \text{ ft} \cdot \dfrac{1 \text{ gridline}}{2 \text{ ft}} = 1\dfrac{1}{2} \text{ gridlines}$

The sofa will fill $3\dfrac{1}{2} \times 1\dfrac{1}{2}$ gridlines.

23)　$\dfrac{13}{20} = \dfrac{n}{325}$

$\dfrac{13}{20} \cdot \dfrac{325}{1} = \dfrac{n}{325} \cdot \dfrac{325}{1}$

$211.25 = n$

$211 \approx n$

Approximately 211 prefer beige paint.

24)　$\dfrac{500}{n} = \dfrac{8}{50}$

$\dfrac{n}{500} = \dfrac{50}{8}$

$\dfrac{500}{1} \cdot \dfrac{n}{500} = \dfrac{500}{1} \cdot \dfrac{50}{8}$

$n = 3,125$

There are approximately 3,125 fish in the pond.

25)　The usual dose of RyPed 200 is 33–50 mg
per kg per day in divided doses.

$$\frac{500 \text{ mg}}{20 \text{ kg}} = \frac{25 \text{ mg}}{\text{kg}}$$

Since 25 mg per kg < 50 mg per kg, the
prescribed dose is considered safe.

26)　The drug concentration of RyPed 200 is 200 mg
per 5 mL.

$$\frac{5 \text{ mL}}{200 \text{ mg}} = \frac{n \text{ mL}}{500 \text{ mg}}$$

$$\frac{5 \text{ mL}}{200 \text{ mg}} \cdot \frac{500 \text{ mg}}{1} = \frac{n \text{ mL}}{500 \text{ mg}} \cdot \frac{500 \text{ mg}}{1}$$

$$12.5 \text{ mL} = n$$

12.5 mL should be administered.

Chapter 5   Percents

5.1   Percents, Fractions, and Decimals

GUIDED PRACTICE

1) $37\% = \dfrac{37}{100}$

2) $63\% = \dfrac{63}{100}$

3) $171\% = \dfrac{171}{100} = 1\dfrac{71}{100}$

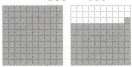

4) $60\% = \dfrac{60\%}{1} \cdot \dfrac{1}{100\%}$

$\quad = \dfrac{60}{100}$

$\quad = \dfrac{3 \cdot 20}{5 \cdot 20}$

$\quad = \dfrac{3}{5}$

5) $16\dfrac{2}{3}\% = \dfrac{50\%}{3}$

$\quad = \dfrac{50\%}{3} \cdot \dfrac{1}{100\%}$

$\quad = \dfrac{50}{3} \cdot \dfrac{1}{100}$

$\quad = \dfrac{50}{3} \cdot \dfrac{1}{2 \cdot 50}$

$\quad = \dfrac{1}{6}$

6) $220\% = \dfrac{220\%}{1} \cdot \dfrac{1}{100\%}$

$\quad = \dfrac{220}{100}$

$\quad = \dfrac{11 \cdot 20}{5 \cdot 20}$

$\quad = \dfrac{11}{5}$ or $2\dfrac{1}{5}$

7) $\dfrac{18}{25} = \dfrac{18}{25} \cdot \dfrac{100\%}{1}$

$\quad = \dfrac{18}{25} \cdot \dfrac{25 \cdot 4\%}{1}$

$\quad = \dfrac{72\%}{1}$

$\quad = 72\%$

8) $\dfrac{3}{8} = \dfrac{3}{8} \cdot \dfrac{100\%}{1}$

$\quad = \dfrac{300\%}{8}$

$\quad = 37\dfrac{1}{2}\%$

9) $1\dfrac{1}{12} = \dfrac{13}{12}$

$\quad = \dfrac{13}{12} \cdot \dfrac{100\%}{1}$

$\quad = \dfrac{13}{3 \cdot 4} \cdot \dfrac{4 \cdot 25\%}{1}$

$\quad = \dfrac{325\%}{3}$

$\quad = 108\dfrac{1}{3}\%$

10) The missing multiplier is 20 because $5 \cdot 20 = 100$.

$\quad \dfrac{4}{5} \cdot \dfrac{20}{20} = \dfrac{80}{100} = 80\%$

11) $\qquad 2\dfrac{6}{25} = 2 + \dfrac{6}{25}$

$2 = 2 \cdot 100\% \qquad \dfrac{6}{25} = \dfrac{6}{25} \cdot \dfrac{4}{4}$

$\quad = 200\% \qquad\qquad = \dfrac{24}{100}$

$\qquad\qquad\qquad\qquad = 24\%$

$\qquad 2\dfrac{6}{25} = 200\% + 24\%$

$\qquad\qquad = 224\%$

12) $62\% = \dfrac{62\%}{1} \cdot \dfrac{1}{100\%}$

$\quad = \dfrac{62}{100}$

$\quad = 0.62$

13) $0.0031 = 0.0031 \cdot 100\%$
$= 0.31\%$

14) $340\% = \dfrac{340\%}{1} \cdot \dfrac{1}{100\%}$
$= \dfrac{340}{100}$
$= 3.4$

15) $10.3 = 10.3 \cdot 100\%$
$= 1,030\%$

16) $\dfrac{3}{5} = 0.6$
$= 0.6 \cdot 100\%$
$= 60\%$

17) $\dfrac{3}{5} = \dfrac{3}{5} \cdot \dfrac{100\%}{1}$
$= \dfrac{3 \cdot \cancel{5} \cdot 20\%}{\cancel{5}}$
$= 60\%$

18) $0.045 = \dfrac{45}{1,000}$
$= \dfrac{45}{1,000} \cdot \dfrac{100\%}{1}$
$= \dfrac{45}{10 \cdot \cancel{100}} \cdot \dfrac{\cancel{100}\%}{1}$
$= \dfrac{45}{10}\%$
$= 4.5\%$

19) $0.045 = 0.045 \cdot 100\%$
$= 4.5\%$

20) a) The red sector represents state funding. The fraction $\dfrac{1}{2}$ is a good estimate for this portion.

$\dfrac{1}{2} \cdot \dfrac{100\%}{1} = \dfrac{100\%}{2} = 50\%$

Approximately 50% of the budget comes from the state funding.

20) b) The sector for local taxes looks like it is more than $\dfrac{1}{4}$ but less than $\dfrac{1}{3}$.

Low Estimate as a Fraction: $\dfrac{1}{4}$

High Estimate as a Fraction: $\dfrac{1}{3}$

Low Estimate as a Percent:

$\dfrac{1}{4} \cdot \dfrac{100\%}{1} = \dfrac{100}{4}\% = 25\%$

High Estimate as a Percent:

$\dfrac{1}{3} \cdot \dfrac{100\%}{1} = 33\dfrac{1}{3}\%$

30% is a good estimate because it is between these values. Approximately 30% of the budget comes from local taxes.

c) The percentages of the local taxes, state funding, and tuition must add to 100%. Subtract the known percents from 100% to find the percent of the budget that comes from tuition.

$100\% - 50\% - 30\% = 50\% - 30\%$
$= 20\%$

Approximately 20% of the budget comes from tuition.

CONCEPT CHECKS AND PRACTICE EXERCISES

A1) $90\% = \dfrac{90}{100}$; f)

A2) $50\% = \dfrac{50}{100}$; h)

A3) $20\% = \dfrac{20}{100}$; e)

A4) $80\% = \dfrac{80}{100}$; b)

A5) $225\% = \dfrac{225}{100} = 2\dfrac{25}{100}$; i)

A6) $325\% = \dfrac{325}{100} = 3\dfrac{25}{100}$; g)

A7) $10\% = \dfrac{10}{100}$

A8) $33\% = \dfrac{33}{100}$

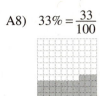

A9) $50\% = \dfrac{50}{100}$

A10) $75\% = \dfrac{75}{100}$

A11) $180\% = \dfrac{180}{100} = 1\dfrac{80}{100}$

A12) $125\% = \dfrac{125}{100} = 1\dfrac{25}{100}$

A13) $290\% = \dfrac{290}{100} = 2\dfrac{90}{100}$

A14) $300\% = \dfrac{300}{100} = 3$

A15) $\dfrac{95}{100} = 95\%$ is shaded.

A16) $\dfrac{20}{100} = 20\%$ is shaded.

A17) $1\dfrac{20}{100} = \dfrac{120}{100} = 120\%$ is shaded.

A18) $2 = \dfrac{200}{100} = 200\%$ is shaded.

B1) Answers may vary. Example: A percent indicates the number of parts out of a whole that is divided into one hundred parts.

B2) Answers may vary. Example: Both fractions and percents represent parts of a whole.

B3) $75\% = \dfrac{75\,\%}{1} \cdot \dfrac{1}{100\,\%}$

$= \dfrac{75}{100}$

$= \dfrac{3 \cdot 25}{4 \cdot 25}$

$= \dfrac{3}{4}$

B4) $20\% = \dfrac{20\,\%}{1} \cdot \dfrac{1}{100\,\%}$

$= \dfrac{20}{100}$

$= \dfrac{1 \cdot 20}{5 \cdot 20}$

$= \dfrac{1}{5}$

B5) $89\% = \dfrac{89\,\%}{1} \cdot \dfrac{1}{100\,\%}$

$= \dfrac{89}{100}$

B6) $8\% = \dfrac{8\,\%}{1} \cdot \dfrac{1}{100\,\%}$

$= \dfrac{8}{100}$

$= \dfrac{2 \cdot 4}{4 \cdot 25}$

$= \dfrac{2}{25}$

B7) $33\dfrac{1}{3}\% = \dfrac{100\%}{3}$

$= \dfrac{100\,\%}{3} \cdot \dfrac{1}{100\,\%}$

$= \dfrac{100}{3} \cdot \dfrac{1}{100}$

$= \dfrac{1}{3}$

B8) $37\frac{1}{2}\% = \frac{75\%}{2}$

$= \frac{75\%}{2} \cdot \frac{1}{100\%}$

$= \frac{75}{2} \cdot \frac{1}{100}$

$= \frac{3 \cdot 25}{2} \cdot \frac{1}{4 \cdot 25}$

$= \frac{3}{8}$

B9) $160\% = \frac{160\%}{1} \cdot \frac{1}{100\%}$

$= \frac{160}{100}$

$= \frac{8 \cdot 20}{5 \cdot 20}$

$= \frac{8}{5}$ or $1\frac{3}{5}$

B10) $115\% = \frac{115\%}{1} \cdot \frac{1}{100\%}$

$= \frac{115}{100}$

$= \frac{5 \cdot 23}{5 \cdot 20}$

$= \frac{23}{20}$ or $1\frac{3}{20}$

C1) The key step used to convert a number to a percent is to multiply by 100%. This step does not change the value of the number because multiplying by 100% is the same as multiplying by one.

C2) Since 100% is equal to 1, a quantity multiplied by a mixed number is larger than the quantity multiplied by 100%.

C3) a) Each value is increased by 100%.

$\frac{1}{4} = 25\%$, $1\frac{1}{4} = 125\%$, $2\frac{1}{4} = 225\%$,

$3\frac{1}{4} = 325\%$, $4\frac{1}{4} = 425\%$

b) The whole number part of the mixed number becomes a percent that is a multiple of 100%.

C4) $\frac{5}{12} = \frac{5}{12} \cdot \frac{100\%}{1}$

$= \frac{500\%}{12}$

$= 41\frac{2}{3}\%$

C5) $\frac{17}{40} = \frac{17}{40} \cdot \frac{100\%}{1}$

$= \frac{1,700\%}{40}$

$= 42\frac{1}{2}\%$

C6) $\frac{1}{8} = \frac{1}{8} \cdot \frac{100\%}{1}$

$= \frac{100\%}{8}$

$= 12\frac{1}{2}\%$

C7) $\frac{5}{6} = \frac{5}{6} \cdot \frac{100\%}{1}$

$= \frac{500\%}{6}$

$= 83\frac{1}{3}\%$

C8) $1\frac{3}{10} = \frac{13}{10}$

$= \frac{13}{10} \cdot \frac{100\%}{1}$

$= \frac{1,300\%}{10}$

$= 130\%$

C9) $2\frac{3}{5} = \frac{13}{5}$

$= \frac{13}{5} \cdot \frac{100\%}{1}$

$= \frac{1,300\%}{5}$

$= 260\%$

C10) $1\frac{1}{2} = \frac{3}{2}$

$= \frac{3}{2} \cdot \frac{100\%}{1}$

$= \frac{300\%}{2}$

$= 150\%$

C11) $3\frac{7}{20} = \frac{67}{20}$

$= \frac{67}{20} \cdot \frac{100\%}{1}$

$= \frac{6,700\%}{20}$

$= 335\%$

FOCUS ON CONVERTING FRACTIONS PRACTICE EXERCISES

1) The missing multiplier is 10 because $10 \cdot 10 = 100$.

$$\frac{9}{10} \cdot \frac{10}{10} = \frac{90}{100} = 90\%$$

2) The missing multiplier is 10 because $10 \cdot 10 = 100$.

$$\frac{3}{10} \cdot \frac{10}{10} = \frac{30}{100} = 30\%$$

3) The missing multiplier is 20 because $5 \cdot 20 = 100$.

$$\frac{1}{5} \cdot \frac{20}{20} = \frac{20}{100} = 20\%$$

4) The missing multiplier is 20 because $5 \cdot 20 = 100$.

$$\frac{4}{5} \cdot \frac{20}{20} = \frac{80}{100} = 80\%$$

5) The missing multiplier is 25 because $4 \cdot 25 = 100$.

$$2\frac{3}{4} = \frac{11}{4} \cdot \frac{25}{25} = \frac{275}{100} = 275\%$$

6) The missing multiplier is 25 because $4 \cdot 25 = 100$.

$$3\frac{1}{4} = \frac{13}{4} \cdot \frac{25}{25} = \frac{325}{100} = 325\%$$

7) The missing multiplier is 5 because $20 \cdot 5 = 100$.

$$2\frac{1}{20} = \frac{41}{20} \cdot \frac{5}{5} = \frac{205}{100} = 205\%$$

8) The missing multiplier is 5 because $20 \cdot 5 = 100$.

$$1\frac{1}{20} = \frac{21}{20} \cdot \frac{5}{5} = \frac{105}{100} = 105\%$$

9) The missing multiplier is 100 because $1 \cdot 100 = 100$.

$$2 = \frac{2}{1} \cdot \frac{100}{100} = \frac{200}{100} = 200\%$$

10) The missing multiplier is 100 because $5 \cdot 100 = 100$.

$$5 = \frac{5}{1} \cdot \frac{100}{100} = \frac{500}{100} = 500\%$$

11) The missing multiplier is 50 because $2 \cdot 50 = 100$.

$$\frac{13}{2} \cdot \frac{50}{50} = \frac{650}{100} = 650\%$$

12) The missing multiplier is 50 because $2 \cdot 50 = 100$.

$$\frac{7}{2} \cdot \frac{50}{50} = \frac{350}{100} = 350\%$$

13) Answers may vary. Example: The whole number portion of a mixed number corresponds to hundred of percent.

D1) a) $\dfrac{1}{100\%}$

b) Answers may vary. Example: This multiplication will divide out the percent symbol.

D2) a) ii is correct; $6.5\% = 0.065$

b) The conversion has the same effect as multiplying by $\dfrac{1}{100\%}$.

D3) $85\% = \dfrac{85\%}{1} \cdot \dfrac{1}{100\%}$

$= \dfrac{85}{100}$

$= 0.85$

D4) $36\% = \dfrac{36\%}{1} \cdot \dfrac{1}{100\%}$

$= \dfrac{36}{100}$

$= 0.36$

D5) $0.92 = 0.92 \cdot 100\%$

$= 92\%$

D6) $0.12 = 0.12 \cdot 100\%$

$= 12\%$

D7) $823\% = \dfrac{823\%}{1} \cdot \dfrac{1}{100\%}$

$= \dfrac{823}{100}$

$= 8.23$

D8) $555\% = \dfrac{555\%}{1} \cdot \dfrac{1}{100\%}$

$= \dfrac{555}{100}$

$= 5.55$

D9) $6.5 = 6.5 \cdot 100\%$

$= 650\%$

D10) $4.2 = 4.2 \cdot 100\%$

$= 420\%$

D11) $0.3\% = \dfrac{0.3\%}{1} \cdot \dfrac{1}{100\%}$

$\quad = \dfrac{0.3}{100}$

$\quad = 0.003$

D12) $0.0087 = 0.0087 \cdot 100\%$

$\quad = 0.87\%$

D13) $0.0012 = 0.0012 \cdot 100\%$

$\quad = 0.12\%$

D14) $0.5\% = \dfrac{0.5\%}{1} \cdot \dfrac{1}{100\%}$

$\quad = \dfrac{0.5}{100}$

$\quad = 0.005$

D15) $1{,}000\% = \dfrac{1{,}000\%}{1} \cdot \dfrac{1}{100\%}$

$\quad = \dfrac{1{,}000}{100}$

$\quad = 10$

D16) $4{,}000 = 4{,}000 \cdot 100\%$

$\quad = 400{,}000\%$

D17) $1{,}000 = 1{,}000 \cdot 100\%$

$\quad = 100{,}000\%$

D18) $4{,}000\% = \dfrac{4{,}000\%}{1} \cdot \dfrac{1}{100\%}$

$\quad = \dfrac{4{,}000}{100}$

$\quad = 40$

SECTION 5.1 EXERCISES

For 1–14, refer to Concept Checks and Practice Exercises.

15) Percent

17) Per

19) $45\% = \dfrac{45}{100}$

21) $78\% = \dfrac{78}{100}$

23) $99\% = \dfrac{99}{100}$

25) Add the percents for the library and school.
$3\% + 54\% = 57\%$

27) $51\% = \dfrac{51\%}{1} \cdot \dfrac{1}{100\%}$

$\quad = \dfrac{51}{100}$

29) $80\% = \dfrac{80\%}{1} \cdot \dfrac{1}{100\%}$

$\quad = \dfrac{80}{100}$

$\quad = \dfrac{4 \cdot 20}{5 \cdot 20}$

$\quad = \dfrac{4}{5}$

31) $100\% = \dfrac{100\%}{1} \cdot \dfrac{1}{100\%}$

$\quad = \dfrac{100}{100}$

$\quad = 1$

33) $33\dfrac{1}{3}\% = \dfrac{100\%}{3}$

$\quad = \dfrac{100\%}{3} \cdot \dfrac{1}{100\%}$

$\quad = \dfrac{100}{3} \cdot \dfrac{1}{100}$

$\quad = \dfrac{1}{3}$

35) $87.5\% = \dfrac{87.5\%}{1} \cdot \dfrac{1}{100\%}$

$= \dfrac{87.5}{100} \cdot \dfrac{10}{10}$

$= \dfrac{875}{1,000}$

$= \dfrac{7 \cdot 125}{8 \cdot 125}$

$= \dfrac{7}{8}$

37) $175\% = \dfrac{175\%}{1} \cdot \dfrac{1}{100\%}$

$= \dfrac{175}{100}$

$= \dfrac{7 \cdot 25}{4 \cdot 25}$

$= \dfrac{7}{4}$ or $1\dfrac{3}{4}$

39) $3,000\% = \dfrac{3,000\%}{1} \cdot \dfrac{1}{100\%}$

$= \dfrac{3,000}{100}$

$= 30$

41) $0.02\% = \dfrac{0.02\%}{1} \cdot \dfrac{1}{100\%}$

$= \dfrac{0.02}{100}$

$= \dfrac{2}{10,000}$

$= \dfrac{1}{5,000}$

43) $55\% = \dfrac{55\%}{1} \cdot \dfrac{1}{100\%}$

$= \dfrac{55}{100}$

$= \dfrac{11 \cdot 5}{20 \cdot 5}$

$= \dfrac{11}{20}$

45) $110\% = \dfrac{110\%}{1} \cdot \dfrac{1}{100\%}$

$= \dfrac{110}{100}$

$= \dfrac{11 \cdot 10}{10 \cdot 10}$

$= \dfrac{11}{10}$ or $1\dfrac{1}{10}$

47) Answers may vary.

49) $\dfrac{3}{10} = \dfrac{3}{10} \cdot \dfrac{100\%}{1}$

$= \dfrac{300\%}{10}$

$= 30\%$

51) $\dfrac{4}{9} = \dfrac{4}{9} \cdot \dfrac{100\%}{1}$

$= \dfrac{400\%}{9}$

$= 44\dfrac{4}{9}\%$

53) $\dfrac{5}{3} = \dfrac{5}{3} \cdot \dfrac{100\%}{1}$

$= \dfrac{500\%}{3}$

$= 166\dfrac{2}{3}\%$

55) $3\dfrac{1}{5} = \dfrac{16}{5} \cdot \dfrac{100\%}{1}$

$= \dfrac{1,600\%}{5}$

$= 320\%$

57) $\dfrac{17}{20} = \dfrac{17}{20} \cdot \dfrac{100\%}{1}$

$= \dfrac{1,700\%}{20}$

$= 85\%$

59) $\dfrac{100}{100} = \dfrac{100}{100} \cdot \dfrac{100\%}{1}$

$= 1 \cdot \dfrac{100\%}{1}$

$= 100\%$

61) $3\dfrac{3}{10} = \dfrac{33}{10} \cdot \dfrac{100\%}{1}$

$= \dfrac{3,300\%}{10}$

$= 330\%$

63) $6 = \dfrac{6}{1} \cdot \dfrac{100\%}{1}$

$= \dfrac{600\%}{1}$

$= 600\%$

65) $\dfrac{14}{56} = \dfrac{14}{56} \cdot \dfrac{100\%}{1}$

$= \dfrac{1400\%}{56}$

$= 25\%$

25% of homes have cable television.

67) $\frac{4}{9} = \frac{4}{9} \cdot \frac{100\%}{1}$
$= \frac{400\%}{9}$
$= 44\frac{4}{9}\%$

The value of the stock account was $44\frac{4}{9}\%$ of the original value.

69) $\frac{0}{5} = \frac{0}{5} \cdot \frac{100\%}{1}$
$= 0\%$
0% wore Bermuda shorts.

71) $3 = \frac{3}{1} \cdot \frac{100\%}{1}$
$= \frac{300\%}{1}$
$= 300\%$

73) $60\% = \frac{60\%}{1} \cdot \frac{1}{100\%}$
$= \frac{60}{100}$
$= 0.6$

75) $0.81 = 0.81 \cdot 100\%$
$= 81\%$

77) $250\% = \frac{250\%}{1} \cdot \frac{1}{100\%}$
$= \frac{250}{100}$
$= 2.5$

79) $0.0076 = 0.0076 \cdot 100\%$
$= 0.76\%$

81) $0.08\% = \frac{0.08\%}{1} \cdot \frac{1}{100\%}$
$= \frac{0.08}{100}$
$= 0.0008$

83) $4.5 = \frac{45}{10}$
$= \frac{45}{10} \cdot \frac{100\%}{1}$
$= \frac{45}{10} \cdot \frac{10 \cdot 10\%}{1}$
$= 450\%$

85) $450\% = \frac{450\%}{1} \cdot \frac{1}{100\%}$
$= \frac{450}{100}$
$= 4.5$

87) $4.00 = 4.00 \cdot 100\%$
$= 400\%$

89) $6\% = \frac{6\%}{1} \cdot \frac{1}{100\%}$
$= \frac{6}{100}$
$= 0.06$

91) $0.082 = 0.082 \cdot 100\%$
$= 8.2\%$

93) $0.001\% = \frac{0.001\%}{1} \cdot \frac{1}{100\%}$
$= \frac{0.001}{100}$
$= 0.00001$

95) a) $\frac{1}{20} = 0.05$
$0.05 \cdot 100\% = 5\%$

   b) $\frac{1}{20} = \frac{1}{20} \cdot \frac{100\%}{1}$
$= \frac{100\%}{20}$
$= 5\%$

97) a) $0.034 = \frac{34}{1,000}$
$\frac{34}{1,000} = \frac{34}{1,000} \cdot \frac{100\%}{1}$
$= \frac{3,400\%}{1,000}$
$= 3.4\%$

   b) $0.034 = 0.034 \cdot 100\%$
$= 3.4\%$

99) a) $220\% = \frac{220\%}{1} \cdot \frac{1}{100\%}$
$= \frac{220}{100}$
$= 2.2$
$2.2 = 2\frac{2}{10} = 2\frac{1}{5}$

99) b) $220\% = \dfrac{220\%}{1} \cdot \dfrac{1}{100\%}$

$= \dfrac{220}{100}$

$= \dfrac{11 \cdot 20}{5 \cdot 20}$

$= \dfrac{11}{5}$ or $2\dfrac{1}{5}$

101) $0.25\% = \dfrac{0.25\%}{1} \cdot \dfrac{1}{100\%}$

$= \dfrac{0.25}{100} \cdot \dfrac{100}{100}$

$= \dfrac{25}{10,000}$

$= \dfrac{1 \cdot 25}{400 \cdot 25}$

$= \dfrac{1}{400}$

$\dfrac{1}{400} = 0.0025$

103) $\dfrac{17}{20} = \dfrac{17}{20} \cdot \dfrac{100\%}{1}$

$= \dfrac{17 \cdot 20 \cdot 5\%}{20 \cdot 1}$

$= 85\%$

$85\% \cdot \dfrac{1}{100\%} = \dfrac{85}{100}$

$= 0.85$

105) To convert a percent to a decimal, multiply by $\dfrac{1}{100\%}$.

107) To convert a fraction to a percent, multiply by 100% and simplify, if necessary.

109) a) About 40% of the minimum payment goes to pay for the principal.

b) $100\% - 40\% = 60\%$
About 60% of the minimum payment goes to pay for interest.

c) Answers may vary. Example: All of the parts of the whole must add to 100%. Since there are only "two parts" in this problem's whole, their sum must equal 100%.

111) a) About 25% of eligible voters did not cast a vote in the 2008 presidential election.

b) About 40% of eligible voters voted for Barack Obama.

c) Answers may vary. Example: These represent only two of the possible outcomes, not all of them.

113) Halves

	Decimal	Fraction	Percent
a)	0.5	$\dfrac{1}{2}$	50%
b)	1	$\dfrac{2}{2}$	100%

115) Fourths

	Decimal	Fraction	Percent
a)	0.25	$\dfrac{1}{4}$	25%
b)	0.5	$\dfrac{2}{4}$	50%
c)	0.75	$\dfrac{3}{4}$	75%
d)	1	$\dfrac{4}{4}$	100%

117) Fifths

	Decimal	Fraction	Percent
a)	0.2	$\dfrac{1}{5}$	20%
b)	0.4	$\dfrac{2}{5}$	40%
c)	0.6	$\dfrac{3}{5}$	60%
d)	0.8	$\dfrac{4}{5}$	80%
e)	1	$\dfrac{5}{5}$	100%

119)

	Decimal	Fraction	Percent
a)	$0.\overline{5}$	$\dfrac{5}{9}$	$55\dfrac{5}{9}\%$
b)	1.28	$1\dfrac{7}{25}$	128%
c)	2.9	$2\dfrac{9}{10}$	290%
d)	0.03125	$\dfrac{1}{32}$	3.125%

5.2   Use Proportions to Solve Percent Exercises

GUIDED PRACTICE

1) The percent is 10%.
   The base is 20.
   The amount is 2.

2) Identify the percent, the base, and the amount.
   To identify the percent, look for the "percent."
   The percent is unknown.
   Let P = percent.
   To identify the base B, look for
   "percent (%) of."
   $B = 20$
   To identify the amount A, look for the word *is*.
   $A = 50$

   Write the percent proportion.
   Substitute the values for P, B, and A.
   $$\frac{P}{100} = \frac{50}{20}$$

3) Identify the percent, the base, and the amount.
   To identify the percent, look for the % symbol.
   $P = 120$
   To identify the base B, look for
   "percent (%) of."
   $B = 35$
   To identify the amount A, look for the word *is*.
   The amount is unknown.
   Let A = amount.

   Write the percent proportion.
   Substitute the values for P, B, and A.
   $$\frac{120}{100} = \frac{A}{35}$$

4) Identify the percent, the base, and the amount.
   To identify the percent, look for the "percent."
   $P = 45$
   To identify the base B, look for
   "percent (%) of."
   The base is unknown.
   Let B = total number of customers.
   To identify the amount A, look for the word *is*.
   In this case, the past tense *was* indicates the amount.
   $A = 10$

   Write the percent proportion.
   Substitute the values for P, B, and A.
   $$\frac{45}{100} = \frac{10}{B}$$

5) The important information in the problem is:
   Henry gave her 232 pages, which was 58% of the paper. How much paper did Henry have before he gave paper to Anais?

   Identify the percent, the base, and the amount.
   To identify the percent, look for the "percent."
   $P = 58$
   To identify the base B, look for
   "percent (%) of."
   The base is unknown.
   Let B = total number of sheets of paper.
   To identify the amount A, look for the word *is*.
   In this case, the key word is *was*.
   $A = 232$

   Write the percent proportion.
   Substitute the values for P, B, and A.
   $$\frac{58}{100} = \frac{232}{B}$$

6) Let P = the percent.
   $B = 40$
   $A = 16$

   $$\frac{P}{100} = \frac{16}{40}$$
   $$\frac{\cancel{100}}{1} \cdot \frac{P}{\cancel{100}} = \frac{100}{1} \cdot \frac{16}{40}$$
   $$P = \frac{100 \cdot 16}{40}$$
   $$P = \frac{10 \cdot \cancel{10} \cdot \cancel{4} \cdot 4}{\cancel{10} \cdot \cancel{4}}$$
   $$P = 40$$
   40% of 40 is 16.

7) $P = 16$
   Let B = the number.
   $A = 64$

   $$\frac{16}{100} = \frac{64}{B}$$
   $$\frac{100}{16} = \frac{B}{64}$$
   $$\frac{100}{16} \cdot \frac{64}{1} = \frac{B}{\cancel{64}} \cdot \frac{\cancel{64}}{1}$$
   $$\frac{100 \cdot 4 \cdot \cancel{16}}{\cancel{16}} = B$$
   $$400 = B$$
   64 is 16% of 400.

8) $P = 40$

  $B = 20$

  Let A = pages to be completed.

$$\frac{40}{100} = \frac{A}{20}$$

$$\frac{40}{100} \cdot \frac{20}{1} = \frac{A}{20} \cdot \frac{20}{1}$$

$$\frac{4 \cdot \cancel{10} \cdot 2 \cdot \cancel{10}}{\cancel{10} \cdot \cancel{10}} = A$$

$$8 = A$$

  Kato must write 8 pages by tomorrow.

9) <u>Identify the question in the application.</u>
  "What percent of the bikes are nonracing bikes?"

  <u>Identify the information needed to answer the question.</u>
  Nonracing bikes = Total bikes − Racing bikes
$$= 125 - 5$$
$$= 120$$

  <u>Write and solve a percent proportion.</u>
  Let P = the percent.
  $B = 125$
  A = nonracing bikes = 120

$$\frac{P}{100} = \frac{120}{125}$$

$$\frac{\cancel{100}}{1} \cdot \frac{P}{\cancel{100}} = \frac{100}{1} \cdot \frac{120}{125}$$

$$P = \frac{2 \cdot 2 \cdot \cancel{5} \cdot \cancel{5} \cdot 2 \cdot 2 \cdot 2 \cdot 3 \cdot \cancel{5}}{\cancel{5} \cdot \cancel{5} \cdot \cancel{5}}$$

$$P = 96$$

  96% are nonracing bikes.

10) <u>Identify the question in the application.</u>
  "How much more does she need to save?"

  <u>Identify the information needed to answer the question.</u>
  Percent to save = 100% − percent saved so far
$$= 100\% - 60\%$$
$$= 40\%$$

  <u>Write and solve a percent proportion.</u>
  $P = 40$
  $B = 1,500$
  Let A = the amount.

$$\frac{40}{100} = \frac{A}{1,500}$$

$$\frac{40}{100} \cdot \frac{1,500}{1} = \frac{A}{1,500} \cdot \frac{1,500}{1}$$

$$\frac{40 \cdot 15 \cdot \cancel{100}}{\cancel{100}} = A$$

$$600 = A$$

Samantha still needs to save \$600.

CONCEPT CHECKS AND PRACTICE EXERCISES

A1) a) "The horses" indicates the base.

  b) "Black" indicates the amount.

A2) a) To identify the percent, look for the word percent or the symbol %.

  b) To identify the base, look for the phrase percent of.

  c) To identify the amount, look for the words *is* or *equal*.

A3) <u>Identify the percent, the base, and the amount.</u>
  To identify the percent, look for the "percent."
    The percent is unknown.
    Let P = percent.
  To identify the base B, look for "percent (%) of."
    $B = 30$
  To identify the amount A, look for the word *is*.
    $A = 18$

  <u>Write the percent proportion.</u>
  Substitute the values for P, B, and A.

$$\frac{P}{100} = \frac{18}{30}$$

A4) <u>Identify the percent, the base, and the amount.</u>
  To identify the percent, look for the "percent."
    The percent is unknown.
    Let P = percent.
  To identify the base B, look for "percent (%) of."
    $B = 40$
  To identify the amount A, look for the word *is*.
    $A = 12$

  <u>Write the percent proportion.</u>
  Substitute the values for P, B, and A.

$$\frac{P}{100} = \frac{12}{40}$$

A5) <u>Identify the percent, the base, and the amount.</u>
To identify the percent, look for the % symbol.
$P = 80$
To identify the base B, look for
"percent (%) of."
$B = 20$
To identify the amount A, look for the word *is*.
The amount is unknown.
Let $A = $ amount.

<u>Write the percent proportion.</u>
Substitute the values for P, B, and A.
$$\frac{80}{100} = \frac{A}{20}$$

A6) <u>Identify the percent, the base, and the amount.</u>
To identify the percent, look for the % symbol.
$P = 20$
To identify the base B, look for
"percent (%) of."
$B = 50$
To identify the amount A, look for the word *is*.
The amount is unknown.
Let $A = $ amount.

<u>Write the percent proportion.</u>
Substitute the values for P, B, and A.
$$\frac{20}{100} = \frac{A}{50}$$

A7) <u>Identify the percent, the base, and the amount.</u>
To identify the percent, look for the % symbol.
$P = 250$
To identify the base B, look for
"percent (%) of."
The base is unknown.
Let $B = $ the base.
To identify the amount A, look for the word *is*.
$A = 90$

<u>Write the percent proportion.</u>
Substitute the values for P, B, and A.
$$\frac{250}{100} = \frac{90}{B}$$

A8) <u>Identify the percent, the base, and the amount.</u>
To identify the percent, look for the % symbol.
$P = 125$
To identify the base B, look for
"percent (%) of."
The base is unknown.
Let $B = $ the base.
To identify the amount A, look for the word *is*.
$A = 50$

<u>Write the percent proportion.</u>
Substitute the values for P, B, and A.
$$\frac{125}{100} = \frac{50}{B}$$

A9) <u>Identify the percent, the base, and the amount.</u>
To identify the percent, look for the % symbol.
$P = 4$
To identify the base B, look for
"percent (%) of."
The base is unknown.
Let $B = $ cost of his book before taxes.
To identify the amount A, look for the word *is*.
In this case, the past tense *was* indicates the amount.
$A = 8.80$

<u>Write the percent proportion.</u>
Substitute the values for P, B, and A.
$$\frac{4}{100} = \frac{8.80}{B}$$

A10) <u>Identify the percent, the base, and the amount.</u>
To identify the percent, look for the % symbol.
$P = 5$
To identify the base B, look for
"percent (%) of."
The base is unknown.
Let $B = $ value of the cars sold.
To identify the amount A, look for the word *is*.
In this case, the past tense *was* indicates the amount.
$A = 3,750$

<u>Write the percent proportion.</u>
Substitute the values for P, B, and A.
$$\frac{5}{100} = \frac{3,750}{B}$$

A11) <u>Identify the percent, the base, and the amount.</u>
To identify the percent, look for the "percent."
The percent is unknown.
Let $P = $ percent.
To identify the base B, look for
"percent (%) of."
$B = 150$
To identify the amount A, look for the word *is*.
$A = 110$

<u>Write the percent proportion.</u>
Substitute the values for P, B, and A.
$$\frac{P}{100} = \frac{110}{150}$$

A12) <u>Identify the percent, the base, and the amount.</u>
    To identify the percent, look for the "percent."
      The percent is unknown.
      Let P = percent.

    To identify the base B, look for
    "percent (%) of."
      $B = 80$

    To identify the amount A, look for the word *is*.
    In this case, the past tense *was* indicates the
    amount.
      $A = 79$

    <u>Write the percent proportion.</u>
    Substitute the values for P, B, and A.
$$\frac{P}{100} = \frac{79}{80}$$

FOCUS ON WRITING PERCENT PROPORTIONS

1) b

2) e

3) a

4) d

5) c

6) f

7) d

8) c

9) f

10) a

11) e

12) b

B1) 24 is 22 percent of some number.

B2) a) Answers may vary. Example: A must be
      more than 30 because 120 is more than 100.
    b) Answers may vary. Example: P must be less
      than 100% because 20 is less than 25.

B3) $P = 30$
    $B = 60$
    Let A = the number.

$$\frac{30}{100} = \frac{A}{60}$$

$$\frac{30}{100} \cdot \frac{60}{1} = \frac{A}{\cancel{60}} \cdot \frac{\cancel{60}}{1}$$

$$\frac{3 \cdot \cancel{10} \cdot 6 \cdot \cancel{10}}{\cancel{10} \cdot \cancel{10}} = A$$

$$18 = A$$

The number is 18.

B4) $P = 110$
    $B = 40$
    Let A = the amount.

$$\frac{110}{100} = \frac{A}{40}$$

$$\frac{110}{100} \cdot \frac{40}{1} = \frac{A}{\cancel{40}} \cdot \frac{\cancel{40}}{1}$$

$$\frac{11 \cdot \cancel{10} \cdot 4 \cdot \cancel{10}}{\cancel{10} \cdot \cancel{10}} = A$$

$$44 = A$$

44 is 110% of 40.

B5) Let P = the percent.
    $B = 80$
    $A = 100$

$$\frac{P}{100} = \frac{100}{80}$$

$$\frac{\cancel{100}}{1} \cdot \frac{P}{\cancel{100}} = \frac{100}{1} \cdot \frac{100}{80}$$

$$P = \frac{\cancel{2} \cdot \cancel{2} \cdot \cancel{5} \cdot 5 \cdot \cancel{2} \cdot \cancel{2} \cdot 5 \cdot 5}{\cancel{2} \cdot \cancel{2} \cdot \cancel{2} \cdot \cancel{2} \cdot \cancel{5}}$$

$$P = 125$$

100 is 125% of 80.

B6) Let P = the percent.
    $B = 45$
    $A = 90$

$$\frac{P}{100} = \frac{90}{45}$$

$$\frac{\cancel{100}}{1} \cdot \frac{P}{\cancel{100}} = \frac{100}{1} \cdot \frac{90}{45}$$

$$P = \frac{100 \cdot 2 \cdot \cancel{45}}{\cancel{45}}$$

$$P = 200$$

90 is 200% of 45.

B7) $P = 25$
      Let B = the number.
      $A = 75$

$$\frac{25}{100} = \frac{75}{B}$$

$$\frac{100}{25} = \frac{B}{75}$$

$$\frac{100}{25} \cdot \frac{75}{1} = \frac{B}{75} \cdot \frac{75}{1}$$

$$\frac{100 \cdot 3 \cdot \cancel{25}}{\cancel{25}} = B$$

$$300 = B$$

75 is 25% of 300.

B8) $P = 15$
      Let B = the number.
      $A = 300$

$$\frac{15}{100} = \frac{300}{B}$$

$$\frac{100}{15} = \frac{B}{300}$$

$$\frac{100}{15} \cdot \frac{300}{1} = \frac{B}{300} \cdot \frac{300}{1}$$

$$\frac{\cancel{5} \cdot 20 \cdot \cancel{3} \cdot 100}{\cancel{3} \cdot \cancel{5}} = B$$

$$2,000 = B$$

300 is 15% of 2,000.

B9) $P = 105$
      $B = 180$
      Let A = the candles made.

$$\frac{105}{100} = \frac{A}{180}$$

$$\frac{105}{100} \cdot \frac{180}{1} = \frac{A}{180} \cdot \frac{180}{1}$$

$$\frac{\cancel{5} \cdot 21 \cdot 9 \cdot \cancel{20}}{\cancel{5} \cdot \cancel{20}} = A$$

$$189 = A$$

Dee made 189 candles.

B10) Let P = the percent.
      $B = 175$
      $A = 140$

$$\frac{P}{100} = \frac{140}{175}$$

$$\frac{\cancel{100}}{1} \cdot \frac{P}{\cancel{100}} = \frac{100}{1} \cdot \frac{140}{175}$$

$$P = \frac{4 \cdot 25 \cdot \cancel{7} \cdot 20}{\cancel{7} \cdot 25}$$

$$P = 80$$

80% of the drivers wore seatbelts.

B11) $P = 80$
      $B = 80$
      Let A = the questions answered correctly.

$$\frac{80}{100} = \frac{A}{80}$$

$$\frac{80}{100} \cdot \frac{80}{1} = \frac{A}{\cancel{80}} \cdot \frac{\cancel{80}}{1}$$

$$\frac{\cancel{8} \cdot 16 \cdot 4 \cdot \cancel{20}}{\cancel{8} \cdot \cancel{20}} = A$$

$$64 = A$$

Ed answered 64 questions correctly.

B12) Let P = the percent.
      $B = 150$
      $A = 90$

$$\frac{P}{100} = \frac{90}{150}$$

$$\frac{\cancel{100}}{1} \cdot \frac{P}{\cancel{100}} = \frac{100}{1} \cdot \frac{90}{150}$$

$$P = \frac{4 \cdot 25 \cdot \cancel{6} \cdot 15}{\cancel{6} \cdot 25}$$

$$P = 60$$

She met 60% of the employees.

SECTION 5.2   EXERCISES

For 1–8, refer to Concept Checks and Practice Exercises.

9)   percent proportion

11) is

13) Identify the percent, the base, and the amount.
     $P = 35$
     $B = 70$
     The amount is unknown.
     Let A = amount.

     Write the percent proportion.
     $$\frac{35}{100} = \frac{A}{70}$$

15) Identify the percent, the base, and the amount.
     $P = 40$
     The base is unknown.
     Let B = the base.
     $A = 60$

     Write the percent proportion.
     $$\frac{40}{100} = \frac{60}{B}$$

17) <u>Identify the percent, the base, and the amount.</u>
 The percent is unknown.
 Let P = percent.
 $B = 50$
 $A = 33$

<u>Write the percent proportion.</u>
 $$\frac{P}{100} = \frac{33}{50}$$

19) <u>Identify the percent, the base, and the amount.</u>
 $P = 1.2$
 $B = 300$
 The amount is unknown.
 Let A = amount.

<u>Write the percent proportion.</u>
 $$\frac{1.2}{100} = \frac{A}{300}$$

21) <u>Identify the percent, the base, and the amount.</u>
 The percent is unknown.
 Let P = percent.
 $B = 450$
 $A = 2$

<u>Write the percent proportion.</u>
 $$\frac{P}{100} = \frac{2}{450}$$

23) <u>Identify the percent, the base, and the amount.</u>
 $P = 40$
 $B = 300$
 Let A = the people who ate a hamburger.

<u>Write the percent proportion.</u>
 $$\frac{40}{100} = \frac{A}{300}$$

25) $P = 300$
 $B = 4$
 Let A = the amount.

$$\frac{300}{100} = \frac{A}{4}$$
$$\frac{4}{1} \cdot \frac{300}{100} = \frac{A}{4} \cdot \frac{4}{1}$$
$$\frac{4 \cdot 25 \cdot 12}{4 \cdot 25} = A$$
$$12 = A$$

27) $P = 0.01$
 Let B = the base.
 $A = 7$

$$\frac{0.01}{100} = \frac{7}{B}$$
$$\frac{100}{0.01} = \frac{B}{7}$$
$$\frac{7}{1} \cdot \frac{100}{0.01} = \frac{B}{7} \cdot \frac{7}{1}$$
$$\frac{700}{0.01} = B$$
$$70,000 = B$$

29) $P = 90$
 Let B = the base.
 $A = 70$

$$\frac{90}{100} = \frac{70}{B}$$
$$\frac{100}{90} = \frac{B}{70}$$
$$\frac{70}{1} \cdot \frac{100}{90} = \frac{B}{70} \cdot \frac{70}{1}$$
$$\frac{7,000}{90} = B$$
$$77.78 \approx B$$

31) Let P = the percent.
 $B = 2$
 $A = 4$

$$\frac{P}{100} = \frac{4}{2}$$
$$\frac{100}{1} \cdot \frac{P}{100} = \frac{4}{2} \cdot \frac{100}{1}$$
$$P = 200$$

33) Let P = the percent.
 $B = 3,000$
 $A = 2,910$

$$\frac{P}{100} = \frac{2,910}{3,000}$$
$$\frac{100}{1} \cdot \frac{P}{100} = \frac{100}{1} \cdot \frac{2,910}{3,000}$$
$$P = \frac{100 \cdot 30 \cdot 97}{30 \cdot 100}$$
$$P = 97$$
 97% of the people had no side effects.

35) Let P = the percent.

$B = 42$

$A = 35$

$$\frac{P}{100} = \frac{35}{42}$$

$$\frac{\cancel{100}}{1} \cdot \frac{P}{\cancel{100}} = \frac{100}{1} \cdot \frac{35}{42}$$

$$P = \frac{\cancel{2} \cdot 50 \cdot 5 \cdot \cancel{7}}{\cancel{2} \cdot 3 \cdot \cancel{7}}$$

$$P = \frac{250}{3}$$

$$P = 83\frac{1}{3} \approx 83.33$$

Bessie's production in March is about 83.33% of her production in April.

37) Let P = the percent.

$B = \$12,000$

$A = \$720$

$$\frac{P}{100} = \frac{720}{12,000}$$

$$\frac{\cancel{100}}{1} \cdot \frac{P}{\cancel{100}} = \frac{100}{1} \cdot \frac{720}{12,000}$$

$$P = \frac{\cancel{100} \cdot 6 \cdot \cancel{120}}{\cancel{100} \cdot \cancel{120}}$$

$$P = 6$$

6% of the base price is the tax.

39) Let P = the percent.

$B = 2,000 + 760 + 760 + 360 + 120 = 4,000$

$A = 120 + 760 + 2,000 = 2,880$

$$\frac{P}{100} = \frac{2880}{4000}$$

$$\frac{\cancel{100}}{1} \cdot \frac{P}{\cancel{100}} = \frac{100}{1} \cdot \frac{2880}{4000}$$

$$P = \frac{\cancel{100} \cdot \cancel{40} \cdot 72}{\cancel{40} \cdot \cancel{100}}$$

$$P = 72$$

72% of energy used to create electricity comes from fossil fuels.

41) Let P = the percent.

$B = 120$

$A = 760$

$$\frac{P}{100} = \frac{760}{120}$$

$$\frac{\cancel{100}}{1} \cdot \frac{P}{\cancel{100}} = \frac{100}{1} \cdot \frac{760}{120}$$

$$P = \frac{\cancel{20} \cdot 5 \cdot \cancel{2} \cdot 380}{\cancel{20} \cdot \cancel{2} \cdot 3}$$

$$P = \frac{1,900}{3} \approx 633$$

The electricity made from nuclear sources is about 633% of the electricity made from oil.

43) Let P = the percent.

$B = 45,000$

$A = 125,000$

$$\frac{P}{100} = \frac{125,000}{45,000}$$

$$\frac{\cancel{100}}{1} \cdot \frac{P}{\cancel{100}} = \frac{100}{1} \cdot \frac{125,000}{45,000}$$

$$P = \frac{100 \cdot \cancel{50} \cdot 2,500}{9 \cdot \cancel{50} \cdot \cancel{100}}$$

$$P = \frac{2,500}{9}$$

$$P = 277\frac{7}{9} \approx 278\%$$

A superintendent's salary is about 278% of a teacher's salary.

45) Let P = the percent.

$B = 50,000$

$A = 80,000$

$$\frac{P}{100} = \frac{80,000}{50,000}$$

$$\frac{\cancel{100}}{1} \cdot \frac{P}{\cancel{100}} = \frac{100}{1} \cdot \frac{80,000}{50,000}$$

$$P = \frac{\cancel{100} \cdot 160 \cdot \cancel{500}}{\cancel{100} \cdot \cancel{500}}$$

$$P = 160$$

A principal's salary is 160% of a librarian's salary.

47) Let P = the percent.

$B = 28$

$A = 36$

$$\frac{P}{100} = \frac{36}{28}$$

$$\frac{100}{1} \cdot \frac{P}{100} = \frac{100}{1} \cdot \frac{36}{28}$$

$$P = \frac{100 \cdot \cancel{4} \cdot 9}{\cancel{4} \cdot 7}$$

$$P = \frac{900}{7}$$

$$P = 128\frac{4}{7} \approx 129$$

A person with an associate degree earns about 129% of someone with a high school diploma.

49) Let P = the percent.

$B = 37$

$A = 88$

$$\frac{P}{100} = \frac{88}{37}$$

$$\frac{100}{1} \cdot \frac{P}{100} = \frac{100}{1} \cdot \frac{88}{37}$$

$$P = \frac{8,800}{37}$$

$$P = 237\frac{31}{37} \approx 238$$

A person with a doctorate degree makes about 238% of the average income.

51) $P = 3$

Let B = the recommended amount of fat.

$A = 2$

$$\frac{3}{100} = \frac{2}{B}$$

$$\frac{100}{3} = \frac{B}{2}$$

$$\frac{100}{3} \cdot \frac{2}{1} = \frac{B}{\cancel{2}} \cdot \frac{\cancel{2}}{1}$$

$$\frac{200}{3} = B$$

$$67 \approx B$$

The daily recommended amount of fat is about 67 grams.

53) Let P = the percent.

$B = 150$

$A = 110$

$$\frac{P}{100} = \frac{110}{150}$$

$$\frac{100}{1} \cdot \frac{P}{100} = \frac{100}{1} \cdot \frac{110}{150}$$

$$P = \frac{2 \cdot \cancel{50} \cdot 110}{3 \cdot \cancel{50}}$$

$$P = \frac{220}{3} = 73\frac{1}{3} \approx 74$$

The Oatloops contribute about 74% of the calories to a bowl of Oatloops with skim milk.

55) $$\frac{210 \text{ mg}}{9\%} = \frac{n \text{ mg}}{12\%}$$

$$210 \cdot 0.12 = 0.09 \cdot n$$

$$25.2 = 0.09n$$

$$280 = n$$

There are 280 mg of sodium in the Oatloops and milk.

$$280 - 210 = 70$$

A half cup of skim milk contains 70 mg of sodium.

57) Since the fiber percentage is the same for both cheerios and cheerios with milk (11%), there is no fiber in a half cup of skim milk.

59) $P = 25$

$B = \$52,000$

Let A = the taxes.

$$\frac{25}{100} = \frac{A}{52,000}$$

$$\frac{25}{100} \cdot \frac{52,000}{1} = \frac{A}{52,000} \cdot \frac{52,000}{1}$$

$$\frac{25 \cdot \cancel{100} \cdot 520}{\cancel{100}} = A$$

$$13,000 = A$$

They paid $13,000 in taxes.

$P = 5$

$B = \$52,000 - \$13,000 = \$39,000$

Let A = the savings.

$$\frac{5}{100} = \frac{A}{39,000}$$

$$\frac{5}{100} \cdot \frac{39,000}{1} = \frac{A}{39,000} \cdot \frac{39,000}{1}$$

$$\frac{5 \cdot \cancel{100} \cdot 390}{\cancel{100}} = A$$

$$1,950 = A$$

They put $1,950 into the savings account.

61) $P = 20$
$B = \$45$
Let $A =$ the tip.

$$\frac{20}{100} = \frac{A}{45}$$

$$\frac{20}{100} \cdot \frac{45}{1} = \frac{A}{45} \cdot \frac{45}{1}$$

$$\frac{20 \cdot 5 \cdot 9}{5 \cdot 20} = A$$

$$9 = A$$

They tipped the waitress $9.

$P = 50$
$B = \$45 + \$9 = \$54$
Let $A =$ the amount that each paid.

$$\frac{50}{100} = \frac{A}{54}$$

$$\frac{50}{100} \cdot \frac{54}{1} = \frac{A}{54} \cdot \frac{54}{1}$$

$$\frac{50 \cdot 2 \cdot 27}{2 \cdot 50} = A$$

$$27 = A$$

Each paid $27.

63) $P = 10$
$B = \$15,000$
Let $A =$ the discount.

$$\frac{10}{100} = \frac{A}{15,000}$$

$$\frac{10}{100} \cdot \frac{15,000}{1} = \frac{A}{15,000} \cdot \frac{15,000}{1}$$

$$\frac{10 \cdot 100 \cdot 150}{100} = A$$

$$1,500 = A$$

The dealer lowered the price by $1,500.

$\$15,000 - \$1,500 - \$1,750 = \$11,750$
Pai was charged $11,750 for the car.

65)   First, find 50% of 76:
$$\frac{50}{100} = \frac{A}{76}$$

$$\frac{76}{1} \cdot \frac{50}{100} = \frac{A}{76} \cdot \frac{76}{1}$$

$$\frac{2 \cdot 38 \cdot 50}{50 \cdot 2} = A$$

$$38 = A$$

Then, find 200% of that number:

$$\frac{200}{100} = \frac{A}{38}$$

$$\frac{38}{1} \cdot \frac{200}{100} = \frac{A}{38} \cdot \frac{38}{1}$$

$$\frac{38 \cdot 100 \cdot 2}{100} = A$$

$$76 = A$$

200% of 50% of 76 is 76.

67) $200\% = 200\% \cdot \dfrac{1}{100\%} = 2$

$50\% = 50\% \cdot \dfrac{1}{100\%} = \dfrac{1}{2}$

Answers may vary. Example: The fractions are reciprocals. Their product is 1.

5.3   Use Equations to Solve Percent Exercises

GUIDED PRACTICE

1)   The percent is 25%.
The base is 200 because it is multiplied by the percent.
The amount is 50 because it is the result of multiplying the percent and the base.

2)   Identify the five parts of the percent equation.
   1.   The percent is 16%.
   2.   "Of" indicates multiplication.
   3.   The points scored is the base since it is multiplied by the percent. It is unknown, so we choose a variable.
         Let $B =$ the number of points scored.
   4.   *Is* indicates an equal sign.
   5.   The amount is 8.

Write the percent equation.
   Percent $\cdot$ Base = Amount
         $16\% \cdot B = 8$

3)   Identify the percent, the base, and the amount.
   To identify the percent, look for the % symbol.
         $P = 40$
   To identify the base, look for "percent (%) of base."
         $B = 200$
   To identify the amount, look for the word *is*.
   The amount is unknown.
   Let $A =$ the number.

Write the percent equation.
   Substitute the values for P, B, and A.
         $P \cdot B = A$
   $40\% \cdot 200 = A$

4) <u>Identify the percent, the base, and the amount.</u>
    To identify the percent, look for the "percent."
       The percent is unknown.
       Let P = percent.
    To identify the base, look for "percent (%) of base."
       $B = 700$
    To identify the amount, look for the word *is*.
       $A = 35$

<u>Write the percent equation.</u>
    Substitute the values for P, B, and A.
$$P \cdot B = A$$
$$P \cdot 700 = 35$$

5) <u>Identify the percent, the base, and the amount.</u>
    To identify the percent, look for the "percent."
       $P = 15$
    To identify the base, look for "percent (%) of base."
       The base is unknown.
       Let B = the weight of the tiger.
    To identify the amount, look for the word *is*.
       $A = 61$

<u>Write the percent equation.</u>
    Substitute the values for P, B, and A.
$$P \cdot B = A$$
$$15\% \cdot B = 61$$

6) <u>Identify the percent, the base, and the amount.</u>
    To identify the percent, look for the % symbol.
       $P = 668$
    To identify the base, look for "percent (%) of base."
       The base is Shaquille O'Neal's earnings.
       $B = 33.4$ million
    Since we know the percent and the base, the amount is Oprah Winfrey's earnings.
       The amount is unknown.
       Let A = Oprah Winfrey's earnings.

<u>Write the percent equation.</u>
    Substitute the values for P, B, and A.
$$P \cdot B = A$$
$$668\% \cdot 33.4 \text{ million} = A$$
$$223.112 = A$$

Oprah earned about $223 million.

7) P is $20\% = 0.2$.
   $B = 50$
   The amount is unknown.
   Let A = "what number."
$$P \cdot B = A$$
$$0.2 \cdot 50 = A$$
$$10 = A$$
The number is 10.

8) P is $6\% = 0.06$.
   The base is unknown.
   Let B = "what number."
   $A = 4.2$
$$P \cdot B = A$$
$$0.06 \cdot B = 4.2$$
$$\frac{\cancel{0.06} \cdot B}{\cancel{0.06}} = \frac{4.2}{0.06}$$
$$B = 70$$
The number is 70.

9) The percent is unknown.
   Let P = "what percent."
   $B = 55$
   $A = 11$
$$P \cdot B = A$$
$$P \cdot 55 = 11$$
$$\frac{P \cdot \cancel{55}}{\cancel{55}} = \frac{11}{55}$$
$$P = 0.2$$
The percent is 20%.

10) P is $40\% = 0.40$.
   $B = 10$
   The amount is unknown.
   Let A = "what number."
$$P \cdot B = A$$
$$0.40 \cdot 10 = A$$
$$4 = A$$
The number is 4.

11) P is $26\% = 0.26$.
   The base is unknown.
   Let B = "what number."
   $A = 104$
$$P \cdot B = A$$
$$0.26 \cdot B = 104$$
$$\frac{\cancel{0.26} \cdot B}{\cancel{0.26}} = \frac{104}{0.26}$$
$$B = 400$$
The number is 400.

12) The percent is unknown.
Let P = "what percent."
$B = 52$
$A = 13$

$$P \cdot B = A$$
$$P \cdot 52 = 13$$
$$\frac{P \cdot \cancel{52}}{\cancel{52}} = \frac{13}{52}$$
$$P = 0.25$$

Angel repaid 25% of her debt.

13) The base is 30.
Shade approximately 10%, and draw an arrow where 25 should be.

Since 25 is not close to the shading for 10% of 30, the answer is not reasonable.

14) The base is 120.
To shade 230%, we need a little more than 2 wholes.

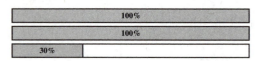

Since each 100% is the same as 120, 2 bases is 240. The difference between 240 and 276 is 36, which is about one-third of 100%. 276 is reasonable.

15) The base is 60.
Shade approximately 45%, and draw an arrow where 40 should be.

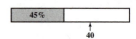

Since 40 is more than half of 60 and 45% is less than half, the statement is not reasonable.

16) The base is 90.
Shade approximately 30%, and draw an arrow where 27 should be.

30% is about one-third of 100%, and 27 is about one-third of 90. The answer is reasonable.

17) P is 30% = 0.3.
$B = 9,200 - 1,400 = 7,800$
The amount is unknown.
Let A = the down payment.

$$P \cdot B = A$$
$$0.30 \cdot 7,800 = A$$
$$2,340 = A$$

He needs $2,340 for the down payment.

CONCEPT CHECKS AND PRACTICE EXERCISES

A1) The word "of" is often used to indicate multiplication in word exercises.

A2) Answers may vary. Examples include is, are, were, equals, and results.

A3) The standard form of a percent equation is $P \cdot B = A$.

A4) <u>Identify the percent, the base, and the amount.</u>
To identify the percent, look for the "percent."
The percent is unknown.
Let P = percent.
To identify the base, look for "percent (%) of base."
$B = 15$
To identify the amount, look for the word *is*.
$A = 12$

<u>Write the percent equation.</u>
Substitute the values for P, B, and A.
$$P \cdot B = A$$
$$P \cdot 15 = 12$$

A5) <u>Identify the percent, the base, and the amount.</u>
To identify the percent, look for the "percent."
The percent is unknown.
Let P = percent.
To identify the base, look for "percent (%) of base."
$B = 20$
To identify the amount, look for the word *is*.
$A = 7$

<u>Write the percent equation.</u>
Substitute the values for P, B, and A.
$$P \cdot B = A$$
$$P \cdot 20 = 7$$

A6) Identify the percent, the base, and the amount.
   To identify the percent, look for the % symbol.
       $P = 30$
   To identify the base, look for "percent (%) of base."
       $B = 70$
   To identify the amount, look for the word *is*.
       The amount is unknown.
       Let $A =$ amount.

   Write the percent equation.
   Substitute the values for P, B, and A.
       $P \cdot B = A$
       $30\% \cdot 70 = A$

A7) Identify the percent, the base, and the amount.
   To identify the percent, look for the % symbol.
       $P = 70$
   To identify the base, look for "percent (%) of base."
       $B = 30$
   To identify the amount, look for the word *is*.
       The amount is unknown.
       Let $A =$ amount.

   Write the percent equation.
   Substitute the values for P, B, and A.
       $P \cdot B = A$
       $70\% \cdot 30 = A$

A8) Identify the percent, the base, and the amount.
   To identify the percent, look for the % symbol.
       $P = 35$
   To identify the base, look for "percent (%) of base."
       The base is unknown.
       Let $B =$ base.
   To identify the amount, look for the word *is*.
       $A = 60$

   Write the percent equation.
   Substitute the values for P, B, and A.
       $P \cdot B = A$
       $35\% \cdot B = 60$

A9) Identify the percent, the base, and the amount.
   To identify the percent, look for the % symbol.
       $P = 25\%$
   To identify the base, look for "percent (%) of base."
       The base is unknown.
       Let $B =$ base.
   To identify the amount, look for the word *is*.
       $A = 60$

Write the percent equation.
   Substitute the values for P, B, and A.
       $P \cdot B = A$
       $25\% \cdot B = 60$

A10) Identify the percent, the base, and the amount.
   To identify the percent, look for the % symbol.
       The percent is the sales tax rate.
       $P = 6$
   To identify the base, look for "percent (%) of base."
       The base is the pretax cost of the car.
       Let $B =$ the pretax cost of the car.
   To identify the amount, look for the word *was*.
       The amount is the sales tax.
       $A = 900$

   Write the percent equation.
   Substitute the values for P, B, and A.
       $P \cdot B = A$
       $6\% \cdot B = 900$

A11) Identify the percent, the base, and the amount.
   To identify the percent, look for the % symbol.
       $P = 20$
   To identify the base, look for "percent (%) of base."
       The base is the cost of her dinner.
       $B = 45.50$
   Since we know the percent and the base, the amount is the tip.
   The amount is unknown.
       Let $A =$ amount of the tip.

   Write the percent equation.
   Substitute the values for P, B, and A.
       $P \cdot B = A$
       $20\% \cdot 45.50 = A$

A12) Identify the percent, the base, and the amount.
   To identify the percent, look for the % symbol.
       The percent is unknown.
       Let $P =$ the discount rate.
   To identify the base, look for "percent (%) of base."
       The base is the regular price of the shoes.
       Let $B = \$120$
   To identify the amount, look for the word *was*.
       The amount is the sales price of the shoes.
       $A = \$90$

   Write the percent equation.
   Substitute the values for P, B, and A.
       $P \cdot B = A$
       $P \cdot 120 = 90$

A13) Identify the percent, the base, and the amount.
        To identify the percent, look for the % symbol.
            The percent is unknown.
            Let P = the percent.

        To identify the base, look for "percent (%) of base."
            The base is the morning lap time.
            Let $B = 40$
        To identify the amount, look for the word *was*.
            The amount is the afternoon lap time.
            $A = 37$

      Write the percent equation.
        Substitute the values for P, B, and A.
            $P \cdot B = A$
            $P \cdot 40 = 37$

FOCUS ON ESTIMATION PRACTICE EXERCISES

1) Fifty percent of forty-hundredths is what?
    f) $0.50 \cdot 0.40 = A$

2) What percent of 40 is 50?
    a) $P \cdot 40 = 50$

3) Forty is what percent of 50?
    d) $P \cdot 50 = 40$

4) Fifty percent of what number is forty?
    e) $0.50 \cdot B = 40$

5) What number is forty percent of fifty?
    b) $0.40 \cdot 50 = A$

6) Four thousand percent of what number is fifty?
    c) $40 \cdot B = 50$

7) Fifty kilograms is four thousand percent of what weight?
    c) $40 \cdot B = 50$

8) What percent of forty pitches is fifty pitches?
    a) $P \cdot 40 = 50$

9) A runner ran forty miles in a week. If that number is fifty percent of his normal distance, how far does he normally run in a week?
    e) $0.50 \cdot B = 40$

10) Originally, there were fifty pamphlets is a display. After two hours, there were forty. What percent of the original amount is the new amount?
    d) $P \cdot 50 = 40$

11) A prescription calls for 0.4 gram of medicine. What is fifty percent of the daily amount?
    f) $0.50 \cdot 0.40 = A$

12) After breaking a neighbor's window, a child had to clean the neighbor's fifty windows. If he has cleaned forty percent of the windows so far, how many has he cleaned?
    b) $0.40 \cdot 50 = A$

B1) a) To perform the calculations for $40\% \cdot B = 60$, convert from a percent to a decimal.

      b) To perform the calculations for $30\% \cdot 20 = A$, convert from a percent to a decimal.

      c) After solving the percent equation $P \cdot 30 = 90$, convert from a decimal to a percent.

B2) When given the equation $0.5 \cdot B = 10$, you must perform division to solve for B.

B3) a) Since 110% is greater than 100%, 110% of a base is more than the base.

      b) Since 90% is less than 100%, 90% of a base is less than the base.

B4) P is $12\% = 0.12$
      $B = 30$
      The amount is unknown.
      Let A = the number.

          $P \cdot B = A$
        $0.12 \cdot 30 = A$
             $3.6 = A$
      The number is 3.6.

B5) P is $90\% = 0.9$.
      $B = 250$
      The amount is unknown.
      Let A = the number.

          $P \cdot B = A$
        $0.9 \cdot 250 = A$
             $225 = A$
      The number is 225.

B6) P is $40\% = 0.4$.
The base is unknown.
Let B = the number.
$A = 20$

$$P \cdot B = A$$
$$0.4 \cdot B = 20$$
$$\frac{\cancel{0.4} \cdot B}{\cancel{0.4}} = \frac{20}{0.4}$$
$$B = 50$$

The number is 50.

B7) P is $20\% = 0.2$.
The base is unknown.
Let B = the number.
$A = 40$

$$P \cdot B = A$$
$$0.2 \cdot B = 40$$
$$\frac{\cancel{0.2} \cdot B}{\cancel{0.2}} = \frac{40}{0.2}$$
$$B = 200$$

The number is 200.

B8)   The percent is unknown.
Let P = the percent.
$B = 50$
$A = 30$

$$P \cdot B = A$$
$$P \cdot 50 = 30$$
$$\frac{P \cdot \cancel{50}}{\cancel{50}} = \frac{30}{50}$$
$$P = 0.6$$

The percent is 60%.

B9)   The percent is unknown.
Let P = the percent.
$B = 120$
$A = 40$

$$P \cdot B = A$$
$$P \cdot 120 = 40$$
$$\frac{P \cdot \cancel{120}}{\cancel{120}} = \frac{40}{120}$$
$$P = 0.\overline{3}$$

The percent is $33\frac{1}{3}\%$.

B10) P is $32\% = 0.32$.
$B = 380$
The amount is unknown.
Let A = the number.

$$P \cdot B = A$$
$$0.32 \cdot 380 = A$$
$$121.60 = A$$

Lourdes paid \$121.60 in tax.

B11) P is $40\% = 0.4$.
$B = 9$
The amount is unknown.
Let A = the number.

$$P \cdot B = A$$
$$0.4 \cdot 9 = A$$
$$3.6 = A$$

She spends 3.6 hours waiting for calls.

SECTION 5.3   EXERCISES

For 1–8, refer to Concept Checks and Practice Exercises.

9)   percent

11)   is

13)   base

15)   Identify the percent, the base, and the amount.
$P = 13$
$B = 312$
The amount is unknown.
Let A = amount.

Write the percent equation.
$$P \cdot B = A$$
$$13\% \cdot 312 = A$$

17)   Identify the percent, the base, and the amount.
The percent is unknown.
Let P = percent.
$B = 20$
$A = 54$

Write the percent equation.
$$P \cdot B = A$$
$$P \cdot 20 = 54$$

19)   Identify the percent, the base, and the amount.
$P = 30$
The base is unknown.
Let B = base.
$A = 30$

Write the percent equation.
$$P \cdot B = A$$
$$30\% \cdot B = 30$$

21) Identify the percent, the base, and the amount.
$$P = 0.05$$
$$B = 10,000$$
The amount is unknown.
Let A = amount.

Write the percent equation.
$$P \cdot B = A$$
$$0.05\% \cdot 10,000 = A$$

23) Identify the percent, the base, and the amount.
The percent is unknown.
Let P = percent.
$$B = 10,000$$
$$A = 2$$

Write the percent equation.
$$P \cdot B = A$$
$$P \cdot 10,000 = 2$$

25) Identify the percent, the base, and the amount.
$$P = 135$$
The base is unknown.
Let B = base.
$$A = 297$$

Write the percent equation.
$$P \cdot B = A$$
$$135\% \cdot B = 297$$

27) Identify the percent, the base, and the amount.
$$P = 150$$
$$B = 70$$
The amount is unknown.
Let A = amount.

Write the percent equation.
$$P \cdot B = A$$
$$150\% \cdot 70 = A$$

29) Identify the percent, the base, and the amount.
The percent is unknown.
Let P = percent.
$$B = 600$$
$$A = 120$$

Write the percent equation.
$$P \cdot B = A$$
$$P \cdot 600 = 120$$

31) Identify the percent, the base, and the amount.
$$P = 350$$
The base is unknown.
Let B = base.
$$A = 21$$

Write the percent equation.
$$P \cdot B = A$$
$$350\% \cdot B = 21$$

33) The percent is unknown.
Let P = the percent.
$$B = 40$$
$$A = 50$$

$$P \cdot B = A$$
$$P \cdot 40 = 50$$
$$\frac{P \cdot \cancel{40}}{\cancel{40}} = \frac{50}{40}$$
$$P = 1.25$$
The percent is 125%.

35) P is 75% = 0.75.
$$B = 212$$
The amount is unknown.
Let A = the number.

$$P \cdot B = A$$
$$0.75 \cdot 212 = A$$
$$159 = A$$
The number is 159.

37) P is 175% = 1.75.
$$B = 7$$
The amount is unknown.
Let A = the number.

$$P \cdot B = A$$
$$1.75 \cdot 7 = A$$
$$12.25 = A$$
The number is 12.25.

39) P is 90% = 0.90.
$$B = 80$$
The amount is unknown.
Let A = the number.

$$P \cdot B = A$$
$$0.90 \cdot 80 = A$$
$$72 = A$$
The number is 72.

41) P is $40\% = 0.40$.
The base is unknown.
Let $B =$ the number.
$A = 70$

$$P \cdot B = A$$
$$0.40 \cdot B = 70$$
$$\frac{0.40 \cdot B}{0.40} = \frac{70}{0.40}$$
$$B = 175$$
The number is 175.

43) The percent is unknown.
Let $P =$ the percent.
$B = 75$
$A = 90$

$$P \cdot B = A$$
$$P \cdot 75 = 90$$
$$\frac{P \cdot 75}{75} = \frac{90}{75}$$
$$P = 1.2$$
The percent is 120%.

45) P is $150\% = 1.50$.
$B = 70$
The amount is unknown.
Let $A =$ the number.

$$P \cdot B = A$$
$$1.50 \cdot 70 = A$$
$$105 = A$$

47) The percent is unknown.
Let $P =$ the percent.
$B = 600$
$A = 120$

$$P \cdot B = A$$
$$P \cdot 600 = 120$$
$$\frac{P \cdot 600}{600} = \frac{120}{600}$$
$$P = 0.20$$
The percent is 20%.

49) P is $350\% = 3.50$.
The base is unknown.
Let $B =$ the number.
$A = 21$

$$P \cdot B = A$$
$$3.50 \cdot B = 21$$
$$\frac{3.50 \cdot B}{3.50} = \frac{21}{3.50}$$
$$B = 6$$

51) The base is 312.
Consider 13%.

Since 42 is slightly more than 10% of 312, the solution $A = 42$ is reasonable.

53) The base is 20.

Since 54 is more than 20, the solution $P = 37\%$ is not reasonable.

55) 30% is approximately $\frac{1}{3}$ and $30 \cdot 3 = 90$.

The solution $B = 100$ is reasonable.

57) The base is 10,000.
Consider 0.05%.

Since 0.05% is less than 1%, and 1% of 10,000 is 100, the solution $A = 500$ is not reasonable.

59) The base is 10,000.

Since 2 is a very small portion of 10,000, the solution $P = 5,000\%$ is not reasonable.

61) 135% is approximately $1\frac{1}{3} = \frac{4}{3}$ and

$297 \cdot \frac{3}{4} = 222.75$.

The solution $B = 220$ is reasonable.

63) Estimate: 35
$$P \cdot B = A$$
$$0.40 \cdot 80 = A$$
$$32 = A$$

65) Estimate: 40%
$$P \cdot B = A$$
$$P \cdot 40 = 15$$
$$\frac{P \cdot 40}{40} = \frac{15}{40}$$
$$P = 0.375$$
The percent is 37.5%.

67) Estimate: 80
$$P \cdot B = A$$
$$0.60 \cdot B = 48$$
$$\frac{0.60 \cdot B}{0.60} = \frac{48}{0.60}$$
$$B = 80$$
The number is 80.

69) Estimate: 25%
$$P \cdot B = A$$
$$P \cdot 2 = 0.5$$
$$\frac{P \cdot \cancel{2}}{\cancel{2}} = \frac{0.5}{2}$$
$$P = 0.25$$
The percent is 25%.

71) Estimate: 17
$$P \cdot B = A$$
$$0.10 \cdot B = 1.7$$
$$\frac{\cancel{0.10} \cdot B}{\cancel{0.10}} = \frac{1.7}{0.10}$$
$$B = 17$$
The number is 17.

73) Estimate: 30
$$P \cdot B = A$$
$$0.10 \cdot B = 3$$
$$\frac{\cancel{0.10} \cdot B}{\cancel{0.10}} = \frac{3}{0.10}$$
$$B = 30$$
The number is 30.

75) P is 5% = 0.05.
$B = 16.60$
The amount is unknown.
Let A = the amount of sales tax.
$$P \cdot B = A$$
$$0.05 \cdot 16.60 = A$$
$$0.83 = A$$
The sales tax is $0.83.

77) The percent is unknown.
Let P = the percent.
$$B = 3\frac{1}{2} = 3.5$$
$$A = 1$$
$$P \cdot B = A$$
$$P \cdot 3.5 = 1$$
$$\frac{P \cdot \cancel{3.5}}{\cancel{3.5}} = \frac{1}{3.5}$$
$$P \approx 0.2857$$
About 29% of her time watching TV was spent watching commercials.

79) $14.00 + $3.50 = $17.50
The total cost is $17.50.
$$P \cdot B = A$$
$$P \cdot 17.50 = 14.00$$
$$\frac{P \cdot \cancel{17.50}}{\cancel{17.50}} = \frac{14.00}{17.50}$$
$$P = 0.80$$
The paper represents 80% of the total cost.

81) $0.06 \cdot 1,020 = 61.20$
The fuel surcharge is $61.20.

$1,020 + $61.20 = $1,081.20
The total cost is $1,081.20. Since this is less than $1,100, Kesta has enough money to buy the ticket.

5.4   Percent Applications

GUIDED PRACTICE

1) Identify the commission earned, commission rate, and sales.
(Amount) Let c = the commission.
(%) The commission rate is 5% = 0.05.
(Base) Sales = $19,000

Write and solve a commission equation.
commission = commission rate · sales
$$c = 0.05 \cdot 19,000$$
$$c = 950$$
The salesperson made $950 in commission.

2) Identify the commission earned, commission rate, and sales.
Commission = $45,000
The commission rate is 5% = 0.05.
Let s = amount of total sales.

Write and solve a commission equation.
commission = commission rate · sales
$$45,000 = 0.05 \cdot s$$
$$\frac{45,000}{0.05} = \frac{\cancel{0.05} \cdot s}{\cancel{0.05}}$$
$$900,000 = s$$
Hank's total sales were $900,000 last year.

3) Identify the commission earned, commission rate, and sales.

Commission = $12,500

Let r = the commission rate.

Savings = $2,500,000

Write and solve a commission equation.

commission = commission rate · savings

$12,500 = r \cdot 2,500,000$

$$\frac{12,500}{2,500,0000} = \frac{r \cdot 2,500,0000}{2,500,0000}$$

$0.005 = r$

Gloria earned a 0.5% commission rate.

4) Identify the interest, principal, interest rate, and time.

Principal = $3,500

Let I = interest.

Interest rate is $1.5\% = 0.015$ per month.

Time = 2 months

Write and solve a simple interest equation.

Interest = Principal · rate · time

$I = P \cdot r \cdot t$

$I = 3,500 \cdot 0.015 \cdot 2$

$I = 52.50 \cdot 2$

$I = 105$

Antonio was charged $105.00 in interest.

5) Identify the interest, principal, interest rate, and time.

Principal = $12,000

Interest = $420

Let r = interest rate per year.

Time = 1 year

Write and solve a simple interest equation.

$I = P \cdot r \cdot t$

$420 = 12,000 \cdot r \cdot 1$

$420 = 12,000 \cdot r$

$$\frac{420}{12,000} = \frac{12,000 \cdot r}{12,000}$$

$0.035 = r$

$3.5\% = r$

Samantha earned an interest rate of 3.5%.

6) Identify the interest, principal, interest rate, and time.

Principal or Balance = $9,500

Let I = interest.

Interest rate = 9% per year

$$= \frac{9\%}{12}$$

$= 0.75\%$

$= 0.0075$ per month

Time = 1 month

Write and solve a simple interest equation.

$I = P \cdot r \cdot t$

$I = 9,500 \cdot 0.0075 \cdot 1$

$I = 71.25$

Andrea owes $71.25 in interest after 1 month.

7) a) When you buy an iPod for $200, you are charged sales tax. The sales tax is a percent increase.

b) A bacteria population that is 75% of its original size has experienced a 25% decrease.

8) Identify the percent change, the change in quantity, and the original quantity.

Percent change is $30\% = 0.30$.

Let c = the change in snow.

The original amount = 36 inches.

Write and solve a percent change equation.

$$\text{Percent Change} = \frac{\text{Change in Amount}}{\text{Original Amount}}$$

$$0.30 = \frac{c}{36}$$

$$0.30 \cdot \frac{36}{1} = \frac{c}{36} \cdot \frac{36}{1}$$

$10.8 = c$

The amount of snow increased by 10.8 inches.

9) Identify the percent change, the change in quantity, and the original quantity.

Let p = the percent change.

The change in production = a decrease of 70,000 balls.

Original amount = 500,000 balls.

Write and solve a percent change equation.

$$\text{Percent Change} = \frac{\text{Change in Production}}{\text{Original Production}}$$

$$p = \frac{70,000}{500,000}$$

$p = 0.14$

$p = 14\%$

Tennis ball production decreased by 14%.

10) Identify the percent change, the change in quantity, and the original quantity.

Let p = the percent decrease.

The change in salary = $\$32,000 - \$28,000$
$$= \$4,000.$$

The original salary = $\$32,000$.

Write and solve a percent change equation.

$$\text{Percent Change} = \frac{\text{Change in Salary}}{\text{Original Salary}}$$
$$p = \frac{4,000}{32,000}$$
$$p = 0.125$$
$$p = 12.5\%$$

Daniel's salary decreased by 12.5%.

11) The principal or starting balance = $\$100$.
The interest rate is $2\% = 0.02$.

$$(1 + r) \cdot \text{Balance} = \text{New Balance}$$
$$(1.02) \cdot \$100.00 = \$102.00$$
$$(1.02) \cdot \$102.00 = \$104.04$$
$$(1.02) \cdot \$104.04 = \$106.12$$

The account balance after three years is $106.12.

12) a) Balance difference after 4 years:
$$\text{Compound} - \text{Simple} \approx \$7,500 - \$7,000$$
$$= \$500$$

b) Balance difference after 16 years:
$$\text{Compound} - \text{Simple} \approx \$23,000 - \$13,000$$
$$\approx \$10,000$$

c) A compound interest account will earn more money than a simple interest account if both have the same interest rate because the interest earned in a compound interest account also earns interest.

CONCEPT CHECKS AND PRACTICE EXERCISES

A1) Commission = Sales · Commission rate

A2) Answers may vary. Example: This would mean that Pat earned more than she sold.

A3) Let c = the commission.
The commission rate is $9\% = 0.09$.
Sales = $\$800$
commission = commission rate · sales
$$c = 0.09 \cdot 800$$
$$c = 72$$
He will earn a $72 commission.

A4) Let c = the commission.
The commission rate is $1\% = 0.01$.
Sales = $\$2,100,000$

commission = commission rate · sales
$$c = 0.01 \cdot 2,100,000$$
$$c = 21,000$$
The manager's commission was $21,000.

A5) Commission = $\$2,125$
Let r = the commission rate.
Sales = $\$42,500$

commission = commission rate · sales
$$2,125 = r \cdot 42,500$$
$$\frac{2,125}{42,500} = \frac{r \cdot \cancel{42,500}}{\cancel{42,500}}$$
$$0.05 = r$$
Lester earned a 5% commission rate.

A6) Commission = $\$4,500$
Let r = the commission rate.
Sales = $\$30,000$

commission = commission rate · sales
$$4,500 = r \cdot 30,000$$
$$\frac{4,500}{30,000} = \frac{r \cdot \cancel{30,000}}{\cancel{30,000}}$$
$$0.15 = r$$
She earned a 15% commission rate.

A7) Let c = the commission.
The commission rate is $0.25\% = 0.0025$.
Sales = $\$1,260,000$

commission = commission rate · sales
$$c = 0.0025 \cdot 1,260,000$$
$$c = 3,150$$
The advisor's commission was $3,150.

A8)   Commission = $1,500
The commission rate $6\% = 0.06$.
Let s = sales.

commission = commission rate · sales
$$1,500 = 0.06 \cdot s$$
$$\frac{1,500}{0.06} = \frac{\cancel{0.06} \cdot s}{\cancel{0.06}}$$
$$25,000 = s$$
The total sales for the month were $25,000.

B1)   Answers may vary. Example: Change a monthly interest rate to a yearly interest rate by multiplying the   monthly interest rate by 12.

B2)   Answers may vary. Example: Change a yearly interest rate to a monthly interest rate by dividing the yearly interest rate by 12.

B3)   Principal = $200,000

Interest = $15,000

Let r = interest rate per year.

Time = 1 year

$$I = P \cdot r \cdot t$$
$$15,000 = 200,000 \cdot r \cdot 1$$
$$15,000 = 200,000 \cdot r$$
$$\frac{15,000}{200,000} = \frac{\cancel{200,000} \cdot r}{\cancel{200,000}}$$
$$0.075 = r$$
$$7.5\% = r$$
This savings account earns an interest rate of 7.5%.

B4)   Principal = $500,000

Interest = $30,000

Let r = interest rate per year.

Time = 1 year

$$I = P \cdot r \cdot t$$
$$30,000 = 500,000 \cdot r \cdot 1$$
$$30,000 = 500,000 \cdot r$$
$$\frac{30,000}{500,000} = \frac{\cancel{500,000} \cdot r}{\cancel{500,000}}$$
$$0.06 = r$$
$$6\% = r$$
This bond account earns an interest rate of 6%.

B5)   Principal = $630

Let I = interest.

The interest rate $1.3\% = 0.013$ per year.

Time = 3 years

Interest = Principal · rate · time
$$I = P \cdot r \cdot t$$
$$I = 630 \cdot 0.013 \cdot 3$$
$$I = 8.19 \cdot 3$$
$$I = 24.57$$
The account earned $24.57 interest in 3 years.

B6)   Principal = $1,580

Let I = interest.
The interest rate $2.5\% = 0.025$ per year.

Time = 4 years

Interest = Principal · rate · time
$$I = P \cdot r \cdot t$$
$$I = 1,580 \cdot 0.025 \cdot 4$$
$$I = 39.50 \cdot 4$$
$$I = 158$$
The account will earn $158 interest in 4 years.

B7)   Let P = Balance.
$I = \$50$
Interest rate = 15% per year
$$= \frac{15\%}{12}$$
$$= 1.25\%$$
$$= 0.0125 \text{ per month}$$
Time = 1 month
$$I = P \cdot r \cdot t$$
$$50 = P \cdot 0.0125 \cdot 1$$
$$50 = P \cdot 0.0125$$
$$\frac{50}{0.0125} = \frac{P \cdot \cancel{0.0125}}{\cancel{0.0125}}$$
$$4,000 = P$$
Hiro's balance was $4,000.

B8)   Let P = Balance.
$I = \$27$
Interest rate = 20% per year
$$= \frac{20\%}{12}$$
$$= 1\frac{2}{3}\%$$
$$= \frac{1}{60} \text{ per month}$$
Time = 1 month

$$I = P \cdot r \cdot t$$
$$27 = P \cdot \frac{1}{60} \cdot 1$$
$$27 \cdot 60 = \frac{P}{\cancel{60}} \cdot \cancel{60}$$
$$1,620 = P$$
Saiesha's balance was $1,620.

C1)   No, everyone in the department does not receive
      the same raise in dollars. Taking 5% of different
      salaries yields different amounts.

C2)   Percent $2.1\% = 0.021$
      Let c = the change in salary.
      Original Salary $= \$35,000$

$$\text{Percent Change} = \frac{\text{Change in Salary}}{\text{Original Salary}}$$

$$0.021 = \frac{c}{35,000}$$

$$0.021 \cdot \frac{35,000}{1} = \frac{c}{35,000} \cdot \frac{35,000}{1}$$

$$735 = c$$

The change in salary is $735.

C3)   Percent $3\% = 0.03$
      Let c = the change in salary.
      Original Salary $= \$38,000$

$$\text{Percent Change} = \frac{\text{Change in Salary}}{\text{Original Salary}}$$

$$0.03 = \frac{c}{38,000}$$

$$0.03 \cdot \frac{38,000}{1} = \frac{c}{38,000} \cdot \frac{38,000}{1}$$

$$1,140 = c$$

The change in salary was $1,140.

C4)   Let p = the percent increase.
      The change in cost = an increase of $207.
      Original Cost $= \$900$

$$\text{Percent Change} = \frac{\text{Change in Cost}}{\text{Original Cost}}$$

$$p = \frac{207}{900}$$

$$p = 0.23$$

$$p = 23\%$$

The cost of Arejo's insurance increased by 23%.

C5)   Let p = the percent increase.
      Change in Cost $= \$3.84 - \$2.40$
      $\qquad\qquad\quad = \$1.44$
      Original Cost $= \$2.40$

$$\text{Percent Change} = \frac{\text{Change in Cost}}{\text{Original Cost}}$$

$$p = \frac{1.44}{2.40}$$

$$p = 0.6$$

$$p = 60\%$$

The price of gasoline has experienced a 60%
increase.

C6)   Let p = the percent decrease.
      The change in salary = a decrease of $5,000.
      Original Salary $= \$40,000$

$$\text{Percent Change} = \frac{\text{Change in Salary}}{\text{Original Salary}}$$

$$p = \frac{5,000}{40,000}$$

$$p = 0.125$$

$$p = 12.5\%$$

A $5,000 cut represents a 12.5% decrease in
salary.

C7)   Let p = the percent decrease.
      The change in number = a decrease of 3.
      Original Number $= 80$

$$\text{Percent Change} = \frac{\text{Change in Number}}{\text{Original Number}}$$

$$p = \frac{3}{80}$$

$$p = 0.0375$$

$$p = 3.75\%$$

A 3.75% decrease in the number of police
officers has occurred.

SECTION 5.4   EXERCISES

For 1–9, refer to Concept Checks and Practice
Exercises.

11)   percent increase

13)   commission rate

15)   interest rate

17)   commission

19)   Let c = the commission.
      The commission rate $11\% = 0.11$.
      Sales $= \$575,000$

$$\text{commission} = \text{commission rate} \cdot \text{sales}$$

$$c = 0.11 \cdot 575,000$$

$$c = 63,250$$

Sigmund's commission is $63,250.

21) Commission = \$3,000
     Let r = the commission rate.
     Sales = \$120,000

     commission = commission rate $\cdot$ sales
     $$3,000 = r \cdot 120,000$$
     $$\frac{3,000}{120,000} = \frac{r \cdot \cancel{120,000}}{\cancel{120,000}}$$
     $$0.025 = r$$
     $$2.5\% = r$$
     Her commission rate was 2.5%.

23) Commission = \$5,500
     The commission rate $2\% = 0.02$.
     Let s = sales.
     commission = commission rate $\cdot$ sales
     $$5,500 = 0.02 \cdot s$$
     $$\frac{5,500}{0.02} = \frac{\cancel{0.02} \cdot s}{\cancel{0.02}}$$
     $$275,000 = s$$
     Marcos's sales last month were \$275,000.

25) commission = commission rate $\cdot$ sales
     $$c = 0.02 \cdot 15,000$$
     $$c = 300$$
     Germaine's commission will be \$300.

27) $0.05 \cdot 125,000 - 0.04 \cdot 124,000 = 6,250 - 4,960$
     $$= 1,290$$
     Her commission increased by \$1,290.

29) commission = commission rate $\cdot$ sales
     $$4,000 = 0.04 \cdot s$$
     $$\frac{4,000}{0.04} = \frac{\cancel{0.04} \cdot s}{\cancel{0.04}}$$
     $$100,000 = s$$
     His car sales were \$100,000.

31) Principal = \$4,500
     Let I = interest.
     The interest rate $2\% = 0.02$ per year.
     Time = 2 years

     Interest = Principal $\cdot$ rate $\cdot$ time
     $$I = P \cdot r \cdot t$$
     $$I = 4,500 \cdot 0.02 \cdot 2$$
     $$I = 90 \cdot 2$$
     $$I = 180$$
     The account will earn \$180 interest in two years.

33) Principal or Balance = \$2,580
     Let I = interest.
     Interest rate = 15% per year
     $$= \frac{15\%}{12}$$
     $$= 1.25\%$$
     $$= 0.0125 \text{ per month}$$
     Time = 1 month

     $$I = P \cdot r \cdot t$$
     $$I = 2,580 \cdot 0.0125 \cdot 1$$
     $$I = 32.25$$
     Her interest charge for one month is \$32.25.

35) Let P = Account Balance.
     I = \$18.53
     Interest rate = 12% per year
     $$= \frac{12\%}{12}$$
     $$= 1\%$$
     $$= 0.01 \text{ per month}$$
     Time = 1 month
     $$I = P \cdot r \cdot t$$
     $$18.53 = P \cdot 0.01 \cdot 1$$
     $$18.53 = P \cdot 0.01$$
     $$\frac{18.53}{0.01} = \frac{P \cdot \cancel{0.01}}{\cancel{0.01}}$$
     $$1,853 = P$$
     Keith's balance was \$1,853.

37) Let p = the percent decrease.
     Change in Cost = \$25.00 − \$8.50
     $$= \$16.50$$
     Original Cost = \$25.00

     Percent Change $= \dfrac{\text{Change in Cost}}{\text{Original Cost}}$
     $$p = \frac{16.50}{25.00}$$
     $$p = 0.66$$
     $$p = 66\%$$
     The markdown represents a 66% decrease in price.

39) Let p = the percent increase.

Change in Cost = $45 - $15 = $30

Original Cost = $15

$$\text{Percent Change} = \frac{\text{Change in Cost}}{\text{Original Cost}}$$

$$p = \frac{30}{15}$$

$$p = 2$$

$$p = 200\%$$

The markup represents an increase of 200%.

41) Let p = the percent increase.

Change in Debt = $178 billion − $160 billion

= $18 billion

Original Debt = $160 billion

$$\text{Percent Change} = \frac{\text{Change in Debt}}{\text{Original Debt}}$$

$$p = \frac{18\text{ billion}}{160\text{ billion}}$$

$$p = 0.1125$$

$$p = 11.25\%$$

This represents an 11.25% increase in the interest paid on the national debt.

43) Percent 4% = 0.04

Let c = the change in salary.

Original Salary = $45,000

$$\text{Percent Change} = \frac{\text{Change in Salary}}{\text{Original Salary}}$$

$$0.04 = \frac{c}{45,000}$$

$$0.04 \cdot \frac{45,000}{1} = \frac{c}{45,000} \cdot \frac{45,000}{1}$$

$$1,800 = c$$

The change in their salary was $1,800.

45) Let p = the percent increase.

Change in Cost = $9 − $5

= $4

Original Cost = $5

$$\text{Percent Change} = \frac{\text{Change in Cost}}{\text{Original Cost}}$$

$$p = \frac{4}{5}$$

$$p = 0.8$$

$$p = 80\%$$

The percent increase is 80%.

47) $p = \frac{40}{50} = 0.8$

The 2008 value was 80% of the 2007 value.

49) $\text{Percent Change} = \frac{\text{Change in Value}}{\text{Original Value}}$

$$p = \frac{50 - 40}{50}$$

$$p = \frac{10}{50}$$

$$p = 0.20$$

$$p = 20\%$$

The value decreased 20% from 2007 to 2008.

51) $p = \frac{72}{50} = 1.44$

The 2010 value was 144% of the 2007 value.

53) $\text{Value} = \frac{1,230}{0.0075}$

$$= 164,000$$

The property is worth $164,000.

55) $\text{Percent Change} = \frac{\text{Change in Lap Time}}{\text{Original Lap Time}}$

$$0.03 = \frac{c}{17.8}$$

$$\frac{0.03}{1} \cdot \frac{17.8}{1} = \frac{c}{17.8} \cdot \frac{17.8}{1}$$

$$0.534 = c$$

Their lap time decreased by 0.534 seconds.

$17.8 - 0.534 = 17.266$

The new lap time is about 17.27 seconds.

57) $38 \div 2 = 19$

Each person's share of the original bill was $19.

$0.20 \cdot 19 = 3.80$

Jennifer gave a tip of $3.80.

$0.15 \cdot 19 = 2.85$

Ben gave a tip of $2.85.

$3.80 + $2.85 = $6.65

The waiter's tip was $6.65.

59) $0.025 \cdot 120,000 = 3,000$

The agent earned a commission of $3,000.

$3,000 − 0.28 \cdot 3000$

$= 3,000 − 840$

$= 2,160$

The agent earned $2,160 after taxes.

61) $10,000,000 + 0.08 \cdot 10,000,000$
$= 10,000,000 + 800,000$
$= 10,800,000$
Before the donation, her net worth was
$10,800,000.

$10,800,000 - 1,080,000 = 9,720,000$
Her new net worth is $9,720,000.

63) $A = s^2 = 2^2 = 4 \text{ cm}^2$
If the sides are increased in length by 100%,
then each side will be of length $2s$.

$A = (2s)^2 = (4)^2 = 16 \text{ cm}^2$

$\text{Percent Change} = \dfrac{\text{Change in Value}}{\text{Original Value}}$

$= \dfrac{16 \text{ cm}^2 - 4 \text{ cm}^2}{4 \text{ cm}^2}$

$= \dfrac{12 \text{ cm}^2}{4 \text{ cm}^2}$

$= 3$

$= 300\%$

This is a 300% increase in the area.

65) $p = \dfrac{4}{20}$
$p = 0.2$
20% of the shapes are circles, but not red.

67) $p = \dfrac{5}{20}$
$p = 0.25$
25% of the shapes are either red or blue squares.

69) $p = \dfrac{0}{20}$
$p = 0$
0% of the shapes are brown circles.

CHAPTER 5   REVIEW EXERCISES

1) % is used in place of the phrases *per hundred* or
out of one hundred.

2) The word cent means "one hundred."

3) a) $\dfrac{1}{2} = 50\%$, $1\dfrac{1}{2} = 150\%$, $2\dfrac{1}{2} = 250\%$,
$3\dfrac{1}{2} = 350\%$, $4\dfrac{1}{2} = 450\%$

b) $1 = 100\%$, $2 = 200\%$, $3 = 300\%$,
$9 = 900\%$, $10 = 1,000\%$

4) 20%

5) 83%

6) 100%

7) 1%

8) 160%

9) 115%
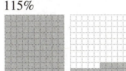

10) $25\% = \dfrac{25\,\%}{1} \cdot \dfrac{1}{100\,\%}$

$= \dfrac{25}{100}$

$= \dfrac{1 \cdot 25}{4 \cdot 25}$

$= \dfrac{1}{4}$

11) $40\% = \dfrac{40\,\%}{1} \cdot \dfrac{1}{100\,\%}$

$= \dfrac{40}{100}$

$= \dfrac{2 \cdot 20}{5 \cdot 20}$

$= \dfrac{2}{5}$

12) $78\% = \dfrac{78\%}{1} \cdot \dfrac{1}{100\%}$

$= \dfrac{78}{100}$

$= \dfrac{2 \cdot 39}{2 \cdot 50}$

$= \dfrac{39}{50}$

13) $82\% = \dfrac{82\%}{1} \cdot \dfrac{1}{100\%}$

$= \dfrac{82}{100}$

$= \dfrac{2 \cdot 41}{2 \cdot 50}$

$= \dfrac{41}{50}$

14) $50\dfrac{1}{4}\% = \dfrac{50.25\%}{1} \cdot \dfrac{1}{100\%}$

$= \dfrac{50.25}{100}$

$= \dfrac{5,025}{10,000}$

$= \dfrac{25 \cdot 201}{25 \cdot 400}$

$= \dfrac{201}{400}$

15) $40\dfrac{1}{2}\% = \dfrac{40.5\%}{1} \cdot \dfrac{1}{100\%}$

$= \dfrac{40.5}{100}$

$= \dfrac{405}{1,000}$

$= \dfrac{5 \cdot 81}{5 \cdot 200}$

$= \dfrac{81}{200}$

16) $180\% = \dfrac{180\%}{1} \cdot \dfrac{1}{100\%}$

$= \dfrac{180}{100}$

$= \dfrac{9 \cdot 20}{5 \cdot 20}$

$= \dfrac{9}{5} = 1\dfrac{4}{5}$

17) $120\% = \dfrac{120\%}{1} \cdot \dfrac{1}{100\%}$

$= \dfrac{120}{100}$

$= \dfrac{6 \cdot 20}{5 \cdot 20}$

$= \dfrac{6}{5} = 1\dfrac{1}{5}$

18) $\dfrac{7}{10} = \dfrac{7}{10} \cdot \dfrac{100\%}{1}$

$= \dfrac{700\%}{10}$

$= 70\%$

19) $\dfrac{11}{20} = \dfrac{11}{20} \cdot \dfrac{100\%}{1}$

$= \dfrac{1,100\%}{20}$

$= 55\%$

20) $\dfrac{3}{8} = \dfrac{3}{8} \cdot \dfrac{100\%}{1}$

$= \dfrac{300\%}{8}$

$= 37.5\%$

21) $\dfrac{5}{8} = \dfrac{5}{8} \cdot \dfrac{100\%}{1}$

$= \dfrac{500\%}{8}$

$= 62.5\%$

22) $4\dfrac{1}{10} = \dfrac{41}{10} \cdot \dfrac{100\%}{1}$

$= \dfrac{4,100\%}{10}$

$= 410\%$

23) $5\dfrac{1}{5} = \dfrac{26}{5} \cdot \dfrac{100\%}{1}$

$= \dfrac{2,600\%}{5}$

$= 520\%$

24) $2\dfrac{1}{9} = \dfrac{19}{9} \cdot \dfrac{100\%}{1}$

$= \dfrac{1,900\%}{9}$

$= 211.\overline{1}\%$

25) $3\dfrac{5}{6} = \dfrac{23}{6} \cdot \dfrac{100\%}{1}$

$= \dfrac{2,300\%}{6}$

$= 383.\overline{3}\%$

26) $57\% = \dfrac{57\%}{1} \cdot \dfrac{1}{100\%}$

$= \dfrac{57}{100}$

$= 0.57$

27) $48\% = \dfrac{48\%}{1} \cdot \dfrac{1}{100\%}$

$= \dfrac{48}{100}$

$= 0.48$

28) $0.67 = 0.67 \cdot 100\%$

$= 67\%$

29) $0.52 = 0.52 \cdot 100\%$

$= 52\%$

30) $235\% = \dfrac{235\%}{1} \cdot \dfrac{1}{100\%}$

$= \dfrac{235}{100}$

$= 2.35$

31) $3.6 = 3.6 \cdot 100\%$

$= 360\%$

32) $8.7 = 8.7 \cdot 100\%$

$= 870\%$

33) $434\% = \dfrac{434\%}{1} \cdot \dfrac{1}{100\%}$

$= \dfrac{434}{100}$

$= 4.34$

34) $0.0054 = 0.0054 \cdot 100\%$

$= 0.54\%$

35) $0.02\% = \dfrac{0.02\%}{1} \cdot \dfrac{1}{100\%}$

$= \dfrac{0.02}{100}$

$= 0.0002$

36) $0.9\% = \dfrac{0.9\%}{1} \cdot \dfrac{1}{100\%}$

$= \dfrac{0.9}{100}$

$= 0.009$

37) $0.0025 = 0.0025 \cdot 100\%$

$= 0.25\%$

38) $6,000\% = \dfrac{6,000\%}{1} \cdot \dfrac{1}{100\%}$

$= \dfrac{6,000}{100}$

$= 60$

39) $7,000 = 7,000 \cdot 100\%$

$= 700,000\%$

40) $3,000 = 3,000 \cdot 100\%$

$= 300,000\%$

41) $3,000\% = \dfrac{3,000\%}{1} \cdot \dfrac{1}{100\%}$

$= \dfrac{3,000}{100}$

$= 30$

	Decimal	Fraction	Percent
42)	0.75	$\dfrac{3}{4}$	75%
43)	$0.\overline{3}$	$\dfrac{1}{3}$	$33\dfrac{1}{3}\%$
44)	0.2	$\dfrac{1}{5}$	20%
45)	1.0	$\dfrac{3}{3}$	100%

	Decimal	Fraction	Percent
46)	0.125	$\dfrac{1}{8}$	12.5%
47)	1.22	$1\dfrac{11}{50}$	122%
48)	2.6	$2\dfrac{3}{5}$	260%
49)	0.7	$\dfrac{7}{10}$	70%

50) $\dfrac{P}{100} = \dfrac{A}{B}$

51) The word *is* is often used to represent equality in word exercises.

52) The word *of* is often used to represent multiplication in word exercises.

53) $P = 30$
$B = 80$
Let A = amount.

$$\frac{30}{100} = \frac{A}{80}$$

$$\frac{30}{100} \cdot \frac{80}{1} = \frac{A}{\cancel{80}} \cdot \frac{\cancel{80}}{1}$$

$$\frac{3 \cdot \cancel{10} \cdot 8 \cdot \cancel{10}}{\cancel{10} \cdot \cancel{10}} = A$$

$$24 = A$$

24 is 30% of 80.

54) $P = 20$
$B = 50$
Let A = amount.

$$\frac{20}{100} = \frac{A}{50}$$

$$\frac{20}{100} \cdot \frac{50}{1} = \frac{A}{\cancel{50}} \cdot \frac{\cancel{50}}{1}$$

$$\frac{\cancel{2} \cdot 10 \cdot \cancel{50}}{\cancel{2} \cdot \cancel{50}} = A$$

$$10 = A$$

10 is 20% of 50.

55) Let P = the percent.
$B = 40$
$A = 100$

$$\frac{P}{100} = \frac{100}{40}$$

$$\frac{\cancel{100}}{1} \cdot \frac{P}{\cancel{100}} = \frac{100}{1} \cdot \frac{100}{40}$$

$$P = \frac{\cancel{10} \cdot 10 \cdot \cancel{4} \cdot 25}{\cancel{4} \cdot \cancel{10}}$$

$$P = 250$$

100 is 250% of 40.

56) Let P = the percent.
$B = 35$
$A = 70$

$$\frac{P}{100} = \frac{70}{35}$$

$$\frac{\cancel{100}}{1} \cdot \frac{P}{\cancel{100}} = \frac{100}{1} \cdot \frac{70}{35}$$

$$P = \frac{100 \cdot 2 \cdot \cancel{35}}{\cancel{35}}$$

$$P = 200$$

70 is 200% of 35.

57) $P = 25$
Let B = the number.
$A = 18$

$$\frac{25}{100} = \frac{18}{B}$$

$$\frac{100}{25} = \frac{B}{18}$$

$$\frac{100}{25} \cdot \frac{18}{1} = \frac{B}{\cancel{18}} \cdot \frac{\cancel{18}}{1}$$

$$\frac{1,800}{25} = B$$

$$72 = B$$

18 is 25% of 72.

58) $P = 15$
Let B = the number.
$A = 15$

$$\frac{15}{100} = \frac{15}{B}$$

$$\frac{100}{15} = \frac{B}{15}$$

$$\frac{100}{\cancel{15}} \cdot \frac{\cancel{15}}{1} = \frac{B}{\cancel{15}} \cdot \frac{\cancel{15}}{1}$$

$$100 = B$$

15 is 15% of 100.

59) Let P = the percent.
$B = 10$
$A = 24 - 10 = 14$

$$\frac{P}{100} = \frac{14}{10}$$

$$\frac{\cancel{100}}{1} \cdot \frac{P}{\cancel{100}} = \frac{100}{1} \cdot \frac{14}{10}$$

$$P = \frac{\cancel{10} \cdot 10 \cdot 14}{\cancel{10}}$$

$$P = 140$$

Apple increased iPhone production by 140%.

60) $P = 85$
$B = 60$
Let A = the questions answered correctly.

$$\frac{85}{100} = \frac{A}{60}$$

$$\frac{85}{100} \cdot \frac{60}{1} = \frac{A}{\cancel{60}} \cdot \frac{\cancel{60}}{1}$$

$$\frac{\cancel{5} \cdot 17 \cdot 3 \cdot \cancel{20}}{\cancel{5} \cdot \cancel{20}} = A$$

$$51 = A$$

You answered 51 questions correctly.

61) $P = 20$
Let $B =$ the purchase price of the car.
$A = 5,300$

$$\frac{20}{100} = \frac{5,300}{B}$$

$$\frac{100}{20} = \frac{B}{5,300}$$

$$\frac{100}{20} \cdot \frac{5,300}{1} = \frac{B}{5,300} \cdot \frac{5,300}{1}$$

$$\frac{5 \cdot \cancel{20} \cdot 5,300}{\cancel{20}} = B$$

$$26,500 = B$$

The purchase price of the car was \$26,500.

62) Let $P =$ the percent.
$B = 150$
$A = 90$

$$\frac{P}{100} = \frac{90}{150}$$

$$\frac{\cancel{100}}{1} \cdot \frac{P}{\cancel{100}} = \frac{100}{1} \cdot \frac{90}{150}$$

$$P = \frac{2 \cdot \cancel{50} \cdot \cancel{3} \cdot 30}{\cancel{3} \cdot \cancel{50}}$$

$$P = 60$$

She met 60% of the employees.

63) $P = 8$
Let $B =$ the recommended amount of sodium.
$A = 200$

$$\frac{8}{100} = \frac{200}{B}$$

$$\frac{100}{8} = \frac{B}{200}$$

$$\frac{100}{8} \cdot \frac{200}{1} = \frac{B}{\cancel{200}} \cdot \frac{\cancel{200}}{1}$$

$$\frac{\cancel{4} \cdot 25 \cdot \cancel{2} \cdot 100}{\cancel{2} \cdot \cancel{4}} = B$$

$$2,500 = B$$

The daily recommended amount of sodium is 2,500 milligrams.

64) Let $P =$ the percent.
$B = 170$
$A = 170 - 130 = 40$

$$\frac{P}{100} = \frac{40}{170}$$

$$\frac{\cancel{100}}{1} \cdot \frac{P}{\cancel{100}} = \frac{100}{1} \cdot \frac{40}{170}$$

$$P = \frac{10 \cdot \cancel{10} \cdot 40}{17 \cdot \cancel{10}}$$

$$P = \frac{400}{17}$$

$$P = 23\frac{9}{17}$$

$$P \approx 24$$

Approximately 24% of the calories come from the milk.

65) $$\frac{30 \text{ mg}}{10\%} = \frac{n \text{ mg}}{12\%}$$
$$30 \cdot 0.12 = 0.10 \cdot n$$
$$3.6 = 0.10n$$
$$36 = n$$
There are 36 mg of carbohydrates in the cheerios and milk.

$36 - 30 = 6$
A half cup of skim milk contains 6 mg of carbohydrates.

66) The word *of* is often used to indicate multiplication in a percent problem

67) The words *is* and *are* are often used to indicate equality in a percent problem.

68) $P \cdot B = A$

69) In a percent equation, the percent describes the portion of the base that is equal to the amount.

70) In a percent equation, the base is the whole quantity.

71) In a percent equation, the amount is the portion that results from multiplying the percent and the base.

72) $P \cdot B = A$
$0.15 \cdot 35 = A$
$5.25 = A$

73) $P \cdot B = A$
$0.80 \cdot 22 = A$
$17.6 = A$

74) $P \cdot B = A$
$0.40 \cdot B = 25$
$\dfrac{\cancel{0.40} \cdot B}{\cancel{0.40}} = \dfrac{25}{0.40}$
$B = 62.5$
The number is 62.5.

75) $P \cdot B = A$
$0.20 \cdot B = 46$
$\dfrac{\cancel{0.20} \cdot B}{\cancel{0.20}} = \dfrac{46}{0.20}$
$B = 230$
The number is 230.

76) $P \cdot B = A$
$P \cdot 15 = 65$
$\dfrac{P \cdot \cancel{15}}{\cancel{15}} = \dfrac{65}{15}$
$P = 4.\overline{3}$
The percent is $433.\overline{3}\%$.

77) $P \cdot B = A$
$P \cdot 12 = 84$
$\dfrac{P \cdot \cancel{12}}{\cancel{12}} = \dfrac{84}{12}$
$P = 7$
The percent is 700%.

78) $P \cdot B = A$
$P \cdot 50 = 42$
$\dfrac{P \cdot \cancel{50}}{\cancel{50}} = \dfrac{42}{50}$
$P = 0.84$
You have used 84% of the storage capacity.

79) $P \cdot B = A$
$0.08 \cdot B = 0.96$
$\dfrac{\cancel{0.08} \cdot B}{\cancel{0.08}} = \dfrac{0.96}{0.08}$
$B = 12$
The strings cost $12.00 before sales tax.

80) $P \cdot B = A$
$0.20 \cdot 109,000 = A$
$21,800 = A$
Her down payment was $21,800.

81)   Commission is the amount of money a salesperson earns as a percent of sales.

82)   An interest rate multiplies the balance and the time period of the investment to determine how much interest is earned.

83)   When money is invested, the original investment is called the principal.

84)   Interest is the amount of money earned as a percent of the balance.

85)   Let c = the commission.
The commission rate 8% = 0.08.
Sales = $1,200

commission = commission rate · sales
$c = 0.08 \cdot 1,200$
$c = 96$
He will earn a $96 commission.

86)   Let c = the commission.
The commission rate 2% = 0.02.
Sales = $1,100,000

commission = commission rate · sales
$c = 0.02 \cdot 1,100,000$
$c = 22,000$
The manager's commission was $22,000.

87)   Commission = $2,250
Let r = the commission rate.
Sales = $25,000

commission = commission rate · sales
$2,250 = r \cdot 25,000$
$\dfrac{2,250}{25,000} = \dfrac{r \cdot \cancel{25,000}}{\cancel{25,000}}$
$0.09 = r$
$9\% = r$
She earned a commission rate of 9%.

88)   Commission = $1,680
Let r = the commission rate.
Sales = $14,000

commission = commission rate · sales
$1,680 = r \cdot 14,000$
$\dfrac{1,680}{14,000} = \dfrac{r \cdot \cancel{14,000}}{\cancel{14,000}}$
$0.12 = r$
$12\% = r$
Her commission rate is 12%.

89) Commission = $650
     The commission rate $0.25\% = 0.0025$.
     Let $s =$ the value of the portfolio.

     commission = rate $\cdot$ value of portfolio
     $$650 = 0.0025 \cdot s$$
     $$\frac{650}{0.0025} = \frac{0.0025 \cdot s}{0.0025}$$
     $$260,000 = s$$
     The value of the portfolio is $260,000.

90) Commission = $580
     The commission rate $4\% = 0.04$.
     Let $s =$ sales.

     commission = commission rate $\cdot$ sales
     $$580 = 0.04 \cdot s$$
     $$\frac{580}{0.04} = \frac{0.04 \cdot s}{0.04}$$
     $$14,500 = s$$
     The sales were $14,500.

91) Principal = $860
     Let $I =$ interest.
     The interest rate $1.5\% = 0.015$ per year.
     Time $= 2$ years

     Interest = Principal $\cdot$ rate $\cdot$ time
     $$I = P \cdot r \cdot t$$
     $$I = 860 \cdot 0.015 \cdot 2$$
     $$I = 12.90 \cdot 2$$
     $$I = 25.80$$
     The account earned $25.80 interest in two years.

92) Principal = $1,080
     Let $I =$ interest.
     The interest rate $2.5\% = 0.025$ per year.
     Time $= 3$ years

     Interest = Principal $\cdot$ rate $\cdot$ time
     $$I = P \cdot r \cdot t$$
     $$I = 1,080 \cdot 0.025 \cdot 3$$
     $$I = 27 \cdot 3$$
     $$I = 81$$
     The account earned $81 interest in three years.

93) $P = \$20,000$
     $I = \$500$
     Let $r =$ interest rate.
     Time $= 1$ year

     $$I = P \cdot r \cdot t$$
     $$500 = 20,000 \cdot r \cdot 1$$
     $$500 = 20,000 \cdot r$$
     $$\frac{500}{20,000} = \frac{20,000 \cdot r}{20,000}$$
     $$0.025 = r$$
     The account earned an interest rate of 2.5%.

94) $P = \$50,000$
     $I = \$800$
     Let $r =$ interest rate.
     Time $= 1$ year
     $$I = P \cdot r \cdot t$$
     $$800 = 50,000 \cdot r \cdot 1$$
     $$800 = 50,000 \cdot r$$
     $$\frac{800}{50,000} = \frac{50,000 \cdot r}{50,000}$$
     $$0.016 = r$$
     The bond account earns an interest rate of 1.6%.

95) Principal or Balance = $1,700
     Let $I =$ interest.
     Interest rate $= 24\%$ per year
     $$= \frac{24\%}{12}$$
     $$= 2\%$$
     $$= 0.02 \text{ per month}$$
     Time $= 1$ month

     $$I = P \cdot r \cdot t$$
     $$I = 1,700 \cdot 0.02 \cdot 1$$
     $$I = 34$$
     He will be charged $34 for the first month.

96) Principal or Balance = $1500
     Let $I =$ interest.
     Interest rate $= 18\%$ per year
     $$= \frac{18\%}{12}$$
     $$= 1.5\%$$
     $$= 0.015 \text{ per month}$$
     Time $= 1$ month

     $$I = P \cdot r \cdot t$$
     $$I = 1,500 \cdot 0.015 \cdot 1$$
     $$I = 22.50$$
     She will be charged $22.50 for the first month.

97)   Percent $3\% = 0.03$
Let c = the change in salary.
Original Salary = $45,000

$$\text{Percent Change} = \frac{\text{Change in Salary}}{\text{Original Salary}}$$

$$0.03 = \frac{c}{45,000}$$

$$0.03 \cdot \frac{45,000}{1} = \frac{c}{45,000} \cdot \frac{45,000}{1}$$

$$1,350 = c$$

The change in her salary is $1,350.

98)   Percent $2\% = 0.02$
Let c = the change in salary.
Original Salary = $36,000

$$\text{Percent Change} = \frac{\text{Change in Salary}}{\text{Original Salary}}$$

$$0.02 = \frac{c}{36,000}$$

$$0.02 \cdot \frac{36,000}{1} = \frac{c}{36,000} \cdot \frac{36,000}{1}$$

$$720 = c$$

The change in salary was $720.

99)   Let p = the percent decrease.
The change in cost = a decrease of $171.
Original Cost = $900

$$\text{Percent Change} = \frac{\text{Change in Cost}}{\text{Original Cost}}$$

$$p = \frac{171}{900}$$

$$p = 0.19$$

$$p = 19\%$$

The cost of Alli's insurance decreased by 19%.

100)   Let p = the percent increase.
The change in salary = an increase of $4,000.
Original Salary = $36,000

$$\text{Percent Change} = \frac{\text{Change in Salary}}{\text{Original Salary}}$$

$$p = \frac{4,000}{36,000}$$

$$p = 0.\overline{1}$$

$$p = 11\frac{1}{9}\%$$

The bonus represents an increase of about 11%.

CHAPTER 5   TEST

1)   Percent is used in place of the phrases *per hundred* or *out of one hundred*.

2)   In a percent equation, the percent tells what portion of the base gives the amount.

3)   In a percent equation, the base is the whole quantity.

4)   In a percent equation, the amount is the portion that results from multiplying the percent and the base.

5)   The word *is* is often used to represent equality in word exercises.

6)   The word *of* is often used to represent multiplication in word exercises.

7)   A commission is the amount of money a salesperson earns as a percent of sales.

8)   When money is invested, the original investment is called the principal.

9)   A percent decrease is the percent of the base that is subtracted to make a new decreased amount.

10)   68%

11)   142%

12) $28\% = \dfrac{28\,\%}{1} \cdot \dfrac{1}{100\,\%}$

$$= \frac{28}{100}$$

$$= \frac{4 \cdot 7}{4 \cdot 25}$$

$$= \frac{7}{25}$$

13) $160\% = \dfrac{160\cancel{\%}}{1} \cdot \dfrac{1}{100\cancel{\%}}$

$= \dfrac{160}{100}$

$= \dfrac{\cancel{2} \cdot 8 \cdot \cancel{10}}{\cancel{2} \cdot 5 \cdot \cancel{10}}$

$= \dfrac{8}{5} = 1\dfrac{3}{5}$

14) $40\dfrac{1}{4}\% = \dfrac{40.25\cancel{\%}}{1} \cdot \dfrac{1}{100\cancel{\%}}$

$= \dfrac{40.25}{100}$

$= \dfrac{4{,}025}{10{,}000}$

$= \dfrac{\cancel{25} \cdot 161}{\cancel{25} \cdot 400}$

$= \dfrac{161}{400}$

15) $\dfrac{9}{10} = \dfrac{9}{10} \cdot \dfrac{100\%}{1}$

$= \dfrac{900\%}{10}$

$= 90\%$

16) $\dfrac{1}{8} = \dfrac{1}{8} \cdot \dfrac{100\%}{1}$

$= \dfrac{100\%}{8}$

$= 12.5\%$

17) $3\dfrac{2}{5} = \dfrac{17}{5} \cdot \dfrac{100\%}{1}$

$= \dfrac{1{,}700\%}{5}$

$= 340\%$

18) $72\% = \dfrac{72\%}{1} \cdot \dfrac{1}{100\%}$

$= \dfrac{72}{100}$

$= 0.72$

19) $0.078 = 0.078 \cdot 100\%$

$= 7.8\%$

20) $0.45 = 0.45 \cdot 100\%$

$= 45\%$

21) $123\% = \dfrac{123\%}{1} \cdot \dfrac{1}{100\%}$

$= \dfrac{123}{100}$

$= 1.23$

22) $P = 60$

$B = 50$

Let $A =$ amount.

$\dfrac{60}{100} = \dfrac{A}{50}$

$\dfrac{60}{100} \cdot \dfrac{50}{1} = \dfrac{A}{\cancel{50}} \cdot \dfrac{\cancel{50}}{1}$

$\dfrac{6 \cdot \cancel{10} \cdot 5 \cdot \cancel{10}}{\cancel{10} \cdot \cancel{10}} = A$

$30 = A$

30 is 60% of 50.

23) Let $P =$ the percent.

$B = 80$

$A = 100$

$\dfrac{P}{100} = \dfrac{100}{80}$

$\dfrac{\cancel{100}}{1} \cdot \dfrac{P}{\cancel{100}} = \dfrac{100}{1} \cdot \dfrac{100}{80}$

$P = \dfrac{\cancel{4} \cdot 25 \cdot 5 \cdot \cancel{20}}{\cancel{4} \cdot \cancel{20}}$

$P = 125$

100 is 125% of 80.

24) $P = 15$

Let $B =$ the number.

$A = 300$

$\dfrac{15}{100} = \dfrac{300}{B}$

$\dfrac{100}{15} = \dfrac{B}{300}$

$\dfrac{100}{15} \cdot \dfrac{300}{1} = \dfrac{B}{\cancel{300}} \cdot \dfrac{\cancel{300}}{1}$

$\dfrac{30{,}000}{15} = B$

$2{,}000 = B$

300 is 15% of 2,000.

25) Let $P =$ the percent.

$B = 18$

$A = 18 - 12 = 6$

$\dfrac{P}{100} = \dfrac{6}{18}$

$\dfrac{\cancel{100}}{1} \cdot \dfrac{P}{\cancel{100}} = \dfrac{100}{1} \cdot \dfrac{6}{18}$

$P = \dfrac{100 \cdot \cancel{6}}{3 \cdot \cancel{6}}$

$P = \dfrac{100}{3} = 33\dfrac{1}{3}$

Microsoft decreased Xbox production
approximately 33%.

26) $P = 95$
$B = 40$
Let $A = $ the questions answered correctly.

$$\frac{95}{100} = \frac{A}{40}$$

$$\frac{95}{100} \cdot \frac{40}{1} = \frac{A}{\cancel{40}} \cdot \frac{\cancel{40}}{1}$$

$$\frac{\cancel{5} \cdot 19 \cdot 2 \cdot \cancel{20}}{\cancel{5} \cdot \cancel{20}} = A$$

$$38 = A$$

You answered 38 questions correctly.

27) $$P \cdot B = A$$
$$0.14 \cdot 44 = A$$
$$6.16 = A$$

28) $$P \cdot B = A$$
$$0.40 \cdot B = 50$$
$$\frac{\cancel{0.40} \cdot B}{\cancel{0.40}} = \frac{50}{0.40}$$
$$B = 125$$
The number is 125.

29) $$P \cdot B = A$$
$$P \cdot 20 = 65$$
$$\frac{P \cdot \cancel{20}}{\cancel{20}} = \frac{65}{20}$$
$$P = 3.25$$
The percent is 325%.

30) $$P \cdot B = A$$
$$P \cdot 160 = 100$$
$$\frac{P \cdot \cancel{160}}{\cancel{160}} = \frac{100}{160}$$
$$P = 0.625$$
You have used about 63% of the storage capacity.

31) $$P \cdot B = A$$
$$0.08 \cdot B = 5.44$$
$$\frac{\cancel{0.08} \cdot B}{\cancel{0.08}} = \frac{5.44}{0.08}$$
$$B = 68$$
The microphone cost $68.00 before sales tax.

32) Let $c = $ the commission.
The commission rate $4\% = 0.04$.
Sales $= \$1,100$

commission $=$ commission rate $\cdot$ sales
$$c = 0.04 \cdot 1,100$$
$$c = 44$$
She will earn a $44 commission.

33) Commission $= \$1,750$
Let $r = $ the commission rate.
Sales $= \$25,000$

commission $=$ commission rate $\cdot$ sales
$$1,750 = r \cdot 25,000$$
$$\frac{1,750}{25,000} = \frac{r \cdot \cancel{25,000}}{\cancel{25,000}}$$
$$0.07 = r$$
$$7\% = r$$
His commission rate was 7%.

34) Commission $= \$680$
The commission rate $5\% = 0.05$.
Let $s = $ sales.

commission $=$ commission rate $\cdot$ sales
$$680 = 0.05 \cdot s$$
$$\frac{680}{0.05} = \frac{\cancel{0.05} \cdot s}{\cancel{0.05}}$$
$$13,600 = s$$
The sales were $13,600.

35) Principal $= \$15,450$
Let $I = $ interest.
The interest rate $1.5\% = 0.015$ per year.
Time $= 3$ years

Interest $=$ Principal $\cdot$ rate $\cdot$ time
$$I = P \cdot r \cdot t$$
$$I = 15,450 \cdot 0.015 \cdot 3$$
$$I = 231.75 \cdot 3$$
$$I = 695.25$$
The account earned $695.25 interest in three years.

36) $P = \$40,000$
$I = \$800$
Let $r = $ interest rate.
Time $= 1$ year
$$I = P \cdot r \cdot t$$
$$800 = 40,000 \cdot r \cdot 1$$
$$800 = 40,000 \cdot r$$
$$\frac{800}{40,000} = \frac{\cancel{40,000} \cdot r}{\cancel{40,000}}$$
$$0.02 = r$$
The bond account earns an interest rate of 2%.

37)   Principal or Balance = $1,300

Let I = interest.

Interest rate = 24% per year

$$= \frac{24\%}{12}$$
$$= 2\%$$
$$= 0.02 \text{ per month}$$

Time = 1 month

$I = P \cdot r \cdot t$
$I = 1,300 \cdot 0.02 \cdot 1$
$I = 26$

He will be charged $26 for the first month.

38)   Percent 2% = 0.02

Let c = the change in salary.

Original Salary = $26,000

$$\text{Percent Change} = \frac{\text{Change in Salary}}{\text{Original Salary}}$$
$$0.02 = \frac{c}{26,000}$$
$$0.02 \cdot \frac{26,000}{1} = \frac{c}{26,000} \cdot \frac{26,000}{1}$$
$$520 = c$$

The change in salary is $520.

39)   Let p = the percent decrease.

The change in cost = a decrease of $150.

Original Cost = $600

$$\text{Percent Change} = \frac{\text{Change in Cost}}{\text{Original Cost}}$$
$$p = \frac{150}{600}$$
$$p = 0.25$$
$$p = 25\%$$

The cost of Elli's insurance decreased by 25%.

Chapter 6   Measurement

6.1   U.S. System Units of Measure

GUIDED PRACTICE

1) Since 12 inches $= 1$ foot, the two unit fractions
   are $\dfrac{12 \text{ in.}}{1 \text{ ft}} = 1$ and $\dfrac{1 \text{ ft}}{12 \text{ in.}} = 1$.

2) 2 pints $= 1$ quart

   $$6 \text{ pt} = \frac{6 \text{ pt}}{1}$$
   $$= \frac{6 \text{ pt}}{1} \cdot \frac{1 \text{ qt}}{2 \text{ pt}}$$
   $$= \frac{6 \text{ qt}}{2}$$
   $$= 3 \text{ quarts}$$

3) 1 pound $= 16$ ounces

   $$6 \text{ lb} = \frac{6 \text{ lb}}{1}$$
   $$= \frac{6 \text{ lb}}{1} \cdot \frac{16 \text{ oz}}{1 \text{ lb}}$$
   $$= \frac{6}{1} \cdot \frac{16 \text{ oz}}{1}$$
   $$= 96 \text{ ounces}$$

4) 1 pint $= 2$ cups

   $$28 \text{ cups} = \frac{28 \text{ cups}}{1}$$
   $$= \frac{28 \text{ cups}}{1} \cdot \frac{1 \text{ pt}}{2 \text{ cups}}$$
   $$= \frac{28}{1} \cdot \frac{1 \text{ pt}}{2}$$
   $$= 14 \text{ pints}$$

5) The first units fraction divides out feet and
   introduces miles.
   The second units fraction divides out minutes
   and introduces hours.

   $$2{,}640 \text{ ft per min} = \frac{2{,}640 \text{ ft}}{1 \text{ min}}$$
   $$= \frac{2{,}640 \text{ ft}}{1 \text{ min}} \cdot \frac{1 \text{ mi}}{5{,}280 \text{ ft}}$$
   $$= \frac{2{,}640 \text{ ft}}{1 \text{ min}} \cdot \frac{1 \text{ mi}}{5{,}280 \text{ ft}} \cdot \frac{60 \text{ min}}{1 \text{ hr}}$$
   $$= \frac{2{,}640 \cdot 2 \cdot 30 \text{ mi}}{2 \cdot 2{,}640 \text{ hr}}$$
   $$= \frac{30 \text{ mi}}{1 \text{ hr}}$$
   $$= 30 \text{ miles per hour}$$

6) $\text{months} = \text{years} \cdot \dfrac{\text{months}}{\text{years}}$

   $$= \frac{1.5 \text{ yr}}{1} \cdot \frac{12 \text{ months}}{1 \text{ yr}}$$
   $$= 18 \text{ months}$$

   $\text{payment} = \text{loan amount} \div \text{months}$
   $$= 1{,}500 \div 18$$
   $$\approx 83.33$$
   The monthly payment is approximately $\$83.33$.

7) $\text{Total} = \text{payment} \cdot \text{months}$

   $$= \frac{\$70}{\text{month}} \cdot \frac{12 \text{ months}}{1 \text{ yr}} \cdot 3 \text{ yr}$$
   $$= \$2{,}520$$

   $\text{Difference} = \text{paid} - \text{loan}$
   $$= 2{,}520 - 2{,}000$$
   $$= 520$$
   $\$520$ will be paid back in total interest.

CONCEPT CHECKS AND PRACTICE EXERCISES

A1) The distance across the state of Georgia may be
   measured in miles.

A2) The length of your notebook may be measured
   in inches.

A3) The time you spend in a typical class meeting
   may be measured in minutes.

A4) The width of the classroom may be measured in
   feet.

A5) The time span of a college semester may be measured in weeks.

A6) a) 1 yard = 3 feet
     b) 1 minute = 60 seconds
     c) 7 days = 1 week
     d) 1 mile = 5,280 feet
     e) 1 year = 365 days
     f) 60 minutes = 1 hour
     g) 1 day = 24 hours
     h) 12 inches = 1 foot

A7) a) 1 d = 24 hr
     b) 365 d = 1 yr
     c) 1 mi = 5,280 ft
     d) 1 min = 60 sec
     e) 12 in. = 1 ft
     f) 1 wk = 7 d
     g) 60 min = 1 h
     h) 1 yd = 3 ft

B1) The weight of the human body may be measured in pounds.

B2) The weight of a pencil may be measured in ounces.

B3) The volume of an aquarium may be measured in gallons.

B4) The volume of a single-serving soda bottle may be measured in fluid ounces.

B5) The weight of an iceberg may be measured in tons.

B6) a) 1 ton = 2,000 pounds
     b) 1 pint = 2 cups
     c) 16 ounces = 1 pound
     d) 1 gallon = 4 quarts
     e) 1 quart = 2 pints
     f) 8 fluid ounces = 1 cup

B7) a) 4 qt = 1 gal
     b) 1 qt = 2 pt
     c) 1 T = 2,000 lb
     d) 1 lb = 16 oz
     e) 1 c = 8 fl oz
     f) 1 pt = 2 c

C1) Since 8 fluid ounces = 1 cup, the two unit fractions are $\dfrac{8\ \text{fl oz}}{1\ \text{c}} = 1$ and $\dfrac{1\ \text{c}}{8\ \text{fl oz}} = 1$.

C2) Since 7 days = 1 week, the two unit fractions are $\dfrac{7\ \text{d}}{1\ \text{wk}} = 1$ and $\dfrac{1\ \text{wk}}{7\ \text{d}} = 1$.

C3) $\dfrac{6\ \cancel{\text{qt}}}{1} \cdot \dfrac{2\ \text{pt}}{1\ \cancel{\text{qt}}} = 12\ \text{pt}$

     6 quarts = 12 pints

C4) $\dfrac{8{,}000\ \cancel{\text{lb}}}{1} \cdot \dfrac{1\ \text{T}}{2{,}000\ \cancel{\text{lb}}} = 4\ \text{T}$

     8,000 pounds = 4 tons

C5) $\dfrac{10\ \cancel{\text{ft}}}{1} \cdot \dfrac{12\ \text{in.}}{1\ \cancel{\text{ft}}} = 120\ \text{in.}$

C6) $\dfrac{10\ \cancel{\text{yd}}}{1} \cdot \dfrac{3\ \text{ft}}{1\ \cancel{\text{yd}}} = 30\ \text{ft}$

C7) $\dfrac{16\ \cancel{\text{c}}}{1} \cdot \dfrac{1\ \text{pt}}{2\ \cancel{\text{c}}} = 8\ \text{pt}$

C8) $\dfrac{14\ \cancel{\text{pt}}}{1} \cdot \dfrac{1\ \text{qt}}{2\ \cancel{\text{pt}}} = 7\ \text{qt}$

C9) $9\ \text{yd} = \dfrac{9\ \cancel{\text{yd}}}{1} \cdot \dfrac{3\ \text{ft}}{1\ \cancel{\text{yd}}}$

     $= 27\ \text{ft}$

     There are 27 feet in 9 yards.

C10) $11\ \text{wk} = \dfrac{11\ \cancel{\text{wk}}}{1} \cdot \dfrac{7\ \text{d}}{1\ \cancel{\text{wk}}}$

      $= 77\ \text{d}$

      There are 77 days in 11 weeks.

C11) $240\ \text{oz} = \dfrac{240\ \cancel{\text{oz}}}{1} \cdot \dfrac{1\ \text{lb}}{16\ \cancel{\text{oz}}}$

      $= 15\ \text{lb}$

      240 ounces is equal to 15 pounds.

C12) $48\ \text{fl oz} = \dfrac{48\ \cancel{\text{fl oz}}}{1} \cdot \dfrac{1\ \text{c}}{8\ \cancel{\text{fl oz}}}$

      $= 6\ \text{c}$

      48 fluid ounces is equal to 6 cups.

C13)   280 yards per minute $= \dfrac{280 \text{ yd}}{1 \text{ min}} \cdot \dfrac{3 \text{ ft}}{1 \text{ yd}} \cdot \dfrac{1 \text{ min}}{60 \text{ s}}$

$= \dfrac{14 \cdot 20 \cdot 3 \text{ ft}}{3 \cdot 20 \text{ s}}$

$= \dfrac{14 \text{ ft}}{\text{s}}$

$= 14$ feet per second

C14)   1,320 ft per min $= \dfrac{1,320 \text{ ft}}{1 \text{ min}} \cdot \dfrac{1 \text{ mi}}{5,280 \text{ ft}} \cdot \dfrac{60 \text{ min}}{1 \text{ h}}$

$= \dfrac{1,320 \cdot 1 \cdot 15 \text{ mi}}{1 \cdot 1,320 \text{ h}}$

$= \dfrac{15 \text{ mi}}{\text{h}}$

$= 15$ miles per hour

SECTION 6.1   EXERCISES

For 1–9, refer to Concept Checks and Practice Exercises.

11) a)   4 quarts $= 1$ gallon

b)   365 days $= 1$ year

c)   1 ton $= 2,000$ pounds

d)   60 minutes $= 1$ hour

e)   1 yard $= 3$ feet

f)   1 pint $= 2$ cups

13) a)   5,280 ft $= 1$ mi

b)   1 ft $= 12$ in.

c)   1 qt $= 2$ pt

d)   1 min $= 60$ sec

e)   1 wk $= 7$ d

f)   1 lb $= 16$ oz

g)   1 c $= 8$ fl oz

15) $\dfrac{6 \text{ d}}{1} \cdot \dfrac{24 \text{ hr}}{1 \text{ d}} = 144$ hours

17) $\dfrac{24 \text{ qt}}{1} \cdot \dfrac{1 \text{ gal}}{4 \text{ qt}} = 6$ gallons

19)   10,560 feet $= \dfrac{10,560 \text{ ft}}{1} \cdot \dfrac{1 \text{ mi}}{5,280 \text{ ft}}$

$= 2$ miles

21)   1,500 seconds $= \dfrac{1,500 \text{ sec}}{1} \cdot \dfrac{1 \text{ min}}{60 \text{ sec}}$

$= 25$ minutes

23)   14 gallons $= \dfrac{14 \text{ gal}}{1} \cdot \dfrac{4 \text{ qt}}{1 \text{ gal}}$

$= 56$ quarts

25)   20,320 feet $= \dfrac{20,320 \text{ ft}}{1} \cdot \dfrac{1 \text{ mi}}{5,280 \text{ ft}}$

≈ 3.8 miles

27) $\dfrac{4 \text{ lb}}{1} \cdot \dfrac{16 \text{ oz.}}{1 \text{ lb}} \cdot \dfrac{1 \text{ batch}}{7 \text{ oz}} = \dfrac{64}{7}$

≈ 9.1 batches

He can make 9 batches.

29)   7 miles $= \dfrac{7 \text{ mi}}{1} \cdot \dfrac{5,280 \text{ ft}}{1 \text{ mi}}$

$= 36,960$ feet

31)   1,400 pounds $= \dfrac{1,400 \text{ lb}}{1} \cdot \dfrac{1 \text{ T}}{2,000 \text{ lb}}$

$= 0.7$ ton

33)   24 cups $= \dfrac{24 \text{ c}}{1} \cdot \dfrac{1 \text{ pt}}{2 \text{ c}}$

$= 12$ pints

35)   8 feet $= \dfrac{8 \text{ ft}}{1} \cdot \dfrac{12 \text{ in.}}{1 \text{ ft}}$

$= 96$ inches

37)   5.5 pounds $= \dfrac{5.5 \text{ lb}}{1} \cdot \dfrac{16 \text{ oz}}{1 \text{ lb}}$

$= 88$ ounces

39)   87 inches $= \dfrac{87 \text{ in.}}{1} \cdot \dfrac{1 \text{ ft}}{12 \text{ in.}}$

$= 7.25$ feet

41) a)   6.2 miles $= \dfrac{6.2 \text{ mi}}{1} \cdot \dfrac{5,280 \text{ ft}}{1 \text{ mi}}$

$= 32,736$ feet

The race is 32,736 feet.

41) b) $32{,}736 \text{ ft} \cdot \dfrac{1 \text{ yd}}{3 \text{ ft}} \cdot \dfrac{1 \text{ step}}{1 \text{ yd}} = 10{,}912 \text{ steps}$

It would take 10,912 steps to walk the Binder Park Road Race.

c) $32{,}736 \text{ ft} \cdot \dfrac{1 \text{ yd}}{3 \text{ ft}} \cdot \dfrac{1 \text{ stride}}{2 \text{ yd}} = 5{,}456 \text{ strides}$

It would take 5,456 strides to jog the Binder Park Road Race.

43) Since an ounce is a smaller unit of measure than a pound, it takes more ounces to measure the same amount of bananas.

45) Since a pint is a smaller unit of measure than a quart, it takes more pints to measure the amount of water in Lake Okeechobee.

47) $45 \text{ miles per hour} = \dfrac{45 \text{ mi}}{1 \text{ hr}} \cdot \dfrac{5{,}280 \text{ ft}}{1 \text{ mi}} \cdot \dfrac{1 \text{ hr}}{60 \text{ min}}$

$= \dfrac{45 \cdot 60 \cdot 88 \text{ ft}}{60 \text{ min}}$

$= \dfrac{3{,}960 \text{ ft}}{\text{min}}$

$= 3{,}960 \text{ feet per minute}$

49) $60 \text{ quarts per day} = \dfrac{60 \text{ qt}}{1 \text{ d}} \cdot \dfrac{1 \text{ gal}}{4 \text{ qt}} \cdot \dfrac{7 \text{ d}}{1 \text{ wk}}$

$= \dfrac{4 \cdot 15 \cdot 7 \text{ gal}}{4 \text{ wk}}$

$= \dfrac{105 \text{ gal}}{\text{wk}}$

$= 105 \text{ gallons per week}$

51) $45 \text{ feet per minute} = \dfrac{45 \text{ ft}}{1 \text{ min}} \cdot \dfrac{12 \text{ in.}}{1 \text{ ft}} \cdot \dfrac{1 \text{ min}}{60 \text{ sec}}$

$= \dfrac{5 \cdot 9 \cdot 12 \text{ in.}}{5 \cdot 12 \text{ sec}}$

$= \dfrac{9 \text{ in.}}{\text{sec}}$

$= 9 \text{ inches per second}$

53) $250 \text{ lb per min} = \dfrac{250 \text{ lb}}{1 \text{ min}} \cdot \dfrac{1 \text{ T}}{2000 \text{ lb}} \cdot \dfrac{60 \text{ min}}{1 \text{ hr}}$

$= \dfrac{250 \cdot 4 \cdot 15 \text{ T}}{2 \cdot 4 \cdot 250 \text{ hr}}$

$= \dfrac{15 \text{ T}}{2 \text{ hr}}$

$= 7.5 \text{ tons per hour}$

55) $5 \text{ yd} + 2 \text{ ft} = 5 \text{ yd} \cdot \dfrac{3 \text{ ft}}{1 \text{ yd}} + 2 \text{ ft}$

$= 15 \text{ ft} + 2 \text{ ft}$

$= 17 \text{ ft}$

The length of the sidewalk is 17 feet.

$17 \text{ ft} \cdot \dfrac{\$19}{1 \text{ ft}} = \$323$

The new sidewalk will cost \$323.

57) $6.2 \text{ mi} \cdot \dfrac{4 \text{ fl oz}}{1 \text{ mi}} \cdot \dfrac{1 \text{ c}}{8 \text{ oz}} = 3.1 \text{ c}$

She will need to drink 3.1 cups of water during the road race.

59) $25.5 \text{ sq miles} = \dfrac{25.5 \text{ sq miles}}{1} \cdot \dfrac{660 \text{ acres}}{1 \text{ sq mile}}$

$= 16{,}830 \text{ acres}$

The fire burned 16,830 acres.

61) $\text{months} = \text{years} \cdot \dfrac{\text{months}}{\text{years}}$

$= \dfrac{30 \text{ yr}}{1} \cdot \dfrac{12 \text{ months}}{1 \text{ yr}}$

$= 360 \text{ months}$

$\text{payment} = \text{loan amount} \div \text{months}$

$= 95{,}400 \div 360$

$= 265$

Each payment is \$265.

63) $\text{Total} = \text{payment} \cdot \text{months}$

$= \dfrac{\$630}{\text{month}} \cdot \dfrac{12 \text{ months}}{1 \text{ yr}} \cdot 30 \text{ yr}$

$= \$226{,}800$

$\text{Difference} = \text{paid} - \text{loan}$

$= 226{,}800 - 95{,}000$

$= 131{,}800$

He will pay \$131,800 in excess of the original purchase price.

6.2   Metric System Units of Measure

GUIDED PRACTICE

1) If we follow the chart, we must move from meters to kilometers. Kilometers are three places to the left of meters. Move the decimal three places to the left.

km hm dam m dm cm mm

$23 \text{ m} = 0.023 \text{ km}$

2) If we follow the chart, we must move from kiloliters to milliliters. Milliliters are six places to the right of kiloliters. Move the decimal six places to the right.

kL hL daL L dL cL mL

$28.2 \text{ kL} = 28,200,000 \text{ mL}$

3) $1.3 \text{ mL per dose} = \dfrac{1.3 \text{ mL}}{\text{dose}} \cdot 8,000 \text{ doses}$
   $= 10,400 \text{ mL}$

$10,400 \text{ mL} = 10.4 \text{ L}$

Since 10.4 L > 10 L, there is not enough glucose to make the necessary doses of placebos.

CONCEPT CHECKS AND PRACTICE EXERCISES

A1) The fundamental unit meter is used to measure length in the metric system.

A2) The fundamental unit liter is used to measure volume in the metric system.

A3) The fundamental unit gram is used to measure weight in the metric system.

A4) King Henry died unceremoniously drinking chocolate milk.

A5) a) centi- hundredths

   b) deci- tenths

   c) hecto- hundreds

   d) deka- tens

   e) milli- thousandths

   f) kilo- thousands

A6) The weight of an apple may be measured in grams.

A7) The weight of a person may be measured in kilograms.

A8) The length of a notebook may be measured in centimeters.

A9) The width of a classroom may be measured in meters.

A10) The perimeter of California may be measured in kilometers.

A11) The thickness of the lead in a pencil may be measured in millimeters.

A12) The volume of a large soda bottle may be measured in liters.

A13) The volume of a syringe may be measured in milliliters.

A14) Answers will vary.

A15) Answers will vary.

A16) Answers will vary.

A17) Answers will vary.

A18) Answers will vary.

A19) Answers will vary.

B1) a) "Kilo" represents the number 1,000. This means 1 km = 1,000 m.

   b) Use a units fraction with those quantities to complete the conversion.
   $$\frac{5.25 \text{ km}}{1} \cdot \frac{1,000 \text{ m}}{1 \text{ km}} = 5,250 \text{ m}$$

   c) To move from kilometers to meters on the metric conversion table, we must move three places to the right.

   d) Move the decimal point three places to the right to convert 5.25 kilometers to meters: 5.25 km = 5,250 m.

   e) Yes, the answers from parts b and d match.

B2) a) "Milli-" represents the number $\dfrac{1}{1,000}$.

   This means 1,000 mg = 1 g.

B2) b) Use a unit fraction with those quantities to complete the conversion.

$$\frac{3{,}253 \ \cancel{mg}}{1} \cdot \frac{1 \ g}{1{,}000 \ \cancel{mg}} = 3.253 \ g$$

c) To move from milligrams to grams on the metric conversion table, we must move three places to the left.

d) Move the decimal point three places to the left to convert 3,253 milligrams to grams: 3,253 mg = 3.253 g.

e) Yes, the answers from parts b and d match.

B3) If we follow the chart, we must move from kilometers to meters. Meters are three places to the right of kilometers. Move the decimal three places to the right.
244 km = 244,000 m

B4) If we follow the chart, we must move from liters to kiloliters. Kiloliters are three places to the left of liters. Move the decimal three places to the left.
110 L = 0.11 kL

B5) If we follow the chart, we must move from milliliters to deciliters. Deciliters are two places to the left of milliliters. Move the decimal two places to the left.
329 mL = 3.29 dL

B6) If we follow the chart, we must move from hectograms to decigrams. Decigrams are three places to the right of hectograms. Move the decimal three places to the right.
564 hg = 564,000 dg

B7) If we follow the chart, we must move from dekagrams to decigrams. Decigrams are two places to the right of dekagrams. Move the decimal two places to the right.
1,264 dag = 126,400 dg

B8) If we follow the chart, we must move from kilometers to millimeters. Millimeters are six places to the right of kilometers. Move the decimal six places to the right.
1,327 km = 1,327,000,000 mm

SECTION 6.2   EXERCISES

For 1–6, refer to Concept Checks and Practice Exercises.

7) Metric prefixes

9) liter

11) a) hecto-   hundreds
     b) kilo-   thousands
     c) centi-   hundredths
     d) deka-   tens
     e) deci-   tenths
     f) milli-   thousandths

13) The six metric prefixes, largest to smallest, are kilo-, hecto-, deka-, deci-, centi-, milli-.

15) The volume of a soda can may be b) 320 milliliters. The other values are too large for the volume of a soda can.

17) The height of a classroom door may be a) 2 meters. The other values are too large or too small for the height of a door.

19) The weight of a tomato may be a) 0.5 kg. Kilograms is the only unit listed that measures weight.

21) The weight of an elephant may be a) 5,000 kg. Kilograms is the only unit listed that measures weight.

23) Liters are three places to the right of kiloliters.
Move the decimal three places to the right.
116 kL = 116,000 L

25) Dekagrams are two places to the left of decigrams.
Move the decimal two places to the left.
32.4 dg = 0.324 dag

27) Millimeters are six places to the right of kilometers.
Move the decimal six places to the right.
1.327 km = 1,327,000 mm

29) Kilograms are six places to the left of milligrams.
Move the decimal six places to the left.
6.8 mg = 0.0000068 kg

31) Meters are three places to the right of kilometers.
Move the decimal three places to the right.
244 km = 244,000 m

33) 22 hL = 220,000 cL

35) 5 m = 0.005 km

37) 12.3 kg = 1,230 dag

39) 4,092 mg = 0.004092 kg

41) 456,678 cm = 45.6678 hm

43) 6,700 cL = 0.67 hL

45) 0.965 hm = 96,500 mm

47) 9 g = 0.09 hg

49) 0.00008 cL = 0.000008 dL

51) $200 \text{ students} \cdot \dfrac{60 \text{ mL}}{1 \text{ student}} \cdot \dfrac{1 \text{ L}}{1,000 \text{ mL}} = 12 \text{ L}$

No, there is not enough solution.

53) $52 \text{ kg} \cdot \dfrac{1,000 \text{ g}}{1 \text{ kg}} \cdot \dfrac{\$6.79}{2,000 \text{ g}} = \$176.54$

This order will cost $176.54.

55) $2 \text{ km} \cdot \dfrac{100,000 \text{ cm}}{1 \text{ km}} \cdot \dfrac{1 \text{ row}}{16 \text{ cm}}$
$= 12,500 \text{ rows}$
There are 12,500 rows of bricks in the road.

57) a) $4,300 \text{ g} = \dfrac{4,300 \text{ g}}{1} \cdot \dfrac{1 \text{ mL}}{1 \text{ g}}$
$= 4,300 \text{ mL}$

b) $4,300 \text{ mL} = \dfrac{4,300 \text{ mL}}{1} \cdot \dfrac{1 \text{ L}}{1,000 \text{ mL}}$
$= 4.3 \text{ L}$

59) a) $8.56 \text{ L} = \dfrac{8.56 \text{ L}}{1} \cdot \dfrac{1,000 \text{ mL}}{1 \text{ L}} \cdot \dfrac{1 \text{ g}}{1 \text{ mL}}$
$= 8,560 \text{ g}$

b) $8,560 \text{ g} = \dfrac{8,560 \text{ g}}{1} \cdot \dfrac{1 \text{ kg}}{1,000 \text{ g}}$
$= 8.56 \text{ kg}$

61) $2,500 \text{ loaves} \cdot \dfrac{180 \text{ mL}}{1 \text{ loaf}} \cdot \dfrac{1 \text{ L}}{1,000 \text{ mL}} = 450 \text{ L}$
450 liters of water are required.

63) $3 \cdot 500 \text{ L} = 1,500 \text{ L}$
The capacity of the tanks is 1,500 liters.

$3,600 \cdot 375 \text{ mL} = 1,350,000 \text{ mL}$

$1,350,000 \text{ mL} \cdot \dfrac{1 \text{ L}}{1,000 \text{ mL}} = 1,350 \text{ L}$

The total volume of the beer is 1,350 liters.
Yes, the three tanks hold enough beer.

6.3   Converting Between the U.S. System and the Metric System

GUIDED PRACTICE

1) $4 \text{ lb} = \dfrac{4 \text{ lb}}{1}$
$\approx \dfrac{4 \text{ lb}}{1} \cdot \dfrac{0.454 \text{ kg}}{1 \text{ lb}}$
$= \dfrac{4}{1} \cdot \dfrac{0.454 \text{ kg}}{1}$
$= 1.816 \text{ kg}$

2) $200 \text{ g} = \dfrac{200 \text{ g}}{1}$
$\approx \dfrac{200 \text{ g}}{1} \cdot \dfrac{0.0353 \text{ oz}}{1 \text{ g}}$
$= \dfrac{200}{1} \cdot \dfrac{0.0353 \text{ oz}}{1}$
$= 7.06 \text{ oz}$

3) $10 \text{ in.} = \dfrac{10 \text{ in.}}{1}$
$= \dfrac{10 \text{ in.}}{1} \cdot \dfrac{2.54 \text{ cm}}{1 \text{ in.}}$
$= \dfrac{10}{1} \cdot \dfrac{2.54 \text{ cm}}{1}$
$= 25.4 \text{ cm}$

4) 120 km per hour $= \dfrac{120 \text{ km}}{1 \text{ hr}}$

$\approx \dfrac{120 \text{ km}}{1 \text{ hr}} \cdot \dfrac{0.62 \text{ mi}}{1 \text{ km}}$

$= \dfrac{120}{1 \text{ hr}} \cdot \dfrac{0.62 \text{ mi}}{1}$

$= \dfrac{74.4 \text{ mi}}{1 \text{ hr}}$

$= 74.4$ miles per hour

5) $F = \dfrac{9}{5} \cdot C + 32$

$= \dfrac{9}{5} \cdot 30 + 32$

$= \dfrac{9}{5} \cdot \dfrac{30}{1} + 32$

$= \dfrac{9}{5} \cdot \dfrac{5 \cdot 6}{1} + 32$

$= 54 + 32$

$F = 86$

$30^\circ \text{C} = 86^\circ \text{F}$

6) $C = \dfrac{5 \cdot F - 160}{9}$

$= \dfrac{5 \cdot 86 - 160}{9}$

$= \dfrac{430 - 160}{9}$

$= \dfrac{270}{9}$

$C = 30$

$86^\circ \text{F} = 30^\circ \text{C}$

7) 1 pound ≈ 0.454 kilogram

Since 1 kilogram is greater than 1 pound, 200 kilograms is greater than 200 pounds.

8) 1 inch ≈ 2.54 centimeters

$12 \text{ in.} = \dfrac{12 \text{ in.}}{1} \cdot \dfrac{2.54 \text{ cm}}{1 \text{ in.}}$

$= 30.48 \text{ cm}$

Since $30.48 < 31$, 12 inches is less than 31 centimeters.

CONCEPT CHECKS AND PRACTICE EXERCISES

A1) 1 mile ≈ 1.61 kilometers

1 mile is longer than 1 kilometer.

A2) 1 quart ≈ 0.946 liter

1 liter has more volume than one quart.

A3) 1 pound ≈ 0.454 kilogram

1 kilogram is heavier than 1 pound.

A4) a) $\dfrac{1.61 \text{ km}}{1 \text{ mi}}$ will be easier to use in calculations because it enables us to multiply by 1.61 rather than divide by 0.62.

A5) b) $\dfrac{3.28 \text{ ft}}{1 \text{ m}}$ will be easier to use in calculations because it enables us to multiply by 3.28 rather than divide by 0.305.

A6) 100 yd $= \dfrac{100 \text{ yd}}{1}$

$\approx \dfrac{100 \text{ yd}}{1} \cdot \dfrac{0.914 \text{ m}}{1 \text{ yd}}$

$= \dfrac{100}{1} \cdot \dfrac{0.914 \text{ m}}{1}$

$= 91.4 \text{ m}$

A7) 10 L $= \dfrac{10 \text{ L}}{1}$

$\approx \dfrac{10 \text{ L}}{1} \cdot \dfrac{0.264 \text{ gal}}{1 \text{ L}}$

$= \dfrac{10}{1} \cdot \dfrac{0.264 \text{ gal}}{1}$

$= 2.64 \text{ gal}$

A8) 5 oz $= \dfrac{5 \text{ oz}}{1}$

$\approx \dfrac{5 \text{ oz}}{1} \cdot \dfrac{28.35 \text{ g}}{1 \text{ oz}}$

$= \dfrac{5}{1} \cdot \dfrac{28.35 \text{ g}}{1}$

$= 141.75 \text{ g}$

A9) 7 in. $= \dfrac{7 \text{ in.}}{1}$

$= \dfrac{7 \text{ in.}}{1} \cdot \dfrac{2.54 \text{ cm}}{1 \text{ in.}}$

$= \dfrac{7}{1} \cdot \dfrac{2.54 \text{ cm}}{1}$

$= 17.78 \text{ cm}$

A10) 3 meters per second $= \dfrac{3 \text{ m}}{1 \text{ sec}}$

$\approx \dfrac{3 \text{ m}}{1 \text{ sec}} \cdot \dfrac{3.28 \text{ ft}}{1 \text{ m}}$

$= \dfrac{3}{1 \text{ sec}} \cdot \dfrac{3.28 \text{ ft}}{1}$

$= \dfrac{9.84 \text{ ft}}{1 \text{ sec}}$

$= 9.84$ feet per second

A11)   4 feet per second $= \dfrac{4 \text{ ft}}{1 \text{ sec}}$

$\approx \dfrac{4 \text{ ft}}{1 \text{ sec}} \cdot \dfrac{0.305 \text{ m}}{1 \text{ ft}}$

$= \dfrac{4}{1 \text{ sec}} \cdot \dfrac{0.305 \text{ m}}{1}$

$= \dfrac{1.22 \text{ m}}{1 \text{ sec}}$

$= 1.22 \text{ meters per second}$

B1)   Water freezes at 0°C.

B2) $F = \dfrac{9}{5} \cdot C + 32$

B3) $F = \dfrac{9}{5} \cdot C + 32$

$= \dfrac{9}{5} \cdot 15 + 32$

$= \dfrac{9}{5} \cdot \dfrac{15}{1} + 32$

$= \dfrac{9}{5} \cdot \dfrac{3 \cdot 5}{1} + 32$

$= 27 + 32$

$F = 59$

$15^\circ \text{C} = 59^\circ \text{F}$

B4) $F = \dfrac{9}{5} \cdot C + 32$

$= \dfrac{9}{5} \cdot 40 + 32$

$= \dfrac{9}{5} \cdot \dfrac{40}{1} + 32$

$= \dfrac{9}{5} \cdot \dfrac{5 \cdot 8}{1} + 32$

$= 72 + 32$

$F = 104$

$40^\circ \text{C} = 104^\circ \text{F}$

B5) $C = \dfrac{5 \cdot F - 160}{9}$

$= \dfrac{5 \cdot 95 - 160}{9}$

$= \dfrac{475 - 160}{9}$

$= \dfrac{315}{9}$

$C = 35$

$95^\circ \text{F} = 35^\circ \text{C}$

B6) $C = \dfrac{5 \cdot F - 160}{9}$

$= \dfrac{5 \cdot 140 - 160}{9}$

$= \dfrac{700 - 160}{9}$

$= \dfrac{540}{9}$

$C = 60$

$140^\circ \text{F} = 60^\circ \text{C}$

SECTION 6.3   EXERCISES

For 1–6, refer to Concept Checks and Practice Exercises.

7)   Most countries that use the U.S. system measure temperature with the Fahrenheit scale.

9)   The lengths from smallest to largest are centimeter, inch, length of a pencil, yard, meter, height of a person, kilometer, and mile. (5, 2, 1, 8, 6, 3, 7, 4)

11)   The weights from smallest to largest are weight of a grain of sand, gram, ounce, pound, kilogram, ton, and weight of a cruise ship. (3, 2, 4, 5, 6, 7, 1)

13)   A quart is a little smaller than a liter.

15)   A kilogram is a little more than twice the weight of a pound.

17) $10 \text{ cm} \approx \dfrac{10 \text{ cm}}{1} \cdot \dfrac{0.394 \text{ in.}}{1 \text{ cm}}$

$= 3.94 \text{ in.}$

$\approx 3.9 \text{ in.}$

19) $8 \text{ in.} = \dfrac{8 \text{ in.}}{1} \cdot \dfrac{2.54 \text{ cm}}{1 \text{ in.}}$

$= 20.32 \text{ cm}$

$\approx 20.3 \text{ cm}$

21) $11 \text{ gal} \approx \dfrac{11 \text{ gal}}{1} \cdot \dfrac{3.79 \text{ L}}{1 \text{ gal}}$

$= 41.69 \text{ L}$

$\approx 41.7 \text{ L}$

23) $14.6 \text{ lb} \approx \dfrac{14.6 \text{ lb}}{1} \cdot \dfrac{0.454 \text{ kg}}{1 \text{ lb}}$
        $= 6.6284 \text{ kg}$
        $\approx 6.6 \text{ kg}$

25) $85 \text{ mi} \approx \dfrac{85 \text{ mi}}{1} \cdot \dfrac{1.61 \text{ km}}{1 \text{ mi}}$
        $= 136.85 \text{ km}$
        $\approx 136.9 \text{ km}$

27) $20 \text{ m} \approx \dfrac{20 \text{ m}}{1} \cdot \dfrac{1.09 \text{ yd}}{1 \text{ m}}$
        $= 21.8 \text{ yd}$

29) $0.3 \text{ oz} \approx \dfrac{0.3 \text{ oz}}{1} \cdot \dfrac{28.35 \text{ g}}{1 \text{ oz}}$
        $= 8.505 \text{ g}$
        $\approx 8.5 \text{ g}$

31) $32 \text{ kg} \approx \dfrac{32 \text{ kg}}{1} \cdot \dfrac{2.2 \text{ lb}}{1 \text{ kg}}$
        $= 70.4 \text{ lb}$

33) $5 \text{ km} \approx \dfrac{5 \text{ km}}{1} \cdot \dfrac{0.62 \text{ mi}}{1 \text{ km}}$
        $= 3.1 \text{ mi}$

35) $100 \text{ g} \approx \dfrac{100 \text{ g}}{1} \cdot \dfrac{0.0353 \text{ oz}}{1 \text{ g}}$
        $= 3.53 \text{ oz}$
        $\approx 3.5 \text{ oz}$

37)   a) $6 \text{ gal} = \dfrac{6 \text{ gal}}{1} \cdot \dfrac{4 \text{ qt}}{1 \text{ gal}}$
              $= 24 \text{ qt}$

        b) $24 \text{ qt} \approx \dfrac{24 \text{ qt}}{1} \cdot \dfrac{0.946 \text{ L}}{1 \text{ qt}}$
              $= 22.704 \text{ L}$
              $\approx 22.7 \text{ L}$

39)   a) $321 \text{ mm} = 32.1 \text{ cm}$

        b) $32.1 \text{ cm} = \dfrac{32.1 \text{ cm}}{1} \cdot \dfrac{0.394 \text{ in.}}{1 \text{ cm}}$
              $= 12.6474 \text{ in.}$
              $\approx 12.6 \text{ in.}$

41) $28.2 \text{ cm} \approx \dfrac{28.2 \text{ cm}}{1} \cdot \dfrac{0.394 \text{ in.}}{1 \text{ cm}}$
        $= 11.1108 \text{ in.}$
        $\approx 11.1 \text{ in.}$
No, this piece will not be long enough. Eino needs a 13-inch board.

43) $35 \text{ gal} \approx \dfrac{35 \text{ gal}}{1} \cdot \dfrac{3.79 \text{ L}}{1 \text{ gal}}$
        $= 132.65 \text{ L}$
        $\approx 132.7 \text{ L}$
Yes, a 35 gallon container will hold 100 liters.

45) $46 \text{ qt} \approx \dfrac{46 \text{ qt}}{1} \cdot \dfrac{0.946 \text{ L}}{1 \text{ qt}}$
        $= 43.516 \text{ L}$
        $\approx 43.5 \text{ L}$
46 quarts equals 43.5 liters.

47)   Costello: $300 \text{ m} \approx \dfrac{300 \text{ m}}{1} \cdot \dfrac{1.09 \text{ yd}}{1 \text{ m}}$
                        $= 327 \text{ yd}$
Antuan: 312 yd

Costello has run the longer distance.

49) $21 \text{ L} \approx \dfrac{21 \text{ L}}{1} \cdot \dfrac{0.264 \text{ gal}}{1 \text{ L}}$
        $= 5.544 \text{ gal}$
        $\approx 5.5 \text{ gal}$
Approximately 5.5 gallons equal 21 liters.

51) $4.5 \text{ L} \approx \dfrac{4.5 \text{ L}}{1} \cdot \dfrac{1.06 \text{ qt}}{1 \text{ L}}$
        $= 4.77 \text{ qt}$
        $\approx 4.8 \text{ qt}$
4.5 liters equals approximately 4.8 quarts.

53) $70 \text{ miles per hour} \approx \dfrac{70 \text{ mi}}{1 \text{ hr}} \cdot \dfrac{1.61 \text{ km}}{1 \text{ mi}}$
        $= \dfrac{112.7 \text{ km}}{1 \text{ hr}}$
        $= 112.7 \text{ km per hour}$

55) $85 \text{ km per hour} \approx \dfrac{85 \text{ km}}{1 \text{ hr}} \cdot \dfrac{0.62 \text{ mi}}{1 \text{ km}}$
        $= \dfrac{52.7 \text{ mi}}{1 \text{ hr}}$
        $= 52.7 \text{ miles per hour}$

57) $4 \text{ L} \approx \dfrac{4 \cancel{\text{L}}}{1} \cdot \dfrac{0.264 \text{ gal}}{1 \cancel{\text{L}}}$

$= 1.056 \text{ gal}$

$\approx 1.1 \text{ gal}$

No, 4 liters cannot be poured into a gallon jug.

59) $19 \text{ miles per hour} \approx \dfrac{19 \cancel{\text{mi}}}{1 \text{ hr}} \cdot \dfrac{1.61 \text{ km}}{1 \cancel{\text{mi}}}$

$= \dfrac{30.59 \text{ km}}{1 \text{ hr}}$

$\approx 30.6 \text{ km per hour}$

Yes, a robin flying 19 miles per hour can catch a pig that is flying 25 miles per hour. (But, only when pigs fly.)

61) Water boils at 100°C.

63) $F = \dfrac{9}{5} \cdot C + 32$

65) $F = \dfrac{9}{5} \cdot C + 32$

$= \dfrac{9}{5} \cdot 105 + 32$

$= \dfrac{9}{\cancel{5}} \cdot \dfrac{\cancel{5} \cdot 21}{1} + 32$

$= 189 + 32$

$F = 221$

$105^{\circ}\text{C} = 221^{\circ}\text{F}$

67) $C = \dfrac{5 \cdot F - 160}{9}$

$= \dfrac{5 \cdot 41 - 160}{9}$

$= \dfrac{205 - 160}{9}$

$= \dfrac{45}{9}$

$C = 5$

$41^{\circ}\text{F} = 5^{\circ}\text{C}$

69) $C = \dfrac{5 \cdot F - 160}{9}$

$= \dfrac{5 \cdot 95 - 160}{9}$

$= \dfrac{475 - 160}{9}$

$= \dfrac{315}{9}$

$C = 35$

The temperature is 35°C.

71) $F = \dfrac{9}{5} \cdot C + 32$

$= \dfrac{9}{5} \cdot 85 + 32$

$= \dfrac{9}{\cancel{5}} \cdot \dfrac{\cancel{5} \cdot 17}{1} + 32$

$= 153 + 32$

$F = 185$

The temperature is 185°F.

73) $100 \text{ km per hour} \approx \dfrac{100 \cancel{\text{km}}}{1 \text{ hr}} \cdot \dfrac{0.62 \text{ mi}}{1 \cancel{\text{km}}}$

$= \dfrac{62 \text{ mi}}{1 \text{ hr}}$

$= 62 \text{ miles per hour}$

No. If the speed limit is 70 miles per hour, Kevin is not speeding.

75) 5 kilograms of steak are needed.

$8 \cdot 20 \text{ oz} = 160 \cancel{\text{oz}} \cdot \dfrac{1 \cancel{\text{lb}}}{16 \cancel{\text{oz}}} \cdot \dfrac{0.454 \text{ kg}}{1 \cancel{\text{lb}}}$

$= 4.54 \text{ kg}$

The total weight of eight steaks is 4.54 kilograms.

No. Eight steaks will not provide enough steak.

77) Hint: Use the conversion 1 gallon $\approx$ 3.79 liters.

a) $20 \text{ gal} \approx \dfrac{20 \cancel{\text{gal}}}{1} \cdot \dfrac{3.79 \text{ L}}{1 \cancel{\text{gal}}}$

$= 75.8 \text{ L}$

Approximately 75.8 liters are needed.

b) $75.8 \cancel{\text{L}} \cdot \dfrac{\$0.69}{1 \cancel{\text{L}}} \approx \$52.30$

The water will cost about $52.30.

CHAPTER 6   REVIEW EXERCISES

1) The distance from New York to Texas may be measured in miles.

2) The distance from your eyes to your lips may be measured in inches.

3) The amount of time needed to do your math homework each week may be measured in hours.

4) The amount of time it takes to get a college degree may be measured in years.

5) The weight of a car may be measured in tons.

6) The weight of a turkey may be measured in pounds.

7) The volume of a soda can may be measured in fluid ounces.

8) The weight of a toad may be measured in ounces.

9) The length of a whale may be measured in yards.

10) The weight of a whale may be measured in tons.

11) The diameter of a quarter may be measured in inches.

12) The volume of a quarter may be measured in fluid ounces.

13) $4 \text{ feet} = \dfrac{4 \text{ ft}}{1} \cdot \dfrac{12 \text{ in.}}{1 \text{ ft}}$
$= 48 \text{ inches}$

14) $1.5 \text{ miles} = \dfrac{1.5 \text{ mi}}{1} \cdot \dfrac{5{,}280 \text{ ft}}{1 \text{ mi}}$
$= 7{,}920 \text{ feet}$

15) $3 \text{ weeks} = \dfrac{3 \text{ wk}}{1} \cdot \dfrac{7 \text{ d}}{1 \text{ wk}} \cdot \dfrac{24 \text{ hr}}{1 \text{ d}}$
$= 504 \text{ hours}$

16) $2.5 \text{ hours} = \dfrac{2.5 \text{ hr}}{1} \cdot \dfrac{60 \text{ min}}{1 \text{ hr}} \cdot \dfrac{60 \text{ s}}{1 \text{ min}}$
$= 9{,}000 \text{ seconds}$

17) $52 \text{ yards} = \dfrac{52 \text{ yd}}{1} \cdot \dfrac{3 \text{ ft}}{1 \text{ yd}}$
$= 156 \text{ feet}$

18) $320 \text{ feet} = \dfrac{320 \text{ ft}}{1} \cdot \dfrac{1 \text{ yd}}{3 \text{ ft}}$
$= 106.\overline{6} \text{ yards}$
$\approx 106.7 \text{ yards}$

19) $2.5 \text{ quarts} = \dfrac{2.5 \text{ qt}}{1} \cdot \dfrac{2 \text{ pt}}{1 \text{ qt}}$
$= 5 \text{ pints}$

20) $3 \text{ lb} = \dfrac{3 \text{ lb}}{1} \cdot \dfrac{16 \text{ oz}}{1 \text{ lb}}$
$= 48 \text{ ounces}$

21) $8 \text{ gallons} = \dfrac{8 \text{ gal}}{1} \cdot \dfrac{4 \text{ qt}}{1 \text{ gal}} \cdot \dfrac{2 \text{ pt}}{1 \text{ qt}}$
$= 64 \text{ pints}$

22) $4.532 \text{ tons} = \dfrac{4.532 \text{ T}}{1} \cdot \dfrac{2{,}000 \text{ lb}}{1 \text{ T}}$
$= 9{,}064 \text{ pounds}$

23) $36 \text{ inches per day} = \dfrac{36 \text{ in.}}{1 \text{ d}} \cdot \dfrac{1 \text{ d}}{24 \text{ hr}}$
$= 1.5 \text{ inches per hour}$

24) $\$231 \text{ per week} = \dfrac{\$231}{1 \text{ wk}} \cdot \dfrac{1 \text{ wk}}{7 \text{ d}}$
$= \$33 \text{ per day}$

25) $60 \text{ miles per hour} = \dfrac{60 \text{ mi}}{1 \text{ hr}} \cdot \dfrac{5{,}280 \text{ ft}}{1 \text{ mi}} \cdot \dfrac{1 \text{ hr}}{60 \text{ min}}$
$= \dfrac{60 \cdot 5{,}280 \text{ ft}}{60 \text{ min}}$
$= \dfrac{5{,}280 \text{ ft}}{\text{min}}$
$= 5{,}280 \text{ feet per minute}$

26) $3 \text{ tons per day} = \dfrac{3 \text{ T}}{1 \text{ d}} \cdot \dfrac{2000 \text{ lb}}{1 \text{ T}} \cdot \dfrac{1 \text{ d}}{24 \text{ hr}}$
$= \dfrac{3 \cdot 8 \cdot 250 \text{ lb}}{3 \cdot 8 \text{ hr}}$
$= \dfrac{250 \text{ lb}}{\text{hr}}$
$= 250 \text{ pounds per hour}$

27) $100 \text{ lb per min} = \dfrac{100 \text{ lb}}{1 \text{ min}} \cdot \dfrac{1 \text{ T}}{2{,}000 \text{ lb}} \cdot \dfrac{60 \text{ min}}{1 \text{ h}} \cdot \dfrac{4 \text{ h}}{1}$
$= \dfrac{100 \cdot 3 \cdot 20 \cdot 4 \text{ T}}{20 \cdot 100}$
$= 12 \text{ tons}$
He will shovel 12 tons in 4 hours.

28)   1 fl oz per min

$$= \frac{1\,\text{fl oz}}{1\,\text{min}} \cdot \frac{1\,\text{c}}{8\,\text{fl oz}} \cdot \frac{1\,\text{pt}}{2\,\text{c}} \cdot \frac{1\,\text{qt}}{2\,\text{pt}} \cdot \frac{1\,\text{gal}}{4\,\text{qt}} \cdot \frac{60\,\text{min}}{1\,\text{hr}} \cdot \frac{24\,\text{hr}}{1\,\text{d}}$$

$$= \frac{4 \cdot 15 \cdot 3 \cdot 8\ \text{gal}}{8 \cdot 2 \cdot 2 \cdot 4\ \text{d}}$$

$$= \frac{11.25\ \text{gal}}{\text{d}}$$

≈ 11.3 gallons per day

The faucet leaks 11.3 gallons of water in one day.

29)   The weight of a person may be measured in kilograms.

30)   The weight of a mouse may be measured in grams.

31)   The distance from Miami to Seattle may be measured in kilometers.

32)   The thickness of a quarter may be measured in millimeters.

33)   The volume of an aquarium may be measured in liters.

34)   The volume of cough syrup in one dose may be measured in milliliters.

35)   The volume of a large sports bottle may be measured in liters.

36)   The distance someone can run in 30 seconds may be measured in meters.

37) $1{,}234\ \text{m} = 1.234\ \text{km}$

38) $0.523\ \text{m} = 52.3\ \text{cm}$

39) $3.89\ \text{L} = 3{,}890\ \text{mL}$

40) $350\ \text{mg} = 0.35\ \text{g}$

41) $349\ \text{mL} = 34.9\ \text{cL}$

42) $750\ \text{g} = 0.75\ \text{kg}$

43) $1.2\ \text{m} = 1{,}200\ \text{mm}$

44) $2\ \text{mg} = 0.000002\ \text{kg}$

45) $50\ \text{mL per min} = \dfrac{50\ \text{mL}}{1\ \text{min}} \cdot \dfrac{1\ \text{L}}{1{,}000\ \text{mL}} \cdot \dfrac{60\ \text{min}}{1\ \text{hr}}$

$$= \frac{50 \cdot 3 \cdot 20\ \text{L}}{20 \cdot 50\ \text{hr}}$$

$$= \frac{3\ \text{L}}{\text{hr}}$$

$= 3$ liters per hour

46) $320\ \text{m per min} = \dfrac{320\ \text{m}}{1\ \text{min}} \cdot \dfrac{1\ \text{km}}{1000\ \text{m}} \cdot \dfrac{60\ \text{min}}{1\ \text{hr}}$

$$= \frac{8 \cdot 40 \cdot 8 \cdot 12\ \text{km}}{8 \cdot 5 \cdot 40\ \text{hr}}$$

$$= \frac{96\ \text{km}}{5\ \text{hr}}$$

$= 19.2$ kilometers per hour

47) $0.5\ \text{L per 15 min} = \dfrac{0.5\ \text{L}}{15\ \text{min}} \cdot \dfrac{1{,}000\ \text{mL}}{1\ \text{L}} \cdot \dfrac{1\ \text{min}}{60\ \text{sec}}$

$$= \frac{8 \cdot 2 \cdot 5 \cdot 10\ \text{mL}}{3 \cdot 8 \cdot 2 \cdot 3 \cdot 10\ \text{sec}}$$

$$= \frac{5\ \text{mL}}{9\ \text{sec}}$$

≈ 0.6 milliliters per second

Approximately 0.6 milliliters must be given per second.

48) $300\ \text{km per hr} = \dfrac{300\ \text{km}}{1\ \text{hr}} \cdot \dfrac{1000\ \text{m}}{1\ \text{km}} \cdot \dfrac{1\ \text{hr}}{60\ \text{min}} \cdot \dfrac{1\ \text{min}}{60\ \text{sec}}$

$$= \frac{8 \cdot 60 \cdot 4 \cdot 250\ \text{m}}{60 \cdot 3 \cdot 4 \cdot 8\ \text{sec}}$$

$$= \frac{250\ \text{m}}{3\ \text{sec}}$$

≈ 83.3 meters per second

The race car travels approximately 83.3 meters per second.

49)   The weights from largest to smallest are 1 kilogram, 1 pound, the weight of an apple, 1 ounce, and 1 gram.

50)   The volumes from largest to smallest are 1 kiloliter, 1 gallon, the volume of a mug of coffee, 1 ounce, and 1 centiliter.

51)   The volumes from largest to smallest are the volume of water in a bathtub, 1 gallon, the volume of a 2-liter soda bottle, 1 liter, 1 quart, 1 pint, 1 cup, 1 fluid ounce, and 1 milliliter.

52) The lengths from largest to smallest are 1 mile, 1 kilometer, the width of a classroom, 1 meter, 1 foot, the width of this page, and 1 inch.

53) $12 \text{ in.} = \dfrac{12 \text{ in.}}{1} \cdot \dfrac{2.54 \text{ cm}}{1 \text{ in.}}$

$= 30.48 \text{ cm}$

$\approx 30.5 \text{ cm}$

54) $5 \text{ m} \approx \dfrac{5 \text{ m}}{1} \cdot \dfrac{3.28 \text{ ft}}{1 \text{ m}}$

$= 16.4 \text{ ft}$

55) $500 \text{ kg} \approx \dfrac{500 \text{ kg}}{1} \cdot \dfrac{2.2 \text{ lb}}{1 \text{ kg}}$

$= 1{,}100 \text{ lb}$

56) $3 \text{ lb} \approx \dfrac{3 \text{ lb}}{1} \cdot \dfrac{16 \text{ oz}}{1 \text{ lb}} \cdot \dfrac{28.35 \text{ g}}{1 \text{ oz}}$

$= 1{,}360.8 \text{ g}$

57) $C = \dfrac{5 \cdot F - 160}{9}$

$= \dfrac{5 \cdot 35 - 160}{9}$

$= \dfrac{175 - 160}{9}$

$= \dfrac{15}{9}$

$C = 1.\overline{6} \approx 1.7$

$5°F \approx 1.7°C$

58) $F = \dfrac{9}{5} \cdot C + 32$

$= \dfrac{9}{5} \cdot 35 + 32$

$= \dfrac{9}{5} \cdot \dfrac{5 \cdot 7}{1} + 32$

$= 63 + 32$

$F = 95$

$35°C = 95°F$

59) $2 \text{ qt} \approx \dfrac{2 \text{ qt}}{1} \cdot \dfrac{0.946 \text{ L}}{1 \text{ qt}}$

$= 1.892 \text{ L}$

$\approx 1.9 \text{ L}$

60) $7 \text{ L} \approx \dfrac{7 \text{ L}}{1} \cdot \dfrac{0.264 \text{ gal}}{1 \text{ L}}$

$= 1.848 \text{ gal}$

$\approx 1.8 \text{ gal}$

61) a) $2 \text{ ft} = \dfrac{2 \text{ ft}}{1} \cdot \dfrac{0.305 \text{ m}}{1 \text{ ft}}$

$= 0.61 \text{ m}$

b) $0.61 \text{ m} = 61 \text{ cm}$

62) a) $2{,}640 \text{ ft} = \dfrac{2{,}640 \text{ ft}}{1} \cdot \dfrac{1 \text{ mi}}{5{,}280 \text{ ft}}$

$= 0.5 \text{ mi}$

b) $0.5 \text{ mi} \approx \dfrac{0.5 \text{ mi}}{1} \cdot \dfrac{1.61 \text{ km}}{1 \text{ mi}}$

$= 0.805 \text{ km}$

$\approx 0.8 \text{ km}$

63) a) $400 \text{ mL} = 0.4 \text{ L}$

b) $0.4 \text{ L} \approx \dfrac{0.4 \text{ L}}{1} \cdot \dfrac{0.264 \text{ gal}}{1 \text{ L}}$

$= 0.1056 \text{ gal}$

$\approx 0.1 \text{ gal}$

64) a) $5 \text{ L} \approx \dfrac{5 \text{ L}}{1} \cdot \dfrac{1.06 \text{ qt}}{1 \text{ L}}$

$= 5.3 \text{ qt}$

b) $5.3 \text{ qt} = \dfrac{5.3 \text{ qt}}{1} \cdot \dfrac{2 \text{ pt}}{1 \text{ qt}} \cdot \dfrac{2 \text{ c}}{1 \text{ pt}}$

$= 21.2 \text{ c}$

65) $\$0.98 \text{ per liter} = \dfrac{\$0.98}{1 \text{ L}} \cdot \dfrac{3.79 \text{ L}}{1 \text{ gal}}$

$\approx \$3.71 \text{ per gallon}$

$\$3.71 - \$3.70 = \$0.01$
Gasoline is $0.01 per gallon more expensive in Canada.

66) $65 \text{ miles per hour} \approx \dfrac{65 \text{ mi}}{1 \text{ hr}} \cdot \dfrac{1.61 \text{ km}}{1 \text{ mi}}$

$= \dfrac{104.65 \text{ km}}{1 \text{ hr}}$

$\approx 104.7 \text{ km per hour}$

$104.7 - 100 = 4.7$
Dale is exceeding the speed limit by 4.7 kilometers per hour.

67) $5 \text{ lb} \approx \dfrac{5 \, \cancel{\text{lb}}}{1} \cdot \dfrac{0.454 \text{ kg}}{1 \, \cancel{\text{lb}}}$

$= 2.27 \text{ kg}$

$\dfrac{\$1.99}{2.27 \text{ kg}} \approx \$0.88 \text{ per kilogram}$

$\dfrac{\$1.49}{2 \text{ kg}} \approx \$0.75 \text{ per kilogram}$

The better buy is 2 kilograms for $1.49.

68) $100 \text{ m} \approx \dfrac{100 \, \cancel{\text{m}}}{1} \cdot \dfrac{1.09 \text{ yd}}{1 \, \cancel{\text{m}}}$

$= 109 \text{ yd}$

$\dfrac{109 \text{ yd}}{15 \text{ sec}} \approx 7.3 \text{ yards per second}$

$\dfrac{100 \text{ yd}}{13 \text{ sec}} \approx 7.7 \text{ yards per second}$

The faster athlete runs 100 yards in 13 seconds.

69) $\$2.90 \text{ per gallon} \approx \dfrac{\$2.90}{1 \, \cancel{\text{gal}}} \cdot \dfrac{1 \, \cancel{\text{gal}}}{3.29 \text{ L}}$

$= \dfrac{\$2.90}{3.29 \text{ L}}$

$\approx \$0.88 \text{ per liter}$

$\$0.88 < \$0.89$

Milk is cheaper in the United States.

70) $2 \text{ pt} \approx \dfrac{2 \, \cancel{\text{pt}}}{1} \cdot \dfrac{1 \, \cancel{\text{qt}}}{2 \, \cancel{\text{pt}}} \cdot \dfrac{0.946 \text{ L}}{1 \, \cancel{\text{qt}}}$

$= 0.946 \text{ L}$

$\dfrac{\$1.25}{0.946 \text{ L}} \approx \$1.32 \text{ per liter}$

$\$1.17 < \$1.32$

The better buy is 1 liter for $1.17.

71) $440 \text{ feet per minute} \approx \dfrac{440 \, \cancel{\text{ft}}}{1 \, \cancel{\text{min}}} \cdot \dfrac{0.305 \text{ m}}{1 \, \cancel{\text{ft}}} \cdot \dfrac{1 \, \cancel{\text{min}}}{60 \text{ sec}}$

$= \dfrac{134.2 \text{ m}}{60 \text{ sec}}$

$\approx 2.24 \text{ meters per second}$

$2.68 > 2.24$

The *Carpe Carp*, at 2.68 meters per second, is traveling faster than *Tranquility*.

72) $25 \text{ km per hour} \approx \dfrac{25 \, \cancel{\text{km}}}{1 \text{ hr}} \cdot \dfrac{0.62 \text{ mi}}{1 \, \cancel{\text{km}}}$

$= \dfrac{15.5 \text{ mi}}{1 \text{ hr}}$

$= 15.5 \text{ miles per hour}$

Shauntel exceeded the speed limit by 15.5 miles per hour.

CHAPTER 6   TEST

1) a) The weight of a large dog may be measured in pounds.

   b) The weight of a large dog may be measured in kilograms.

2) a) The amount of liquid in a small bottle may be measured in fluid ounces.

   b) The amount of liquid in a small bottle may be measured in milliliters.

3) a) The height of an oak tree may be measured in feet.

   b) The height of an oak tree may be measured in meters.

4) a) The weight of a ping-pong ball may be measured in ounces.

   b) The weight of a ping-pong ball may be measured in grams.

5) $8 \text{ yd} = \dfrac{8 \, \cancel{\text{yd}}}{1} \cdot \dfrac{3 \text{ ft}}{1 \, \cancel{\text{yd}}}$

   $= 24 \text{ ft}$

6) $32 \text{ mm} = 0.032 \text{ m}$

7) $1.4 \text{ kL} = 1{,}400 \text{ L}$

8) $13 \text{ fl oz} = \dfrac{13 \, \cancel{\text{fl oz}}}{1} \cdot \dfrac{1 \text{ c}}{8 \, \cancel{\text{fl oz}}}$

   $= 1.625 \text{ c}$

   $\approx 1.6 \text{ c}$

9) $C = \dfrac{5 \cdot F - 160}{9}$

$= \dfrac{5 \cdot 85 - 160}{9}$

$= \dfrac{425 - 160}{9}$

$= \dfrac{265}{9}$

$C = 19.\overline{4} \approx 19.4$

$85°F = 29.4°C$

10) $F = \dfrac{9}{5} \cdot C + 32$

$= \dfrac{9}{5} \cdot 115 + 32$

$= \dfrac{9}{\cancel{5}} \cdot \dfrac{\cancel{5} \cdot 23}{1} + 32$

$= 207 + 32$

$F = 239$

$115°C = 239°F$

11) $43 \text{ cg} = 0.43 \text{ g}$

12) $2.5 \text{ pounds} = \dfrac{2.5 \,\cancel{lb}}{1} \cdot \dfrac{16 \text{ oz}}{1 \,\cancel{lb}}$

$= 40 \text{ ounces}$

13) $15 \text{ ft per sec} = \dfrac{15\,\cancel{ft}}{1\,\text{sec}} \cdot \dfrac{1 \text{ mi}}{5280\,\cancel{ft}} \cdot \dfrac{60\,\cancel{sec}}{1\,\cancel{min}} \cdot \dfrac{60\,\cancel{min}}{1\,h}$

$= \dfrac{15 \cdot \cancel{60} \cdot \cancel{4} \cdot 15 \text{ mi}}{\cancel{60} \cdot \cancel{4} \cdot 22 \text{ h}}$

$= \dfrac{225 \text{ mi}}{22 \text{ h}}$

$\approx 10.2 \text{ miles per hour}$

14) $31 \text{ km/hr} = \dfrac{31\,\cancel{km}}{1\,\cancel{hr}} \cdot \dfrac{1{,}000 \text{ m}}{1\,\cancel{km}} \cdot \dfrac{1\,\cancel{hr}}{60\,\cancel{min}} \cdot \dfrac{1\,\cancel{min}}{60 \text{ sec}}$

$= \dfrac{31 \cdot \cancel{2} \cdot 5 \cdot \cancel{10} \cdot \cancel{10} \text{ m}}{\cancel{2} \cdot 3 \cdot \cancel{10} \cdot 6 \cdot \cancel{10} \text{ sec}}$

$= \dfrac{155 \text{ m}}{18 \text{ sec}}$

$\approx 8.6 \text{ meters per second}$

15) $25 \text{ mL/sec} = \dfrac{25\,\cancel{mL}}{1\,\text{sec}} \cdot \dfrac{1 \text{ L}}{1{,}000\,\cancel{mL}} \cdot \dfrac{60\,\cancel{sec}}{1\,\cancel{min}} \cdot \dfrac{60\,\cancel{min}}{1 \text{ hr}}$

$= \dfrac{\cancel{5} \cdot 5 \cdot \cancel{2} \cdot 3 \cdot \cancel{10} \cdot 6 \cdot \cancel{10} \text{ L}}{\cancel{2} \cdot \cancel{5} \cdot \cancel{10} \cdot \cancel{10} \text{ hr}}$

$= \dfrac{90 \text{ L}}{\text{hr}}$

$= 90 \text{ liters per hour}$

16) $8 \text{ oz per hr} = \dfrac{8\,\cancel{oz}}{1\,\text{hr}} \cdot \dfrac{1\,\cancel{c}}{8\,\cancel{oz}} \cdot \dfrac{1\,\cancel{pt}}{2\,\cancel{c}} \cdot \dfrac{1\,\cancel{qt}}{2\,\cancel{pt}} \cdot \dfrac{1 \text{ gal}}{4\,\cancel{qt}} \cdot \dfrac{24\,\cancel{h}}{1 \text{ d}}$

$= \dfrac{\cancel{8} \cdot \cancel{2} \cdot 3 \cdot \cancel{4} \text{ gal}}{\cancel{8} \cdot \cancel{2} \cdot 2 \cdot \cancel{4} \text{ d}}$

$= \dfrac{3 \text{ gal}}{2 \text{ d}}$

$= 1.5 \text{ gallons per day}$

17) $400 \text{ m} \approx \dfrac{400\,\cancel{m}}{1} \cdot \dfrac{1.09 \text{ yd}}{1\,\cancel{m}}$

$= 436 \text{ yd}$

18) $100 \text{ mi} \approx \dfrac{100\,\cancel{mi}}{1} \cdot \dfrac{1.61 \text{ km}}{1\,\cancel{mi}}$

$= 161 \text{ km}$

19) $10 \text{ ft} \approx \dfrac{10\,\cancel{ft}}{1} \cdot \dfrac{0.305 \text{ m}}{1\,\cancel{ft}}$

$= 3.05 \text{ m}$

$\approx 3.1 \text{ m}$

20) $30 \text{ km} \approx \dfrac{30\,\cancel{km}}{1} \cdot \dfrac{0.62 \text{ mi}}{1\,\cancel{km}}$

$= 18.6 \text{ mi}$

21) a) $3 \text{ in.} = \dfrac{3\,\cancel{in.}}{1} \cdot \dfrac{2.54 \text{ cm}}{1\,\cancel{in.}}$

$= 7.62 \text{ cm}$

   b) $7.62 \text{ cm} = 76.2 \text{ mm}$

22) a) $250 \text{ g} = 0.25 \text{ kg}$

   b) $0.25 \text{ kg} \approx \dfrac{0.25\,\cancel{kg}}{1} \cdot \dfrac{2.2 \text{ lb}}{1\,\cancel{kg}}$

$= 0.55 \text{ lb}$

$\approx 0.6 \text{ lb}$

23) a) $1 \text{ qt} = \dfrac{1\,\cancel{qt}}{1} \cdot \dfrac{1 \text{ gal}}{4\,\cancel{qt}}$

$= 0.25 \text{ gal}$

   b) $0.25 \text{ gal} \approx \dfrac{0.25\,\cancel{gal}}{1} \cdot \dfrac{3.79 \text{ L}}{1\,\cancel{gal}}$

$= 0.9475 \text{ L}$

$\approx 0.9 \text{ L}$

24) a) $1,000 \text{ yd} \approx \dfrac{1,000 \text{ yd}}{1} \cdot \dfrac{0.914 \text{ m}}{1 \text{ yd}}$

$= 914 \text{ m}$

b) $914 \text{ m} = 0.914 \text{ km} \approx 0.9 \text{ km}$

25) $480 \text{ ft per hour} = \dfrac{480 \text{ ft}}{1 \text{ hr}} \cdot \dfrac{1 \text{ hr}}{60 \text{ min}}$

$= \dfrac{8 \text{ ft}}{1 \text{ min}}$

$= 8 \text{ feet per minute}$

The tortoise walks 8 feet per minute.

26) $50 \text{ times per second} = \dfrac{50 \text{ times}}{1 \text{ sec}} \cdot \dfrac{60 \text{ sec}}{1 \text{ min}}$

$= \dfrac{3,000 \text{ times}}{1 \text{ min}}$

$= 3,000 \text{ times per minute}$

The hummingbird flaps its wings 3,000 times in one minute.

27) $\$0.58 \text{ per pound} \approx \dfrac{\$0.58}{1 \text{ lb}} \cdot \dfrac{2.2 \text{ lb}}{1 \text{ kg}}$

$= \dfrac{\$1.276}{1 \text{ kg}}$

$\approx \$1.28 \text{ per kilogram}$

$\$1.25 < \$1.28$

The better buy is $1.25 per kilogram.

28) $\$4.23 \text{ per gallon} \approx \dfrac{\$4.23}{1 \text{ gal}} \cdot \dfrac{1 \text{ gal}}{3.79 \text{ L}}$

$= \dfrac{\$4.23}{3.79 \text{ L}}$

$\approx \$1.12 \text{ per liter}$

$\$1.12 < \$1.22$

The better buy is $4.23 per gallon.

29) $120 \text{ km per hr} = \dfrac{120 \text{ km}}{1 \text{ hr}} \cdot \dfrac{1000 \text{ m}}{1 \text{ km}} \cdot \dfrac{1 \text{ hr}}{60 \text{ min}} \cdot \dfrac{1 \text{ min}}{60 \text{ sec}}$

$= \dfrac{100 \text{ m}}{3 \text{ sec}}$

$\approx 33.3 \text{ meters per second}$

No. A snowmobile traveling at 25 meters per second cannot outrun the avalanche.

30) $12 \text{ ft per sec} = \dfrac{12 \text{ ft}}{1 \text{ sec}} \cdot \dfrac{1 \text{ mi}}{5280 \text{ ft}} \cdot \dfrac{60 \text{ sec}}{1 \text{ min}} \cdot \dfrac{60 \text{ min}}{1 \text{ hr}}$

$= \dfrac{90 \text{ mi}}{11 \text{ hr}}$

$\approx 8.2 \text{ miles per hour}$

Yes. A firefighter running 10 miles per hour can outrun the fire.

31) $22 \text{ T} \approx \dfrac{22 \text{ T}}{1} \cdot \dfrac{2,000 \text{ lb}}{1 \text{ T}} \cdot \dfrac{0.454 \text{ kg}}{1 \text{ lb}}$

$= 19,976 \text{ kg}$

22 tons is approximately 19,976 kilograms.

32) $17 \text{ gal per second} \approx \dfrac{17 \text{ gal}}{1 \text{ sec}} \cdot \dfrac{3.79 \text{ L}}{1 \text{ gal}} \cdot \dfrac{60 \text{ sec}}{1 \text{ min}}$

$= \dfrac{3,865.8 \text{ L}}{1 \text{ min}}$

$= 3,865.8 \text{ liters per minute}$

Approximately 3,865.8 liters of water go over the waterfall in a minute.

Chapter 7   Geometry

7.1   Angles

GUIDED PRACTICE

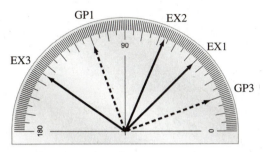

1) To draw GP1 at $110°$, we count eleven big tick marks from zero.

2) GP1 is drawn at $110°$. EX3 is drawn at $145°$. The angle between them is the difference in those values.

    The angle has a measure of $145° - 110° = 35°$.

3) An angle could be drawn either to the right or left of GP1.

    Since $90°$ isn't shown to the left of GP1, it must be drawn to the right.

    Draw GP3 at $110° - 90° = 20°$.

4) The $50°$ angle can be named either $\angle FGH$ or $\angle HGF$. The angle is an acute angle.

5) $m\angle HGJ = 90°$, a right angle.

6) $m\angle HGK = m\angle HGJ + m\angle JGK$
    $= 90° + 40°$
    $= 130°$
    $\angle HGK$ is an obtuse angle.

7) $m\angle FGJ = m\angle FGH + m\angle HGJ$
    $= 50° + 90°$
    $= 140°$
    $\angle FGJ$ is an obtuse angle.

8) $m\angle FGK = m\angle FGH + m\angle HGJ + m\angle JGK$
    $= 50° + 90° + 40°$
    $= 180°$
    $\angle FGK$ is a straight angle.

9) $\angle a$ is supplementary to $141°$.
    $m\angle a + 141° = 180°$
    $m\angle a = 180° - 141°$
    $m\angle a = 39°$

10) $\angle a$ is complementary to $21°$.
    $m\angle a + 21° = 90°$
    $m\angle a = 90° - 21°$
    $m\angle a = 69°$

11) There are no $90°$ or $180°$ angles, so we won't be able to use complements or supplements.

    However, the angle $123°$ has the same measure as the angles $88°$ and a added together. The smaller angles added = the larger angle.
    $88° + m\angle a = 123°$
    $m\angle a = 123° - 88°$
    $m\angle a = 35°$

CONCEPT CHECKS AND PRACTICE EXERCISES

A1) A ray has one endpoint and extends forever in one direction. A line has no endpoints and extends forever in both directions.

A2) The point at which two rays meet to form an angle is called the vertex of the angle.

A3) A protractor is used to measure angles.

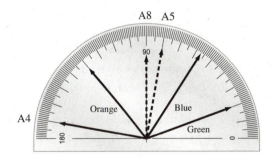

A4) Refer to the figure.

A5) Refer to the figure.

A6) $130° - 57° = 73°$

A7) $170° - 20° = 150°$

A8) $20° + 70° = 90°$

Draw the ray A8 to make a $90°$ angle.

A9) $170° - 90° = 80°$

B1) Answers may vary. Example: There are three
angles at C. It would be hard to discern which
angle was being referred to.

B2) $\angle BCD$ is acute.

B3) $\angle BCA$ is obtuse.

B4) A square is often drawn on a right angle to show
that it is $90°$.

B5) $m\angle NOL = 65°$

B6) $m\angle LOM = 20°$

B7) $m\angle JON = 90° - 65°$
$= 25°$

B8) $m\angle MOK = 90° - 20°$
$= 70°$

B9) $m\angle JOM = m\angle JON + m\angle NOL + m\angle LOM$
$= 25° + 65° + 20°$
$= 110°$

B10) $m\angle LOK = m\angle JOK - m\angle JOL$
$= 180° - 90°$
$= 90°$

B11) The five acute angles in the figure are $\angle JON$,
$\angle NOL$, $\angle LOM$, $\angle MOK$, and $\angle NOM$.

B12) The two obtuse angles in the figure are $\angle JOM$
and $\angle NOK$.

B13) The straight angle in the figure is $\angle JOK$.

B14) The two right angles in the figure are $\angle JOL$
and $\angle LOK$.

C1) The measure of an angle and its supplement add
to $180°$.

C2) The measure of an angle and its complement
add to $90°$.

C3) $180° - 35° = 145°$

C4) $90° - 35° = 55°$

C5) $90° - 45° = 45°$

C6) $180° - 90° = 90°$

C7) $90° - m\angle NOL = 90° - 72°$
$= 18°$

C8) $m\angle MOK = 90° - 23°$
$= 67°$

C9) $180° - m\angle NOK$
$= 180° - (m\angle NOL + m\angle LOM + m\angle MOK)$
$= 180° - (72° + 23° + 67°)$
$= 180° - 162°$
$= 18°$

C10) $180° - m\angle MON$
$= 180° - (m\angle NOL + m\angle LOM)$
$= 180° - (72° + 23°)$
$= 180° - 95°$
$= 85°$

C11) $m\angle a = 81° - (20° + 38°)$
$= 81° - 58°$
$= 23°$

C12) $m\angle a = 142° - 93°$
$= 49°$

SECTION 7.1   EXERCISES

For 1–9, refer to Concept Checks and Practice
Exercises.

11) rays; vertex

13) right angles

15) Acute angles; obtuse angles

17) complementary angles

19) See B in figure.   21) See D in figure.

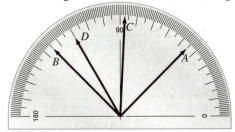

23) $m\angle AOD = 120° - 45°$
            $= 75°$

25) $m\angle BOC = 135° - 87°$
            $= 48°$

27) a) The measure of the smallest angle formed
          between any two of the rays A, B, C or D is
          $135° - 120° = 15°$.

     b) The name of the angle is $\angle BOD$.

29) Two obtuse angles that do not include ray B are
     $\angle AOD$ and $\angle COE$.

31) Two acute angles that include ray B are
     $\angle AOB$ and $\angle BOC$.

33) The straight angle is $\angle AOE$.

35) $90° - m\angle AOB = 90° - \left(180° - m\angle BOE\right)$
                  $= 90° - \left(180° - 133°\right)$
                  $= 90° - 47°$
                  $= 43°$

37) $180° - m\angle BOD = 180° - \left(m\angle BOE - m\angle DOE\right)$
                  $= 180° - \left(133° - 20°\right)$
                  $= 180° - 113°$
                  $= 67°$

39) a) $m\angle COD = 180° - \left(90° + 50°\right)$
              $= 180° - 140°$
              $= 40°$

     b) $m\angle AOB = 180° - 90°$
              $= 90°$

     c) $180° - m\angle COD = 180° - 40°$
                    $= 140°$

41) a) $m\angle BOC = 180° - \left(20° + 90°\right)$
              $= 180° - 110°$
              $= 70°$

     b) $m\angle DOE = 90° - 20°$
              $= 70°$

     c) $m\angle BOE = 180° - 20°$
              $= 160°$

     d) $m\angle BOD = m\angle BOC + m\angle COD$
              $= 70° + 20°$
              $= 90°$
          The names of the three right angles are
          $\angle AOC$, $\angle COE$, and $\angle BOD$.

43) a) $m\angle AOB = 140° - \left(40° + 30°\right)$
              $= 140° - 70°$
              $= 70°$

     b) $m\angle AOC = m\angle AOB + m\angle BOC$
              $= 70° + 40°$
              $= 110°$

     c) $m\angle BOC = 40°$

43) d) $90° - m\angle AOB = 90° - 70°$
                    $= 20°$

The complement of $\angle AOB$ is $20°$.

$90° - m\angle BOC = 90° - 40°$
                    $= 50°$

The complement of $\angle BOC$ is $50°$.

$\angle AOB$ has the smaller complement.

45) $4a = 180° - 90°$
     $4a = 90°$
      $a = 22.5°$

The angle is $22.5°$.

47) The statement is sometimes true.

True example: If $m\angle A = 45°$, the complement is $90° - m\angle A = 90° - 45° = 45°$, and the angle is equal to its complement.

False example: If $m\angle A = 30°$, the complement is $90° - m\angle A = 90° - 30° = 60°$, and the angle is not equal to its complement.

49) This statement is sometimes true.

True example: $m\angle A = 130°$ and $m\angle B = 50°$

False example: $m\angle A = 90°$ and $m\angle B = 90°$

51) The statement is always true. Answers may vary. Example: Choose any angle, say $100°$.

The supplement of $100°$ is $180° - 100° = 80°$.

The supplement of $80°$ is $180° - 80° = 100°$, or the original angle. This is true for any angle chosen.

7.2   Polygons

GUIDED PRACTICE

1) There are six sides.
    The angles between sides are different.
    The figure is a hexagon.

2) There are five sides.
    The angles between sides are the same and the lengths of the sides are the same.
    The figure is a regular pentagon.

3) There are eight sides.
    The angles between sides are the same and the lengths of the sides are different.
    The figure is an octagon.

4) $P = 10 + 5 + 9 + 4 + 6$
     $= 15 + 9 + 4 + 6$
     $= 24 + 4 + 6$
     $= 28 + 6$
     $= 34$ m

5) Figures d) and f) are rhombuses because all sides are equal.

6) Figures a), c), d), and f) are parallelograms because opposite sides are parallel.

7) Figures a) and f) are rectangles because all angles are $90°$.

8) Figures a) and c) are scalene triangles because no sides are equal.

9) Figure d) is an equilateral triangle because all sides are equal. Also, all angles are equal.

10) Only figures a) and c) are scalene triangles. Of those two, only figure a) has an angle of $90°$. Therefore, only figure a) is a right scalene triangle.

11) In an obtuse triangle, one of the angles will be greater than $90°$. This is true only of figure c). Figure c) is an obtuse triangle.

12) None of the sides are of the same length. The triangle is not isosceles.
     There is not a right angle. The triangle is not a right triangle.
     The triangle is acute because all angles measure less than $90°$.
     The triangle is an acute scalene triangle.

CONCEPT CHECKS AND PRACTICE EXERCISES

A1) Answers may vary. Example: A polygon is regular when all the sides are the same length and all the angles are the same measure.

A2) a) A quadrilateral has four sides.

b) A pentagon has five sides.

c) An octagon has eight sides.

d) A hexagon has six sides.

A3) There are three sides.
The angles between sides are different.
The shape is a triangle.

A4) There are four sides.
Opposite sides are parallel. The angles between sides are 90°, but the sides are not the same length.
The shape is a quadrilateral.

A5) There are six sides.
The angles between sides are the same and the lengths of the sides are the same.
The shape is a regular hexagon.

A6) There are eight sides.
The angles between sides are the same and the lengths of the sides are the same.
The shape is a regular octagon.

A7) There are six sides.
The angles between sides are not the same.
The shape is a hexagon.

A8) There are five sides.
The angles between sides are different.
The shape is a pentagon.

A9) There are five sides.
The angles between sides are the same and the lengths of the sides are the same.
The shape is a regular pentagon.

A10) There are three sides.
The angles between sides are the same and the lengths of the sides are the same.
The shape is a regular triangle.

A11) $P = 5 + 2.5 + 5 + 2.5$
$= 7.5 + 5 + 2.5$
$= 12.5 + 2.5$
$= 15$ in.

A12) $P = 1 + 1 + 1.5 + 1 + 1 + 1.5$
$= 2 + 1.5 + 1 + 1 + 1.5$
$= 3.5 + 1 + 1 + 1.5$
$= 4.5 + 1 + 1.5$
$= 5.5 + 1.5$
$= 7$ cm

B1) Yes, a rectangle is also a parallelogram. Answers may vary. Example: In a parallelogram, opposite sides are equal and parallel. This is also true for a rectangle.

B2) No, a parallelogram does not have to be a rectangle. Answers may vary. Example: In a rectangle, the angles must measure 90°. This is not true of a parallelogram.

B3) Yes, a rectangle can be a rhombus. A square can be classified as both a rectangle (a parallelogram with 90° angles) and a rhombus (a parallelogram with equal sides). Thus, the special rectangle called a square is also a rhombus.

B4) Answers may vary. Example: A rectangle is a polygon with four sides. This is the definition of a quadrilateral.

B5) The shape is a rectangle because it is a parallelogram with all angles equal to 90°.

B6) The shape is a trapezoid because only one pair of sides is parallel.

B7) The shape is a square because it is a parallelogram with all angles equal to 90° and all sides of the same length.

B8) The shape is a parallelogram because opposite sides are parallel.

B9) The shape is a rhombus because it is a parallelogram with all sides of the same length.

B10) The shape is a square because it is a parallelogram with all angles equal to 90° and all sides of the same length.

B11) The shape is a parallelogram because opposite sides are parallel.

B12) The shape is a quadrilateral because no sides are parallel.

B13) The shape is a quadrilateral.

B14) The shape is a parallelogram.

B15) The shape is a rectangle.

B16) The shape is a trapezoid.

C1)   Answers may vary. Example: In an equilateral triangle, all three sides are equal in length and all three angles have the same measure. In an isosceles triangle, this is true of only two sides and two angles.

C2)   Answers may vary. Example: A right scalene triangle will have one angle measuring 90° and two unequal acute angles. None of the sides will be equal.

C3)   Answers may vary. Example: In an equilateral triangle, all angles must be equal to 60°. No angle will be a right angle.

C4)   Two sides are equal. Two angles are equal. All angles are acute. The shape is an acute isosceles triangle.

C5)   Three sides are equal. Three angles are equal. The shape is an equilateral triangle.

C6)   Two sides are equal. Two angles are equal. One angle is a right angle. The shape is a right isosceles triangle.

C7)   Two sides are equal. Two angles are equal. All angles are acute. The shape is an acute isosceles triangle.

C8)   No sides are equal. No angles are equal. There is one obtuse angle. The shape is an obtuse scalene triangle.

C9)   Two sides are equal. Two angles are equal. There is a right angle. The shape is a right isosceles triangle.

C10) Three sides are equal. Three angles are equal. The shape is an equilateral triangle.

C11) No sides are equal. No angles are equal. There is a right angle. The shape is a right scalene triangle.

SECTION 7.2   EXERCISES

For 1–9, refer to Concept Checks and Practice Exercises.

11)   regular polygon

13)   hexagon

15)   False. A polygon cannot have a curved side.

17)   There are eight sides.
      The lengths of the sides are different.
      The shape is an octagon.

19)   There are five sides.
      The angles between sides are different.
      The shape is a pentagon.

21)   There are eight sides.
      The angles between sides are the same and the lengths of the sides are the same.
      The shape is a regular octagon.

23)   There are four sides.
      The angles between sides are different.
      The shape is a quadrilateral.

25)   There are three sides.
      None of the sides are of equal length.
      The angles between sides are different.
      One angle is obtuse.
      The shape is an obtuse scalene triangle.

27)   There are three sides.
      Two sides are equal.
      Two angles are equal.
      All angles are acute.
      The shape is an acute isosceles triangle.

29)   There are six sides.
      The angles between sides are the same and the lengths of the sides are the same.
      The shape is a regular hexagon.

31)   There are four sides.
      One pair of opposite sides is parallel.
      The shape is a trapezoid.

33)   There are three sides.
      No sides and no angles are equal.
      There is a right angle.
      The shape is a right scalene triangle.

35)   The sides do not form a closed figure, so the shape is not a polygon.

37)   The triangles in 27) and 36) are acute triangles.

39)   The shape is a rhombus.

41)   The shape is an isosceles triangle.

43)   The shape is a parallelogram.

45)   The shape is a hexagon.

47) The shapes are equilateral triangles.

49) The shapes are regular hexagons.

51) The shapes are equilateral triangles.

53) A triangle best approximates the shape. Answers may vary. Example: The giraffe's head is a three-sided figure.

55) A square best approximates the shape. Answers may vary. Example: The mallet's end is a four-sided polygon that appears to have sides that are the same length with right angles at the corners.

57) A square best approximates the shape. Answers may vary. Example: The wheels are four-sided polygons whose angles appear to be right angles and whose sides appear to be equal in length.

59) A regular octagon best approximates the shape. Answers may vary. Example: The plate has eight sides and each side and angle appears to be the same.

61) $P = 4 \cdot 2.2$ ft
     $= 8.8$ ft

63) $P = 8 \cdot 2.5$ in.
     $= 20$ in.

7.3   Perimeter and Area

GUIDED PRACTICE

1) $P = 18 + 8 + 12 + 20 + 10$
    $= 26 + 12 + 20 + 10$
    $= 38 + 20 + 10$
    $= 58 + 10$
    $= 68$ m

2) The missing horizontal length is $4 + 6 = 10$.
    The missing vertical length is $8 - 5 = 3$.

    $P = 10 + 8 + 6 + 5 + 4 + 3$
    $= 18 + 6 + 5 + 4 + 3$
    $= 24 + 5 + 4 + 3$
    $= 29 + 4 + 3$
    $= 33 + 3$
    $= 36$ ft

3) Counting, there are a total of 12 squares. The area is 12 square meters.

    Number of squares per row $= 4$
    Number of rows $= 3$

    Area $= ($squares per row$) \cdot ($rows$)$
    $\quad\;\; = 4 \cdot 3$
    $\quad\;\; = 12$ square meters

4) From the diagram, the width is 5 feet and the length is 12 feet.

    $A = l \cdot w$
    $\;\; = (12)(5)$
    $\;\; = 60$ square feet

5) **Lower Triangle**   **Upper Triangle**
    $b = 22$ in. $b = 18$ in.
    $h = 10$ in. $h = 10$ in.

    $A_{\text{triangle}} = \frac{1}{2} b \cdot h$ $A_{\text{triangle}} = \frac{1}{2} b \cdot h$

    $\quad\quad\;\; = \frac{1}{2} \cdot 22 \cdot 10$ $\quad\quad\;\; = \frac{1}{2} \cdot 18 \cdot 10$

    $\quad\quad\;\; = 11 \cdot 10$ $\quad\quad\;\; = 9 \cdot 10$

    $\quad\quad\;\; = 110$ sq in. $\quad\quad\;\; = 90$ sq in.

    Area $= 110 + 90$
    $\quad\quad\; = 200$ square inches

6) $A = \frac{1}{2} \cdot b \cdot h$

$\quad = \frac{1}{2} \cdot 5 \cdot 2$

$\quad = \frac{5}{2} \cdot 2$

$\quad = 5$

The area is 5 m^2.

7) $A = b \cdot h$

$\quad = 2 \cdot 3$

$\quad = 6$

The area is 6 in.^2

8) $A = \frac{1}{2} \cdot (B + b) \cdot h$

$\quad = \frac{1}{2} \cdot (20 + 14) \cdot 7$

$\quad = \frac{1}{2} \cdot (34) \cdot 7$

$\quad = 17 \cdot 7$

$\quad = 119$

The area is 119 ft^2.

CONCEPT CHECKS AND PRACTICE EXERCISES

A1) Answers may vary. Example: The perimeter is the distance around the outside of the object.

A2) Answers may vary. Examples: inches, feet, centimeters.

A3) $P = 3 + 2 + 3 + 4 + 3$

$\quad = 5 + 3 + 4 + 3$

$\quad = 8 + 4 + 3$

$\quad = 12 + 3$

$\quad = 15 \text{ ft}$

A4) $P = 8 + 3 + 8 + 3$

$\quad = 11 + 8 + 3$

$\quad = 19 + 3$

$\quad = 22 \text{ ft}$

A5) $P = 4 + 6 + 4 + 6$

$\quad = 10 + 4 + 6$

$\quad = 14 + 6$

$\quad = 20 \text{ ft}$

A6) $P = 9 + 3 + 4 + 9 + 4 + 3$

$\quad = 12 + 4 + 9 + 4 + 3$

$\quad = 16 + 9 + 4 + 3$

$\quad = 25 + 4 + 3$

$\quad = 29 + 3$

$\quad = 32 \text{ ft}$

A7) The missing vertical length is
$20 \text{ ft} - 12 \text{ ft} = 8 \text{ ft}$.

$\quad P = 7 + 20 + 24 + 12 + 17 + 8$

$\quad = 27 + 24 + 12 + 17 + 8$

$\quad = 51 + 12 + 17 + 8$

$\quad = 63 + 17 + 8$

$\quad = 80 + 8$

$\quad = 88 \text{ ft}$

A8) The missing horizontal length is
$21 \text{ ft} - 10 \text{ ft} - 4 \text{ ft} = 7 \text{ ft}$.
The missing vertical length is 6 ft.

$\quad P = 21 + 13 + 4 + 6 + 7 + 6 + 10 + 13$

$\quad = 34 + 4 + 6 + 7 + 6 + 10 + 13$

$\quad = 38 + 6 + 7 + 6 + 10 + 13$

$\quad = 44 + 7 + 6 + 10 + 13$

$\quad = 51 + 6 + 10 + 13$

$\quad = 57 + 10 + 13$

$\quad = 67 + 13$

$\quad = 80 \text{ ft}$

A9) The missing horizontal length is
$18 \text{ ft} - 6 \text{ ft} - 6 \text{ ft} = 6 \text{ ft}$.
The missing vertical length is 6 ft.

$\quad P = 6 + 6 + 6 + 20 + 18 + 20 + 6 + 6$

$\quad = 12 + 6 + 20 + 18 + 20 + 6 + 6$

$\quad = 18 + 20 + 18 + 20 + 6 + 6$

$\quad = 38 + 18 + 20 + 6 + 6$

$\quad = 56 + 20 + 6 + 6$

$\quad = 76 + 6 + 6$

$\quad = 82 + 6$

$\quad = 88 \text{ ft}$

A10) The missing horizontal length is
18 m − 5 m = 13 m.
The missing vertical lengths are 7 m and 5 m.

$$P = 18 + 7 + 5 + 5 + 5 + 13 + 7$$
$$= 25 + 5 + 5 + 5 + 13 + 7$$
$$= 30 + 5 + 5 + 13 + 7$$
$$= 35 + 5 + 13 + 7$$
$$= 40 + 13 + 7$$
$$= 53 + 7$$
$$= 60 \text{ m}$$

A11) $P = 2 \cdot 2.5 + 2 \cdot 5$
$= 5 + 10$
$= 15 \text{ ft}$

A12) $P = 4 \cdot 1 + 2 \cdot 1.5$
$= 4 + 3$
$= 7 \text{ ft}$

B1) Answers may vary. Example: To find area, the shape must be "covered." Linear measures are only one-dimensional. To cover a region, a two-dimensional measure is needed. Square units are two-dimensional.

B2) There are 15 squares inside the object. The area is 15 cm^2.

B3) Answers may vary. Example: Perimeter is the distance around a region and is measured with a one-dimensional measure (i.e., inches). Area is the amount needed to "cover" the region and is measured with a two-dimensional measure (i.e., square inches).

B4) Answers may vary. Example: By drawing a diagonal, the rectangle has been "cut in half," forming two triangles. Thus, the area of a triangle is half that of the corresponding rectangle.

B5) $A = l \cdot w$
$= (4)(2)$
$= 8 \text{ square feet}$

B6) $A = l \cdot w$
$= (3)(3)$
$= 9 \text{ square inches}$

B7) $A = \frac{1}{2} \cdot b \cdot h$
$= \frac{1}{2} \cdot 8 \cdot 6$
$= 4 \cdot 6$
$= 24 \text{ square meters}$

B8) $A = \frac{1}{2} \cdot b \cdot h$
$= \frac{1}{2} \cdot 4 \cdot 6$
$= 2 \cdot 6$
$= 12 \text{ square centimeters}$

B9) The shape is made of two rectangles.

B10) The shape is made of two rectangles.

Answers may vary.

B11) The shape is made of two triangles.

Answers may vary.

B12) The shape is made of a rectangle and a triangle.

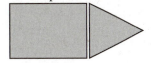

B13) The shape is made of two rectangles.

Answers may vary.

The missing vertical length is 17 m − 8 m = 9 m.

$$A = l_1 \cdot w_1 + l_2 \cdot w_2$$
$$= (10)(8) + (12)(9)$$
$$= 80 + 108$$
$$= 188 \text{ square meters}$$

B14)   The shape is made of two triangles.

Answers may vary.

$$A = \frac{1}{2} \cdot b_1 \cdot h_1 + \frac{1}{2} \cdot b_2 \cdot h_2$$
$$= \frac{1}{2} \cdot 20 \cdot 12 + \frac{1}{2} \cdot 20 \cdot 12$$
$$= 120 + 120$$
$$= 240 \text{ square centimeters}$$

B15)   The shape is made of two triangles.

$$A = \frac{1}{2} \cdot b_1 \cdot h_1 + \frac{1}{2} \cdot b_2 \cdot h_2$$
$$= \frac{1}{2} \cdot 2 \cdot 6 + \frac{1}{2} \cdot 6 \cdot 4$$
$$= 6 + 12$$
$$= 18 \text{ square feet}$$

B16)   The shape is made of two rectangles.

Answers may vary.

The missing horizontal length is
12 in. − 5 in. = 7 in.
The missing vertical length is
16 in. − 14 in. = 2 in.

$$A = l_1 \cdot w_1 + l_2 \cdot w_2$$
$$= (14)(7) + (16)(5)$$
$$= 98 + 80$$
$$= 178 \text{ square inches}$$

C1)   a)   Triangle: $A = \frac{1}{2} \cdot b \cdot h$

b)   Rectangle: $A = l \cdot w$

c)   Parallelogram: $A = b \cdot h$

d)   Trapezoid: $A = \frac{1}{2} \cdot (B + b) \cdot h$

C2)   A rectangle that measures 4 feet by 5 feet has an
area of 20 square feet.

C3) $A = b \cdot h$
$$= 24 \cdot 12$$
$$= 288 \text{ in.}^2$$

C4) $A = b \cdot h$
$$= 40 \cdot 15$$
$$= 600 \text{ cm}^2$$

C5) $A = \frac{1}{2} \cdot (B + b) \cdot h$
$$= \frac{1}{2} \cdot (30 + 20) \cdot 8$$
$$= \frac{1}{2} \cdot (50) \cdot 8$$
$$= 25 \cdot 8$$
$$= 200 \text{ m}^2$$

C6) $A = \frac{1}{2} \cdot (B + b) \cdot h$
$$= \frac{1}{2} \cdot (26 + 14) \cdot 3$$
$$= \frac{1}{2} \cdot (40) \cdot 3$$
$$= 20 \cdot 3$$
$$= 60 \text{ yd}^2$$

C7) $A = \frac{1}{2} \cdot (B + b) \cdot h$
$$= \frac{1}{2} \cdot (10 + 8) \cdot 7$$
$$= \frac{1}{2} \cdot (18) \cdot 7$$
$$= 9 \cdot 7$$
$$= 63 \text{ ft}^2$$

C8) $A = b \cdot h$
$$= 7 \cdot 6$$
$$= 42 \text{ m}^2$$

C9) $A = \frac{1}{2} \cdot (B + b) \cdot h$
$$= \frac{1}{2} \cdot (8 + 6) \cdot 3$$
$$= \frac{1}{2} \cdot (14) \cdot 3$$
$$= 7 \cdot 3$$
$$= 21 \text{ in.}^2$$

C10) $A = \frac{1}{2} \cdot (B + b) \cdot h$
$$= \frac{1}{2} \cdot (6 + 3) \cdot 8$$
$$= \frac{1}{2} \cdot (9) \cdot 8$$
$$= \frac{9}{2} \cdot 8$$
$$= 36 \text{ yd}^2$$

SECTION 7.3   EXERCISES

For 1–9, refer to Concept Checks and Practice Exercises.

11) area

13) lengths^2

15) $P = 10 + 6 + 10 + 6$
     $= 16 + 10 + 6$
     $= 26 + 6$
     $= 32$ in.

17) $P = 14 + 13 + 5$
     $= 27 + 5$
     $= 32$ km

19) The missing horizontal length is
     6 ft $- 5$ ft $= 1$ ft.
     The missing vertical length is
     3 ft $- 1$ ft $= 2$ ft.

     $P = 5 + 2 + 1 + 1 + 6 + 3$
     $= 7 + 1 + 1 + 6 + 3$
     $= 8 + 1 + 6 + 3$
     $= 9 + 6 + 3$
     $= 15 + 3$
     $= 18$ ft

21) $P = 10 \cdot 4$ m
     $= 40$ m

23) $A = l \cdot w$
     $= 13 \cdot 8$
     $= 104$ in.^{2}

25) $A = \frac{1}{2} \cdot b \cdot h$
     $= \frac{1}{2} \cdot 12 \cdot 7$
     $= 6 \cdot 7$
     $= 42$ ft^2

27) The shape is made of two rectangles.

     The missing horizontal length is
     25 ft $- 15$ ft $= 10$ ft.
     The missing vertical length is
     12 ft $- 6$ ft $= 6$ ft.

     $A = l_1 \cdot w_1 + l_2 \cdot w_2$
     $= (15)(12) + (10)(6)$
     $= 180 + 60$
     $= 240$ ft^2

29) The shape is made of a rectangle and a triangle.

     The missing height of the triangle is
     20 ft $- 10$ ft $= 10$ ft.

     $A = l \cdot w + \frac{1}{2} \cdot b \cdot h$
     $= 30 \cdot 10 + \frac{1}{2} \cdot 30 \cdot 10$
     $= 300 + 150$
     $= 450$ ft^2

31) The shape is made of two rectangles.

     The missing horizontal length is
     6 ft $- 5$ ft $= 1$ ft.
     The missing vertical length is
     3 ft $- 1$ ft $= 2$ ft.

     $A = l_1 \cdot w_1 + l_2 \cdot w_2$
     $= (5)(3) + (1)(1)$
     $= 15 + 1$
     $= 16$ ft^2

33) The shape is made of two rectangles.

     The missing horizontal length is
     14 cm $- 8$ cm $= 6$ cm.
     The missing vertical length is
     5 cm $- 3$ cm $= 2$ cm.

     $A = l_1 \cdot w_1 + l_2 \cdot w_2$
     $= (8)(5) + (6)(3)$
     $= 40 + 18$
     $= 58$ cm^2

35) $A = l \cdot w$
     $= (10)(3)$
     $= 30$ cm^2

37) $A = \frac{1}{2} \cdot b \cdot h$
     $= \frac{1}{2} \cdot 18 \cdot 9$
     $= 9 \cdot 9$
     $= 81$ in.^{2}

39) $A = b \cdot h$
     $= 3 \cdot 15$
     $= 45$ m^2

41) $A = \frac{1}{2} \cdot (B + b) \cdot h$

$= \frac{1}{2} \cdot (20 + 12) \cdot 7$

$= \frac{1}{2} \cdot (32) \cdot 7$

$= 16 \cdot 7$

$= 112 \text{ cm}^2$

43) $A = \frac{1}{2} \cdot b \cdot h$

$= \frac{1}{2} \cdot 8 \cdot 3$

$= 4 \cdot 3$

$= 12 \text{ in.}^2$

45) $A = l \cdot w$

$= \left(\frac{1}{2}\right)\left(\frac{3}{8}\right)$

$= \frac{3}{16} \text{ mi}^2$

47) a) Counting, we see a total of 8 squares. Each square has a length of 3 feet, so the total width of the windows is $8 \cdot 3 \text{ ft} = 24 \text{ ft}$.

$A = 24 \text{ ft} \cdot 5 \text{ ft}$

$= 120 \text{ ft}^2$

The combined area of all the windows is 120 square feet.

b) $120 \text{ ft}^2 \cdot \frac{\$2}{\text{ft}^2} = \$240$

The total cost of installing curtains is $240.

49) a) Counting, we see a total of 8 squares. Each square has a length of 3 feet, so the total width of the windows is $8 \cdot 3 \text{ ft} = 24 \text{ ft}$.

Since there are six windows, the total height of the windows is $6 \cdot 5 \text{ ft} = 30 \text{ ft}$.

$P = 2 \cdot 24 \text{ ft} + 2 \cdot 30 \text{ ft}$

$= 48 \text{ ft} + 60 \text{ ft}$

$= 108 \text{ ft}$

The combined perimeter of all the windows is 108 feet.

b) $108 \text{ ft} \cdot \frac{\$2.25}{\text{ft}} = \$243$

It will cost $243 to install trim around all the windows.

51) $A = \frac{1}{2} \cdot b_1 \cdot h_1 + \frac{1}{2} \cdot b_2 \cdot h_2$

$= \frac{1}{2} \cdot 12 \cdot 3 + \frac{1}{2} \cdot 12 \cdot 3$

$= 18 + 18$

$= 36 \text{ ft}^2$

53) $A = \frac{1}{2} \cdot b_1 \cdot h_1 + \frac{1}{2} \cdot b_2 \cdot h_2$

$= \frac{1}{2} \cdot 8 \cdot 5 + \frac{1}{2} \cdot 4 \cdot 5$

$= 20 + 10$

$= 30 \text{ m}^2$

55) a) $A = \frac{1}{2} \cdot b \cdot h$

$= \frac{1}{2} \cdot 1 \cdot 1.3$

$= 0.65 \text{ in.}^2$

The area of each triangle is 0.65 square inches.

$8 \cdot 0.65 \text{ in.}^2 = 5.2 \text{ in.}^2$

The estimate for the area of the circle is 5.2 square inches.

b) Answers may vary. Example: The estimate with the red triangles is more accurate. The length of the base is shorter and will create less "waste" within the circle.

7.4   Circles

GUIDED PRACTICE

1) $r = 5 \text{ ft}$

$C = 2\pi r$

$= 2 \cdot \pi \cdot 5$

$\approx 10 \cdot 3.14$

$\approx 31.4 \text{ ft}$

2) $d = 40 \text{ mi}$

$C = \pi d$

$= \pi \cdot 40$

$\approx 3.14 \cdot 40$

$= 125.6 \text{ mi}$

3) To find the radius, divide the diameter by 2.
    Because the diameter is 12 mm, the radius is
    $12 \div 2 = 6$ mm.

    $A = \pi r^2$
    $ = \pi \cdot 6^2$
    $ \approx 3.14 \cdot 36$
    $ = 113.04$ mm^2

4) The diameter is the same as the height of the
    rectangle, so the diameter is 20 ft.
    The radius is 10 ft.

    Rectangle   **Circle**
    $A = l \cdot w$ $A = \pi r^2$
    $ = 70 \cdot 20$ $ = \pi \cdot 10^2$
    $ = 1{,}400$ ft^2 $ \approx 3.14 \cdot 100$
     $ = 314$ ft^2

    Area = ☐ − ◯

    Area $= 1{,}400 - 314$
    $ = 1{,}086$ ft^2

5) $r = \dfrac{1}{2} \cdot 4$ ft
    $ = 2$ ft

    Triangle   **Half Circle**
    $A = \dfrac{1}{2} \cdot b \cdot h$ $A = \dfrac{1}{2}\pi r^2$
    $ = \dfrac{1}{2} \cdot 4 \cdot 8$ $ \approx 0.5 \cdot 3.14 \cdot 2^2$
    $ = 16$ ft^2 $ \approx 0.5 \cdot 3.14 \cdot 4$
     $ \approx 6.28$ ft^2

    Area = ▷ + ◖

    Area $\approx 6.28 + 16$
    $ \approx 22.28$ ft^2

6) Perimeter $= \dfrac{1}{2}$ circumference $+ 3$ sides of trapezoid

    Half-Circle   **3 sides of the Trapezoid**
    $P = \dfrac{1}{2} \cdot \pi d$ $P = 5 + 8 + 5$
    $ \approx 0.5 \cdot 3.14 \cdot 6$ $ = 18$ ft
    $ = 9.42$ ft

    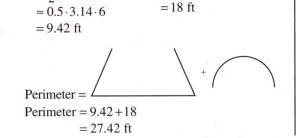

    Perimeter =
    Perimeter $= 9.42 + 18$
    $ = 27.42$ ft

7) $r = \dfrac{1}{2} \cdot 6$ ft
    $ = 3$ ft

    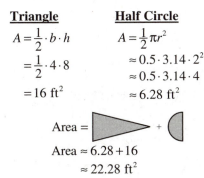

    Area =

    Area $= A_{\text{trapezoid}} - \dfrac{1}{2} A_{\text{circle}}$
    $ = \dfrac{1}{2} \cdot (10 + 6) \cdot 7 - \dfrac{1}{2} \cdot \pi \cdot (3)^2$
    $ \approx \dfrac{1}{2} \cdot 16 \cdot 7 - \dfrac{1}{2} \cdot 3.14 \cdot 9$
    $ \approx 56 - 14.13$
    $ \approx 41.87$ ft^2

8) Answers will vary. Example: The perimeter of
    both figures includes the lengths of the same
    three sides of the trapezoid and half the
    circumference of the circle.

CONCEPT CHECKS AND PRACTICE EXERCISES

A1) If the radius of a circle is 4 miles, the diameter
    is twice that measure, or 8 miles.

A2) The name of the measure from the center of a
    circle to its outer edge is the radius.

A3) The circumference is the name used for the
    perimeter of a circle.

A4) $r = 12$ cm

    $C = 2\pi r$
    $ = 2 \cdot \pi \cdot 12$
    $ \approx 24 \cdot 3.14$
    $ = 75.36$ cm

A5) $d = 3$ m

    $C = \pi d$
    $ = \pi \cdot 3$
    $ \approx 3.14 \cdot 3$
    $ = 9.42$ m

A6) $d = 8$ ft

    $C = \pi d$
    $ = \pi \cdot 8$
    $ \approx 3.14 \cdot 8$
    $ = 25.12$ ft

A7) $r = 9$ yd

$C = 2\pi r$
$\quad = 2 \cdot \pi \cdot 9$
$\quad \approx 18 \cdot 3.14$
$\quad = 56.52$ yd

B1)

B2)   If the radius of a circle is given in inches, then the circle's area will have units of square inches.

B3) $r = 18$ in.

$A = \pi r^2$
$\quad = \pi \cdot 18^2$
$\quad \approx 3.14 \cdot 324$
$\quad = 1{,}017.36$ in.^{2}

B4) $r = \frac{1}{2} \cdot 4$ mm $= 2$ mm

$A = \pi r^2$
$\quad = \pi \cdot 2^2$
$\quad \approx 3.14 \cdot 4$
$\quad = 12.56$ mm^2

B5) $r = \frac{1}{2} \cdot 12$ yd $= 6$ yd

$A = \pi r^2$
$\quad = \pi \cdot 6^2$
$\quad \approx 3.14 \cdot 36$
$\quad = 113.04$ yd^2

B6) $r = 3$ m

$A = \pi r^2$
$\quad = \pi \cdot 3^2$
$\quad \approx 3.14 \cdot 9$
$\quad = 28.26$ m^2

B7)   10 m

$r = \frac{1}{2} \cdot 10$ m $= 5$ m

Area $= A_{\text{square}} - A_{\text{circle}}$
$\quad = 10 \cdot 10 - \pi \cdot 5^2$
$\quad \approx 10 \cdot 10 - 3.14 \cdot 25$
$\quad \approx 100 - 78.5$
$\quad \approx 21.5$ m^2

B8)   9 in.

$h = 4$ in.

$r = \frac{1}{2} \cdot 4$ in. $= 2$ in.

Area $= A_{\text{parallelogram}} - A_{\text{circle}}$
$\quad = 9 \cdot 4 - \pi \cdot 2^2$
$\quad \approx 9 \cdot 4 - 3.14 \cdot 4$
$\quad = 36 - 12.56$
$\quad = 23.44$ in.^{2}

B9)   5 mm

$h = 3$ mm

7 mm

$r = \frac{1}{2} \cdot 3$ mm $= \frac{3}{2}$ mm $= 1.5$ mm

Area $= A_{\text{trapezoid}} - A_{\text{circle}}$
$\quad = \frac{1}{2} \cdot (7 + 5) \cdot 3 - \pi \cdot 1.5^2$
$\quad = \frac{1}{2} \cdot 12 \cdot 3 - \pi \cdot 2.25$
$\quad \approx \frac{1}{2} \cdot 12 \cdot 3 - 3.14 \cdot 2.25$
$\quad \approx 18 - 7.065$
$\quad \approx 10.935$ mm^2

B10)   5 in.

3 in.

$r = \frac{1}{2} \cdot 3$ in. $= \frac{3}{2}$ in. $= 1.5$ in.

Area $= A_{\text{rectangle}} - A_{\text{circle}}$
$\quad = 5 \cdot 3 - \pi \cdot (1.5)^2$
$\quad = 5 \cdot 3 - \pi \cdot 2.25$
$\quad \approx 5 \cdot 3 - 3.14 \cdot 2.25$
$\quad = 15 - 7.065$
$\quad = 7.935$ in.^{2}

C1)

C2)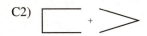

C3)

C4) Area $= A_{square} + A_{parallelogram}$
$= 3 \cdot 3 + 4 \cdot 6$
$= 9 + 24$
$= 33 \text{ in.}^2$

C5) $r = \frac{1}{2} \cdot 2 \text{ m} = 1 \text{ m}$

Area $= \frac{1}{2} A_{circle} + A_{trapezoid}$
$= \frac{1}{2} \cdot \pi \cdot 1^2 + \frac{1}{2} \cdot (4+2) \cdot 3$
$\approx \frac{1}{2} \cdot 3.14 \cdot 1^2 + \frac{1}{2} \cdot 6 \cdot 3$
$\approx 1.57 + 9$
$\approx 10.57 \text{ m}^2$

C6) $r = \frac{1}{2} \cdot 4 \text{ ft} = 2 \text{ ft}$

Area $= A_{circle} - A_{square}$
$= \pi \cdot 2^2 - 2 \cdot 2$
$\approx 3.14 \cdot 4 - 2 \cdot 2$
$\approx 12.56 - 4$
$\approx 8.56 \text{ ft}^2$

C7) $r = \frac{1}{2} \cdot 12 \text{ cm} = 6 \text{ cm}$

Area $= A_{square} - A_{circle}$
$= 12 \cdot 12 - \pi \cdot 6^2$
$\approx 12 \cdot 12 - 3.14 \cdot 36$
$= 144 - 113.04$
$= 30.96 \text{ cm}^2$

C8) The shape is made of two rectangles.

The missing horizontal length is
7 ft − 5 ft = 2 ft.

$P = 7 + 2 + 2 + 3 + 5 + 3 + 2$
$= 9 + 2 + 3 + 5 + 3 + 2$
$= 11 + 3 + 5 + 3 + 2$
$= 14 + 5 + 3 + 2$
$= 19 + 3 + 2$
$= 22 + 2$
$= 24 \text{ ft}$

C9) The shape is made of a half circle and a triangle.

$r = \frac{1}{2} \cdot 5 \text{ in.} = \frac{5}{2} \text{ in.} = 2.5 \text{ in.}$

Perimeter $= \frac{1}{2}$ circumference + 2 sides of triangle

$P = \frac{1}{2} \cdot 2\pi r + 3 + 4$
$\approx 0.5 \cdot 2 \cdot 3.14 \cdot 2.5 + 3 + 4$
$\approx 7.85 + 7$
$\approx 14.85 \text{ in.}$

C10) The shape is made of a half circle and a trapezoid.

$r = \frac{1}{2} \cdot 2 \text{ ft} = 1 \text{ ft}$

Perimeter
$= \frac{1}{2}$ circumference + 3 sides of trapezoid

$P = \frac{1}{2} \cdot 2\pi r + 3.4 + 4 + 3.2$
$\approx 0.5 \cdot 2 \cdot 3.14 \cdot 1 + 3.4 + 4 + 3.2$
$= 3.14 + 10.6$
$= 13.74 \text{ m}$

C11) The shape is made of a square and a parallelogram. Let s be the length of a side of the square.

Perimeter $= P_{parallelogram} + P_{square} - 2 \cdot s$
$= (2 \cdot 6 + 2 \cdot 3) + 4 \cdot 3 - 2 \cdot 3$
$= (12 + 6) + 12 - 6$
$= 18 + 12 - 6$
$= 30 - 6$
$= 24 \text{ in.}$

SECTION 7.4 EXERCISES

For 1–9, refer to Concept Checks and Practice Exercises.

11) Pi

13) radius

15) $C = 2\pi r$; $C = \pi d$

17) $r = 7 \text{ m}$
$d = 2 \cdot 7 \text{ m} = 14 \text{ m}$

19) $d = 7 \text{ cm}$
$r = \frac{1}{2} \cdot 7 \text{ cm} = \frac{7}{2} \text{ cm} = 3.5 \text{ cm}$

21) $r = 8$ m

$C = 2\pi r$
$\quad = 2 \cdot \pi \cdot 8$
$\quad \approx 16 \cdot 3.14$
$\quad \approx 50.24$ m

23) $d = 20$ cm

$C = \pi d$
$\quad = \pi \cdot 20$
$\quad \approx 3.14 \cdot 20$
$\quad \approx 62.8$ cm

25) $r = 5$ cm

$A = \pi r^2$
$\quad = \pi \cdot 5^2$
$\quad \approx 3.14 \cdot 25$
$\quad \approx 78.5$ cm^2

27) $r = \frac{1}{2} \cdot 8$ km $= 4$ km

$A = \pi r^2$
$\quad = \pi \cdot 4^2$
$\quad \approx 3.14 \cdot 16$
$\quad \approx 50.24$ km^2

29)   9 ft

7 ft

$r = \frac{1}{2} \cdot 7$ ft $= \frac{7}{2}$ ft $= 3.5$ ft

Area $= A_{\text{rectangle}} - A_{\text{circle}}$
$\quad = 7 \cdot 9 - \pi \cdot (3.5)^2$
$\quad = 7 \cdot 9 - \pi \cdot 12.25$
$\quad \approx 7 \cdot 9 - 3.14 \cdot 12.25$
$\quad \approx 63 - 38.465$
$\quad \approx 24.535$ ft^2

31)   10 cm

$h = 3$ cm

12 cm

$r = \frac{1}{2} \cdot 3$ cm $= \frac{3}{2}$ cm $= 1.5$ cm

Area $= A_{\text{trapezoid}} - A_{\text{circle}}$
$\quad = \frac{1}{2} \cdot (12 + 10) \cdot 3 - \pi \cdot 1.5^2$
$\quad = \frac{1}{2} \cdot (12 + 10) \cdot 3 - \pi \cdot 2.25$
$\quad = \frac{1}{2} \cdot 22 \cdot 3 - \pi \cdot 2.25$
$\quad \approx \frac{1}{2} \cdot 22 \cdot 3 - 3.14 \cdot 2.25$
$\quad = 33 - 7.065$
$\quad = 25.935$ cm^2

33) a)

  b) $r = \frac{1}{2} \cdot 9$ in. $= 4.5$ in.

Perimeter $= \frac{1}{2}$ circumference $+$
                    3 sides of trapezoid

$P = \frac{1}{2} \cdot 2\pi r + 6 + 12 + 6$
$\quad \approx 0.5 \cdot 2 \cdot 3.14 \cdot 4.5 + 6 + 12 + 6$
$\quad \approx 14.13 + 24$
$\quad \approx 38.13$ in.

35) a)

  b) Note that the perimeter of the shape in the same as in exercise 33. Refer to Guided Practice 8.

$P = 22.28$ cm

37) a)

  b) The shape is made of a half circle and a parallelogram.

$r = \frac{1}{2} \cdot 7$ m $= \frac{7}{2}$ m $= 3.5$ m

Perimeter
$= \frac{1}{2}$ circumference $+ 3$ sides of parallelogram

$P = \frac{1}{2} \cdot 2\pi r + 3.4 + 4 + 3.2$
$\quad \approx 0.5 \cdot 2 \cdot 3.14 \cdot 3.5 + 10 + 7 + 10$
$\quad \approx 10.99 + 27$
$\quad \approx 37.99$ m

39) a) $\left(\ \ +\ \right)\ +\ \right)$

b) The shape is made of a large half circle and two smaller half circles, which combine to make one smaller whole circle.

$$r_{\text{large circle}} = 2 \text{ cm}$$

$$r_{\text{small circle}} = \frac{1}{2} \cdot 2 \text{ cm} = 1 \text{ cm}$$

$$\text{Perimeter} = \frac{1}{2} \text{circumference}_{\text{large circle}}$$
$$+ \text{circumference}_{\text{small circle}}$$

$$P = \frac{1}{2} \cdot 2\pi r_{\text{large circle}} + 2\pi r_{\text{small circle}}$$
$$\approx 0.5 \cdot 2 \cdot 3.14 \cdot 2 + 2 \cdot 3.14 \cdot 1$$
$$\approx 6.28 + 6.28$$
$$\approx 12.56 \text{ cm}$$

41) a) ◖ + ◼

b) $r = \frac{1}{2} \cdot 9 \text{ in.} = \frac{9}{2} \text{ in.} = 4.5 \text{ in.}$

$$\text{Area} = A_{\text{trapezoid}} + \frac{1}{2} A_{\text{circle}}$$
$$= \frac{1}{2}(9+12) \cdot 10 + \frac{1}{2}\pi(4.5)^2$$
$$\approx \frac{1}{2} \cdot 21 \cdot 10 + \frac{1}{2} \cdot 3.14 \cdot 20.25$$
$$\approx 105 + 31.7925$$
$$\approx 136.7925 \text{ in.}^2$$

43) a) ◼ − ◡

b) $r = \frac{1}{2} \cdot 9 \text{ in.} = \frac{9}{2} \text{ in.} = 4.5 \text{ in.}$

$$\text{Area} = A_{\text{trapezoid}} - \frac{1}{2} A_{\text{circle}}$$
$$= \frac{1}{2}(9+12) \cdot 10 - \frac{1}{2}\pi(4.5)^2$$
$$\approx \frac{1}{2} \cdot 21 \cdot 10 - \frac{1}{2} \cdot 3.14 \cdot 20.25$$
$$\approx 105 - 31.7925$$
$$\approx 73.2075 \text{ in.}^2$$

45) a) $\text{Perimeter} = 8+6+8+\frac{1}{2}\pi(6)$
$$\approx 8+6+8+\frac{1}{2}(3.14)(6)$$
$$\approx 22+9.42$$
$$\approx 31.42 \text{ ft}$$
The gasket must be 31.42 feet long.

b) The shape is made of a half circle and a rectangle.

$$r = \frac{1}{2} \cdot 6 \text{ ft} = 3 \text{ ft}$$

$$\text{Area} = A_{\text{rectangle}} + \frac{1}{2} A_{\text{circle}}$$
$$= 8 \cdot 6 + \frac{1}{2}\pi(3)^2$$
$$\approx 8 \cdot 6 + \frac{1}{2} \cdot 3.14 \cdot 9$$
$$= 48 + 14.13$$
$$= 62.13 \text{ ft}^2$$

$62.13 \text{ ft}^2 \cdot 3 = 186.39 \text{ ft}^2$
186.39 square feet of fabric is needed.

47) $r = \frac{1}{2} \cdot 6 \text{ m} = 3 \text{ m}$

$$\text{Area} = A_{\text{circle}} - A_{\text{triangle}}$$
$$= \pi \cdot 3^2 - \frac{1}{2} \cdot 6 \cdot 2$$
$$\approx 3.14 \cdot 9 - \frac{1}{2} \cdot 6 \cdot 2$$
$$= 28.26 - 6$$
$$= 22.26 \text{ m}^2$$

49) $r = \frac{1}{2} \cdot 1 \text{ m} = \frac{1}{2} \text{ m} = 0.5 \text{ m}$

$$\text{Area} = \frac{1}{2} \cdot A_{\text{circle}}$$
$$= 0.5 \cdot \pi \cdot (0.5)^2$$
$$\approx 0.5 \cdot 3.14 \cdot 0.25$$
$$= 0.3925 \text{ m}^2$$

51) a) $d = 0.5 \text{ ft}$

$$C = \pi d$$
$$= \pi \cdot 0.5$$
$$\approx 3.14 \cdot 0.5$$
$$= 1.57 \text{ ft}$$
The part moves 1.57 feet for every revolution of the wheel.

b) $\frac{1 \text{ revolution}}{1.57 \text{ ft}} \cdot 15.7 \text{ ft} = 10 \text{ revolutions}$
Ten revolutions must be made in one minute.

c) Since the dial is set at 15 revolutions per minute, the conveyor belt is set too fast.

53) a) The resulting shape is a circle.

b) $A = \pi r^2$

53) c) Answers may vary. Example: Use the formula for the area of an ellipse with a and b equal to the same value; for instance, 5. Then the area is $\pi \cdot 5 \cdot 5$, or $\pi \cdot 25$. Using the formula for the area of a circle with the radius equal to 5, we get $\pi \cdot 5^2$, or $\pi \cdot 25$, which is the same as the area of the ellipse with $a = b$.

7.5   Volume

GUIDED PRACTICE

1) $8 \text{ m} + 7 \text{ m} + 9 \text{ m} = 24 \text{ m}$

Perimeter is a 1-dimensional measure. We measure perimeter with line segments, so the units are lengths such as meters.

2) $1 \text{ ft}^2 + 1 \text{ ft}^2 + 1 \text{ ft}^2 + 1 \text{ ft}^2 + 1 \text{ ft}^2 + 1 \text{ ft}^2 = 6 \text{ ft}^2$

Area is a 2-dimensional measure. We measure area with squares, so area has units of square length such as square inches or in.^{2}.

3) $1 \text{ m}^3 + 1 \text{ m}^3 + 1 \text{ m}^3 = 3 \text{ m}^3$

Volume is a 3-dimensional measure. We measure volume with cubes, so volume has units of cubic length such as cubic meters, or m^3.

4)   Since 15 cubes are needed to fill the figure, the volume is 15 in.^{3}, or 15 cubic inches.

5)   height $= 4$ m
     width $= 3$ m
     length $= 5$ m

To find the number of cubes, we multiply the length, width, and height.

$$\text{Volume} = \text{length} \cdot \text{width} \cdot \text{height}$$
$$= 5 \cdot 3 \cdot 4$$
$$= 15 \cdot 4$$
$$= 60$$

The volume is 60 m^3, or 60 cubic meters.

6) $V = \frac{1}{3} \cdot l \cdot w \cdot h$
$$= \frac{1}{3} \cdot 4 \cdot 4 \cdot 12$$
$$= \frac{1}{3} \cdot 192$$
$$= 64$$

The volume of the steeple is 64 ft^3.

7) $r = 3$ cm

$$V = \frac{4}{3} \cdot \pi \cdot r^3$$
$$\approx \frac{4}{3} \cdot 3.14 \cdot 3^3$$
$$= \frac{4}{3} \cdot 3.14 \cdot 27$$
$$= 36 \cdot 3.14$$
$$= 113.04$$

The ball's volume is about 113.04 milliliters.

CONCEPT CHECKS AND PRACTICE EXERCISES

A1)   The units used to measure perimeter, area, and volume are lengths, square lengths, and cubic lengths, respectively.

A2)   Answers may vary. Example: Volume is measured using a three-dimensional unit or cube.

A3)   Since there are 20 cubes, the volume is 20 ft^3 or 20 cubic feet.

A4)   Since there are 15 cubes, the volume is 15 cm^3 or 15 cubic centimeters.

A5)   a)   length $= 3$ m

      b)   width $= 4$ m

      c)   height $= 2$ m

      d)   Volume $= \text{length} \cdot \text{width} \cdot \text{height}$
           $$= 3 \cdot 4 \cdot 2$$
           $$= 12 \cdot 2$$
           $$= 24$$
      The volume is 24 m^3 or 24 cubic meters.

A6)   a)   length $= 6$ yd

      b)   width $= 4$ yd

      c)   height $= 4$ yd

      d)   Volume $= \text{length} \cdot \text{width} \cdot \text{height}$
           $$= 6 \cdot 4 \cdot 4$$
           $$= 24 \cdot 4$$
           $$= 96$$
      The volume is 96 yd^3 or 96 cubic yards.

A7) a) True. Answers may vary. Example: There are no "indentations" in the top layer to indicate that any layer has fewer cubes than the top layer.

   b) True. Answers may vary. Example: When the number of cubes in the layer is counted, the answer is the same as multiplying the number of cubes in the length by the number of cubes in the width.

   c) True. Answers may vary. Example: When the number of cubes in the entire object is counted, the answer is the same as multiplying the number of layers by the number of cubes in each layer.

   d) Answers may vary. Example: The number of cubes in a layer is the same as multiplying the number of cubes in the width of the layer by the number of cubes in the length. Multiplying the number of cubes in the layer by the number of layers is the same as multiplying the layer by the height of the object.

B1) The decimal approximation used for π is 3.14.

B2) Find the radius by multiplying the diameter by $\frac{1}{2}$ or by dividing the diameter by 2.

B3) Answers may vary. Example: Writing the formula may enable her to easily see what values must be substituted into the formula.

B4) $r = \frac{1}{2} \cdot 2 \text{ ft} = 1 \text{ ft}$

$V = \pi \cdot r^2 \cdot h$
$\approx 3.14 \cdot 1^2 \cdot 6$
$\approx 3.14 \cdot 1 \cdot 6$
$\approx 3.14 \cdot 6$
$\approx 18.84 \text{ ft}^3$

B5) $V = \frac{1}{3} \cdot l \cdot w \cdot h$
$= \frac{1}{3} \cdot 6 \cdot 4 \cdot 5$
$= \frac{1}{3} \cdot 120$
$= 40 \text{ ft}^3$

B6) $V = l \cdot w \cdot h$
$= 14 \cdot 9 \cdot 3$
$= 126 \cdot 3$
$= 378 \text{ in.}^3$
The volume of the roaster is 378 cubic inches.

B7) $r = \frac{1}{2} \cdot 20 \text{ ft} = 10 \text{ ft}$

$V = \pi \cdot r^2 \cdot h$
$\approx 3.14 \cdot 10^2 \cdot 50$
$\approx 3.14 \cdot 100 \cdot 50$
$\approx 3.14 \cdot 5,000$
$\approx 15,700 \text{ ft}^3$
The volume of the tank is about 15,700 cubic feet.

B8) $r = \frac{1}{2} \cdot 2 \text{ m} = 1 \text{ m}$

$V = \frac{1}{3} \cdot \pi \cdot r^2 \cdot h$
$\approx \frac{1}{3} \cdot 3.14 \cdot 1^2 \cdot 1.5$
$\approx \frac{1}{3} \cdot 3.14 \cdot 1 \cdot 1.5$
$\approx 3.14 \cdot 0.5$
$\approx 1.57 \text{ m}^3$
About 1.57 cubic meters of grain fits in the cone.

B9) $r = \frac{1}{2} \cdot 2 \text{ m} = 1 \text{ m}$

$V = \pi \cdot r^2 \cdot h$
$\approx 3.14 \cdot 1^2 \cdot 5$
$\approx 3.14 \cdot 1 \cdot 5$
$\approx 3.14 \cdot 5$
$\approx 15.7 \text{ m}^3$
About 15.7 cubic meters of grain fits in the cylinder.

B10) $1.57 \text{ m}^3 + 15.7 \text{ m}^3 = 17.27 \text{ m}^3$
The total volume of the hopper is 17.27 cubic meters.

SECTION 7.5   EXERCISES

For 1–6, refer to Concept Checks and Practice Exercises.

7) cone

9) pyramid

11) cylinder

13) Since there are 15 cubes, the volume is 15 ft^3 or 15 cubic feet.

15) Since there are 12 cubes, the volume is 12 cm^3 or 12 cubic centimeters.

17) Since there are 19 cubes, the volume is 19 mm^3 or 19 cubic millimeters.

19) Volume = length · width · height
$$= 5 \cdot 3 \cdot 2$$
$$= 15 \cdot 2$$
$$= 30 \text{ ft}^3$$

21) $V = \frac{1}{3} \cdot \pi \cdot r^2 \cdot h$
$$\approx \frac{1}{3} \cdot 3.14 \cdot 6^2 \cdot 15$$
$$\approx \frac{1}{3} \cdot 3.14 \cdot 36 \cdot 15$$
$$\approx 3.14 \cdot 180$$
$$\approx 565.2 \text{ in.}^3$$

23) $r = \frac{1}{2} \cdot 12 \text{ in.} = 6 \text{ in.}$

$V = \frac{4}{3} \cdot \pi \cdot r^3$
$$\approx \frac{4}{3} \cdot 3.14 \cdot 6^3$$
$$\approx \frac{4}{3} \cdot 3.14 \cdot 216$$
$$\approx 288 \cdot 3.14$$
$$\approx 904.32 \text{ in.}^3$$

25) $V = \frac{4}{3} \cdot \pi \cdot r^3$
$$\approx \frac{4}{3} \cdot 3.14 \cdot 1{,}000^3$$
$$\approx \frac{4}{3} \cdot 3.14 \cdot 1{,}000{,}000{,}000$$
$$\approx 4{,}186{,}666{,}667 \text{ mi}^3$$
The volume of the moon is approximately 4,186,666,667 cubic miles.

27) $r = \frac{1}{2} \cdot 6 \text{ ft} = 3 \text{ ft}$

$V = \pi \cdot r^2 \cdot h$
$$\approx 3.14 \cdot 3^2 \cdot 8$$
$$= 3.14 \cdot 9 \cdot 8$$
$$= 3.14 \cdot 72$$
$$= 226.08 \text{ ft}^3$$
The volume of the oil tank is approximately 226.08 cubic feet.

29) $V = \frac{1}{3} \cdot l \cdot w \cdot h$
$$= \frac{1}{3} \cdot 750 \cdot 750 \cdot 450$$
$$= \frac{1}{3} \cdot 253{,}125{,}000$$
$$= 84{,}375{,}000 \text{ ft}^3$$
The approximate volume of the great pyramid at Giza is 84,375,000 cubic feet.

31) $V = l \cdot w \cdot h$
$$= 150 \cdot 150 \cdot 630$$
$$= 22{,}500 \cdot 630$$
$$= 14{,}175{,}000 \text{ ft}^3$$
The approximate volume of the skyscraper is 14,175,000 cubic feet.

33) a) The shape of the juice container is best approximated as a rectangular solid.

   b) $V = l \cdot w \cdot h$
$$= 4 \cdot 4 \cdot 9$$
$$= 16 \cdot 9$$
$$= 144 \text{ in.}^3$$
The volume of the juice container is approximately 144 cubic inches.

35) a) $V = \pi \cdot r^2 \cdot h$
$$\approx 3.14 \cdot 1^2 \cdot 4$$
$$\approx 3.14 \cdot 1 \cdot 4$$
$$\approx 3.14 \cdot 4$$
$$\approx 12.56 \text{ in.}^3$$

   b) $V = \frac{1}{3} \cdot \pi \cdot r^2 \cdot h$
$$\approx \frac{1}{3} \cdot 3.14 \cdot 1^2 \cdot 4$$
$$\approx \frac{1}{3} \cdot 3.14 \cdot 1 \cdot 4$$
$$\approx 4.18\overline{6}$$
$$\approx 4.19 \text{ in.}^3$$

   c) $\dfrac{V_{\text{cylinder}}}{V_{\text{cone}}} = \dfrac{\pi \cdot r^2 \cdot h}{\frac{1}{3} \cdot \pi \cdot r^2 \cdot h} = \dfrac{12.56}{4.19} \approx 3$

You would have to pour three cones full of water into the cylinder to fill the cylinder.

   d) Answers may vary. Example: The volume of the cone is one-third that of the cylinder.

   e) Answers may vary. Example: The formula for the volume of a cone is the formula for the volume of a cylinder multiplied by one-third.

7.6   Square Roots and the Pythagorean Theorem

GUIDED PRACTICE

1) $\sqrt{49} = 7$ because $7 \cdot 7 = 49$.

2) $\sqrt{64} = 8$ because $8 \cdot 8 = 64$.

3) $3 \cdot 3 = 9$
    $2 \cdot 2 = 4$

$\sqrt{\dfrac{9}{4}} = \dfrac{3}{2}$ because $\dfrac{3}{2} \cdot \dfrac{3}{2} = \dfrac{9}{4}$

4) This is an order of operations exercise. First, evaluate the square roots. Then subtract.

$\sqrt{100} - \sqrt{64} = 10 - 8 = 2$

5) Since $5^2 = 25$ and $6^2 = 36$, $\sqrt{33}$ is between 5 and 6.

6) From the square root table, $\sqrt{11} \approx 3.317$.

7) $\sqrt{6} \cdot \sqrt{4} \approx 2.449 \cdot 2$
    $= 4.898$

8) a) The hypotenuse is opposite the right angle.

Hypotenuse = x
1st leg = 1 m
2nd leg = 4 m

b) Since the hypotenuse in unknown, use the formula

hypotenuse $= \sqrt{(\text{leg})^2 + (\text{leg})^2}$

Substitute the values:

$x = \sqrt{(1)^2 + (4)^2}$

9) Since the triangle is a right triangle, we may use the Pythagorean theorem.

One leg = 5 m
Hypotenuse = 6 m

leg $= \sqrt{(\text{hypotenuse})^2 - (\text{known leg})^2}$
$= \sqrt{6^2 - 5^2}$
$= \sqrt{36 - 25}$
$= \sqrt{11}$
≈ 3.317 m

10) Since the triangle is a right triangle, we may use the Pythagorean theorem.

1st leg = 7 in.
2nd leg = 11 in.

hypotenuse $= \sqrt{\left(1^{st}\ \text{leg}\right)^2 + \left(2^{nd}\ \text{leg}\right)^2}$
$= \sqrt{7^2 + 11^2}$
$= \sqrt{49 + 121}$
$= \sqrt{170}$
≈ 13.038 in.

11) 1st leg = 27 yd
    2nd leg = unknown
    Hypotenuse = 30 yd

leg $= \sqrt{(\text{hypotenuse})^2 - (\text{known leg})^2}$
$= \sqrt{30^2 - 27^2}$
$= \sqrt{900 - 729}$
$= \sqrt{171}$
≈ 13.1 yd

12) $a = 5$, b = unknown, $c = 13$

$a^2 + b^2 = c^2$
$5^2 + b^2 = 13^2$
$25 + b^2 = 169$
$25 - 25 + b^2 = 169 - 25$
$b^2 = 144$
$b = \sqrt{144}$
$b = 12$ m

13) $a = 9$, $b = 10$, c = unknown

$a^2 + b^2 = c^2$
$9^2 + 10^2 = c^2$
$81 + 100 = c^2$
$181 = c^2$
$\sqrt{181} = c$
13.454 ft $\approx c$

CONCEPT CHECKS AND PRACTICE EXERCISES

A1) Answers may vary. Example: The square root of a number is a value, which, when multiplied by itself, gives the original number.

A2) Answers may vary. Example: To find the square root of a fraction, you must find the square root of the numerator and the square root of the denominator.

A3)

Number, n	1	2	3	4	5
Perfect square, n^2	1	4	9	16	25

Number, n	6	7	8	9	10
Perfect square, n^2	36	49	64	81	100

Number, n	11	12	13	14	15
Perfect square, n^2	121	144	169	196	225

A4) Square roots must be evaluated during the exponent step of the order of operations.

A5) $\sqrt{9} = 3$ because $3 \cdot 3 = 9$.

A6) $\sqrt{4} = 2$ because $2 \cdot 2 = 4$.

A7) $\sqrt{0} = 0$ because $0 \cdot 0 = 0$.

A8) $\sqrt{1} = 1$ because $1 \cdot 1 = 1$.

A9) $7 \cdot 7 = 49$
     $6 \cdot 6 = 36$
     $$\sqrt{\frac{49}{36}} = \frac{7}{6}$$

A10) $1 \cdot 1 = 1$
     $8 \cdot 8 = 64$
     $$\sqrt{\frac{1}{64}} = \frac{1}{8}$$

A11) $15 \cdot 15 = 225$
     $7 \cdot 7 = 49$
     $$\sqrt{\frac{225}{49}} = \frac{15}{7}$$

A12) $14 \cdot 14 = 196$
     $5 \cdot 5 = 25$
     $$\sqrt{\frac{196}{25}} = \frac{14}{5}$$

A13) $\sqrt{4} + \sqrt{9} = 2 + 3 = 5$

A14) $\sqrt{144} - \sqrt{36} = 12 - 6 = 6$

A15) $\sqrt{25} \cdot \sqrt{16} = 5 \cdot 4 = 20$

A16) $\sqrt{49} \cdot \sqrt{1} = 7 \cdot 1 = 7$

B1) The square root of a number that is a perfect square will have no decimal part.

B2) If a number is not a perfect square, its square root will have a decimal part.

B3) $(1)^2 = 1$ $(2)^2 = 4$ $(3)^2 = 9$
     $(4)^2 = 16$ $(5)^2 = 25$ $(6)^2 = 36$
     $(7)^2 = 49$ $(8)^2 = 64$ $(9)^2 = 81$

B4) Answers may vary. Example: Determine the square roots of the perfect squares just less than 31 and just greater than 31. The square root of 31 will lie between these two values.

B5) Since $1^2 = 1$ and $2^2 = 4$, $\sqrt{3}$ is between 1 and 2.

B6) Since $4^2 = 16$ and $5^2 = 25$, $\sqrt{19}$ is between 4 and 5.

B7) Since $4^2 = 16$ and $5^2 = 25$, $\sqrt{17}$ is between 4 and 5.

B8) Since $2^2 = 4$ and $3^2 = 9$, $\sqrt{5}$ is between 2 and 3.

B9) $\sqrt{3} \approx 1.732$

B10) $\sqrt{19} \approx 4.359$

B11) $\sqrt{17} \approx 4.123$

B12) $\sqrt{5} \approx 2.236$

B13) $\sqrt{36} + \sqrt{3} \approx 6 + 1.732 = 7.732$

B14) $\sqrt{49} - \sqrt{20} \approx 7 - 4.472 = 2.528$

B15) $\sqrt{12} - \sqrt{3} \approx 3.464 - 1.732 = 1.732$

B16) $\sqrt{8} + \sqrt{8} \approx 2.828 + 2.828 = 5.656$

B17) $\sqrt{18} \cdot \sqrt{2} = \sqrt{18 \cdot 2} = \sqrt{36} = 6$

B18) $\sqrt{5} \cdot \sqrt{3} = \sqrt{5 \cdot 3} = \sqrt{15} \approx 3.873$

C1) The Pythagorean theorem can be used only with right triangles.

C2) The longest side of a right triangle is called the hypotenuse.

C3) a) $\text{hypotenuse} = \sqrt{(\text{leg})^2 + (\text{leg})^2}$

     b) $\text{leg} = \sqrt{(\text{hypotenuse})^2 - (\text{leg})^2}$

C4) One leg $= 6$ ft
     Hypotenuse $= 10$ ft

$$\begin{aligned}\text{leg} &= \sqrt{(\text{hypotenuse})^2 - (\text{known leg})^2} \\ &= \sqrt{10^2 - 6^2} \\ &= \sqrt{100 - 36} \\ &= \sqrt{64} \\ &= 8 \text{ ft}\end{aligned}$$

C5) One leg $= 12$ in.
     Hypotenuse $= 13$ in.

$$\begin{aligned}\text{leg} &= \sqrt{(\text{hypotenuse})^2 - (\text{known leg})^2} \\ &= \sqrt{13^2 - 12^2} \\ &= \sqrt{169 - 144} \\ &= \sqrt{25} \\ &= 5 \text{ in.}\end{aligned}$$

C6) 1st leg $= 6$ m
     2nd leg $= 6$ m

$$\begin{aligned}\text{hypotenuse} &= \sqrt{(1^{\text{st}} \text{ leg})^2 + (2^{\text{nd}} \text{ leg})^2} \\ &= \sqrt{6^2 + 6^2} \\ &= \sqrt{36 + 36} \\ &= \sqrt{72} \\ &\approx 8.5 \text{ m}\end{aligned}$$

C7) 1st leg $= 8$ yd
     2nd leg $= 11$ yd

$$\begin{aligned}\text{hypotenuse} &= \sqrt{(1^{\text{st}} \text{ leg})^2 + (2^{\text{nd}} \text{ leg})^2} \\ &= \sqrt{8^2 + 11^2} \\ &= \sqrt{64 + 121} \\ &= \sqrt{185} \\ &\approx 13.6 \text{ yd}\end{aligned}$$

C8) 1st leg $= 7$ ft
     2nd leg $= 12$ ft

$$\begin{aligned}\text{hypotenuse} &= \sqrt{(1^{\text{st}} \text{ leg})^2 + (2^{\text{nd}} \text{ leg})^2} \\ &= \sqrt{7^2 + 12^2} \\ &= \sqrt{49 + 144} \\ &= \sqrt{193} \\ &\approx 13.9 \text{ ft}\end{aligned}$$

C9) One leg $= 8$ km
     Hypotenuse $= 16$ km

$$\begin{aligned}\text{leg} &= \sqrt{(\text{hypotenuse})^2 - (\text{known leg})^2} \\ &= \sqrt{16^2 - 8^2} \\ &= \sqrt{256 - 64} \\ &= \sqrt{192} \\ &\approx 13.9 \text{ km}\end{aligned}$$

C10) One leg $= 6$ cm
      Hypotenuse $= 15$ cm

$$\begin{aligned}\text{leg} &= \sqrt{(\text{hypotenuse})^2 - (\text{known leg})^2} \\ &= \sqrt{15^2 - 6^2} \\ &= \sqrt{225 - 36} \\ &= \sqrt{189} \\ &\approx 13.7 \text{ cm}\end{aligned}$$

C11) 1st leg $= 12$ in.
      2nd leg $= 14$ in.

$$\begin{aligned}\text{hypotenuse} &= \sqrt{(1^{\text{st}} \text{ leg})^2 + (2^{\text{nd}} \text{ leg})^2} \\ &= \sqrt{12^2 + 14^2} \\ &= \sqrt{144 + 196} \\ &= \sqrt{340} \\ &\approx 18.4 \text{ in.}\end{aligned}$$

FOCUS ON ALGEBRA EXERCISES

1) $a = 8$, $b = \text{unknown}$, $c = 10$

$$\begin{aligned}a^2 + b^2 &= c^2 \\ 8^2 + b^2 &= 10^2 \\ 64 + b^2 &= 100 \\ 64 - 64 + b^2 &= 100 - 64 \\ b^2 &= 36 \\ b &= \sqrt{36} \\ b &= 6 \text{ in.}\end{aligned}$$

2) $a = 12$, $b = $ unknown, $c = 13$

$$a^2 + b^2 = c^2$$
$$12^2 + b^2 = 13^2$$
$$144 + b^2 = 169$$
$$144 - 144 + b^2 = 169 - 144$$
$$b^2 = 25$$
$$b = \sqrt{25}$$
$$b = 5 \text{ ft}$$

3) $a = 8$, $b = $ unknown, $c = 9$

$$a^2 + b^2 = c^2$$
$$8^2 + b^2 = 9^2$$
$$64 + b^2 = 81$$
$$64 - 64 + b^2 = 81 - 64$$
$$b^2 = 17$$
$$b = \sqrt{17}$$
$$b \approx 4.123 \text{ m}$$

4) $a = 7$, $b = $ unknown, $c = 8$

$$a^2 + b^2 = c^2$$
$$7^2 + b^2 = 8^2$$
$$49 + b^2 = 64$$
$$49 - 49 + b^2 = 64 - 49$$
$$b^2 = 15$$
$$b = \sqrt{15}$$
$$b \approx 3.873 \text{ cm}$$

5) $a = 3$, $b = 5$, $c = $ unknown

$$a^2 + b^2 = c^2$$
$$3^2 + 5^2 = c^2$$
$$9 + 25 = c^2$$
$$34 = c^2$$
$$\sqrt{34} = c$$
$$5.831 \text{ ft} \approx c$$

6) $a = 7$, $b = $ unknown, $c = 12$

$$a^2 + b^2 = c^2$$
$$7^2 + b^2 = 12^2$$
$$49 + b^2 = 144$$
$$49 - 49 + b^2 = 144 - 49$$
$$b^2 = 95$$
$$b = \sqrt{95}$$
$$b = 9.747 \text{ in.}$$

7) $a = 4$, $b = 4$, $c = $ unknown

$$a^2 + b^2 = c^2$$
$$4^2 + 4^2 = c^2$$
$$16 + 16 = c^2$$
$$32 = c^2$$
$$\sqrt{32} = c$$
$$5.657 \text{ cm} \approx c$$

8) $a = 5$, $b = 8$, $c = $ unknown

$$a^2 + b^2 = c^2$$
$$5^2 + 8^2 = c^2$$
$$25 + 64 = c^2$$
$$89 = c^2$$
$$\sqrt{89} = c$$
$$9.434 \text{ m} \approx c$$

SECTION 7.6   EXERCISES

For 1–11, refer to Concept Checks and Practice Exercises.

13) perfect square

15) square root

17) Answers will vary. Example: $5^2 = 25$ and $6^2 = 36$. Since 33 lies between 25 and 36, the square root of 33 will lie between 5 and 6.

19) $\sqrt{49} = 7$

21) $\sqrt{144} = 12$

23) $\sqrt{\dfrac{64}{25}} = \dfrac{8}{5}$

25) $\sqrt{\dfrac{4}{169}} = \dfrac{2}{13}$

27) Since $2^2 = 4$ and $3^2 = 9$, $\sqrt{7}$ is between 2 and 3.

29) Since $7^2 = 49$ and $8^2 = 64$, $\sqrt{55}$ is between 7 and 8.

31) Since $8^2 = 64$ and $9^2 = 81$, $\sqrt{65}$ is between 8 and 9.

33) Since $8^2 = 64$ and $9^2 = 81$, $\sqrt{77}$ is between 8 and 9.

35) $A = 16 \text{ ft}^2$
$s = \sqrt{16 \text{ ft}^2} = 4 \text{ ft}$

37) $A = 64 \text{ cm}^2$
$s = \sqrt{64 \text{ cm}^2} = 8 \text{ cm}$

39) $A = 8 \text{ mi}^2$
$s = \sqrt{8 \text{ mi}^2} \approx 2.828 \text{ mi}$

41) $A = 48 \text{ in.}^2$
$s = \sqrt{48 \text{ in.}^2} \approx 6.928 \text{ in.}$

43) $3^2 = 9$
$4^2 = 16 \qquad 9 + 16 = 25$
$5^2 = 25$

45) $\sqrt{3} \approx 1.732$

47) $\sqrt{30} \approx 5.477$

49) $\sqrt{83} \approx 9.110$

51) $\sqrt{8} \approx 2.828$

53) $\sqrt{4} + \sqrt{36} = 2 + 6 = 8$

55) $\sqrt{22} - \sqrt{2} \approx 4.690 - 1.414 = 3.276$

57) $a = 5$, $b = $ unknown, $c = 13$

$$a^2 + b^2 = c^2$$
$$5^2 + b^2 = 13^2$$
$$25 + b^2 = 169$$
$$25 - 25 + b^2 = 169 - 25$$
$$b^2 = 144$$
$$b = \sqrt{144}$$
$$b = 12 \text{ in.}$$

59) $a = 24$, $b = $ unknown, $c = 25$

$$a^2 + b^2 = c^2$$
$$24^2 + b^2 = 25^2$$
$$576 + b^2 = 625$$
$$576 - 576 + b^2 = 625 - 576$$
$$b^2 = 49$$
$$b = \sqrt{49}$$
$$b = 7 \text{ m}$$

61) $a = 3$, $b = 6$, $c = $ unknown

$$a^2 + b^2 = c^2$$
$$3^2 + 6^2 = c^2$$
$$9 + 36 = c^2$$
$$45 = c^2$$
$$\sqrt{45} = c$$
$$6.71 \text{ ft} \approx c$$

63) $a = 5$, $b = 5$, $c = $ unknown

$$a^2 + b^2 = c^2$$
$$5^2 + 5^2 = c^2$$
$$25 + 25 = c^2$$
$$50 = c^2$$
$$\sqrt{50} = c$$
$$7.07 \text{ cm} \approx c$$

65) $a = 6$, $b = $ unknown, $c = 10$

$$a^2 + b^2 = c^2$$
$$6^2 + b^2 = 10^2$$
$$36 + b^2 = 100$$
$$36 - 36 + b^2 = 100 - 36$$
$$b^2 = 64$$
$$b = \sqrt{64}$$
$$b = 8 \text{ ft}$$
The ladder must be placed 8 feet from the base.

67) a) $a = 5 \text{ ft}$, $b = 12 \text{ ft}$, $c = $ unknown

$$a^2 + b^2 = c^2$$
$$5^2 + 12^2 = c^2$$
$$25 + 144 = c^2$$
$$169 = c^2$$
$$\sqrt{169} = c$$
$$13 \text{ ft} = c$$
Each side of the roof has a length of 40 feet and a width of 13 feet.

67) b) $A = 2 \cdot 40 \cdot 13$
$= 1{,}040 \text{ ft}^2$

The area of the roof is 1,040 square feet.

69) a) Area $= A_{\text{square}} + A_{\text{triangle}}$
$= 2 \cdot 2 + \frac{1}{2} \cdot 2 \cdot 2$
$= 4 + 2$
$= 6 \text{ ft}^2$

The area of the glass used in the window is 6 square feet.

b) $a = 2$ ft, $b = 2$ ft, $c = $ unknown

$a^2 + b^2 = c^2$
$2^2 + 2^2 = c^2$
$4 + 4 = c^2$
$8 = c^2$
$\sqrt{8} = c$
$2.828 \text{ ft} \approx c$

The hypotenuse of the triangular window is approximately 2.828 feet.

Perimeter
$= 3 \text{ sides of square} + 2 \text{ sides of triangle}$

$P \approx 3 \cdot 2 + 2 + 2.828$
$= 6 + 2 + 2.828$
$= 8 + 2.828$
$= 10.828 \text{ ft}$

10.828 feet of trim must be used on the perimeter of the window.

71) a) $a = 1$ m, $b = 1$ m, $c = $ unknown

$a^2 + b^2 = c^2$
$1^2 + 1^2 = c^2$
$1 + 1 = c^2$
$2 = c^2$
$\sqrt{2} = c$
$1.414 \text{ m} \approx c$

The diameter of the circle is $\sqrt{2} \approx 1.414$ meters.

b) $r = \frac{1}{2} \cdot \sqrt{2} = \frac{\sqrt{2}}{2}$ m

$A = \pi \cdot r^2$
$= \pi \cdot \left(\frac{\sqrt{2}}{2} \right)^2$
$= \pi \cdot \frac{1}{2}$
$\approx 3.14 \cdot \frac{1}{2}$
$= 1.570 \text{ m}^2$

The area of the circle is approximately 1.570 square meters.

c) Area $= A_{\text{circle}} - A_{\text{square}}$
$= 1.570 - 1 \cdot 1$
$= 1.570 - 1$
$= 0.570 \text{ m}^2$

The area inside the circle but outside the square is approximately 0.570 square meters.

7.7   Similarity

GUIDED PRACTICE

1) a) Since the triangle is isosceles, it can only be similar to the isosceles triangles, a), c), or f). The longest side is about 4 times the shortest side. This matches f).

The triangle is similar to f).

b) T corresponds to C.
R corresponds to B.
S corresponds to A.

2) a) Since the triangle is obtuse, it can only be similar to the other obtuse triangles, e) or g). The longest side is much longer than the shortest side. This matches g).

The triangle is similar to g).

b) Shortest sides: N corresponds to B.
Longest sides: M corresponds to C.
Medium sides: P corresponds to A.

3) a) Since the triangle is obtuse, it can only be similar to the other obtuse triangles, e) or g). The two shortest sides are close in length. This matches e).

The triangle is similar to e).

b) C corresponds to K.
B corresponds to L.
A corresponds to J.

4) Form ratios using corresponding sides.

Find the ratio of the lengths: $\dfrac{13}{x}$

Find the ratio of the widths: $\dfrac{6}{2}$

Write the proportion: $\dfrac{13}{x} = \dfrac{6}{2}$

5) $\dfrac{y}{13} = \dfrac{2}{6}$

   $\dfrac{y}{\cancel{13}} \cdot \dfrac{\cancel{13}}{1} = \dfrac{2}{6} \cdot \dfrac{13}{1}$

   $y = \dfrac{26}{6}$

   $y = 4\dfrac{1}{3}$ in. ≈ 4.33 in.

6) $\dfrac{h}{5} = \dfrac{20}{2}$

   $\dfrac{h}{\cancel{5}} \cdot \dfrac{\cancel{5}}{1} = \dfrac{20}{2} \cdot \dfrac{5}{1}$

   $h = \dfrac{100}{2}$

   $h = 50$ ft

   The building is 50 feet tall.

7) $\dfrac{w}{4} = \dfrac{14}{8}$

   $\dfrac{w}{\cancel{4}} \cdot \dfrac{\cancel{4}}{1} = \dfrac{14}{8} \cdot \dfrac{4}{1}$

   $w = \dfrac{56}{8}$

   $w = 7$ ft

   $A = l \cdot w$

   $ = 14 \cdot 7$

   $ = 98$ ft^2

8) $\dfrac{x}{4} = \dfrac{18}{8}$

   $8x = 72$

   $\dfrac{8x}{8} = \dfrac{72}{8}$

   $x = 9$ ft

   $A = \dfrac{1}{2} \cdot b \cdot h$

   $ = \dfrac{1}{2} \cdot 9 \cdot 18$

   $ = 81$ ft^2

CONCEPT CHECKS AND PRACTICE EXERCISES

A1) When two figures are similar, the corresponding sides are located directly opposite the corresponding angles.

A2) Similar figures may have different sizes, but they must have the same shape.

A3) Since this is a right triangle, it can only be similar to the right triangle *E*.

A4) Since this is a rectangle, it can only be similar to the rectangles *A* or *C*. The length is about two and a half times the width. This matches *A*.

A5) Since this is a rectangle, it can only be similar to the rectangle *A* or *C*. The length is about three and a half times the width. This matches *C*.

A6) Since this is an obtuse triangle, it can only be similar to the obtuse triangle *F*.

A7) Since this is a square, it can only be similar to the square *B*.

A8) Since this is a regular pentagon, it can only be similar to the regular pentagon *D*.

A9)   *A* corresponds to *D*.
     B corresponds to *E*.
     C corresponds to *F*.

A10) *A* corresponds to *D*.
     B corresponds to *E*.
     C corresponds to *F*.

A11) *A* corresponds to *E*.
     B corresponds to *F*.
     C corresponds to *D*.

A12) *A* corresponds to *F*.
     B corresponds to *E*.
     C corresponds to *D*.

B1) The corresponding sides of similar figures form an equivalent ratio.

B2) A proportion is an equation stating that two ratios or rates are equal.

B3) $\dfrac{B}{10} = \dfrac{40}{6}$

   $\dfrac{B}{\cancel{10}} \cdot \dfrac{\cancel{10}}{1} = \dfrac{40}{6} \cdot \dfrac{10}{1}$

   $B = \dfrac{400}{6}$

   $B = 66\dfrac{2}{3}$ cm ≈ 66.7 cm

B4)
$$\frac{D}{2} = \frac{30}{7}$$
$$\frac{D}{\cancel{2}} \cdot \frac{\cancel{2}}{1} = \frac{30}{7} \cdot \frac{2}{1}$$
$$D = \frac{60}{7}$$
$$D = 8\frac{4}{7} \text{ in.} \approx 8.6 \text{ in.}$$

B5)

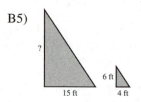

B6)

B7)
$$\frac{h}{6} = \frac{15}{4}$$
$$\frac{h}{\cancel{6}} \cdot \frac{\cancel{6}}{1} = \frac{15}{4} \cdot \frac{6}{1}$$
$$h = \frac{90}{4}$$
$$h = 22\frac{1}{2} \text{ ft}$$
The flagpole is 22.5 feet tall.

B8)
$$\frac{h}{20} = \frac{120}{25}$$
$$\frac{h}{\cancel{20}} \cdot \frac{\cancel{20}}{1} = \frac{120}{25} \cdot \frac{20}{1}$$
$$h = \frac{2400}{25}$$
$$h = 96 \text{ ft}$$
The radio tower is 96 feet tall.

C1) If an object is measured in meters, then square meters are used when writing the object's area.

C2) $\frac{x}{9} = \frac{9}{12}$; $\frac{x}{9} = \frac{6}{8}$; $\frac{x}{9} = \frac{3}{4}$

C3)
$$\frac{h}{3} = \frac{20}{8}$$
$$\frac{h}{\cancel{3}} \cdot \frac{\cancel{3}}{1} = \frac{20}{8} \cdot \frac{3}{1}$$
$$h = \frac{60}{8}$$
$$h = 7.5 \text{ ft}$$

$$A = \frac{1}{2} \cdot b \cdot h$$
$$= \frac{1}{2} \cdot 20 \cdot 7.5$$
$$= 75 \text{ ft}^2$$

C4)
$$\frac{h}{12} = \frac{3}{24}$$
$$\frac{h}{\cancel{12}} \cdot \frac{\cancel{12}}{1} = \frac{3}{24} \cdot \frac{12}{1}$$
$$h = \frac{36}{24}$$
$$h = 1.5 \text{ ft}$$

$$A = \frac{1}{2} \cdot b \cdot h$$
$$= \frac{1}{2} \cdot 3 \cdot 1.5$$
$$= 2.25 \text{ ft}^2 \approx 2.3 \text{ ft}^2$$

C5)
$$\frac{h}{2} = \frac{1}{2} \qquad\qquad \frac{B}{5} = \frac{1}{2}$$
$$\frac{h}{\cancel{2}} \cdot \frac{\cancel{2}}{1} = \frac{1}{2} \cdot \frac{2}{1} \qquad \frac{B}{\cancel{5}} \cdot \frac{\cancel{5}}{1} = \frac{1}{2} \cdot \frac{5}{1}$$
$$h = 1 \text{ cm} \qquad\qquad B = \frac{5}{2}$$
$$B = 2.5 \text{ cm}$$

$$A = \frac{1}{2} \cdot (B + b) \cdot h$$
$$= \frac{1}{2} \cdot (2.5 + 1) \cdot 1$$
$$= \frac{1}{2} \cdot (3.5) \cdot 1$$
$$= 1.75 \text{ cm}^2 \approx 1.8 \text{ cm}^2$$

C6)
$$\frac{b}{20} = \frac{2}{8} \qquad\qquad \frac{B}{24} = \frac{2}{8}$$
$$\frac{b}{\cancel{20}} \cdot \frac{\cancel{20}}{1} = \frac{2}{8} \cdot \frac{20}{1} \qquad \frac{B}{\cancel{24}} \cdot \frac{\cancel{24}}{1} = \frac{2}{8} \cdot \frac{24}{1}$$
$$b = \frac{40}{8} \qquad\qquad B = \frac{48}{8}$$
$$b = 5 \text{ m} \qquad\qquad B = 6 \text{ m}$$

$$A = \frac{1}{2} \cdot (B + b) \cdot h$$
$$= \frac{1}{2} \cdot (6 + 5) \cdot 2$$
$$= \frac{1}{2} \cdot (11) \cdot 2$$
$$= 11 \text{ m}^2$$

C7) $\dfrac{w}{7} = \dfrac{50}{12}$

$12w = 350$

$\dfrac{12w}{12} = \dfrac{350}{12}$

$w = 29\dfrac{1}{6}$ ft

$A = l \cdot w$

$= 50 \cdot 29\dfrac{1}{6}$

$= 1{,}458\dfrac{1}{3}$ ft$^2 \approx 1{,}458.3$ ft^2

The area of the room is about 1,458.3 square feet.

C8) $\dfrac{h}{6} = \dfrac{20}{8}$

$8h = 120$

$\dfrac{8h}{h} = \dfrac{120}{8}$

$h = 15$ ft

$A = \dfrac{1}{2} \cdot b \cdot h$

$= \dfrac{1}{2} \cdot 20 \cdot 15$

$= 150$ ft^2

The area of the patio is 150 square feet.

SECTION 7.7   EXERCISES

For 1–9, refer to Concept Checks and Practice Exercises.

11) similar objects

13) proportion

15) a) The shape is similar to a).

b) A corresponds to K.
B corresponds to L.
C corresponds to J.

17) a) The shape is similar to d).

b) A corresponds to J.
B corresponds to K.

19) a) The shape is similar to c).

b) A corresponds to M.
B corresponds to J.
C corresponds to K.
D corresponds to L.

21) $\dfrac{x}{4} = \dfrac{8}{5}$

$\dfrac{x}{\cancel{4}} \cdot \dfrac{\cancel{4}}{1} = \dfrac{8}{5} \cdot \dfrac{4}{1}$

$x = \dfrac{32}{5}$

$x = 6.4$ in.

23) $\dfrac{x}{13} = \dfrac{8}{5}$

$\dfrac{x}{\cancel{13}} \cdot \dfrac{\cancel{13}}{1} = \dfrac{8}{5} \cdot \dfrac{13}{1}$

$x = \dfrac{104}{5}$

$x = 20.8$ in.

25) $\dfrac{h}{8} = \dfrac{35}{7}$

$\dfrac{h}{\cancel{8}} \cdot \dfrac{\cancel{8}}{1} = \dfrac{35}{7} \cdot \dfrac{8}{1}$

$h = \dfrac{280}{7}$

$h = 40$ ft

The height of the flagpole is 40 feet.

27) $\dfrac{h}{9} = \dfrac{8}{5}$

$\dfrac{h}{\cancel{9}} \cdot \dfrac{\cancel{9}}{1} = \dfrac{8}{5} \cdot \dfrac{9}{1}$

$h = \dfrac{72}{5}$

$h = 14.4$ m

$A = \dfrac{1}{2} \cdot b \cdot h$

$= \dfrac{1}{2} \cdot 8 \cdot 14.4$

$= 57.6$ m^2

29) $\dfrac{h}{7} = \dfrac{24}{15}$

$\dfrac{h}{\cancel{7}} \cdot \dfrac{\cancel{7}}{1} = \dfrac{24}{15} \cdot \dfrac{7}{1}$

$h = \dfrac{168}{15}$

$h = 11.2$ yd

$A = \dfrac{1}{2} \cdot b \cdot h$

$= \dfrac{1}{2} \cdot 24 \cdot 11.2$

$= 134.4$ yd^2

31) $\dfrac{l}{7} = \dfrac{7}{4}$

$4l = 49$

$\dfrac{4l}{4} = \dfrac{49}{4}$

$l = 12.25$ ft ≈ 12.3 ft

The van is about 12.3 feet long.

33) $\dfrac{w}{5} = \dfrac{15}{6}$

$6w = 75$

$\dfrac{6w}{6} = \dfrac{75}{6}$

$w = 12.5$ ft

The room is 12.5 feet wide.

$A = l \cdot w$

$ = 15 \cdot 12.5$

$ = 187.5$ ft^2

187.5 square feet of carpet are needed.

35) $\dfrac{d}{20} = \dfrac{1.3}{3}$

$3d = 26$

$\dfrac{3d}{3} = \dfrac{26}{3}$

$d \approx 8.7$ mi

The island is about 8.7 miles away.

37) $\dfrac{d}{1} = \dfrac{40,000,000,000,000}{400,000}$

$d = 100,000,000$ cm

$ = 1,000$ km

In the model, the star must be 1,000 kilometers from the sun.

39) a) The ratio of the lengths of the sides of the two squares is 3:1.

b) The perimeter of the larger square is $3+3+3+3 = 12$ ft.
The perimeter of the smaller square is $1+1+1+1 = 4$ ft.
The ratio of the perimeters of the two squares is 12:4 or 3:1.

c) The area of the larger square is 9 ft^2.
The area of the smaller square is 1 ft^2.
The ratio of the areas of the two squares is 9:1.

41) a) The ratio of the lengths of the sides of the two squares is 5:1.

b) The perimeter of the larger square is $5+5+5+5 = 20$ ft.
The perimeter of the smaller square is $1+1+1+1 = 4$ ft.
The ratio of the perimeters of the two squares is 20:4 or 5:1.

c) The area of the smaller square is 1 ft^2.
The area of the larger square is 25 ft^2.
The ratio of the areas of the two squares is 25:1.

43) The area will be multiplied by a factor of n^2.

45) a) $\dfrac{5.4}{3} = 1.8$

The ratio of the sides of the two shapes is 1:1.8.

b) The area will be multiplied by the factor $1.8^2 = 3.24$.

$3.24 \cdot 4.8$ ft$^2 = 15.552$ ft^2
The area of the enlarged figure is 15.552 square feet.

47) If the "one-fourth-size" photo is one-fourth as wide and one-fourth as tall, then 16 one-fourth-size photos can be printed on one sheet of paper.

Note: $\left(\dfrac{1}{4}\right)^2 = \dfrac{1}{16}$.

CHAPTER 7   REVIEW EXERCISES

1) $m\angle AOD = 170° - 10°$
$ = 160°$

2) $m\angle BOD = 105° - 10°$
$ = 95°$

3) The three acute angles are $\angle AOB$, $\angle BOC$ and $\angle COD$.

4) The two obtuse angles are $\angle AOD$ and $\angle BOD$.

5) The right angle is $\angle AOC$.

6) Since $m\angle BOE = 100°$, and ray B is drawn at 105°, then ray E must be drawn at 5°. Refer to the figure below.

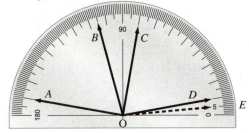

7) $m\angle DOE = 180° - \left(20° + 90°\right)$
$$= 180° - 110°$$
$$= 70°$$

8) $m\angle BOE = 180° - 40°$
$$= 140°$$

9) $m\angle AOD = 180° - m\angle DOE$
$$= 180° - 70°$$
$$= 110°$$

10) $m\angle BOD = m\angle AOD - m\angle AOB$
$$= 110° - 40°$$
$$= 70°$$

11) The shape is an equilateral triangle.

12) The shape is a right scalene triangle.

13) The shape is a rectangle.

14) The shape is a parallelogram.

15) The shape is a regular hexagon.

16) The shape is a trapezoid.

17) The shape is not a polygon. A circle is not a polygon.

18) The shape is an acute isosceles triangle.

19) The shape is a regular octagon.

20) The shape is an acute scalene triangle.

21) $P = 5 + 8 + 5 + 8$
$$= 13 + 5 + 8$$
$$= 18 + 8$$
$$= 26 \text{ in.}$$

22) $P = 3 + 12 + 9$
$$= 15 + 9$$
$$= 24 \text{ cm}$$

23) The missing horizontal length is $15 \text{ cm} - 8 \text{ cm} = 7 \text{ cm}$. The missing vertical length is $6 \text{ cm} - 2 \text{ cm} = 4 \text{ cm}$.

$P = 15 + 2 + 7 + 4 + 8 + 6$
$$= 17 + 7 + 4 + 8 + 6$$
$$= 24 + 4 + 8 + 6$$
$$= 28 + 8 + 6$$
$$= 36 + 6$$
$$= 42 \text{ cm}$$

24) The missing horizontal length is $9 \text{ ft} - 7 \text{ ft} = 2 \text{ ft}$. The missing vertical length is $5 \text{ ft} - 2 \text{ ft} = 3 \text{ ft}$.

$P = 7 + 3 + 2 + 2 + 9 + 5$
$$= 10 + 2 + 2 + 9 + 5$$
$$= 12 + 2 + 9 + 5$$
$$= 14 + 9 + 5$$
$$= 23 + 5$$
$$= 28 \text{ ft}$$

25) $A = b \cdot h$
$$= 15 \cdot 9$$
$$= 135 \text{ yd}^2$$

26) $A = \frac{1}{2} \cdot (B + b) \cdot h$
$$= \frac{1}{2} \cdot (18 + 10) \cdot 6$$
$$= \frac{1}{2} \cdot (28) \cdot 6$$
$$= 14 \cdot 6$$
$$= 84 \text{ cm}^2$$

27) The shape is made of two rectangles.

The missing horizontal length is
20 ft − 12 ft = 8 ft.
The missing vertical length is
10 ft − 5 ft = 5 ft.

$$A = l_1 \cdot w_1 + l_2 \cdot w_2$$
$$= (12)(5) + (20)(5)$$
$$= 60 + 100$$
$$= 160 \text{ ft}^2$$

28) The shape is made of a rectangle and a triangle.

The missing height of the triangle is
16 in. − 8 in. = 8 in.

$$A = l \cdot w + \frac{1}{2} \cdot b \cdot h$$
$$= 20 \cdot 8 + \frac{1}{2} \cdot 20 \cdot 8$$
$$= 160 + 80$$
$$= 240 \text{ in.}^2$$

29) $r = 5$ in.

$$A = \pi r^2$$
$$= \pi \cdot 5^2$$
$$\approx 3.14 \cdot 25$$
$$\approx 78.5 \text{ in.}^2$$

30) $r = \frac{1}{2} \cdot 6$ in. $= 3$ in.

$$A = \pi r^2$$
$$= \pi \cdot 3^2$$
$$\approx 3.14 \cdot 9$$
$$\approx 28.26 \text{ in.}^2$$

31) $d = 20$ ft

$$C = \pi d$$
$$= \pi \cdot 20$$
$$\approx 3.14 \cdot 20$$
$$\approx 62.8 \text{ ft}$$

32) $r = 8$ m

$$C = 2\pi r$$
$$= 2 \cdot \pi \cdot 8$$
$$\approx 16 \cdot 3.14$$
$$\approx 50.24 \text{ m}$$

33) $d = 12$ in.

$$C = \frac{1}{2} \cdot \pi d + 12$$
$$= \frac{1}{2} \cdot \pi \cdot 12 + 12$$
$$\approx 6 \cdot 3.14 + 12$$
$$\approx 18.84 + 12$$
$$\approx 30.84 \text{ in.}$$

34) $r = \frac{1}{2} \cdot 14$ mm $= 7$ mm

$$A = \pi r^2$$
$$= \pi \cdot 7^2$$
$$\approx 3.14 \cdot 49$$
$$\approx 153.86 \text{ mm}^2$$

35) $r = 7$ cm

$$\text{Area} = A_{circle} - A_{square}$$
$$= \pi \cdot 7^2 - 2 \cdot 2$$
$$\approx 3.14 \cdot 49 - 2 \cdot 2$$
$$\approx 153.86 - 4$$
$$\approx 149.86 \text{ cm}^2$$

36) The shape is made of a half circle and a rectangle.

$$r = \frac{1}{2} \cdot 8 \text{ ft} = 4 \text{ ft}$$

$$\text{Perimeter} = \frac{1}{2} \text{circumference} + 3 \text{ sides of rectangle}$$

$$P = \frac{1}{2} \cdot 2\pi r + 12 + 8 + 12$$
$$\approx 0.5 \cdot 2 \cdot 3.14 \cdot 4 + 12 + 8 + 12$$
$$\approx 12.56 + 32$$
$$\approx 44.56 \text{ ft}$$

37) The shape is made of a half circle and a triangle.

$$r = \frac{1}{2} \cdot 8 \text{ cm} = 4 \text{ cm}$$

$$\text{Perimeter} = \frac{1}{2} \text{circumference} + 2 \text{ sides of triangle}$$

$$P = \frac{1}{2} \cdot 2\pi r + 15 + 15$$
$$\approx 0.5 \cdot 2 \cdot 3.14 \cdot 4 + 15 + 15$$
$$\approx 12.56 + 30$$
$$\approx 42.56 \text{ cm}$$

38) $r = \frac{1}{2} \cdot 8 \text{ ft} = 4 \text{ ft}$

$$\text{Area} = A_{\text{rectangle}} + \frac{1}{2} A_{\text{circle}}$$
$$= 12 \cdot 8 + \frac{1}{2} \pi (4)^2$$
$$\approx 12 \cdot 8 + \frac{1}{2} \cdot 3.14 \cdot 16$$
$$\approx 96 + 25.12$$
$$\approx 121.12 \text{ ft}^2$$

39) $r = \frac{1}{2} \cdot 8 \text{ cm} = 4 \text{ cm}$

$$\text{Area} = A_{\text{triangle}} + \frac{1}{2} A_{\text{circle}}$$
$$= \frac{1}{2} \cdot 8 \cdot 12 + \frac{1}{2} \pi (4)^2$$
$$\approx \frac{1}{2} \cdot 8 \cdot 12 + \frac{1}{2} \cdot 3.14 \cdot 16$$
$$\approx 48 + 25.12$$
$$\approx 73.12 \text{ cm}^2$$

40) Since there are 17 cubes, the volume is 17 ft^3 or 17 cubic feet.

41) Since there are 27 cubes, the volume is 27 cm^3 or 27 cubic centimeters.

42) $r = \frac{1}{2} \cdot 3 \text{ ft} = \frac{3}{2} \text{ ft} = 1.5 \text{ ft}$

$$V = \pi \cdot r^2 \cdot h$$
$$\approx 3.14 \cdot (1.5)^2 \cdot 40$$
$$\approx 3.14 \cdot 2.25 \cdot 40$$
$$\approx 3.14 \cdot 90$$
$$\approx 282.6 \text{ ft}^3$$
The volume of one column is 282.6 cubic feet.

43) $r = \frac{1}{2} \cdot 20 \text{ cm} = 10 \text{ cm}$

$$V = \frac{4}{3} \cdot \pi \cdot r^3$$
$$\approx \frac{4}{3} \cdot 3.14 \cdot 10^3$$
$$\approx \frac{4}{3} \cdot 3.14 \cdot 1,000$$
$$\approx 4,186 \frac{2}{3} \text{ cm}^3$$
$$\approx 4,187 \text{ cm}^3$$
The volume of the sphere is approximately 4,187 cubic centimeters.

44) $V = \frac{1}{3} \cdot l \cdot w \cdot h$
$$= \frac{1}{3} \cdot 140 \cdot 140 \cdot 120$$
$$= \frac{1}{3} \cdot 2,352,000$$
$$= 784,000 \text{ ft}^3$$
The volume of the pyramid is 784,000 cubic feet.

45) $r = \frac{1}{2} \cdot 40 \text{ yd} = 20 \text{ yd}$

$$V = \frac{1}{3} \cdot \pi \cdot r^2 \cdot h$$
$$\approx \frac{1}{3} \cdot 3.14 \cdot 20^2 \cdot 15$$
$$= \frac{1}{3} \cdot 3.14 \cdot 400 \cdot 15$$
$$= 3.14 \cdot 2,000$$
$$= 6,280 \text{ yd}^3$$
There are 6,280 cubic yards of sand.

46) Since $3^2 = 9$ and $4^2 = 16$, $\sqrt{13}$ is between 3 and 4.

47) Since $5^2 = 25$ and $6^2 = 36$, $\sqrt{29}$ is between 5 and 6.

48) Since $9^2 = 81$ and $10^2 = 100$, $\sqrt{85}$ is between 9 and 10.

49) Since $7^2 = 49$ and $8^2 = 64$, $\sqrt{56}$ is between 7 and 8.

50) $\sqrt{36} = 6$

51) $\sqrt{49} = 7$

52) $\sqrt{\frac{4}{81}} = \frac{2}{9}$

53) $\sqrt{\frac{64}{25}} = \frac{8}{5}$

54) $a = 12$, $b = $ unknown, $c = 13$

$$a^2 + b^2 = c^2$$
$$12^2 + b^2 = 13^2$$
$$144 + b^2 = 169$$
$$144 - 144 + b^2 = 169 - 144$$
$$b^2 = 25$$
$$b = \sqrt{25}$$
$$b = 5 \text{ in.}$$

$$A = \frac{1}{2} \cdot b \cdot h$$
$$= \frac{1}{2} \cdot 12 \cdot 5$$
$$= 30 \text{ in.}^2$$

55) $a = 3$, $b = $ unknown, $c = 5$

$$a^2 + b^2 = c^2$$
$$3^2 + b^2 = 5^2$$
$$9 + b^2 = 25$$
$$9 - 9 + b^2 = 25 - 9$$
$$b^2 = 16$$
$$b = \sqrt{16}$$
$$b = 4 \text{ cm}$$

$$A = \frac{1}{2} \cdot b \cdot h$$
$$= \frac{1}{2} \cdot 4 \cdot 3$$
$$= 6 \text{ cm}^2$$

56) $a = 3$, $b = 8$, $c = $ unknown

$$a^2 + b^2 = c^2$$
$$3^2 + 8^2 = c^2$$
$$9 + 64 = c^2$$
$$73 = c^2$$
$$\sqrt{73} = c$$
$$8.5 \text{ m} \approx c$$

$$P \approx 8 + 3 + 8.5$$
$$\approx 11 + 8.5$$
$$\approx 19.5 \text{ m}$$

57) $a = 4$, $b = 7$, $c = $ unknown

$$a^2 + b^2 = c^2$$
$$4^2 + 7^2 = c^2$$
$$16 + 49 = c^2$$
$$65 = c^2$$
$$\sqrt{65} = c$$
$$8.1 \text{ ft} \approx c$$

$$P \approx 7 + 4 + 8.1$$
$$\approx 11 + 8.1$$
$$\approx 19.1 \text{ ft}$$

58) $$\frac{h}{6} = \frac{50}{15}$$
$$\frac{h}{6} \cdot \frac{6}{1} = \frac{50}{15} \cdot \frac{6}{1}$$
$$h = \frac{300}{15}$$
$$h = 20 \text{ ft}$$

The height of the flagpole is 20 feet.

59) $$\frac{h}{4} = \frac{9}{1.5}$$
$$\frac{h}{4} \cdot \frac{4}{1} = \frac{9}{1.5} \cdot \frac{4}{1}$$
$$h = \frac{36}{1.5}$$
$$h = 24 \text{ ft}$$

The height of the wall is 24 feet.

CHAPTER 7   PRACTICE TEST

1) a) $m\angle BOC = 120° - 70°$
$$= 50°$$

   b) The angle is acute.

2) a) $m\angle AOC = 160° - 70°$
$$= 90°$$

   b) The angle is a right angle.

3) a) $m\angle AOD = 160° - 30°$
$$= 130°$$

   b) The angle is obtuse.

4) $90° - m\angle BOC = 90° - 50°$
$$= 40°$$

5) $180° - m\angle BOC = 180° - 50°$
$$= 130°$$

6) The shape is a regular octagon.

7) The shape is a trapezoid.

8) The shape is a rhombus.

9) The shape is a right scalene triangle.

10) $a = 4$ m, $b = 9$ m, $c = $ unknown

$$a^2 + b^2 = c^2$$
$$4^2 + 9^2 = c^2$$
$$16 + 81 = c^2$$
$$97 = c^2$$
$$\sqrt{97} = c$$
$$9.8 \text{ m} \approx c$$

$$P \approx 4 + 9 + 9.8$$
$$\approx 13 + 9.8$$
$$\approx 22.8 \text{ m}$$

11) $P = 11 + 4 + 11 + 4$
$$= 15 + 11 + 4$$
$$= 26 + 4$$
$$= 30 \text{ ft}$$

12) $P = 8 \cdot 7$ cm
$$= 56 \text{ cm}$$

13) $P = 7 + 4 + 7 + 4$
$$= 11 + 7 + 4$$
$$= 18 + 4$$
$$= 22 \text{ ft}$$

14) The shape is made of a rectangle and a triangle.

$$A = l \cdot w + \frac{1}{2} \cdot b \cdot h$$
$$= 16 \cdot 10 + \frac{1}{2} \cdot 16 \cdot 8$$
$$= 160 + 64$$
$$= 224 \text{ ft}^2$$

15) The shape is made of two rectangles.

The missing horizontal length is
5 ft $-$ 4 ft $= 1$ ft.
The missing vertical length is
3 ft $-$ 1 ft $= 2$ ft.

$$A = l_1 \cdot w_1 + l_2 \cdot w_2$$
$$= (4)(3) + (1)(1)$$
$$= 12 + 1$$
$$= 13 \text{ ft}^2$$

16) $A = \frac{1}{2} \cdot (B + b) \cdot h$
$$= \frac{1}{2} \cdot (11 + 7) \cdot 4$$
$$= \frac{1}{2} \cdot (18) \cdot 4$$
$$= 9 \cdot 4$$
$$= 36 \text{ in.}^2$$

17) $a = 8$ in., $b = $ unknown, $c = 10$ in.

$$a^2 + b^2 = c^2$$
$$8^2 + b^2 = 10^2$$
$$64 + b^2 = 100$$
$$64 - 64 + b^2 = 100 - 64$$
$$b^2 = 36$$
$$b = \sqrt{36}$$
$$b = 6 \text{ in.}$$

$$A = \frac{1}{2} \cdot b \cdot h$$
$$= \frac{1}{2} \cdot 8 \cdot 6$$
$$= 24 \text{ in.}^2$$

18) Since there are 22 cubes, the volume is 22 m^3 or 22 cubic meters.

19) $r = \frac{1}{2} \cdot 4$ in. $= 2$ in.

$$V = \frac{4}{3} \cdot \pi \cdot r^3$$
$$\approx \frac{4}{3} \cdot 3.14 \cdot 2^3$$
$$\approx \frac{4}{3} \cdot 3.14 \cdot 8$$
$$\approx 33.5 \text{ in.}^3$$
The volume of the sphere is 33.5 cubic inches.

20) $r = \frac{1}{2} \cdot 10$ m $= 5$ m

$$V = \pi \cdot r^2 \cdot h$$
$$\approx 3.14 \cdot (5)^2 \cdot 5$$
$$\approx 3.14 \cdot 25 \cdot 5$$
$$\approx 3.14 \cdot 125$$
$$\approx 392.5 \text{ m}^3$$
The volume of the cylinder is approximately 392.5 cubic meters.

21) $V = \frac{1}{3} \cdot l \cdot w \cdot h$

$\quad\quad = \frac{1}{3} \cdot 20 \cdot 14 \cdot 12$

$\quad\quad = \frac{1}{3} \cdot 3{,}360$

$\quad\quad = 1{,}120 \text{ ft}^3$

The volume of the pyramid is $1{,}120 \text{ ft}^3$.

22) Since $8^2 = 64$ and $9^2 = 81$, $\sqrt{73}$ is between
8 and 9.

23) Since $1^2 = 1$ and $2^2 = 4$, $\sqrt{3}$ is between
1 and 2.

24) $\sqrt{\frac{9}{25}} = \frac{3}{5}$

25) $\sqrt{\frac{81}{16}} = \frac{9}{4}$

26) $\sqrt{31} \approx 5.57$

27) $\sqrt{21} \approx 4.58$

28) $\quad\quad \frac{h}{8} = \frac{27}{16}$

$\quad\quad \frac{h}{\cancel{8}} \cdot \frac{\cancel{8}}{1} = \frac{27}{16} \cdot \frac{8}{1}$

$\quad\quad\quad h = \frac{216}{16}$

$\quad\quad\quad h = 13.5 \text{ ft}$

The height of the flagpole is 13.5 feet.

29) $\quad\quad \frac{b}{2.2} = \frac{20}{5}$

$\quad\quad \frac{b}{\cancel{2.2}} \cdot \frac{\cancel{2.2}}{1} = \frac{20}{5} \cdot \frac{2.2}{1}$

$\quad\quad\quad b = \frac{44}{5}$

$\quad\quad\quad b = 8.8 \text{ ft}$

The measure of b is 8.8 feet.

Chapter 8   Statistics

8.1   Reading Graphs

GUIDED PRACTICE

1) Since the smallest pie-shaped sector of the circle is labeled "Other," that field of study has the fewest students.

2) The number next to the transfer sector is 3,000. That is the number of students planning to transfer.

3) 4,700 out of 10,000 students are studying nursing.

$$\frac{\text{nursing students}}{\text{total students}} = \frac{4,700}{10,000}$$
$$= \frac{47 \cdot \cancel{100}}{100 \cdot \cancel{100}}$$
$$= \frac{47}{100}$$
$$= 47\%$$

47% of the students are studying nursing.

4) Number of manufacturing students: 1,100
Number of nursing students: 4,700

$$\frac{\text{manufacturing students}}{\text{nursing students}} = \frac{1,100}{4,700}$$
$$= \frac{11 \cdot \cancel{100}}{47 \cdot \cancel{100}}$$
$$= \frac{11}{47}$$

The ratio of manufacturing students to nursing students is 11 to 47. For every 11 manufacturing students, there are 47 nursing students.

5) Since the largest sector of the circle is labeled "South", that region contributed the most viewers. The South contributed 39% of the overall viewers.

6) North Central: 21%
South: 39%

$21\% + 39\% = 60\%$
60% of the viewers were from the North Central or South regions.

7) 39% of 1,045 people were from the South.

$$\text{percent} \cdot \text{base} = \text{amount}$$
$$0.39 \cdot 1,045 = 407.55$$
$$\approx 408$$

408 of the people polled were from the South.

8) The yellow bar has the greatest height. Looking at the graph, we see that the yellow bar represents Janet Jackson. Janet Jackson has earned 26 gold singles in the United States.

9) Mariah Carey has 25 gold singles.
The Temptations have 10 gold singles.

$25 - 10 = 15$
Mariah Carey has 15 more gold singles than The Temptations.

10) Since the highest point on the graph occurred in April, that was the month with the most customers.

11) March had approximately 4,000 customers. April had approximately 6,000 customers.

$6,000 - 4,000 = 2,000$
There was an increase of 2,000 customers from March to April.

12) The line goes down at the steepest angle between the months of April and May, so this represents the largest decrease in customers.

13)

Grade/Score	Tally	Frequency
A 90–99	‖	2
B 80–89	卌	5
C 70–79	‖‖	4
D 60–69	∣	1
F 50–59		0

Histogram of Grades/Scores

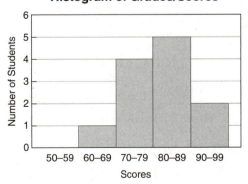

14) Because the class interval 80–89 has the highest frequency, the grade of B was assigned the most. There were more Bs than any other grade.

15)

Work Bonus	Tally	Frequency
101–150	|||	3
151–200	||	2
201–250	|||||	5
251–300	||	2
301–350	|	1

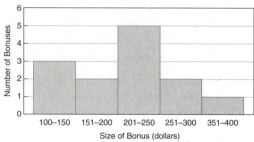

Histogram of Raphael's Work Bonuses

16) The class frequency is highest for the class interval of $201 to $250. Raphael was more likely to get a bonus in this class interval than in any other.

CONCEPT CHECKS AND PRACTICE EXERCISES

A1) Answers may vary. Example: The reader can clearly see that the parts, when "glued" together, will be a whole.

A2) The pie shaped pieces that make up a circle graph are called sectors.

A3) $36 + 36 + 28 = 100$
The U.S. won 100 medals

A4) $36 - 28 = 8$
There were 8 more silver medals won than bronze medals.

A5) 36 out of 100 medals were silver.

$$\frac{\text{silver medals}}{\text{total medals}} = \frac{36}{100} = 0.36$$
36% of the medals were silver.

A6) 28 out of 100 medals were bronze.

$$\frac{\text{bronze medals}}{\text{total medals}} = \frac{28}{100} \approx 0.28$$
28% of the medals were bronze.

A7) Number of bronze medals: 28
Number of total medals: 100

$$\frac{\text{bronze medals}}{\text{total medals}} = \frac{28}{100}$$
$$= \frac{\cancel{4} \cdot 7}{\cancel{4} \cdot 25}$$
$$= \frac{7}{25}$$

The ratio of bronze medals to total medals is 7 to 25. For every 7 bronze medals, there are 25 total medals.

A8) Number of gold medals: 36
Number of bronze medals: 28

$$\frac{\text{gold medals}}{\text{bronze medals}} = \frac{36}{28}$$
$$= \frac{\cancel{4} \cdot 9}{\cancel{4} \cdot 7}$$
$$= \frac{9}{7}$$

The ratio of gold medals to bronze medals is 9 to 7. For every 9 gold medals, there are 7 bronze medals.

A9) Randy Jackson: 26%
Paula Abdul: 6%

$26\% + 6\% = 32\%$
32% of the total viewers were influenced by either Randy Jackson or Paula Abdul.

A10) Simon Cowell: 58%
Paula Abdul: 6%

$58\% - 6\% = 52\%$
52% more viewers were influenced by Simon Cowell than Paula Abdul.

A11) 6% of 1,045 people surveyed said they were influenced by Paula Abdul.

$$\text{percent} \cdot \text{base} = \text{amount}$$
$$0.06 \cdot 1,045 = 62.7$$
$$\approx 63$$
About 63 of the people surveyed said they were influenced by Paula Abdul.

A12) 10% of 1,045 people surveyed said the judges did not influence their opinion.

$$\text{percent} \cdot \text{base} = \text{amount}$$
$$0.10 \cdot 1045 = 104.5$$
$$\approx 105$$
About 105 of the people surveyed said the judges did not influence their opinion.

B1)   A line graph includes data points connected by lines to show a trend.

B2)   Answers may vary. Example: Enrollment at several area colleges.

B3)   The average salary for a nurse working in Illinois is $55,000.

B4)   The average salary for a nurse working in Florida is $52,000.

B5)   The average nursing salary is Texas is $53,500. The average nursing salary is Georgia is $52,000.

$$53,500 - 52,000 = 1,500$$
The average nursing salary is $1,500 more in Texas than Georgia.

B6)   The average nursing salary is Michigan is $53,000.
The average nursing salary is North Carolina is $45,000.

$$53,000 - 45,000 = 8,000$$
The average nursing salary is $8,000 more in Michigan than North Carolina.

B7)   The highest point on the graph occurred for California.
The lowest point on the graph occurred for North Carolina.

California and North Carolina have the largest difference in nursing salaries.

B8)   Florida and Georgia both have average nursing salaries of $52,000.

Florida and Georgia have the smallest difference in nursing salaries.

B9)   The approximate outside temperature at 6 P.M. was $10°$ F.

B10)   The approximate inside temperature at 6 P.M. was $62°$ F.

B11) $50°F - 0°F = 50°F$
The inside temperature was approximately $50°$ F warmer than the outside temperature at midnight.

B12) $45°F - 5°F = 40°F$
The inside temperature was approximately $40°$ F warmer than the outside temperature at 6 a.m.

C1)   The statement, "A class interval of 6–10 includes five whole numbers" is true because there are five whole numbers in the interval. They are 6, 7, 8, 9 and 10.

C2)   The statement, "A class interval of 7–12 includes five whole numbers" is false because there are six whole numbers in the interval. They are 7, 8, 9, 10, 11, and 12.

C3)   There are 30 whole numbers in the interval of 31–60, including the endpoints 31 and 60.

$$30 \div 10 = 3$$
To make a histogram with class intervals of size 10, three class intervals will be needed.

C4)   The lowest value is 31.
The highest value is 60.

C5)

Intervals	Tally	Frequency
31–36	|||	3
37–42	|||	3
43–48	||||	4
49–54	||	2
55–60	|||	3

C6)

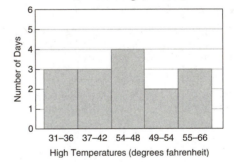

Histogram of Daily High Temperatures in Anchorage, Alaska

SECTION 8.1   EXERCISES

For 1–9, refer to Concept Checks and Practice Exercises.

11) circle graph

13) histogram

15) sectors

17) Valentine's Day has the third most floral sales.

19) $75 - 45 = 30$
30 million more floral sales take place around Christmas than Valentine's Day.

21) $\dfrac{45}{75+55+45+38+15+9} = \dfrac{45}{237}$
≈ 0.19
Approximately 19% of floral sales take place around Valentine's Day.

23) Most of the penalty fees are in the late fees category.

25) $70\% + 15\% = 85\%$
85% of the penalty fees were charged for late fees and over-the-limit fees combined.

27) $\text{percent} \cdot \text{base} = \text{amount}$
$0.70 \cdot 17.1 \text{ billion} = 11.97 \text{ billion}$
$\approx 12.0 \text{ billion}$
In 2006, approximately $12 billion was collected in late fees.

29) $45 - 26 = 19$
The hybrid Honda Civic gets 19 more miles per gallon than the standard Ford Escape.

31) Since the difference in bar heights is the greatest, the Honda Civic standard and hybrid models have the greatest difference in fuel economy.

33) a) $495 \div 34 \approx 14.56 \approx 15$
It would take about 15 gallons of gas to drive the hybrid Ford Escape 495 miles.

   b) $495 \div 26 \approx 19.04 \approx 19$
It would take approximately 19 gallons of gas to drive the standard Ford Escape 495 miles.

35) 49% of respondents said a cell phone is a necessity.

37) $98\% - 35\% = 63\%$
There was a difference of 63% between respondents who owned a TV and respondents who said a TV is a luxury.

39) Looking at the highest bars, we see that the top three items that respondents thought were necessities are a car, a clothes washer, and a TV set.

41) Answers may vary.

43) $2:26:11$
$= 2 \cancel{\text{h}} \cdot \dfrac{3{,}600 \text{ s}}{1 \cancel{\text{h}}} + 26 \cancel{\text{min}} \cdot \dfrac{60 \text{ s}}{1 \cancel{\text{min}}} + 11 \text{ s}$
$= 7{,}200 \text{ s} + 1{,}560 \text{ s} + 11 \text{ s}$
$= 8{,}771 \text{ s}$

$2:22:43$
$= 2 \cancel{\text{h}} \cdot \dfrac{3{,}600 \text{ s}}{1 \cancel{\text{h}}} + 22 \cancel{\text{min}} \cdot \dfrac{60 \text{ s}}{1 \cancel{\text{min}}} + 43 \text{ s}$
$= 7{,}200 \text{ s} + 1{,}320 \text{ s} + 43 \text{ s}$
$= 8{,}563 \text{ s}$

$8{,}771 \text{ s} - 8{,}563 \text{ s} = 208 \text{ s}$

$208 \text{ s} \cdot \dfrac{1 \text{ min}}{60 \text{ s}} = 3\dfrac{7}{15} \text{ min}$
$= 3 \text{ min} + \dfrac{7}{15} \text{ min} \cdot \dfrac{60 \text{ s}}{1 \text{ min}}$
$= 3 \text{ min } 28 \text{ s}$
Joan Benoit reduced her best marathon time by 3 minutes, 28 seconds.

45) Since the line graph goes down from left to right, there is a decrease in energy-efficiency spending from 1996 to 1998.

47) $1.4 - 1.2 = 0.2$
The approximate change in spending from 1996 to 2004 was $0.2 billion.

49) The approximate rainfall is Seattle during the month of September was about 2 inches.

51) $4.3 - 0.3 = 4.0$
In July, about 4 more inches of rain fell in Indianapolis than Seattle.

53) The rainfall in Indianapolis was greater than the rainfall in Seattle for seven months.

55) Five students are age 28 to 33.

57) $6 + 8 + 5 = 19$
     19 students are less than 34 years old.

59) a) The lowest value is 60.
         The highest value is 77.
         The range of the data values is 60 to 77.

     b) The six intervals of 3 inches each are as
         follows.
         60–62
         63–65
         66–68
         69–71
         72–74
         75–77

Height	Tally	Frequency
60–62	||||	4
63–65	||||	4
66–68	|||| |	6
69–71	||||	5
72–74	|||	3
75–77	||	2

     c) c.
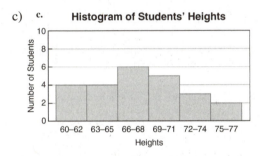

     d) Based on the highest bar, the range of
         heights 66–68 inches describes the most
         students.

8.2   Mean, Median, and Mode

GUIDED PRACTICE

1) $\dfrac{\text{Sum of Values}}{\text{Number of Values}} = \dfrac{2+8+3+6+3+2}{6}$
    $= \dfrac{24}{6}$
    $= 4$
    The mean of the set of data is 4.

2) $\dfrac{\text{Sum of Values}}{\text{Number of Values}} = \dfrac{5+7+4+2+2}{5}$
    $= \dfrac{20}{5}$
    $= 4$
    The mean of the set of data is 4.

3) $\dfrac{\text{Sum of Values}}{\text{Number of Values}} = \dfrac{8+6+11+7+1+2+2}{7}$
    $= \dfrac{37}{7}$
    ≈ 5.29
    The mean rounded to the nearest tenth is 5.3.

4) Arrange in numerical order. $\Rightarrow$ 2, 3, 3, 6, 8
        Find the middle. $\Rightarrow$ 2̸, 3̸, 3, 6̸, 8̸
    The median of the set of data is 3.

5) Arrange in numerical order. $\Rightarrow$ 0, 2, 4, 11, 15, 17
        Find the middle. $\Rightarrow$ 0̸, 2̸, 4, 11, 1̸5̸, 1̸7̸

    Calculate the mean of the two middle numbers.
    $\dfrac{4+11}{2} = \dfrac{15}{2} = 7.5$
    The median is 7.5.

6) Since 3 occurs most often, the mode is 3.

7) {6, 3, 6, 2, 3, 3} = {2, 3, 3, 3, 6, 6}
    Since 3 occurs most often, the mode is 3.

8) {17, 0, 4, 2, 0, 17} = {0, 0, 2, 4, 17, 17}
    The data has two modes, 0 and 17.

9) {1, 0, 7, 4} = {0, 1, 4, 7}
    The data has no mode since every value appears
    the same number of times.

CONCEPT CHECKS AND PRACTICE EXERCISES

A1) Use the word "average" to remember the
     definition of mean.

A2) The mean must be between the values of 42 and
     80.

A3) Answers may vary. Example:
     Step 1: Find the sum of the values.
     Step 2: Divide the sum by the number of
     values.

A4) $\dfrac{\text{Sum of Values}}{\text{Number of Values}} = \dfrac{5+7+12+8}{4}$
     $= \dfrac{32}{4}$
     $= 8$
     The mean of the set of data is 8.

A5) $\dfrac{\text{Sum of Values}}{\text{Number of Values}} = \dfrac{22+26+29}{3}$

$\qquad\qquad\qquad\quad = \dfrac{77}{3}$

$\qquad\qquad\qquad\quad = 25.\overline{6}$

$\qquad\qquad\qquad\quad \approx 25.7$

The mean rounded to the nearest tenth is 25.7.

A6) $\dfrac{\text{Sum of Values}}{\text{Number of Values}} = \dfrac{8+5+10+0+3}{5}$

$\qquad\qquad\qquad\quad = \dfrac{26}{5}$

$\qquad\qquad\qquad\quad = 5.2$

The mean of the set of data is 5.2.

A7) $\dfrac{\text{Sum of Values}}{\text{Number of Values}} = \dfrac{12+4+7+2+1+6}{6}$

$\qquad\qquad\qquad\quad = \dfrac{32}{6}$

$\qquad\qquad\qquad\quad = 5.\overline{3}$

$\qquad\qquad\qquad\quad \approx 5.3$

The mean rounded to the nearest tenth is 5.3.

A8) $\dfrac{\text{Sum of Values}}{\text{Number of Values}} = \dfrac{16+9+5+13+7}{5}$

$\qquad\qquad\qquad\quad = \dfrac{50}{5}$

$\qquad\qquad\qquad\quad = 10$

The mean of the set of data is 10.

A9) $\dfrac{\text{Sum of Values}}{\text{Number of Values}} = \dfrac{24+2+5+35+0+16+3}{7}$

$\qquad\qquad\qquad\quad = \dfrac{85}{7}$

$\qquad\qquad\qquad\quad \approx 12.1$

The mean rounded to the nearest tenth is 12.1.

B1) Use the word "middle" to remember the definition of median.

B2) If there are two middle numbers, the median is the mean of the two middle numbers.

B3) Arrange in numerical order. $\Rightarrow$ 2, 5, 7, 8, 12

Find the middle. $\Rightarrow \not{2}, \not{5}, 7, \not{8}, \not{12}$

The median of the set of data is 7.

B4) Arrange in numerical order. $\Rightarrow$ 5, 7, 9, 13, 16

Find the middle. $\Rightarrow \not{5}, \not{7}, 9, \not{13}, \not{16}$

The median of the set of data is 9.

B5) Arrange in numerical order. $\Rightarrow$ 0, 3, 3, 5, 8, 10

Find the middle. $\Rightarrow \not{0}, \not{3}, 3, 5, \not{8}, \not{10}$

Calculate the mean of the two middle numbers.

$\dfrac{3+5}{2} = \dfrac{8}{2} = 4$

The median is 4.

B6) Arrange in numerical order. $\Rightarrow$ 1, 2, 2, 4, 6, 7

Find the middle. $\Rightarrow \not{1}, \not{2}, 2, 4, \not{6}, \not{7}$

Calculate the mean of the two middle numbers.

$\dfrac{2+4}{2} = \dfrac{6}{2} = 3$

The median is 3.

C1) Use the word "most" to remember the definition of mode.

C2) If each number appears equally often, there is no mode.

C3) Since 5 occurs most often, the mode is 5.

C4) Since 7 occurs most often, the mode is 7.

C5) The data has two modes, 3 and 8.

C6) The data has two modes, 2 and 8.

C7) The data has no mode since every value appears the same number of times.

C8) The data has no mode since every value appears the same number of times.

SECTION 8.2   EXERCISES

For 1–9, refer to Concept Checks and Practice Exercises.

11) median

13) $\dfrac{\text{Sum of Values}}{\text{Number of Values}} = \dfrac{2+7+8+8+5}{5}$

$\qquad\qquad\qquad\quad = \dfrac{30}{5}$

$\qquad\qquad\qquad\quad = 6$

The mean of the set of data is 6.

15) $\dfrac{\text{Sum of Values}}{\text{Number of Values}} = \dfrac{54+68+32+91}{4}$

$\qquad\qquad\qquad\quad = \dfrac{245}{4}$

$\qquad\qquad\qquad\quad = 61.25$

$\qquad\qquad\qquad\quad \approx 61.3$

The mean rounded to the nearest tenth is 61.3.

17) $\left(\dfrac{1}{4}+\dfrac{3}{8}+\dfrac{1}{8}\right)\div 3=\left(\dfrac{2}{8}+\dfrac{3}{8}+\dfrac{1}{8}\right)\cdot\dfrac{1}{3}$

$=\dfrac{6}{8}\cdot\dfrac{1}{3}$

$=\dfrac{3}{4}\cdot\dfrac{1}{3}$

$=\dfrac{1}{4}$

The mean of the set of data is $\dfrac{1}{4}$.

19) Arrange in numerical order. $\Rightarrow 2, 4, 5, 8, 8$

Find the middle. $\Rightarrow \cancel{2}, \cancel{4}, 5, \cancel{8}, \cancel{8}$

The median of the set of data is 5.

21) Arrange in numerical order.

$\Rightarrow 5, 7, 43, 64, 123, 634$

Find the middle.

$\Rightarrow \cancel{5}, \cancel{7}, 43, 64, \cancel{123}, \cancel{634}$

Calculate the mean of the two middle numbers.

$\dfrac{43+64}{2}=\dfrac{107}{2}=53.5$

The median is 53.5.

23) Arrange in numerical order.

$\Rightarrow 21, 23, 25, 34, 67, 84, 84$

Find the middle.

$\Rightarrow \cancel{21}, \cancel{23}, \cancel{25}, 34, \cancel{67}, \cancel{84}, \cancel{84}$

The median of the set of data is 34.

25) Since 1 occurs most often, the mode is 1.

27) The data has no mode since every value appears the same number of times.

29) The data has two modes, 1 and 6.

31) $\dfrac{\text{Sum of Values}}{\text{Number of Values}}=\dfrac{10+10+16+13+11}{5}$

$=\dfrac{60}{5}$

$=12$

The mean of the set of data is 12.

Arrange in numerical order. $\Rightarrow 10, 10, 11, 13, 16$

Find the middle. $\Rightarrow \cancel{10},\cancel{10}, 11,\cancel{13},\cancel{16}$

The median of the set of data is 11.

Since 10 occurs most often, the mode is 10.

33) a) $\dfrac{\text{Total Pizzas}}{\text{Number of Days}}$

$=\dfrac{32+20+17+24+26+38+37}{7}$

$=\dfrac{194}{7}$

≈ 27.7

The mean number of pizzas delivered per day was about 27.7.

b) Arrange in numerical order.

$\Rightarrow 17, 20, 24, 26, 32, 37, 38$

Find the middle.

$\Rightarrow \cancel{17}, \cancel{20}, \cancel{24}, 26, \cancel{32}, \cancel{37}, \cancel{38}$

The median number of pizzas delivered per day was 26.

35) $\dfrac{\text{Sum of Salaries}}{\text{Number of Financial Officers}}$

$=\dfrac{135,000+112,000+72,500+83,000+72,700+65,000}{6}$

$=\dfrac{540,200}{6}$

$\approx 90,033$

The mean of these salaries is approximately $90,033.

37) a) $\left(\begin{array}{l}88+77+98+87+82+75+45+99\\+92+77+59+62+81+85+83+90\end{array}\right)\div 16$

$=1,280\div 16$

$=80$

The mean of the set of data is 80.

Arrange in numerical order.

$\Rightarrow 45, 59, 62, 75, 77, 77, 81, 82,$
$83, 85, 87, 88, 90, 92, 98, 99$

Find the middle.

$\Rightarrow \cancel{45}, \cancel{59}, \cancel{62}, \cancel{75}, \cancel{77}, \cancel{77}, \cancel{81}, 82,$
$83, \cancel{85}, \cancel{87}, \cancel{88}, \cancel{90}, \cancel{92}, \cancel{98}, \cancel{99}$

Calculate the mean of the two middle numbers.

$\dfrac{82+83}{2}=\dfrac{165}{2}=82.5$

The median is 82.5.

Since 77 occurs most often, the mode is 77.

37) b)

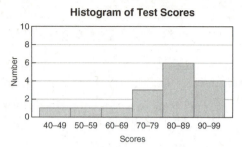

Histogram of Test Scores

c) Answers may vary.

39) a) Day 1: $252 \div 9 = 28$ mpg

Day 2: $286 \div 11 = 26$ mpg

Day 3: $42 \div 2 = 21$ mpg

Day 4: $290 \div 10 = 29$ mpg

b) $\dfrac{\text{Sum of mpg Values}}{\text{Number of Values}} = \dfrac{28+26+21+29}{4}$

$= \dfrac{104}{4}$

$= 26$

The average of the daily gas mileages is 26.

c) $\dfrac{\text{Total Miles Driven}}{\text{Total Gallons Used}} = \dfrac{252+286+42+290}{9+11+2+10}$

$= \dfrac{870}{32}$

≈ 27.2

The actual gas mileage is approximately 27.2 miles per gallon.

41) a) Day 1: $252 \div 9 = 28$ mpg

Day 2: $286 \div 11 = 26$ mpg

Day 3: $210 \div 10 = 21$ mpg

Day 4: $290 \div 10 = 29$ mpg

b) $\dfrac{\text{Sum of mpg Values}}{\text{Number of Values}} = \dfrac{28+26+21+29}{4}$

$= \dfrac{104}{4}$

$= 26$

The average of the daily gas mileages is 26.

c) $\dfrac{\text{Total Miles Driven}}{\text{Total Gallons Used}} = \dfrac{252+286+210+290}{9+11+10+10}$

$= \dfrac{1038}{40}$

$= 25.95$

≈ 26.0

The actual gas mileage is approximately 26.0 miles per gallon.

43) Answers may vary. Example: In Exercise 41, the miles driven on day three increases, raising its influence on the mean. Since the mileage of day three is low (21 mpg), it lowers the overall average.

45) a) $\dfrac{\text{Sum of Hourly Wages}}{\text{Number of Values}}$

$= \dfrac{\$9.50+\$8.75+\$6.85+\$6.85+\$6.85+\$23.54}{6}$

$= \dfrac{\$62.34}{6}$

$= \$10.39$

The mean wage is \$10.39.

Arrange in numerical order.

$\Rightarrow \$6.85, \$6.85, \$6.85, \$8.75, \$9.50, \$23.54$

Find the middle.

$\Rightarrow \cancel{\$6.85},\ \cancel{\$6.85},\ \$6.85,\ \$8.75,\ \cancel{\$9.50},\ \cancel{\$23.54}$

Calculate the mean of the two middle numbers.

$\dfrac{\$6.85+\$8.75}{2} = \dfrac{\$15.60}{2} = \$7.80$

The median wage is \$7.80.

The mode is \$6.85.

b) Answers may vary. Example: The most meaningful measure is the mode. An individual is more likely to be hired as a server than any other position.

47) a) Mean $= \dfrac{\left(\begin{array}{c}65{,}079+70{,}093+60{,}447+59{,}741\\+72{,}639+60{,}755+87{,}256\end{array}\right)}{7}$

$= \dfrac{476{,}010}{7}$

$\approx 68{,}001.43$

Arrange in numerical order.

$\Rightarrow 59{,}741; 60{,}447; 60{,}755; 65{,}079;$
$\quad 70{,}093; 72{,}639; 87{,}256$

Find the middle.

$\Rightarrow \cancel{59{,}741}; \cancel{60{,}447}; \cancel{60{,}755}; 65{,}079;$
$\quad \cancel{70{,}093}; \cancel{72{,}639}; \cancel{87{,}256}$

The median is \$65,079.

No salary appears more than once, so there is no mode.

b) Answers may vary.

49) Mean decreases because the total of the data points decreases.
Median stays the same because the order has not changed.
Mode stays the same because the value of the most data points has not changed.

51) Mean decreases because the total of the data points decreases.
Median increases because the new value is smaller than the old value, thereby affecting the order of the values.
There is no longer a mode because the changed data element was the mode.

53) Answers may vary. Example: The mode will change if the change is either with a value that is the current mode or if the change duplicates a value in the set of numbers.

55) a) Salaries A:
$35,000, $35,000, $35,000, $35,000, $35,000, $35,000, $35,000, $35,000, $35,000, $35,000, $90,000

Salaries B:
$40,000, $40,000, $40,000, $40,000, $40,000, $40,000, $40,000, $40,000, $40,000, $40,000, $40,000

b) The median changes because the middle value is now $40,000.

c) The mean doesn't change because the total value of the salaries is unchanged.

CHAPTER 8   REVIEW EXERCISES

1) The most common response to the survey question, "How many HDTVs do you own?" is 0.

2) $33.3\% - 9.1\% = 24.2\%$
24.2% more households own one high-def TV than two high-def TVs.

3) $9.1\% + 2.0\% + 1.0\% = 12.1\%$
12.1% of homes in the U.S. have two or more high-def TVs.

4) $54.5\% + 33.3\% = 87.8\%$
87.8% of homes in the U.S. have fewer than two high-def TVs.

5) percent · base = amount
$0.333 \cdot 1,059 \approx 353$
Approximately 353 television owners surveyed had one high-definition television.

6) percent · base = amount
$0.091 \cdot 1,059 \approx 96$
Approximately 96 television owners surveyed had two high-definition televisions.

7) The average attendance of a Kentucky basketball game was 23,421.

8) The average attendance of a North Carolina basketball game was 20,693.

9) $21,516 - 19,661 = 1,855$
The average attendance at a Syracuse basketball game was 1,855 more than a Tennessee basketball game.

10) $20,693 - 18,488 = 2,205$
The average attendance at a North Carolina basketball game was 2,205 more than a Louisville basketball game.

11) Syracuse and North Carolina had the smallest difference in attendance.
$21,516 - 20,693 = 823$
The difference in attendance was 823.

12) Kentucky and Louisville had the largest difference in attendance.
$23,421 - 18,488 = 4,933$
The difference in attendance was 4,933.

13) The least number of accidents/spinouts occurred during 2002.

14) The greatest increase in accidents/spinouts occurred between 2002 and 2003.

15) $7 - 6 = 1$
There was approximately one more accident/spinout per race in 2005 than in 2004.

16) The average number of accidents/spinouts is increasing slightly.

17) The range of salaries that describes the most part-time employees is $2,101–$4,290.

18) $23+15+8 = 46$
46 part-time employees make between $6,481 and $13,050.

19) $46+10+23+15+8+2 = 104$
104 part-time employees make less than $15,241.

20) $8+2+2 = 12$
12 part-time employees make more than $10,880.

21) Use the word "average" to remember the definition of mean.

22) Use the word "middle" to remember the definition of median.

23) Use the word "most" to remember the definition of mode.

24) $\dfrac{\text{Sum of Values}}{\text{Number of Values}} = \dfrac{10+5+12+5}{4}$
$= \dfrac{32}{4}$
$= 8$
The mean of the set of data is 8.

25) $\dfrac{\text{Sum of Values}}{\text{Number of Values}} = \dfrac{8+7+7+2+1}{5}$
$= \dfrac{25}{5}$
$= 5$
The mean of the set of data is 5.

26) $\dfrac{\text{Sum of Values}}{\text{Number of Values}} = \dfrac{13+15+12+12+10}{5}$
$= \dfrac{62}{5}$
$= 12.4$
The mean of the set of data is 12.4.

27) $\dfrac{\text{Sum of Values}}{\text{Number of Values}} = \dfrac{18+10+11+14}{4}$
$= \dfrac{53}{4}$
$= 13.25$
The mean is 13.25.

28) $(\$145+\$198+\$230+\$210+\$176+\$138)\div 6$
$= \$1,097 \div 6$
$= \$182.83$
The mean of the bills is $182.83.

29) $\left(\begin{array}{l}\$28,500+\$29,300+\$21,690+\$35,000 \\ +\$37,000+\$43,600+\$45,300+\$38,600\end{array}\right)\div 8$
$= \$278,990 \div 8$
$= \$34,873.75$
$\approx \$34,874$
The mean vehicle cost is approximately $34,874.

30) Arrange in numerical order.
$\Rightarrow 3, 6, 11, 12, 15$
Find the middle.
$\Rightarrow \cancel{3}, \cancel{6}, 11, \cancel{12}, \cancel{15}$
The median of the set of data is 11.

31) Arrange in numerical order.
$\Rightarrow 8, 12, 14, 19, 24$
Find the middle.
$\Rightarrow \cancel{8}, \cancel{12}, 14, \cancel{19}, \cancel{24}$
The median of the set of data is 14.

32) Arrange in numerical order.
$\Rightarrow 3, 5, 6, 7, 35, 35, 40, 43$
Find the middle.
$\Rightarrow \cancel{3}, \cancel{5}, \cancel{6}, 7, 35, \cancel{35}, \cancel{40}, \cancel{43}$

Calculate the mean of the two middle numbers.
$\dfrac{7+35}{2} = \dfrac{42}{2} = 21$
The median age is 21.

33) Arrange in numerical order.
$\Rightarrow 11.1,\ 11.2,\ 11.6,\ 11.8,\ 12.1,\ 13$
Find the middle.
$\Rightarrow \cancel{11.1}, \cancel{11.2}, 11.6, 11.8, \cancel{12.1}, \cancel{13}$

Calculate the mean of the two middle numbers.
$\dfrac{11.6+11.8}{2} = \dfrac{23.4}{2} = 11.7$
The median of the times is 11.7.

34) The data has two modes, 3 and 5.

35) The data has no mode since every value appears the same number of times.

36) The data has no mode since every value appears the same number of times.

37) The data has two modes, 17 and 18.

38) $\dfrac{\text{Sum of Values}}{\text{Number of Values}} = \dfrac{25+31+15+11+12+11}{6}$

$= \dfrac{105}{6}$

$= 17.5$

The mean of the data set is 17.5.

Arrange in numerical order.
    $\Rightarrow 11,\ 11,\ 12,\ 15,\ 25,\ 31$
Find the middle.
    $\Rightarrow \cancel{11},\ \cancel{11},\ 12,\ 15,\ \cancel{25},\ \cancel{31}$

Calculate the mean of the two middle numbers.
$\dfrac{12+15}{2} = \dfrac{27}{2} = 13.5$
The median is 13.5.

The mode is 11.

39) $\dfrac{\text{Sum of Values}}{\text{Number of Values}} = \dfrac{18+19+18+14+24}{5}$

$= \dfrac{93}{5}$

$= 18.6$

The mean of the data set is 18.6.

Arrange in numerical order.
    $\Rightarrow 14,\ 18,\ 18,\ 19,\ 24$
Find the middle.
    $\Rightarrow \cancel{14},\ \cancel{18},\ 18,\ \cancel{19},\ \cancel{24}$
The median of the data set is 18.

The mode is 18.

40) $\dfrac{\text{Total Pizzas Delivered}}{\text{Number of Days}}$

$= \dfrac{21+16+15+19+24+13+18}{7}$

$= \dfrac{126}{7}$

$= 18$

The mean number of pizzas delivered is 18.

Arrange in numerical order.
    $\Rightarrow 13,\ 15,\ 16,\ 18,\ 19,\ 21,\ 24$
Find the middle.
    $\Rightarrow \cancel{13},\ \cancel{15},\ \cancel{16},\ 18,\ \cancel{19},\ \cancel{21},\ \cancel{24}$
The median number of pizzas delivered is 18.

There is no mode.

41) $\dfrac{\text{Total Lunch Customers}}{\text{Number of Days}}$

$= \dfrac{36+46+25+28+34+20+17+28}{8}$

$= \dfrac{234}{8}$

$= 29.25$

≈ 29.3

The mean number of lunch customers is 29.3.

Arrange in numerical order.
    $\Rightarrow 17,\ 20,\ 25,\ 28,\ 28,\ 34,\ 36,\ 46$
Find the middle.
    $\Rightarrow \cancel{17},\ \cancel{20},\ \cancel{25},\ 28,\ 28,\ \cancel{34},\ \cancel{36},\ \cancel{46}$

Calculate the mean of the two middle numbers.
$\dfrac{28+28}{2} = \dfrac{56}{2} = 28$

The median number of lunch customers is 28.

The mode is 28.

42)
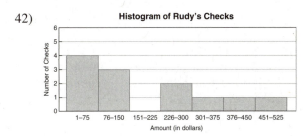

43)
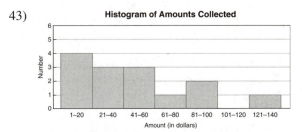

CHAPTER 8   TEST

1) 3% of people answered that they don't know how they eat an apple.

2) $47\% - 11\% = 36\%$
    36% more people answered "bite into it" than "peel it."

3) $39\% + 11\% = 50\%$
    50% of people answered either "cut it into slices" or "peel it."

4) $0.11 \cdot 2{,}000 = 220$
    220 people responded that they peel their apple.

5) Bill Elliot had 44 wins in 2007.

6) $35 - 33 = 2$
Mark Martin had two more wins than
Jimmie Johnson in 2007.

7) $\dfrac{33}{81} = \dfrac{\cancel{3} \cdot 11}{\cancel{3} \cdot 27} = \dfrac{11}{27}$
The ratio that compares Jimmie Johnson's wins
to Jeff Gordon's wins in 2007 is 11 to 27, or
$\dfrac{11}{27}$.

8) $\dfrac{81}{44} \approx 1.84$
Jeff Gordon's total is approximately 184% of
Bill Elliott's.

9) The number of ski resorts first dropped below
700 in 1987.

10) A slight increase occurred in 1995 and 1998.

11) $727 - 600 = 127$
There were approximately 127 fewer ski resorts
in 1990 than in 1985.

12) $\dfrac{485}{727} \approx 0.667$
The number of ski resorts in 2007 was
approximately 66.7% of the resorts in operation
during 1985.

13) The range of salaries that describes the most
employees is $26,001–$29,000.

14) $23 + 15 + 8 + 7 = 53$
53 employees make between $26,001 and
$38,000.

15) $23,000 - 20,001 + 1 = 2,999 + 1$
$= 3,000$
The size of the class intervals is $3,000.

16) The class interval $35,001–$38,000 has the
lowest frequency.

17) $\text{GPA} = \dfrac{4+3+4+4}{4}$
$= \dfrac{15}{4}$
$= 3.75$
Terry's GPA is 3.75.

18) a) $\dfrac{\text{Sum of Scores}}{\text{Number of Scores}} = \dfrac{74+87+72+93+94}{5}$
$= \dfrac{420}{5}$
$= 84$
The score 87 is above the mean.

b) Arrange in numerical order.
$\Rightarrow 72, 74, 87, 93, 94$
Find the middle.
$\Rightarrow \cancel{72}, \cancel{74}, 87, \cancel{93}, \cancel{94}$
The score 87 is equal to the median.

c) Answers may vary.

19) Arrange in numerical order.
$\Rightarrow 13, 16, 35, 40, 43, 50, 72$
Find the middle.
$\Rightarrow \cancel{13}, \cancel{16}, \cancel{35}, 40, \cancel{43}, \cancel{50}, \cancel{72}$
The median age is 40.

20) Arrange in numerical order.
$\Rightarrow 52.1, 53.2, 56.0, 57.6, 60.1, 61.3$
Find the middle.
$\Rightarrow \cancel{52.1}, \cancel{53.2}, 56.0, 57.6, \cancel{60.1}, \cancel{61.3}$

Calculate the mean of the two middle numbers.
$\dfrac{56.0+57.6}{2} = \dfrac{113.6}{2} = 56.8$
The median time is 56.8 seconds.

21) The data has two modes, 64° and 73°.

22) The mode is 18.

23) $\dfrac{\text{Sum of Values}}{\text{Number of Values}} = \dfrac{15+11+15+24+4}{5}$
$= \dfrac{69}{5}$
$= 13.8$
The mean of the data set is 13.8.

Arrange in numerical order.
$\Rightarrow 4, 11, 15, 15, 24$
Find the middle.
$\Rightarrow \cancel{4}, \cancel{11}, 15, \cancel{15}, \cancel{24}$
The median of the data set is 15.

The mode is 15.

24) $\dfrac{\text{Sum of Values}}{\text{Number of Values}} = \dfrac{8+35+18+23+24}{5}$

$= \dfrac{108}{5}$

$= 21.6$

The mean of the data set is 21.6.

Arrange in numerical order.
$\Rightarrow 8,\ 18,\ 23,\ 24,\ 35$
Find the middle.
$\Rightarrow \cancel{8},\ \cancel{18},\ 23,\ \cancel{24},\ \cancel{35}$
The median of the data set is 23.

There is no mode.

25) $\dfrac{\text{Total Cars Sold}}{\text{Number of Days}} = \dfrac{8+11+10+0+7+13}{6}$

$= \dfrac{49}{6}$

≈ 8.2

The mean number of cars sold is 8.2.

Arrange in numerical order.
$\Rightarrow 0,\ 7,\ 8,\ 10,\ 11,\ 13$
Find the middle.
$\Rightarrow \cancel{0},\ \cancel{7},\ 8,\ 10,\ \cancel{11},\ \cancel{13}$

Calculate the mean of the two middle numbers.
$\dfrac{8+10}{2} = \dfrac{18}{2} = 9$
The median number of cars sold is 9.

There is no mode.

26) $\dfrac{\text{Total Lunch Customers}}{\text{Number of Days}}$

$= \dfrac{48+47+35+38+39+47}{6}$

$= \dfrac{254}{6}$

≈ 42.3

The mean number of lunch customers is 42.3.

Arrange in numerical order.
$\Rightarrow 35,\ 38,\ 39,\ 47,\ 47,\ 48$
Find the middle.
$\Rightarrow \cancel{35},\ \cancel{38},\ 39,\ 47,\ \cancel{47},\ \cancel{48}$

Calculate the mean of the two middle numbers.
$\dfrac{39+47}{2} = \dfrac{86}{2} = 43$
The median number of lunch customers is 43.

The mode is 47.

27)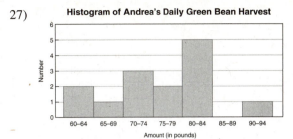

Histogram of Andrea's Daily Green Bean Harvest

$\dfrac{\text{Daily Harvest}}{\text{Number of Days}}$

$= \dfrac{\left(\begin{array}{c}65+71+78+84+83+72+61\\ +75+74+81+90+60+83+84\end{array}\right)}{14}$

$= \dfrac{1{,}061}{14}$

≈ 75.8

The mean daily harvest is about 75.8 pounds.

Arrange in numerical order.
$\Rightarrow 60, 61, 65, 71, 72, 74, 75,$
$78, 81, 83, 83, 84, 84, 90$
Find the middle.
$\Rightarrow \cancel{60}, \cancel{61}, \cancel{65}, \cancel{71}, \cancel{72}, \cancel{74}, 75,$
$78, \cancel{81}, \cancel{83}, \cancel{83}, \cancel{84}, \cancel{84}, \cancel{90}$

Calculate the mean of the two middle numbers.
$\dfrac{75+78}{2} = \dfrac{153}{2} = 76.5$
The median daily havest is 76.5 pounds.

The modes are 83 pounds and 84 pounds.

Chapter 9   Signed Numbers

9.1   Understanding Signed Numbers

GUIDED PRACTICE

1) Since deleting songs decreases the amount of music on an MP3 player, use a negative number.

   Deleting 317 songs can be represented with the signed number -317.

2) Since adding RAM is an increase, use a positive number.

   Adding 2.5 GB of RAM can be represented with the signed number 2.5.

3) The range goes beyond the largest value, 3, and the smallest value, -2.5.

4)
$$-9 > -12$$
-9 is greater than -12.

5) $\dfrac{5}{12} = \dfrac{5}{12}$ $\dfrac{1}{4} \cdot \dfrac{3}{3} = \dfrac{3}{12}$

$$-\frac{5}{12} < -\frac{1}{4}$$
$-\dfrac{5}{12}$ is less than $-\dfrac{1}{4}$.

6) $-(-4)$ is read as "the opposite of negative four."
$$-(-4) = 4$$

7) Since 4 is 4 units from zero, $|4| = 4$. Since the absolute value is a measure of distance, the answer is never negative.

8) $|-12|$ is read as "the absolute value of negative twelve."
$|-12| = 12$, because -12 is 12 units from zero.

9) $|45|$ is read as "the absolute value of forty-five."
$|45| = 45$, because 45 is 45 units from zero.

10) $\left|-\dfrac{2}{3}\right| = \dfrac{2}{3}$, because $-\dfrac{2}{3}$ is $\dfrac{2}{3}$ unit from zero.

11) $-|-6|$ is read as "the opposite of the absolute value of negative six."
$$-|-6| = -(6)$$
$$= -6$$
Since we had to evaluate the absolute value before finding the opposite, this exercise required 2 steps.

12) $-(-6)$ is read as "the opposite of -6."
$$-(-6) = 6$$
Since grouping is not an operation, we find the opposite of negative six in one step.

13) $|-6| = 6$; $-1 = -1$; $-(-4) = 4$;
$0 = 0$; $-|-4| = -4$

14) The two numbers are 7 units to the left and right of 11.
$$11 - 7 = 4$$
$$11 + 7 = 18$$

11 and 4 are $11 - 4 = 7$ units apart.
11 and 18 are $18 - 11 = 7$ units apart.
Both 4 and 18 are 7 units from 11.

CONCEPT CHECKS AND PRACTICE EXERCISES

A1) Answers will vary.

A2) a) -48
     b) 48
     c) Yes, it is possible for the same event to be described with either a positive or a negative number.
     d) Answers will vary.

A3) -15

A4) 500

A5) $-3\dfrac{1}{2}$

A6) $-5\dfrac{1}{2}$

A7) -25.34

283

A8) 267.15

A9) 4.56

A10) −3.56

B1) Zero is neither positive nor negative.

B2) Answers may vary. Example: If a number line is drawn without labels, there would be no indication of the location of the points on the line.

B3)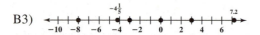

B4)

B5)

B6)

B7)

B8)

C1) Answers may vary. Example: A negative number is always further to the left on the number line than a positive number.

C2) Answers may vary. Example: Zero is always further to the right on the number line than a negative number.

C3) $-5 < 2$

C4) $-13 < 8$

C5) $10 > -12$

C6) $3 > -15$

C7) $-0.054 < -0.05$

C8) $0.093 > 0.09$

C9) $-\dfrac{7}{16} < -\dfrac{3}{8}$

C10) $\dfrac{5}{3} > \dfrac{14}{9}$

D1) a) $-(-5)$ is read "the opposite of negative five"; so $-(-5) = 5$.

   b) $-(2)$ is read "the opposite of two"; so $-(2) = -2$.

   c) $-\left(\dfrac{1}{2}\right)$ is read "the opposite of one half"; so $-\left(\dfrac{1}{2}\right) = -\dfrac{1}{2}$.

D2) $-(-2) = 2$

D3) $-(+4) = -4$

D4) $-(+5) = -5$

D5) $-(-6) = 6$

D6) $-(11) = -11$

D7) $-(-14) = 14$

D8) $-(-13.5) = 13.5$

D9) $-(2.59) = -2.59$

E1) Answers may vary. Example: The absolute value of a number indicates its distance from 0 on the number line.

E2) Answers may vary. Example: The absolute value of a number cannot be negative since distance is always measured with a positive value or zero.

E3) a) two
   b) $-|7| = -(7) = -7$

E4) a) two
   b) $-|-3| = -(3) = -3$

E5) a) one
   b) $-(-7) = 7$

E6) a) one
   b) $-(3) = 3$

E7) $\left|-5\right| = 5$

E8) $\left|16\right| = 16$

E9) $\left|\dfrac{7}{9}\right| = \dfrac{7}{9}$

E10) $\left|\dfrac{-1}{10}\right| = \dfrac{1}{10}$

E11) $-(-5) = 5$

E12) $-(-31) = 31$

E13) $-\left|-42\right| = -(42)$
          $= -42$

E14) $-\left|-15\right| = -(15)$
          $= -15$

SECTION 9.1 EXERCISES

For 1–15, refer to Concept Checks and Practice Exercises.

17) opposites

19) negative number

21) 89

23) −15

25) 1,453

27) −13

29) 1

31)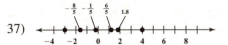

33)

35)

37)

39) $-8 < 3$

41) $-6 < -3$

43) $6 < 18$

45) $42 > -50$

47) $-\dfrac{1}{3} < -\dfrac{1}{4}$

49) $-4.18 > -4.2$

51) Answers may vary. Example: All negative numbers are to the left of all positive numbers on the number line.

53) Answers may vary. Example: Zero is greater than every negative number and less than every positive number because it lies to the right of all negative numbers and to the left of all positive numbers on the number line.

55) a) $\left|-12\right|$ is read as "the absolute value of negative twelve."
     b) $\left|-12\right| = 12$, because −12 is 12 units from 0.

57) a) $-(-3)$ is read as "the opposite of negative three."
     b) $-(-3) = 3$

59) a) $-(7)$ is read as "the opposite of seven."
     b) $-(7) = -7.$

61) a) $-\left|-10\right|$ is read as "the opposite of the absolute value of negative ten."
     b) $-\left|-10\right| = -(10)$
              $= -10$

63) $\left|-8\right| = 8$

65) $(-5) = -5$

67) $-(-7) = 7$

69) $-|-13| = -(13)$
$\qquad = -13$

71) $-(91) = -91$

73) $-|75| = -(75)$
$\qquad = -75$

75) Zero and all positive numbers are equal to their own absolute value.

77) Zero is the only number that is neither positive nor negative.

79) $|25| = 25$
$(-33) = -33$
$|25| > (-33)$

81) $|-45| = 45$
$|-100| = 100$
$|-45| < |-100|$

83) $-|2| = -2$
$-|-8| = -8$
$-|2| > -|-8|$

85) $-(-2) = 2$
$-(3) = -3$
$-(-2) > -(3)$

87) $|-10| = 10;\ -(10) = -10$
$-|-4| = -4;\ |-4| = 4$
$-(10) < -|-4| < |-4| < |-10|$

89) $-(-12) = 12;\ -|-12| = -12$
$-(15) = -15;\ |-15| = 15$
$-(15) < -|-12| < -(-12) < |-15|$

91) $-3 - 5 = -8$
$-3 + 5 = 2$

-8 and 2 are 5 units from -3.

93) $-5 - 3 = -8$
$-5 + 3 = -2$

-8 and -2 are 3 units from -5.

95) $|-7| = 7$
$|7| = 7$

$n = -7, 7$

9.2   Adding and Subtracting Signed Numbers

GUIDED PRACTICE

1) Draw 4 negative discs for -4.
Draw 2 negative discs for -2.

$(-4) + (-2)$

$\boxed{-1}\qquad\boxed{-1}$
$\boxed{-1}\qquad\boxed{-1}$
$\boxed{-1}$
$\boxed{-1}$

$(-4) + (-2) = -6$
Together, there are a total of 6 negative discs.

2) Add the absolute values of the two numbers.
$|-34| = 34$ and $|-21| = 21$
$\qquad 34 + 21 = 55$

The sign of the answer matches the sign of the original numbers.
$(-34) + (-21) = -55$

When adding two negative numbers, the answer is negative.

3) Add the numerators.

3 negative discs	+	4 negative discs	=	7 negative discs

$$\frac{-3}{18} + \frac{-4}{18} = \frac{-7}{18}$$

Since both numbers are negative, the answer is negative.

Add the absolute values of the two numbers.
$\left|\dfrac{-3}{18}\right| = \dfrac{3}{18}$ and $\left|\dfrac{-4}{18}\right| = \dfrac{4}{18}$

$$\frac{3}{18} + \frac{4}{18} = \frac{7}{18}$$

The sign of the answer matches the sign of the original numbers.

$$\frac{-3}{18} + \frac{-4}{18} = \frac{-7}{18}$$

4) Draw 4 negative discs and 2 positive discs.
$1 + -1 = 0$, so two pairs of discs are removed.

$(-4) + 2$

$(-4) + 2 = -2$

Two negative discs remain.

5)

Zero Pair	Reason
4 and -4	$4 + (-4) = 0$
12 and -12	$12 + (-12) = 0$
231 and $-.231$	$231 + (-231) = 0$

6) The largest zero pair is 8 and –8.
The answer will be negative since there are more negatives than positives.

$(-14) + (8) = -6$

Subtract $14 - 8 = 6$ to find how many negatives remain.

7) $(-14) + (30) = 16$

Subtract $30 - 14 = 16$ to find how many positives remain.

Find the difference between the absolute values of the numbers.
$|-14| = 14$ and $|30| = 30$
$\qquad 30 - 14 = 16$

Use the sign of the number with the larger absolute value.
$(-14) + (30) = 16$

Because $|30|$ is larger than $|-14|$, the answer is positive.

8) To add, build like fractions.

$-\dfrac{4}{15} + \dfrac{1}{3} = -\dfrac{4}{15} + \dfrac{1}{3}\left(\dfrac{5}{5}\right)$

$\qquad\qquad = -\dfrac{4}{15} + \dfrac{5}{15}$

$\left(-\dfrac{4}{15}\right) + \left(\dfrac{5}{15}\right) = \dfrac{1}{15}$

Subtract $5 - 4 = 1$ to find how many positives remain.

Find the difference between the absolute values of the numbers.
$\left|-\dfrac{4}{15}\right| = \dfrac{4}{15}$ and $\left|\dfrac{1}{3}\right| = \dfrac{1}{3} = \dfrac{5}{15}$
$\qquad \dfrac{5}{15} - \dfrac{4}{15} = \dfrac{1}{15}$

Use the sign of the number with the larger absolute value.

$\left(-\dfrac{4}{15}\right) + \left(\dfrac{1}{3}\right) = \dfrac{1}{15}$

Because $\left|\dfrac{1}{3}\right|$ is larger than $\left|-\dfrac{4}{15}\right|$, the answer is positive.

9) $3 - (-2)$ Three subtract negative two.

$9 + (-5)$ Nine plus negative five.

$-1 - (+3)$ Negative one subtract positive three.

10) $-13 - 17 = -13 + (-17)$
$\qquad\qquad = -30$

11) $-32 - (-40) = -32 + (+40)$
$\qquad\qquad\quad = 8$

12) a) Opposite signs, subtract. $3 + (-5) = 3 - 5$

b) Opposite signs, subtract. $-7 - (+5) = -7 - 5$

c) Same sign, add. $2 - (-3) = 2 + 3$

13) Graph 13.
To subtract 19, move 19 units to the left.

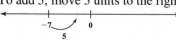

We do pass 0, so the answer is negative.
$19 - 13$ is 6.
$13 - 19 = -6$

14) Graph -7.
To add 5, move 5 units to the right.

We do not pass 0, so the answer is negative.
$7 - 5$ is 2.
$-7 + 5 = -2$

15) $30 + (-51) = 30 - 51$
$$= -21$$

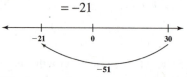

We do pass 0, so the answer is negative.

16) $(-13) + (+20) = -13 + 20$
$$= 7$$

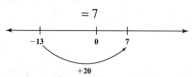

We do pass 0, so the answer is positive.

17) $(21) - (-30) = 21 + 30$
$$= 51$$

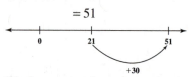

We do not pass 0, so the answer is positive.

18) $(-61) - (+75) = -61 - 75$
$$= -136$$

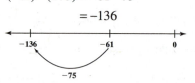

We do not pass 0, so the answer is negative.

19) $(-3) - |-6| + 8 = (-3) - 6 + 8$
$$= -9 + 8$$
$$= -1$$

20) $5 + (-17) - 3 + 4 - 2 - (-5)$
$= 5 + (-17) + (-3) + 4 + (-2) + 5$
$= (5 + 4 + 5) + ((-17) + (-3) + (-2))$
$= 14 + (-22)$
$= -8$

21)

Change	Altitude
Start	3,729 ft
+250	3,979 ft
−490	3,489 ft
+172	3,661 ft
−100	3,561 ft
+82	3,643 ft
−215	3,428 ft

The hiker must descend 215 feet to get to her car.

CONCEPT CHECKS AND PRACTICE EXERCISES

A1) a) True. Answers may vary. When adding numbers with the same sign, the total number of negatives or positives will increase, so the sign remains the same.

b) True. Answers may vary. When adding numbers with the same sign, the total number of negatives or positives will increase, so the sign remains the same.

A2) Draw 8 negative discs for -8.
Draw 9 negative discs for -9.

$(-8) + (-9)$

−1	−1
−1	−1
−1	−1
−1	−1
−1	−1
−1	−1
−1	−1
−1	−1
	−1

$(-8) + (-9) = -17$
Together, there are a total of 17 negative discs.

A3) Draw 5 negative discs for −5.
Draw 6 negative discs for −6.

$(-5)+(-6)$

−1		−1
−1		−1
−1		−1
−1		−1
−1		−1
		−1

$(-5)+(-6)=-11$

Together, there are a total of 11 negative discs.

A4) Add the absolute values of the two numbers.
$|-12|=12$ and $|-16|=16$
$12+16=28$

The sign of the answer matches the sign of the original numbers.
$-12+-16=-28$

A5) Add the absolute values of the two numbers.
$|-19|=19$ and $|-8|=8$
$19+8=27$

The sign of the answer matches the sign of the original numbers.
$-19+-8=-27$

A6) Add the absolute values of the two numbers.
$|-17.3|=17.3$ and $|-3.2|=3.2$
$17.3+3.2=20.5$

The sign of the answer matches the sign of the original numbers.
$(-17.3)+(-3.2)=-20.5$

A7) Add the absolute values of the two numbers.
$|-8.7|=8.7$ and $|-9.5|=9.5$
$8.7+9.5=18.2$

The sign of the answer matches the sign of the original numbers.
$(-8.7)+(-9.5)=-18.2$

A8) Add the numerators.

7 negative discs	+	12 negative discs	=	19 negative discs

$$\frac{-7}{28}+\frac{-12}{28}=\frac{-19}{28}$$

Since both numbers are negative, the answer is negative.

Add the absolute values of the two numbers.
$$\left|\frac{-7}{28}\right|=\frac{7}{28} \text{ and } \left|\frac{-12}{28}\right|=\frac{12}{28}$$
$$\frac{7}{28}+\frac{12}{28}=\frac{19}{28}$$

The sign of the answer matches the sign of the original numbers.
$$\frac{-7}{28}+\frac{-12}{28}=\frac{-19}{28}$$

A9) Add the numerators.

3 negative discs	+	4 negative discs	=	7 negative discs

$$\frac{-3}{10}+\frac{-4}{10}=\frac{-7}{10}$$

Since both numbers are negative, the answer is negative.

Add the absolute values of the two numbers.
$$\left|\frac{-3}{10}\right|=\frac{3}{10} \text{ and } \left|\frac{-4}{10}\right|=\frac{4}{10}$$
$$\frac{3}{10}+\frac{4}{10}=\frac{7}{10}$$

The sign of the answer matches the sign of the original numbers.
$$\frac{-3}{10}+\frac{-4}{10}=\frac{-7}{10}$$

B1) Answers may vary. Example: When adding numbers with opposite signs, the answer will have the same sign as the number with the larger absolute value because there will be "more" of that type of number (i.e. positive or negative).

B2) Answers may vary. Example: Subtracting tells us how many positives or negatives remain.

B3) Draw 6 negative discs and 3 positive discs.
$1 + -1 = 0$, so three pairs of discs are removed.

$$(-6) + (3) = -3$$

Three negative discs remain.

B4) Draw 5 positive discs and 2 negative discs.
$1 + -1 = 0$, so two pairs of discs are removed.

$$(5) + (-2) = 3$$

Three positive discs remain.

B5) Find the difference between the absolute values of the numbers.
$|-11| = 11$ and $|21| = 21$
$$21 - 11 = 10$$

Use the sign of the number with the larger absolute value.
$$(-11) + (21) = 10$$

B6) Find the difference between the absolute values of the numbers.
$|50| = 50$ and $|-43| = 43$
$$50 - 43 = 7$$
Use the sign of the number with the larger absolute value.
$$(50) + (-43) = 7$$

B7) Find the difference between the absolute values of the numbers.
$|-19| = 19$ and $|15| = 15$
$$19 - 15 = 4$$
Use the sign of the number with the larger absolute value.
$$(-19) + (15) = -4$$

B8) Find the difference between the absolute values of the numbers.
$|-29| = 29$ and $|16| = 16$
$$29 - 16 = 13$$

Use the sign of the number with the larger absolute value.
$$(-29) + (16) = -13$$

B9) Find the difference between the absolute values of the numbers.
$$\left|\frac{4}{5}\right| = \frac{4}{5}\left(\frac{2}{2}\right) = \frac{8}{10} \text{ and } \left|-\frac{7}{10}\right| = \frac{7}{10}$$
$$\frac{8}{10} - \frac{7}{10} = \frac{1}{10}$$

Use the sign of the number with the larger absolute value.
$$\left(\frac{4}{5}\right) + \left(-\frac{7}{10}\right) = \frac{1}{10}$$

B10) Find the difference between the absolute values of the numbers.
$$\left|-\frac{3}{8}\right| = \frac{3}{8} \text{ and } \left|\frac{1}{2}\right| = \frac{1}{2}\left(\frac{4}{4}\right) = \frac{4}{8}$$
$$\frac{4}{8} - \frac{3}{8} = \frac{1}{8}$$

Use the sign of the number with the larger absolute value.
$$\left(-\frac{3}{8}\right) + \left(\frac{1}{2}\right) = \frac{1}{8}$$

B11) Find the difference between the absolute values of the numbers.
$|4.9| = 4.9$ and $|-1.5| = 1.5$
$$4.9 - 1.5 = 3.4$$

Use the sign of the number with the larger absolute value.
$$(4.9) + (-1.5) = 3.4$$

B12) Find the difference between the absolute values of the numbers.
$|-15.3| = 15.3$ and $|21.8| = 21.8$
$$21.8 - 15.3 = 6.5$$

Use the sign of the number with the larger absolute value.
$$(-15.3) + (21.8) = 6.5$$

FOCUS ON PERSONAL FINANCE

Date	Payment Debit	Fee (if any)	Deposit Credit	Balance
1 – Jan				$512.73
1 – Jan	$375.00			$137.73
5 – Jan	$ 20.00	$ 3.00		$114.73
				$114.73
8 – Jan	$ 78.95			$ 35.78
8 – Jan	$ 20.00	$ 3.00		$ 12.78
8 – Jan			$438.24	$451.02
10 – Jan	$100.00			$351.02
10 – Jan	$ 59.99			$291.03
10 – Jan	$ 48.73			$242.30
11 – Jan	$ 20.00	$ 3.00		$219.30
13 – Jan	$ 20.00	$ 3.00		$196.30
15 – Jan	$185.00			$ 11.30
16 – Jan	$150.00			–$138.70
16 – Jan		$20.00		–$158.70
18 – Jan		$30.00		–$188.70

1) Answers may vary. Example: In the first half of the month, Alfonso paid $12.00 in ATM fees. If he continues to withdraw money in the same manner, the total for the month will be approximately $24.00. He should try to determine the amount of money he needs for a couple of weeks and withdraw it all at the same time.

2) Answers may vary.

3) $\dfrac{\$50}{3\,\text{min}} \cdot \dfrac{60\,\text{min}}{h} = \dfrac{\$50}{\cancel{3}\,\cancel{\text{min}}} \cdot \dfrac{\cancel{3} \cdot 20\,\cancel{\text{min}}}{h}$

$= \dfrac{\$1{,}000}{h}$

Alfonso's lapse in bookkeeping cost him $1,000 per hour.

C1) a) $10-16=10+(-16)$

b) $-12-5=-12+(-5)$

c) $15-(-2)=15+2$

d) $-4-(-8)=-4+8$

C2) Answers may vary. Example: Both problems have the same answer

C3) $16-20=16+(-20)$
$\qquad = -4$

C4) $15-19=15+(-19)$
$\qquad = -4$

C5) $-12-13=-12+(-13)$
$\qquad = -25$

C6) $-15-10=-15+(-10)$
$\qquad = -25$

C7) $8-(-3)=8+(+3)$
$\qquad = 11$

C8) $6-(-2)=6+(+2)$
$\qquad = 8$

C9) $-3-(-6)=-3+(+6)$
$\qquad = 3$

C10) $-2-(-8)=-2+(+8)$
$\qquad = 6$

D1) Answers may vary. Example: Subtraction is the opposite of addition. A negative value is the opposite of a positive value. By changing both the operation and the sign of the number to its opposite, the same problem has been maintained. It is important to remember that *BOTH* the operation and the sign must be changed to the opposite.

D2) a) You need to move three units to the left of the starting point to reach zero.

b) The largest zero pair that can be made from these numbers is 3 and -3.

c) Answers may vary.

d) Answers may vary.

D3) a) Opposite signs, subtract.
$\qquad 10-(+16)=10-16$

b)

c) $10-(+16)=10-16$
$\qquad = -6$

D4) a) Opposite signs, subtract.
$\qquad 15-(+18)=15-18$

b)

c) $15-(+18)=15-18$
$\qquad = -3$

D5) a) Opposite signs, subtract.
$$-12-(-13)=-12+13$$

b)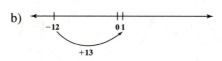

c) $-12-(-13)=-12+13$
$$=1$$

D6) a) Opposite signs, subtract.
$$-7-(-5)=-7+5$$

b)

c) $-7-(-5)=-7+5$
$$=-2$$

D7) $5-(-6)=5+6$
$$=11$$

D8) $4-(-8)=4+8$
$$=12$$

D9) $-6-(-9)=-6+9$
$$=3$$

D10) $-40-(-50)=-40+50$
$$=10$$

D11) $51-(+16)=51-16$
$$=35$$

D12) $40+(-65)=40-65$
$$=-25$$

D13) $-19-(-30)=-19+30$
$$=11$$

D14) $-50+(+75)=-50+75$
$$=25$$

SECTION 9.2   EXERCISES

For 1–14, refer to Concept Checks and Practice Exercises.

15) zero pairs

17) $(-4)+(-6)=-10$

19) $73+28=101$

21) $14+(-9)=14-9$
$$=5$$

23) $(-6)+5=-1$

25) $18+(-23)=18-23$
$$=-5$$

27) $(-8)+20=12$

29) $(-2.5)+3.8=1.3$

31) $(-2.2)+(-8.8)=-11$

33) $\left(\dfrac{3}{5}\right)+\left(-\dfrac{7}{10}\right)=\dfrac{3}{5}-\dfrac{7}{10}$
$$=\dfrac{3}{5}\left(\dfrac{2}{2}\right)-\dfrac{7}{10}$$
$$=\dfrac{6}{10}-\dfrac{7}{10}$$
$$=-\dfrac{1}{10}$$

35) $\left(-\dfrac{8}{9}\right)+\left(-\dfrac{2}{9}\right)=-\dfrac{10}{9}$
$$=-1\dfrac{1}{9}$$

37) $4-12=-8$

39) $21-8=13$

41) $(-17)-14=-31$

43) $38-(-12)=38+12$
$$=50$$

45) $(-13)-(-18)=-13+18$
$$=5$$

47) $(-71)-(-86)=-71+86$
$$=15$$

49) $\frac{3}{5} - \left(-\frac{1}{10}\right) = \frac{3}{5} + \frac{1}{10}$

$= \frac{3}{5}\left(\frac{2}{2}\right) + \frac{1}{10}$

$= \frac{6}{10} + \frac{1}{10}$

$= \frac{7}{10}$

51) $(-5.5) - (-6.3) = -5.5 + 6.3$

$= 0.8$

53) a) $17 - 32 = -15$

b) $32 - 17 = 15$

c) Subtraction is not commutative.

55) a) $(15 + 2) + -8 = 17 + -8$

$= 17 - 8$

$= 9$

b) $15 + (2 + -8) = 15 + (2 - 8)$

$= 15 + (-6)$

$= 15 - 6$

$= 9$

55) c) Addition is associative.

57) $7 - 3 + 5 = 4 + 5$

$= 9$

59) $8 + 3 - 6 = 11 - 6$

$= 5$

61) $-4 + |-3| - (-2) = -4 + 3 - (-2)$

$= -4 + 3 + 2$

$= -1 + 2$

$= 1$

63) $3 - (-3) - 4 = 3 + 3 - 4$

$= 6 - 4$

$= 2$

65) $7 + (-5) - 9 + (-15) = 7 - 5 - 9 - 15$

$= 2 - 9 - 15$

$= -7 - 15$

$= -22$

67) $2.1 - (-3.8) + 4.5 + (-1.1) = 2.1 + 3.8 + 4.5 - 1.1$

$= 5.9 + 4.5 - 1.1$

$= 10.4 - 1.1$

$= 9.3$

69) $1 + 2 - 3 + 4 - 5 + 6 - 7 = 3 - 3 + 4 - 5 + 6 - 7$

$= 0 + 4 - 5 + 6 - 7$

$= 4 - 5 + 6 - 7$

$= -1 + 6 - 7$

$= 5 - 7$

$= -2$

71) $|5 - (-3) - 15| = |5 + 3 - 15|$

$= |8 - 15|$

$= |-7|$

$= 7$

73) $-3,149 + 2,987 = -162$

The weight of the freight decreased by 162 pounds.

75) $43°F - 85°F = -42°F$

The change is temperature from day to night is $-42°F$.

77) $9°F - \left(-3°F\right) = 9°F + 3°F$

$= 12°F$

The change is temperature from New Year's Eve to New Year's Day is $12°F$.

79) $-282 + 170 = -112$

The altitude of the Caspian Sea is -112 feet.

81) $113,200 + 18,700 - 34,700 - 6,300$

$= 131,900 - 34,700 - 6,300$

$= 97,200 - 6,300$

$= 90,900$

The company had $90,900 at the end of March.

83) $-34,700 - 6,300$

$= -41,000$

There was a $41,000 loss from the first of February to the end of March.

85) a) Chante:

$1 + 2 + 1 + 1 + 1 - 1 + 2 - 1 - 1 = 5$

Angelica:

$-1 + 1 + 1 - 1 - 1 - 1 + 1 - 2 + 1 - 1 - 1 - 1 = -5$

Sergio:

$1 - 1 + 1 + 2 + 1 + 1 + 1 + 1 + 2 - 1 = 8$

Mike:

$-1 - 1 + 1 - 2 + 2 + 2 - 1 - 1 + 2 + 2 = 3$

85) b) First Place: Angelica, 5 below par
Second Place: Mike, 3 above par
Third Place: Chante, 5 above par
Fourth Place: Sergio, 8 above par

9.3   Multiplying and Dividing Signed Numbers

GUIDED PRACTICE

1) $(-9)\cdot(8) = -(9\cdot 8)$
$= -72$

2) $\left(-\dfrac{3}{4}\right)\cdot\left(-\dfrac{8}{1}\right) = +\left(\dfrac{3}{4}\right)\cdot\left(\dfrac{8}{1}\right)$
$= \dfrac{3\cdot 8}{4\cdot 1}$
$= \dfrac{3\cdot 2\cdot \cancel{4}}{\cancel{4}\cdot 1}$
$= 6$

3) $\boxed{\text{Opposite Signs}}\ \boxed{\text{Negative Product}}$
$(-3)\cdot(2) = -6$

4) $\boxed{\text{Same Signs}}\ \boxed{\text{Positive Product}}$
$(-1.6)\cdot(-1) = 1.6$

5) $\boxed{\text{Opposite Signs}}\ \boxed{\text{Negative Quotient}}$
$(-20)\div(5) = -4$

6) $\boxed{\text{Same Signs}}\ \boxed{\text{Positive Quotient}}$
$(-2.7)\div(-9) = 0.3$

7) $\boxed{\text{Same Signs}}\ \boxed{\text{Positive Quotient}}$
$\left(-\dfrac{5}{4}\right)\div\left(-\dfrac{35}{8}\right) = +\left(\dfrac{5}{4}\right)\cdot\left(\dfrac{8}{35}\right)$
$= \dfrac{\cancel{5}\cdot 2\cdot \cancel{4}}{\cancel{4}\cdot \cancel{5}\cdot 7}$
$= \dfrac{2}{7}$

8) There is an odd number of negatives, so the answer is negative.
$(-14)\div(-2)\cdot(-7) = -(14)\div(2)\cdot(7)$
$= -7\cdot 7$
$= -49$

9) There is an even number of negatives, so the answer is positive.
$(4)\cdot(-3)\cdot(2)\div(3)\div(-5)$
$= +(4)\cdot(3)\cdot(2)\div(3)\div(5)$
$= (12)\cdot(2)\div(3)\div(5)$
$= (24)\div(3)\div(5)$
$= (8)\div(5)$
$= 1\dfrac{3}{5}$

10) The exponent "touches" the grouping. The base includes the negative sign.

The exponent is 2.
The base is –4.
$(-4)^2 = (-4)\cdot(-4)$

11) The exponent "touches" only the 5, not the negative sign. The base does not include the negative sign.

The exponent is 2.
The base is 5.
$-5^2 = -(5\cdot 5)$

12) $(-1)^3 = (-1)\cdot(-1)\cdot(-1)$
$= -(1\cdot 1\cdot 1)$
$= -1$

13) $-3^2 = -(3\cdot 3)$
$= -9$

14) $(-7)^2 = (-7)\cdot(-7)$
$= +(7\cdot 7)$
$= 49$

15) $-7^2 = -(7\cdot 7)$
$= -49$

16) Average $= \dfrac{\text{Sum of the data values}}{\text{Number of data values}}$

$= \dfrac{(-10.2)+(-8.9)+(-4.0)+(4.5)}{4}$
$= \dfrac{-18.6}{4}$
$= -4.65$

The average monthly temperature is -4.65°C.

17) List all the factor pairs whose product is 6.
Add the factor pairs and look for a sum of -5.

Products that $= 6$	Sums of factors
$1 \cdot 6$	$1 + 6 = 7$
$2 \cdot 3$	$2 + 3 = 5$
$(-1) \cdot (-6)$	$(-1) + (-6) = -7$
$(-2) \cdot (-3)$	$(-2) + (-3) = -5$

The two numbers are (-2) and (-3).

Check: $(-2)(-3) = 6$ and $(-2) + (-3) = -5$.

18) List all the factor pairs whose product is -12.
Add the factor pairs and look for a sum of -4.

Products that $= -12$	Sums of factors
$1 \cdot (-12)$	$1 + (-12) = -11$
$2 \cdot (-6)$	$2 + (-6) = -4$
$3 \cdot (-4)$	$3 + (-4) = -1$
$(-1) \cdot 12$	$(-1) + 12 = 11$
$(-2) \cdot 6$	$(-2) + 6 = 4$
$(-3) \cdot 4$	$(-3) + 4 = 1$

The two numbers are (2) and (-6).

Check: $(2)(-6) = -12$ and $(2) + (-6) = -4$.

CONCEPT CHECKS AND PRACTICE EXERCISES

A1) Answers may vary. Example:
$(-4) + (-4) + (-4) = -12$. Therefore, adding -4 three times gives -12.

A2) a) $1 \cdot (-2) = -2$
$0 \cdot (-2) = 0$
$(-1) \cdot (-2) = 2$
$(-2) \cdot (-2) = 4$
$(-3) \cdot (-2) = 6$

b) Based on the pattern, two negatives result in a positive product.

A3) $13 \cdot (-8) = -104$

A4) $21 \cdot (-20) = -420$

A5) $(-7) \cdot (-15) = 105$

A6) $(-8) \cdot (-22) = 176$

A7) $(-3.3) \cdot (2) = -6.6$

A8) $(-5.5) \cdot (-10) = 55$

A9) $(-1.1) \cdot (-100) = 110$

A10) $(-7.1) \cdot (3) = -21.3$

B1) The product or quotient of two numbers that have the same sign will be positive.

B2) The product or quotient of two numbers that have opposite signs will be negative.

B3) $(27) \div (-3) = -9$

B4) $(14) \div (-7) = -2$

B5) $(-30) \div (-12) = \dfrac{30}{12}$
$= \dfrac{5 \cdot 6}{2 \cdot 6}$
$= \dfrac{5}{2} = 2\dfrac{1}{2}$

B6) $(-40) \div (-15) = \dfrac{40}{15}$
$= \dfrac{5 \cdot 8}{3 \cdot 5}$
$= \dfrac{8}{3} = 2\dfrac{2}{3}$

B7) $\left(\dfrac{5}{8}\right) \div \left(\dfrac{-3}{4}\right) = -\left(\dfrac{5}{8}\right) \cdot \left(\dfrac{4}{3}\right)$
$= -\dfrac{5 \cdot 4}{2 \cdot 4 \cdot 3}$
$= -\dfrac{5}{6}$

B8) $\left(-\dfrac{21}{1}\right) \div \left(-\dfrac{1}{2}\right) = +\left(\dfrac{21}{1}\right) \cdot \left(\dfrac{2}{1}\right)$
$= 42$

B9) $(-4.4) \div (0.02) = -220$

B10) $(-6.4) \div (-0.8) = 8$

C1) Answers may vary. Example: The product will be negative when there is an odd number of negative signed numbers.

C2) If there are no grouping symbols, perform the operations in the order encountered from left to right.

C3) $(-3)\cdot(2)\cdot 1\cdot(-1) = +(3)\cdot(2)\cdot 1\cdot(1)$
$= +(6)\cdot 1\cdot(1)$
$= +(6)\cdot(1)$
$= 6$

C4) $(-3)\cdot(-2)\cdot 2\cdot(-1) = -(3)\cdot(2)\cdot 2\cdot(1)$
$= -(6)\cdot 2\cdot(1)$
$= -(12)\cdot(1)$
$= -12$

C5) $(-2)\cdot(3)\cdot(3) = -(2)\cdot(3)\cdot(3)$
$= -(6)\cdot(3)$
$= -18$

C6) $6\cdot(-4)\div(-2)\cdot 3 = +6\cdot(4)\div(2)\cdot 3$
$= +(24)\div(2)\cdot 3$
$= +(12)\cdot 3$
$= 36$

C7) $(-12)\cdot 2\div 4\cdot(-2) = +(12)\cdot 2\div 4\cdot(2)$
$= +(24)\div 4\cdot(2)$
$= +(6)\cdot(2)$
$= 12$

C8) $18\div(-3)\cdot(-3) = +18\div(3)\cdot(3)$
$= +(6)\cdot(3)$
$= 18$

C9) $\dfrac{2\cdot(-3)\cdot 5}{(-20)} = +\dfrac{2\cdot(3)\cdot 5}{(20)}$
$= +\dfrac{30}{20}$
$= \dfrac{3}{2} = 1\dfrac{1}{2}$

C10) $3\cdot\left(-\dfrac{8}{9}\right)\cdot(-6) = +3\cdot\left(\dfrac{8}{9}\right)\cdot(6)$
$= +\dfrac{8}{3}\cdot(6)$
$= 16$

D1) a) 3^2 has a base of 3 and an exponent of 2.
$3^2 = 9$

b) $(-5)^3$ has a base of -5 and an exponent of 3.
$(-5)^3 = -125$

c) -8^2 has a base of 8 and an exponent of 2.
$-8^2 = -64$

d) $(-1)^9$ has a base of -1 and an exponent of 9.
$(-1)^9 = -1$

D2) $(5)^3 = (5)\cdot(5)\cdot(5)$
$= 125$

D3) $10^3 = (10)\cdot(10)\cdot(10)$
$= 1,000$

D4) $(-3)^2 = (-3)\cdot(-3)$
$= 9$

D5) $(-7)^2 = (-7)\cdot(-7)$
$= 49$

D6) $-2^5 = -(2)\cdot(2)\cdot(2)\cdot(2)\cdot(2)$
$= -32$

D7) $-2^4 = -(2)\cdot(2)\cdot(2)\cdot(2)$
$= -16$

D8) $(-2)^5 = (-2)\cdot(-2)\cdot(-2)\cdot(-2)\cdot(-2)$
$= -32$

D9) $(-2)^4 = (-2)\cdot(-2)\cdot(-2)\cdot(-2)$
$= 16$

SECTION 9.3   EXERCISES

For 1–12, refer to Concept Checks and Practice Exercises.

13) quotient

15) exponent

17) $-2\cdot 12 = -24$

19) $(-13)\cdot(-3) = 39$

21) $(2.5)\cdot(8) = 20$

23) $(-10.1)\cdot(9) = -90.9$

25) $\left(-\dfrac{3}{5}\right)\cdot\left(-\dfrac{10}{9}\right)=+\left(\dfrac{3}{5}\right)\cdot\left(\dfrac{10}{9}\right)$

$=\dfrac{\cancel{3}\cdot2\cdot\cancel{5}}{\cancel{5}\cdot\cancel{3}\cdot3}$

$=\dfrac{2}{3}$

27) $\left(-\dfrac{3}{12}\right)\cdot\left(\dfrac{4}{17}\right)=-\left(\dfrac{3}{12}\right)\cdot\left(\dfrac{4}{17}\right)$

$=-\dfrac{\cancel{3}\cdot\cancel{4}}{\cancel{3}\cdot\cancel{4}\cdot17}$

$=-\dfrac{1}{17}$

29) $50\div(-25)=-2$

31) $(-65)\div(-13)=5$

33) $(-1.6)\div(4)=-0.4$

35) $(-0.35)\div(-0.07)=5$

37) $\dfrac{-100}{-20}=5$

39) $\left(-\dfrac{3}{5}\right)\div\left(-\dfrac{9}{10}\right)=+\left(\dfrac{3}{5}\right)\cdot\left(\dfrac{10}{9}\right)$

$=\dfrac{\cancel{3}\cdot2\cdot\cancel{5}}{\cancel{5}\cdot\cancel{3}\cdot3}$

$=\dfrac{2}{3}$

41) $(-14)\cdot(2)\div(-7)\cdot(2)=+(14)\cdot(2)\div(7)\cdot(2)$

$=+(28)\div(7)\cdot(2)$

$=+(4)\cdot(2)$

$=8$

43) $(60)\div(-2)\div(3)\cdot(10)=-(60)\div(2)\div(3)\cdot(10)$

$=-(30)\div(3)\cdot(10)$

$=-(10)\cdot(10)$

$=-100$

45) $\left(-\dfrac{3}{4}\right)\cdot\left(-\dfrac{1}{3}\right)\div(-7)=-\left(\dfrac{3}{4}\right)\cdot\left(\dfrac{1}{3}\right)\div(7)$

$=-\left(\dfrac{1}{4}\right)\div(7)$

$=-\left(\dfrac{1}{4}\right)\cdot\left(\dfrac{1}{7}\right)$

$=-\dfrac{1}{28}$

47) $(16)\div(8)\cdot(2)\div(-0.1)=-(16)\div(8)\cdot(2)\div(0.1)$

$=-(2)\cdot(2)\div(0.1)$

$=-(4)\div(0.1)$

$=-40$

49) $(13)\cdot(-2)\div(-26)\cdot(0)=+(13)\cdot(2)\div(26)\cdot(0)$

$=+(26)\div(26)\cdot(0)$

$=+(1)\cdot(0)$

$=0$

51) $(-100)\div(-25)\div(2)=+(100)\div(25)\div(2)$

$=+(4)\div(2)$

$=2$

53) $9^3=(9)\cdot(9)$

$=81$

55) $(-3)^3=(-3)\cdot(-3)\cdot(-3)$

$=-27$

57) $-2^4=-(2)\cdot(2)\cdot(2)\cdot(2)$

$=-16$

59) $(6)^2=(6)\cdot(6)$

$=36$

61) $\left(-\dfrac{3}{5}\right)^2=\left(-\dfrac{3}{5}\right)\cdot\left(-\dfrac{3}{5}\right)$

$=\dfrac{9}{25}$

63) $-\left(\dfrac{2}{3}\right)^3=-\left(\dfrac{2}{3}\right)\cdot\left(\dfrac{2}{3}\right)\cdot\left(\dfrac{2}{3}\right)$

$=-\dfrac{8}{27}$

65) $0^4=0$

67) $(-1)^{10}=1$

69)

Products that $=3$	Sums of factors
$1\cdot3$	$1+3=4$
$(-1)\cdot(-3)$	$(-1)+(-3)=-4$

The two numbers are 1 and 3.

71)

Products that $=-8$	Sums of factors
$1\cdot(-8)$	$1+(-8)=-7$
$2\cdot(-4)$	$2+(-4)=-2$
$(-1)\cdot 8$	$(-1)+8=7$
$(-2)\cdot 4$	$(-2)+4=2$

The two numbers are -2 and 4.

73)

Products that $=-18$	Sums of factors
$1\cdot(-18)$	$1+(-18)=-17$
$2\cdot(-9)$	$2+(-9)=-7$
$3\cdot(-6)$	$3+(-6)=-3$
$(-1)\cdot 18$	$(-1)+18=17$
$(-2)\cdot 9$	$(-2)+9=7$
$(-3)\cdot 6$	$(-3)+6=3$

The two numbers are 2 and -9.

75)

Products that $=40$	Sums of factors
$1\cdot 40$	$1+40=41$
$2\cdot 20$	$2+20=22$
$4\cdot 10$	$4+10=14$
$5\cdot 8$	$5+8=13$
$(-1)\cdot(-40)$	$(-1)+(-40)=-41$
$(-2)\cdot(-20)$	$(-2)+(-20)=-22$
$(-4)\cdot(-10)$	$(-4)+(-10)=-14$
$(-5)\cdot(-8)$	$(-5)\cdot(-8)=-13$

The two numbers are -5 and -8.

77) Answers may vary. Example: Each of these fractions is equal to negative three.

79) 0 and 1 are both correct answers to the question.

81) $\dfrac{-30\text{ ft}}{1\ \cancel{s}}\cdot\dfrac{60\ \cancel{s}}{1\text{ min}}=-1{,}800\text{ feet}$

The jet descends 1,800 feet in one minute.

83) $80(\$1.50)+30(-\$2.20)=\$120-\$66$
$=\$54$
The value of Greta's stocks increased by \$54.

85) $13\cdot(-2)=-26$
13 oxide ions have a total charge of -26.

87) $7\cdot(1)+6\cdot(2)=7+12$
$=19$
The total charge is 19.

89) $-6\cdot(-2)=12$
The change in the charge of the remaining substance is 12.

91) Three aluminum and one magnesium ion have a total charge of $+11$.
$3\cdot(3)+1\cdot(2)=9+2$
$=11$

93) $\text{Average}=\dfrac{\text{Sum of the data values}}{\text{Number of data values}}$
$=\dfrac{(20)+(-30)+(-40)+(10)}{4}$
$=\dfrac{-40}{4}$
$=-10$

95) $\text{Average}=\dfrac{\text{Sum of the data values}}{\text{Number of data values}}$
$=\dfrac{(1)+(-3)+(5)+(-7)+(9)+(-11)}{6}$
$=\dfrac{-6}{6}$
$=-1$

97) $\text{Average}=\dfrac{\text{Sum of the data values}}{\text{Number of data values}}$
$=\dfrac{(3)+(7)+(30)+(-12)+(-8)+(-20)}{6}$
$=\dfrac{0}{6}$
$=0$

9.4   The Order of Operations and Signed Numbers

GUIDED PRACTICE

1) $(-3)-(4)+(-7)=(-3)+(-4)+(-7)$
$=-7+(-7)$
$=-14$

2) $35-15\cdot(-2)=35+30$
$=65$

3) $\dfrac{-42\div(-21)}{6-(-8)}=\dfrac{2}{14}$

$\qquad\qquad\quad =\dfrac{\cancel{2}}{\cancel{2}\cdot 7}$

$\qquad\qquad\quad =\dfrac{1}{7}$

4) $(1-3)^2+(5-7)^3\div 2\cdot(-4)$

$=(-2)^2+(-2)^3\div 2\cdot(-4)$

$=4+(-8)\div 2\cdot(-4)$

$=4+(-4)\cdot(-4)$

$=4+16$

$=20$

5) money left $=$ original amount $+$ change

change $=$ number of games $\cdot$ cost per game

change $=7\cdot(-\$0.25)$

money left $=$ original amount $+$ change

$\qquad\qquad\quad =\$3.45+7\cdot(-\$0.25)$

6) $16\div 2-6\overset{?}{=}-4$

$\qquad 8-6\overset{?}{=}-4$

$\qquad\quad 2\overset{?}{=}-4$

Not true.

$(16\div 2)-6\overset{?}{=}-4$ $\qquad$ $16\div(2-6)\overset{?}{=}-4$

The calculation order $\qquad$ The calculation order
does not change, so do $\qquad$ changes, so check
not check this equation. $\qquad$ this equation.

$16\div(2-6)\overset{?}{=}-4$

$\quad 16\div(-4)\overset{?}{=}-4$

$\qquad\qquad -4=-4$

True.

The equation $16\div(2-6)=-4$ is a true

statement.

CONCEPT CHECKS AND PRACTICE EXERCISES

A1) Perform all operations inside grouping symbols
and calculate absolute values.
Perform operations with exponents.
Multiply/divide in order from left to right.
Add/subtract in order from left to right.

A2) $-7-3+2=(-7)+(-3)+2$

$\qquad\qquad\quad =-10+2$

$\qquad\qquad\quad =-8$

A3) $-5-8+4=-5+(-8)+4$

$\qquad\qquad\quad =-13+4$

$\qquad\qquad\quad =-9$

A4) $-10\div 2\cdot 5=-5\cdot 5$

$\qquad\qquad\quad =-25$

A5) $-6\div 2\cdot 3=-3\cdot 3$

$\qquad\qquad\quad =-9$

A6) $(-3)+(-2)\cdot 6-5=(-3)+(-12)-5$

$\qquad\qquad\qquad\qquad =-15-5$

$\qquad\qquad\qquad\qquad =-20$

A7) $(4)-(6)\div(-2)+(-3)=(4)-(-3)+(-3)$

$\qquad\qquad\qquad\qquad\qquad =4+3+(-3)$

$\qquad\qquad\qquad\qquad\qquad =7+(-3)$

$\qquad\qquad\qquad\qquad\qquad =4$

A8) $(-1)^3-(-1)^2+(-5)\cdot 2$

$=-1-1+(-5)\cdot 2$

$=-1-1+(-10)$

$=-2+(-10)$

$=-12$

A9) $(-1)^4+(-1)^3-(-3)\cdot(-4)$

$=(-1)^4+(-1)^3-(-3)\cdot(-4)$

$=1+(-1)-(12)$

$=0-12$

$=-12$

A10) $\dfrac{1}{2}\div\dfrac{1+2}{4}=\dfrac{1}{2}\div\dfrac{3}{4}$

$\qquad\qquad\quad =\dfrac{1}{2}\cdot\dfrac{4}{3}$

$\qquad\qquad\quad =\dfrac{1}{\cancel{2}}\cdot\dfrac{\cancel{2}\cdot 2}{3}$

$\qquad\qquad\quad =\dfrac{2}{3}$

A11) $\dfrac{3}{4}\div\dfrac{5+(-2)}{8}=\dfrac{3}{4}\div\dfrac{3}{8}$

$\qquad\qquad\qquad =\dfrac{3}{4}\cdot\dfrac{8}{3}$

$\qquad\qquad\qquad =\dfrac{\cancel{3}}{\cancel{4}}\cdot\dfrac{2\cdot\cancel{4}}{\cancel{3}}$

$\qquad\qquad\qquad =2$

A12) $\left(\dfrac{5-(7-12)}{(-5)+(7-12)}\right)^2 = \left(\dfrac{5-(-5)}{(-5)+(-5)}\right)^2$

$\qquad\qquad\qquad = \left(\dfrac{10}{-10}\right)^2$

$\qquad\qquad\qquad = (-1)^2$

$\qquad\qquad\qquad = 1$

A13) $\left(\dfrac{(-4)+(13-8)}{4-(13-8)}\right)^2 = \left(\dfrac{(-4)+(5)}{4-(5)}\right)^2$

$\qquad\qquad\qquad = \left(\dfrac{1}{-1}\right)^2$

$\qquad\qquad\qquad = (-1)^2$

$\qquad\qquad\qquad = 1$

SECTION 9.4   EXERCISES

For 1–3, refer to Concept Checks and Practice Exercises.

5) $(-5)+12-(8) = 7-(8)$
$\qquad\qquad\quad = -1$

7) $3\cdot(-8)\div(4) = -24\div(4)$
$\qquad\qquad\qquad = -6$

9) $15\cdot(-2)\div(-1) = 15\cdot(-2)\div(-1)$
$\qquad\qquad\qquad\quad = -30\div(-1)$
$\qquad\qquad\qquad\quad = 30$

11) $-7\cdot(-7)+3\cdot(-8) = 49+3\cdot(-8)$
$\qquad\qquad\qquad\qquad = 49+(-24)$
$\qquad\qquad\qquad\qquad = 25$

13) $5-(7)\cdot(3)+(-35) = 5-21+(-35)$
$\qquad\qquad\qquad\qquad = -16+(-35)$
$\qquad\qquad\qquad\qquad = -51$

15) $\dfrac{3}{4}\cdot\dfrac{1+(-7)}{5} = \dfrac{3}{4}\cdot\dfrac{-6}{5}$
$\qquad\qquad\qquad = -\dfrac{9}{10}$

17) $28\div(-8-6)+2\cdot(-3) = 28\div(-14)+2\cdot(-3)$
$\qquad\qquad\qquad\qquad\quad = -2+2\cdot(-3)$
$\qquad\qquad\qquad\qquad\quad = -2+(-6)$
$\qquad\qquad\qquad\qquad\quad = -8$

19) $(6^2-2^3)\div 4\cdot 2 = (36-8)\div 4\cdot 2$
$\qquad\qquad\qquad\quad = (28)\div 4\cdot 2$
$\qquad\qquad\qquad\quad = 7\cdot 2$
$\qquad\qquad\qquad\quad = 14$

21) $3(6-8)+5 = 3(-2)+5$
$\qquad\qquad\qquad = -6+5$
$\qquad\qquad\qquad = -1$

23) $|15-17|-2+(-12)\div(6) = |-2|-2+(-12)\div(6)$
$\qquad\qquad\qquad\qquad\qquad = 2-2+(-12)\div(6)$
$\qquad\qquad\qquad\qquad\qquad = 2-2+(-2)$
$\qquad\qquad\qquad\qquad\qquad = 0+(-2)$
$\qquad\qquad\qquad\qquad\qquad = -2$

25) $6-7+(9-5)^2 = 6-7+(4)^2$
$\qquad\qquad\qquad\quad = 6-7+16$
$\qquad\qquad\qquad\quad = -1+16$
$\qquad\qquad\qquad\quad = 15$

27) $10-(-30-17) = 10-(-47)$
$\qquad\qquad\qquad = 10+47$
$\qquad\qquad\qquad = 57$

29) $\dfrac{(-3)\cdot 4-(-2)}{-2+4} = \dfrac{-12-(-2)}{-2+4}$
$\qquad\qquad\qquad = \dfrac{-12+2}{-2+4}$
$\qquad\qquad\qquad = \dfrac{-10}{2}$
$\qquad\qquad\qquad = -5$

31) $\dfrac{(-9)\div 3\cdot 4}{4-(-8)} = \dfrac{(-3)\cdot 4}{4+8}$
$\qquad\qquad\qquad = \dfrac{-12}{12}$
$\qquad\qquad\qquad = -1$

33) $\dfrac{(-7)-5\cdot 2+13}{(-1)^5+9-8\cdot 2} = \dfrac{(-7)-5\cdot 2+13}{-1+9-8\cdot 2}$
$\qquad\qquad\qquad\qquad = \dfrac{(-7)-10+13}{-1+9-16}$
$\qquad\qquad\qquad\qquad = \dfrac{-17+13}{8-16}$
$\qquad\qquad\qquad\qquad = \dfrac{-4}{-8}$
$\qquad\qquad\qquad\qquad = \dfrac{1}{2}$

35) $\dfrac{3^2-5^2}{6^2-8^2} = \dfrac{9-25}{36-64}$

$\phantom{\dfrac{3^2-5^2}{6^2-8^2}} = \dfrac{-16}{-28}$

$\phantom{\dfrac{3^2-5^2}{6^2-8^2}} = \dfrac{4}{7}$

37) $\dfrac{5\cdot 4\div(-2)+15}{-13+8\cdot 1} = \dfrac{20\div(-2)+15}{-13+8}$

$\phantom{\dfrac{5\cdot 4\div(-2)+15}{-13+8\cdot 1}} = \dfrac{-10+15}{-13+8}$

$\phantom{\dfrac{5\cdot 4\div(-2)+15}{-13+8\cdot 1}} = \dfrac{5}{-5}$

$\phantom{\dfrac{5\cdot 4\div(-2)+15}{-13+8\cdot 1}} = -1$

39) $\dfrac{64+(-36)+8}{-21+3} = \dfrac{28+8}{-21+3}$

$\phantom{\dfrac{64+(-36)+8}{-21+3}} = \dfrac{36}{-18}$

$\phantom{\dfrac{64+(-36)+8}{-21+3}} = -2$

41) $\text{Average} = \dfrac{\text{Differences in points}}{\text{Number of games}}$

$\phantom{\text{Average}} = \dfrac{(8)+(-12)+(7)+(9)+(-1)+(7)}{6}$

$\phantom{\text{Average}} = \dfrac{18}{6}$

$\phantom{\text{Average}} = 3$

The average difference in points over the six games was 3.

43) $\text{Average} = \dfrac{\text{Rates of return}}{\text{Number of investments}}$

$\phantom{\text{Average}} = \dfrac{(12)+(5)+(-3)+(2)}{4}$

$\phantom{\text{Average}} = \dfrac{16}{4}$

$\phantom{\text{Average}} = 4$

The average rate of return was 4%.

45) $(-3)^2 \div \dfrac{9}{2} + \dfrac{-1}{2} = 9 \div \dfrac{9}{2} + \dfrac{-1}{2}$

$\phantom{(-3)^2 \div \dfrac{9}{2} + \dfrac{-1}{2}} = \dfrac{9}{1}\cdot\dfrac{2}{9} + \dfrac{-1}{2}$

$\phantom{(-3)^2 \div \dfrac{9}{2} + \dfrac{-1}{2}} = \dfrac{2}{1} - \dfrac{1}{2}$

$\phantom{(-3)^2 \div \dfrac{9}{2} + \dfrac{-1}{2}} = \dfrac{4}{2} - \dfrac{1}{2}$

$\phantom{(-3)^2 \div \dfrac{9}{2} + \dfrac{-1}{2}} = \dfrac{3}{2} = 1\dfrac{1}{2}$

47) $(-4)\cdot\dfrac{3}{20} + \dfrac{-3}{10} = -\dfrac{4}{1}\cdot\dfrac{3}{20} + \dfrac{-3}{10}$

$\phantom{(-4)\cdot\dfrac{3}{20} + \dfrac{-3}{10}} = -\dfrac{3}{5} - \dfrac{3}{10}$

$\phantom{(-4)\cdot\dfrac{3}{20} + \dfrac{-3}{10}} = -\dfrac{6}{10} - \dfrac{3}{10}$

$\phantom{(-4)\cdot\dfrac{3}{20} + \dfrac{-3}{10}} = -\dfrac{9}{10}$

49) $\left(\dfrac{1}{4}+\dfrac{-1}{3}\right) + \left(\dfrac{1}{8}-\dfrac{1}{6}\right) = \left(\dfrac{3}{12}-\dfrac{4}{12}\right) + \left(\dfrac{3}{24}-\dfrac{4}{24}\right)$

$\phantom{\left(\dfrac{1}{4}+\dfrac{-1}{3}\right) + \left(\dfrac{1}{8}-\dfrac{1}{6}\right)} = \left(-\dfrac{1}{12}\right) + \left(-\dfrac{1}{24}\right)$

$\phantom{\left(\dfrac{1}{4}+\dfrac{-1}{3}\right) + \left(\dfrac{1}{8}-\dfrac{1}{6}\right)} = -\dfrac{2}{24} - \dfrac{1}{24}$

$\phantom{\left(\dfrac{1}{4}+\dfrac{-1}{3}\right) + \left(\dfrac{1}{8}-\dfrac{1}{6}\right)} = -\dfrac{3}{24}$

$\phantom{\left(\dfrac{1}{4}+\dfrac{-1}{3}\right) + \left(\dfrac{1}{8}-\dfrac{1}{6}\right)} = -\dfrac{1}{8}$

51) $(0.3)^2 - \dfrac{9}{100} = 0.09 - \dfrac{9}{100}$

$\phantom{(0.3)^2 - \dfrac{9}{100}} = \dfrac{9}{100} - \dfrac{9}{100}$

$\phantom{(0.3)^2 - \dfrac{9}{100}} = 0$

53) $-(6)^2 + 24 \div (-4) = -36 + 24 \div (-4)$

$ = -36 + (-6)$

$ = -42$

55) $-(-1)^6 + (12-14)^3 = -(-1)^6 + (-2)^3$

$ = -1-8$

$ = -9$

57) $150\cdot(\$3.21) + 200\cdot(-\$1.75)$

59) $\$11,000 + 12(-\$850)$

61) $(2-3)\cdot 4 = (-1)\cdot 4 = -4$

63) Parentheses not needed.
$1-2+3 = -1+3 = 2$

65) $18 \div (3\cdot 2) = 18 \div (6) = 3$

67) Parentheses not needed.
$28 \div 4\cdot 3 = 7\cdot 3 = 21$

69) $(8+8) \div 4 = (16) \div 4 = 4$

71) The sum of twice ten and four
c) $2(10)+4$

73) The product of two and negative four,
    multiplied by ten

    b) $2(-4) \cdot 10$

75) The quotient of ten and two, decreased by four

    i) $\dfrac{10}{2} - 4$

77) The quotient of ten and negative four, increased
    by two

    d) $\dfrac{10}{-4} + 2$

79) Negative ten, decreased by half of negative four

    h) $-10 - \dfrac{-4}{2}$

CHAPTER 9   REVIEW EXERCISES

1) -75

2) 175

3) $7{,}400$

4) -48.50

5) $-11 > -21$

6) $-24 < -19$

7) $-7.89 < -7.8$

8) $-3.4 > -3.46$

9) $-(24) = -24$

10) $-(-52) = 52$

11) $-|-6| = -6$

12) $-|56| = -56$

13)
$-9 < -4$

14)
$-7 < 3$

15) $-(-9) = 9;\ |-8| = 8;\ -5$
$-5 < |-8| < -(-9)$

16) $|-3| = 3;\ -(-2) = 2;\ 5$
$-(-2) < |-3| < 5$

17) All negative numbers are to the left of all
    positive numbers on the number line.

18) All positive numbers are to the right of all
    negative numbers on the number line.

19) $-5 + -6 = -11$

20) $-3 + -2 = -5$

21) $-5 + 7 = 2$

22) $2+(-6)=-4$

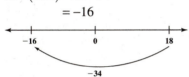

23) $-7+(-5)=-12$

24) $18-(+34)=18-34$
$\qquad\qquad\quad =-16$

25) $-5-10=-15$

26) $-4-7=-11$

27) $-12+(-4)=-16$

28) $-6+-18=-24$

29) $10-16=-6$

30) $8-24=-16$

31) $-11-(-9)=-11+9$
$\qquad\qquad\quad\; =-2$

32) $-13-(-2)=-13+2$
$\qquad\qquad\quad\; =-11$

33) $37+(-24)=37-24$
$\qquad\qquad\quad =13$

34) $-28+24=-4$

35) $-\dfrac{3}{8}+\dfrac{3}{2}=-\dfrac{3}{8}+\dfrac{12}{8}$
$\qquad\qquad =\dfrac{9}{8}=1\dfrac{1}{8}$

36) $\dfrac{2}{3}-\left(-\dfrac{3}{2}\right)=\dfrac{2}{3}+\dfrac{3}{2}$
$\qquad\qquad\quad =\dfrac{4}{6}+\dfrac{9}{6}$
$\qquad\qquad\quad =\dfrac{13}{6}=2\dfrac{1}{6}$

37) $|21|+(-3)-30-(-2)=21+(-3)-30-(-2)$
$\qquad\qquad\qquad\qquad\quad =18-30-(-2)$
$\qquad\qquad\qquad\qquad\quad =-12-(-2)$
$\qquad\qquad\qquad\qquad\quad =-12+2$
$\qquad\qquad\qquad\qquad\quad =-10$

38) $|-15|-17+(-4)-(-8)=15-17+(-4)-(-8)$
$\qquad\qquad\qquad\qquad\quad =-2+(-4)-(-8)$
$\qquad\qquad\qquad\qquad\quad =-6-(-8)$
$\qquad\qquad\qquad\qquad\quad =-6+8$
$\qquad\qquad\qquad\qquad\quad =2$

39)

Type	Date	Desc	Payment/Debit (−)	Fee (if any)	Deposit/Credit (+)	Balance
	4-May	Bal Fwd				$132.85
212	5-May	Credit card	$48.24			$84.61
ATM	8-May	Needed cash	$20.00	$3.00		$61.61
Dep	17-May	Pay day!			$324.62	$386.23
213	18-May	Bill	$117.43			$268.80
ATM	24-May	Needed cash	$40.00	$2.00		$226.80

40) $4\cdot(-3)=-12$

41) $(-5)\cdot(-7)=35$

42) $-15\div3=-5$

43) $-20\div(-4)=5$

44) $3\cdot(-2)\cdot(-8)=(-6)\cdot(-8)$
$\qquad\qquad\qquad =48$

45) $-2\cdot(-14)\div(-4)=(28)\div(-4)$
$\qquad\qquad\qquad\quad =-7$

46) $(-3)^2=9$

47) $-4^2=-16$

48) $\frac{4}{15} \cdot (-3)^2 = \frac{4}{15} \cdot \frac{9}{1}$

$\quad = \frac{4}{\cancel{3} \cdot 5} \cdot \frac{\cancel{3} \cdot 3}{1}$

$\quad = \frac{12}{5} = 2\frac{2}{5}$

49) $\frac{-3}{14} \cdot (-7)^2 = \frac{-3}{14} \cdot \frac{49}{1}$

$\quad = -\frac{3}{2 \cdot \cancel{7}} \cdot \frac{\cancel{7} \cdot 7}{1}$

$\quad = -\frac{21}{2} = -10\frac{1}{2}$

50) $\left(3 \cdot \frac{-2}{3}\right)^2 = (-2)^2$

$\quad = 4$

51) $\left(-2 \div \frac{1}{3}\right)^2 = (-2 \cdot 3)^2$

$\quad = (-6)^2$

$\quad = 36$

52)

Products that $= -20$	Sums of factors
$1 \cdot (-20)$	$1 + (-20) = -19$
$2 \cdot (-10)$	$2 + (-10) = -8$
$4 \cdot (-5)$	$4 + (-5) = -1$
$(-1) \cdot 20$	$(-1) + 20 = 19$
$(-2) \cdot 10$	$(-2) + 10 = 8$
$(-4) \cdot 5$	$(-4) + 5 = 1$

The two numbers are 2 and –10.

53)

Products that $= 14$	Sums of factors
$1 \cdot 14$	$1 + 14 = 15$
$2 \cdot 7$	$2 + 7 = 9$
$(-1) \cdot (-14)$	$(-1) + (-14) = -15$
$(-2) \cdot (-7)$	$(-2) + (-7) = -9$

The two numbers are –2 and –7.

54) $-8 + 12 \div 4 = -8 + 3$

$\quad = -5$

55) $18 - 9 \div 3 = 18 - 3$

$\quad = 15$

56) $2 \cdot (-3)^2 - 6 = 2 \cdot 9 - 6$

$\quad = 18 - 6$

$\quad = 12$

57) $-10 - 3^2 \cdot 2 = -10 - 9 \cdot 2$

$\quad = -10 - 18$

$\quad = -28$

58) $3 - 2(4 - 6) = 3 - 2(-2)$

$\quad = 3 - (-4)$

$\quad = 3 + 4$

$\quad = 7$

59) $4 + 3(5 - 8) = 4 + 3(-3)$

$\quad = 4 + (-9)$

$\quad = 4 - 9$

$\quad = -5$

60) $\frac{2 - 7}{8 + (-6)} = \frac{-5}{2}$

$\quad = -\frac{5}{2} = -2\frac{1}{2}$

61) $\frac{3 + 2}{15 - 8} = \frac{5}{7}$

62) $\frac{4 - 7 + 3}{7^2} = \frac{4 - 7 + 3}{49}$

$\quad = \frac{0}{49}$

$\quad = 0$

63) $\frac{5 - 8 + 3}{5^2} = \frac{5 - 8 + 3}{25}$

$\quad = \frac{0}{25}$

$\quad = 0$

64) Average $= \dfrac{\text{Sum of the data values}}{\text{Number of data values}}$

$\quad = \dfrac{(-10) + (4) + (-8) + (3) + (-6) + (5)}{6}$

$\quad = \dfrac{-12}{6}$

$\quad = -2$

65) Average $= \dfrac{\text{Sum of the data values}}{\text{Number of data values}}$

$\quad = \dfrac{(20) + (-33) + (15) + (-17) + (-2) + (7)}{6}$

$\quad = \dfrac{-10}{6}$

$\quad = -\frac{5}{3} = -1\frac{2}{3}$

66) $\$8.45 + 18 \cdot (-\$0.25)$

67) $\$450 + 12 \cdot (-\$20)$

CHAPTER 9   TEST

1) -54

2) 2.5

3) $-18 > -125$

4) $-5.61 < -5.6$

5) $|-4| = 4,\ -(-8) = 8,\ -5$

$-5 < |-4| < -(-8)$

6) $-4,\ -(-3) = 3,\ |-2| = 2$

$-4 < |-2| < -(-3)$

7) $5 + (-12) = 5 - 12$
$\quad\quad\quad\quad = -7$

8) $-4.8 + (-12) = -16.8$

9) $(-5)(-6) = 30$

10) $21 \div (-7) = -3$

11) $-\dfrac{3}{4} + \dfrac{7}{12} = -\dfrac{9}{12} + \dfrac{7}{12}$
$\quad\quad\quad\quad\ = -\dfrac{2}{12}$
$\quad\quad\quad\quad\ = -\dfrac{1}{6}$

12) $130 - 175 = -45$

13) $-3 - (-9) = -3 + 9$
$\quad\quad\quad\quad = 6$

14) $-(-3)^3 = -(-27)$
$\quad\quad\quad\ = 27$

15) $-\$193 - \$284 = -\$477$
The change in balance is $-\$477$.

16) $\dfrac{3.4 \text{ feet}}{1 \text{ s}} \cdot 8 \text{ s} = 27.2 \text{ feet}$
The concrete will flow 27.2 feet in eight seconds.

17) $(-12) \div 2 + 4 = (-6) + 4$
$\quad\quad\quad\quad\quad\quad = -2$

18) $7 - 9 + 11 = -2 + 11$
$\quad\quad\quad\quad\ = 9$

19) $10 - 3^2 = 10 - 9$
$\quad\quad\quad\ = 1$

20) $12 - 4^2 = 12 - 16$
$\quad\quad\quad\ = -4$

21) $5(-7 + 3) = 5(-4)$
$\quad\quad\quad\quad = -20$

22) $\dfrac{-4 + 5}{-15 - 10} = \dfrac{1}{-25}$
$\quad\quad\quad\quad = -\dfrac{1}{25}$

23) $(-2)(-1) + (4 - 7)^2 = (-2)(-1) + (-3)^2$
$\quad\quad\quad\quad\quad\quad\quad\quad\ = (-2)(-1) + 9$
$\quad\quad\quad\quad\quad\quad\quad\quad\ = 2 + 9$
$\quad\quad\quad\quad\quad\quad\quad\quad\ = 11$

24) $-\dfrac{1}{8} - \dfrac{1}{2} \cdot \dfrac{1}{4} = -\dfrac{1}{8} - \dfrac{1}{8}$
$\quad\quad\quad\quad\quad\ = -\dfrac{2}{8}$
$\quad\quad\quad\quad\quad\ = -\dfrac{1}{4}$

25) Average $= \dfrac{\text{Sum of the data values}}{\text{Number of data values}}$
$\quad\quad\quad\ = \dfrac{(-10) + (4) + (-9)}{3}$
$\quad\quad\quad\ = \dfrac{-15}{3}$
$\quad\quad\quad\ = -5$

26) Average $= \dfrac{\text{Sum of the data values}}{\text{Number of data values}}$
$\quad\quad\quad\ = \dfrac{(20) + (-3) + (15) + (-4)}{4}$
$\quad\quad\quad\ = \dfrac{28}{4}$
$\quad\quad\quad\ = 7$

27) Average $= \dfrac{\text{Sum of temperatures}}{\text{Number of days}}$

$= \dfrac{(16)+(19)+(-5)+(-10)+(15)}{5}$

$= \dfrac{35}{5}$

$= 7$

The average high temperature from Monday to

Friday was $7°$F.

28) a) $15-(-10)=15+10=25$

The greatest increase in high temperature

from one day to the next was $25°$F.

b) $-5-19=-24$

The greatest decrease in high temperature

from one day to the next was $-24°$F.

c) The greatest change in high temperature

from one day to the next was $25°$F.

29) $-10-32=-42$

The high temperature on Thursday was $42°$F
below freezing.

30) Average $= \dfrac{\text{Sum of temperatures}}{\text{Number of days}}$

$= \dfrac{(-5)+(-10)+(15)}{3}$

$= \dfrac{0}{3}$

$= 0$

Wednesday, Thursday, and Friday had an

average high temperature of $0°$F.

Chapter 10 Introduction to Algebra

10.1 Introduction to Variables

GUIDED PRACTICE

1) Since the number of lollipops in a bag is unknown, we let l = the number of lollipops per bag.

   There are 2 bags and 4 lollipops. We can describe the number of lollipops using the variable expression $2l + 4$.

2) Let b = the number of batteries per case.

   Number of batteries $= 3b + 4$

3) Let c = the number of cans of corn in each case.

   Number of cans of corn $= 3c - 14$

4) $3x - 8 = 3(4) - 8$
   $= 12 - 8$
   $= 4$

5) $8 - y = 8 - (-7)$
   $= 8 + 7$
   $= 15$

6) $\dfrac{5x - y}{y} = \dfrac{5(8) - (-5)}{(-5)}$
   $= \dfrac{40 - (-5)}{-5}$
   $= \dfrac{40 + 5}{-5}$
   $= \dfrac{45}{-5}$
   $= -9$

7) $\boxed{5z}\ \boxed{-2y}\ \boxed{-3}\ \boxed{+z}$

   There are four terms.
   The terms are $5z$, $-2y$, -3, and z.

8) $\boxed{2y}\ \boxed{-3}\ \boxed{+15}\ \boxed{-x}$

   Of the four terms, only -3 and 15 have no variable. The constant terms are -3 and 15.

9) $\boxed{4y^2}\ \boxed{-6}\ \boxed{+5yz}\ \boxed{-3x}$

   The variable terms are $4y^2$, $5yz$, and $-3x$

because they are the terms with at least one variable.

10) $\boxed{-y^2}\ \boxed{+5xy}\ \boxed{-3}$

   $-y^2$ Coefficient $= -1$
          Variable Part $= y^2$

   $5xy$ Coefficient $= 5$
          Variable Part $= xy$

   -3 Coefficient $= -3$
          -3 is a constant term so there is no variable part.

11) $\boxed{-xy^2}\ \boxed{+5xy}\ \boxed{+2}\ \boxed{-2x}\ \boxed{-3xy}\ \boxed{-7}$

   Identical variable part, xy
   $5xy$ and $-3xy$ are like terms.

   Constant terms have no variable parts.
   2 and -7 are like terms.

   $-xy^2$ and $-2x$ are unlike terms because the variable parts are different.

12) a) Let c = cost per month.
   Let m = minutes over 450.

   Cost equals $60 plus $0.45 per minute over 450.
   $c = 60 + 0.45m$

   b) $m = 500 - 450$
   $= 50$

   $c = 60 + 0.45m$
   $= 60 + 0.45(50)$
   $= 60 + 22.5$
   $= 82.5$
   It will cost Juanita $82.50 to talk 500 minutes on this plan.

CONCEPT CHECKS AND PRACTICE EXERCISES

A1) Answers may vary. Example: A variable is a symbol or letter representing a numerical quantity.

A2) Answers may vary. Example: b represents the number of batteries per box, and there are four boxes. To find the total number of batteries, multiply the number of boxes (4) by the number of batteries (b) in each box.

A3) Let $c =$ the number of pieces of chalk in one box.

Number of pieces of chalk $= 3c + 6$

A4) Let $b =$ the number of candy bars in one box.

Number of candy bars $= 4b + 5$

A5) Let $n =$ the number of notebooks in one box.

Number of notebooks $= 6n + 31$

A6) Let $b =$ the number of beef jerky sticks per box.

Number of beef jerky sticks $= 8b + 14$

A7) Let a represent the amount in the account.
$a - 1{,}500$

A8) Let a represent the amount in the account.
$A + 3{,}500$

B1) Answers may vary. Example: Evaluating an expression means finding the value of a variable expression by replacing each variable with its value.

B2) Expression b) is correct. When a coefficient sits directly to the left of a variable with no spaces or other signs of punctuation or operation, it indicates multiplication.

B3) Answers may vary. Example: By placing the ( ) in the expression, the operation can easily be identified.

B4) $2x + 5 = 2(4) + 5$
$= 8 + 5$
$= 13$

B5) $3x + 4 = 3(4) + 4$
$= 12 + 4$
$= 16$

B6) $10 - z = 10 - (-2)$
$= 10 + 2$
$= 12$

B7) $25 - z = 25 - (-2)$
$= 25 + 2$
$= 27$

B8) $\dfrac{2x+y}{y} = \dfrac{2(4)+(8)}{(8)}$
$= \dfrac{8+8}{8}$
$= \dfrac{16}{8}$
$= 2$

B9) $\dfrac{3x-y}{y} = \dfrac{3(4)-(8)}{(8)}$
$= \dfrac{12-8}{8}$
$= \dfrac{4}{8}$
$= \dfrac{1}{2}$

B10) $\dfrac{4xz}{y} = \dfrac{4(4)(-2)}{(8)}$
$= \dfrac{16(-2)}{8}$
$= \dfrac{-32}{8}$
$= -4$

B11) $\dfrac{4y}{xz} = \dfrac{4(8)}{(4)(-2)}$
$= \dfrac{32}{-8}$
$= -4$

B12) $y^2 - 2z = (8)^2 - 2(-2)$
$= 64 - 2(-2)$
$= 64 + 4$
$= 68$

B13) $x^2 - 3z = (4)^2 - 3(-2)$
$= 16 - 3(-2)$
$= 16 + 6$
$= 22$

B14) $\dfrac{4x}{y} - (2x+1) = \dfrac{4(4)}{(8)} - (2(4)+1)$
$= \dfrac{16}{8} - (8+1)$
$= \dfrac{16}{8} - (9)$
$= 2 - 9$
$= -7$

B15) $\dfrac{3y}{x} - (4y+6) = \dfrac{3(8)}{(4)} - (4(8)+6)$

$\qquad\qquad\qquad = \dfrac{24}{4} - (32+6)$

$\qquad\qquad\qquad = \dfrac{24}{4} - (38)$

$\qquad\qquad\qquad = 6 - 38$

$\qquad\qquad\qquad = -32$

C1) Answers may vary. Example: A variable term has a variable part; a constant term does not.

C2) Answers may vary. Example: The lines clearly identify the terms.

C3) Answers may vary. Example: Like terms have identical variable parts.

C4) a) There are four terms.

   b) The constant term is 3.

   c) The variable terms are $2x$, $-5y$, and $-6x$.

   d) The numerical coefficients are $2, -5, -6$, and 3.

   e) $2x$ and $-6x$ are like terms.

C5) a) There are four terms.

   b) The constant term is 5.

   c) The variable terms are $3r$, $-8y$, and $-6r$.

   d) The numerical coefficients are $3, -8, -6$, and 5.

   e) $3r$ and $-6r$ are like terms.

C6) a) There are four terms.

   b) The constant term is -7.

   c) The variable terms are $3y$, $-y^2$, and $-4y$.

   d) The numerical coefficients are $-7, 3, -1$, and -4.

   e) $3y$ and $-4y$ are like terms.

C7) a) There are four terms.

   b) The constant term is 14.

   c) The variable terms are $-z$, $4z^2$, and $-2z^2$.

   d) The numerical coefficients are $-1, 4, -2$, and 14.

   e) $4z^2$ and $-2z^2$ are like terms.

SECTION 10.1   EXERCISES

For 1–9, refer to Concept Checks and Practice Exercises.

11) coefficient

13) Like terms

15) constant term

17) term

19) Let $c =$ the number of foam cups per box.

   Number of foam cups $= 18c + 40$

21) $b + 900$

23) a) Let $k =$ the number of candy bars in one box.

   Number of candy bars $= 3k + 2$

   b) $3k + 2 = 3(36) + 2$

   $\qquad\qquad = 108 + 2$

   $\qquad\qquad = 110$

   There is a total of 110 candy bars.

25) $4x - 5 = 4(3) - 5$

   $\qquad\quad = 12 - 5$

   $\qquad\quad = 7$

27) $y - z = 9 - (-4)$

   $\qquad\quad = 9 + 4$

   $\qquad\quad = 13$

29) $5yz = 5(9)(-4)$

   $\qquad\quad = 45(-4)$

   $\qquad\quad = -180$

31) $\dfrac{2y}{3x+y} = \dfrac{2(9)}{3(3)+(9)}$

   $\qquad\quad = \dfrac{18}{9+9}$

   $\qquad\quad = \dfrac{18}{18}$

   $\qquad\quad = 1$

33) $7(x-4)-8 = 7((3)-4)-8$
$$= 7(-1)-8$$
$$= -7-8$$
$$= -15$$

35) $3z^2 = 3(-4)^2$
$$= 3(16)$$
$$= 48$$

37) $7-2(3y-4) = 7-2(3(9)-4)$
$$= 7-2(27-4)$$
$$= 7-2(23)$$
$$= 7-46$$
$$= -39$$

39) $\dfrac{24x}{yz} = \dfrac{24(3)}{(9)(-4)}$
$$= \dfrac{72}{-36}$$
$$= -2$$

41) $\dfrac{1}{4}y + \dfrac{1}{4} = \dfrac{1}{4}(9) + \dfrac{1}{4}$
$$= \dfrac{9}{4} + \dfrac{1}{4}$$
$$= \dfrac{10}{4}$$
$$= \dfrac{5}{2}$$

43) $2y-(y+6) = 2(9)-(9+6)$
$$= 2(9)-(15)$$
$$= 18-15$$
$$= 3$$

45) a) There are four terms.

   b) The constant term is 5.

   c) The variable terms are $4r$, y, and r.

   d) The numerical coefficients are 4, 1, 1, and 5.

   e) $4r$ and r are like terms.

47) a) There are four terms.

   b) The constant terms are 4 and 14.

   c) The variable terms are $3x^3$ and $-x$.

   d) The numerical coefficients are 3, 4, −1, and 14.

   e) 4 and 14 are like terms.

49) a) There are four terms.

   b) There is no constant term.

   c) The variable terms are
      $10m$, $-4m^3$, $6m$, and m^3.

   d) The numerical coefficients are
      10, − 4, 6, and 1.

   e) $-4m^3$ and m^3 are like terms.
      $10m$ and $6m$ are like terms.

51) a) There are five terms.

   b) The constant term is −8.

   c) The variable terms are
      $10l$, $-4m^3$, $6l$, and m.

   d) The numerical coefficients are
      10, − 4, 6, 1, and −8.

   e) $10l$ and $6l$ are like terms.

53) $A = l \cdot w$
$$= 22 \cdot 29$$
$$= 638 \text{ cm}^2$$
The area of the laptop screen is 132 square centimeters.

55) $P = 2l + 2w$
$$= 2(21) + 2(13)$$
$$= 42 + 26$$
$$= 68 \text{ ft}$$
They will need 68 feet of fencing.

57) $d = rt$
$$= 35 \cdot 9$$
$$= 315 \text{ mi}$$
Ralucca drove 315 miles.

59) a) Let $c =$ cost for rental.
      Let $d =$ number of days.
      Let $m =$ miles over limit.

      Clifton's: $c = 32d + 0.35m$
      Kyle's: $c = 37d + 0.15m$

59) b) $m = 140 - 100 = 40$
$d = 1$

Clifton's: $c = 32d + 0.35m$
$= 32(1) + 0.35(40)$
$= 32 + 14$
$= 46$

Kyle's: $c = 37d + 0.15m$
$= 37(1) + 0.15(40)$
$= 37 + 6$
$= 43$

c) Kyle's car rental is a better buy for the customer.

10.2 Operations with Variable Expressions

GUIDED PRACTICE

1) $6x + 14 + 4x + 9 = 6x + 4x + 14 + 9$
$= 10x + 23$

2) $\frac{1}{3}x + 11 + \frac{5}{6}x + 18 = \frac{1}{3}x + \frac{5}{6}x + 11 + 18$
$= \frac{2}{6}x + \frac{5}{6}x + 11 + 18$
$= \frac{7}{6}x + 29$

3) $7z - 2 - 3z + 5 = 7z + (-3z) + (-2) + 5$
$= 4z + 3$

4) $-8 - 8t + 4 + 6t = -8t + 6t + (-8) + 4$
$= -2t + (-4)$
$= -2t - 4$

5) $4(2x + 7) = 4 \cdot 2x + 4 \cdot 7$
$= 8x + 28$

6) $8(7x) = (8 \cdot 7)x$
$= 56x$

7) $11z \cdot 6 = 11 \cdot 6 \cdot z$
$= 66z$

8) $4(5x - 1) = 4 \cdot 5x + 4 \cdot (-1)$
$= 20x - 4$

9) $-3(x - 8) = (-3) \cdot x + (-3) \cdot (-8)$
$= -3x + 24$

10) $4(2x) + 6(3x) - 114 = 8x + 18x - 114$
$= 26x - 114$

11) $\frac{10z}{5} + 28z = \frac{2 \cdot \cancel{5}z}{\cancel{5}} + 28z$
$= \frac{2z}{1} + 28z$
$= 2z + 28z$
$= 30z$

12) $-2(n - 8) + 5(6) = -2n + 16 + 30$
$= -2n + 46$

13) $P = \text{Side}_1 + \text{Side}_2 + \text{Side}_3$
$P = (4x - 3) + (x + 6) + (14)$
$P = 4x - 3 + x + 6 + 14$
$P = 4x + x - 3 + 6 + 14$
$P = 5x + 17 \text{ units}$

14) $A = \text{length} \cdot \text{width}$
$A = 8(5y + 2)$
$A = 40y + 16 \text{ square units}$

CONCEPT CHECKS AND PRACTICE EXERCISES

A1) a) 3 pencils + 2 pencils = 5 pencils

b) $\frac{3}{5} + \frac{1}{5} = \frac{4}{5}$

c) $5x - 2x = 3x$

A2) *Like* indicates the same type of item; that is, pencils, fifths, or x's.

A3) Answers may vary. Example: $3y$ and $3x$; the variable parts are not identical.

A4) $2x + 3x = (2 + 3)x$
$= 5x$

A5) $7z - 3z = (7 + (-3))z$
$= 4z$

A6) $6 - 5y - 10 = -5y + 6 + (-10)$
$= -5y + (-4)$
$= -5y - 4$

A7) $4 - 7t - 26 = -7t + 4 + (-26)$
$= -7t + (-22)$
$= -7t - 22$

A8) $-6z + 9z - 14 = 3z - 14$

A9) $-2x + 17 + 4x = -2x + 4x + 17$
$\qquad\qquad\quad = 2x + 17$

A10) $11 + 7y - y = 7y - y + 11$
$\qquad\qquad = 6y + 11$

A11) $15 + 8n - n = 8n - n + 15$
$\qquad\qquad = 7n + 15$

A12) $3 + 2x - 3x + 13 = 2x + (-3x) + 3 + 13$
$\qquad\qquad\qquad\quad = (-x) + 26$
$\qquad\qquad\qquad\quad = -x + 26$

A13) $4 + 5z - 6z + 4 = 5z + (-6z) + 4 + 4$
$\qquad\qquad\qquad = -z + 8$

A14) $13y - 9 + y - 14 = 13y + y + (-9) + (-14)$
$\qquad\qquad\qquad\quad = 14y + (-23)$
$\qquad\qquad\qquad\quad = 14y - 23$

A15) $7h - 8 + h - 17 = 7h + h + (-8) + (-17)$
$\qquad\qquad\qquad\quad = 8h + (-25)$
$\qquad\qquad\qquad\quad = 8h - 25$

A16) $-\dfrac{1}{4}x + \dfrac{3}{4}x - 3 = \dfrac{2}{4}x - 3$
$\qquad\qquad\qquad\quad = \dfrac{1}{2}x - 3$

A17) $\dfrac{4}{5}y - \dfrac{2}{5}y - 23 = \dfrac{2}{5}y - 23$

A18) $2.4r - 10s + 3.2s + 4r$
$\quad = 2.4r + 4r + (-10s) + 3.2s$
$\quad = 6.4r + (-6.8s)$
$\quad = 6.4r - 6.8s$

A19) $10t - 4.5v - 3.6t + 5v$
$\quad = 10t + (-3.6t) + (-4.5v) + 5v$
$\quad = 6.4t + 0.5v$

B1) We cannot perform the addition inside the parentheses because the terms are not like terms.

B2) When a factor multiplies a grouping, it must multiply every term in the grouping.

B3) a) $2(3 + 4) = 2(7) = 14$

b) $2(3 + 4) = 2 \cdot 3 + 2 \cdot 4 = 6 + 8 = 14$

c) They are the same.

B4) $2(4x) = (2 \cdot 4)x$
$\qquad\quad = 8x$

B5) $5(7y) = (5 \cdot 7)y$
$\qquad\quad = 35y$

B6) $3x \cdot 3 = 3 \cdot 3 \cdot x$
$\qquad\quad = 9x$

B7) $12z \cdot 4 = 12 \cdot 4 \cdot z$
$\qquad\quad = 48z$

B8) $2(x + 6) = 2 \cdot x + 2 \cdot 6$
$\qquad\qquad = 2x + 12$

B9) $4(y + 3) = 4 \cdot y + 4 \cdot 3$
$\qquad\qquad = 4y + 12$

B10) $7(4x - 7) = 7 \cdot 4x + 7 \cdot (-7)$
$\qquad\qquad\quad = 28x - 49$

B11) $8(3x - 6) = 8 \cdot 3x + 8 \cdot (-6)$
$\qquad\qquad\quad = 24x - 48$

B12) $-6(6x + 1) = (-6) \cdot 6x + (-6) \cdot 1$
$\qquad\qquad\qquad = -36x - 6$

B13) $-5(2x + 2) = (-5) \cdot 2x + (-5) \cdot 2$
$\qquad\qquad\qquad = -10x - 10$

B14) $-8(3x - 1) = (-8) \cdot 3x + (-8) \cdot (-1)$
$\qquad\qquad\qquad = -24x + 8$

B15) $-1(4x - 1) = (-1) \cdot 4x + (-1) \cdot (-1)$
$\qquad\qquad\qquad = -4x + 1$

C1) In the order of operations, the distributive property is performed at the multiplication step.

C2) In the order of operations, we combine like terms at the addition/subtraction step.

C3) $6(3t)+5=18t+5$

C4) $6(5y)+10=30y+10$

C5) $\dfrac{4z}{2}-3z=\dfrac{2\cdot\cancel{2}z}{\cancel{2}}-3z$

$=\dfrac{2z}{1}-3z$

$=2z-3z$

$=-z$

C6) $\dfrac{8x}{4}+3x=\dfrac{2\cdot\cancel{4}x}{\cancel{4}}+3x$

$=\dfrac{2x}{1}+3x$

$=2x+3x$

$=5x$

C7) $2(n+8)+4n=2n+16+4n$

$=2n+4n+16$

$=6n+16$

C8) $4(n+2)+7n=4n+8+7n$

$=4n+7n+8$

$=11n+8$

C9) $-6(y-4)+5(6)+4y=-6y+24+30+4y$

$=-6y+4y+24+30$

$=-2y+54$

C10) $-3(m-9)+9(3)+7m=-3m+27+27+7m$

$=-3m+7m+27+27$

$=4m+54$

C11) $2(n-8)-5(n+1)=2n-16-5n-5$

$=2n-5n-16-5$

$=-3n-21$

C12) $7(t^2-2)-4(t^2+1)=7t^2-14-4t^2-4$

$=7t^2-4t^2-14-4$

$=3t^2-18$

Section 10.2   Exercises

For 1–9, refer to Concept Checks and Practice
Exercises.

11) coefficient

13) distributive property

15) Like terms

17) a)   8 pencils $-$ 2 pencils $=$ 6 pencils

b)   5 teachers $-$ 4 pens; Unlike

c) $9x-4x=5x$

d) $6z-1t$; Unlike

19) $8z-4z=(8-4)z$

$=4z$

21) $4t-7t=(4-7)t$

$=-3t$

23) $4z+2z-z+4z=(4+2-1+4)z$

$=9z$

25) $5h-6+2h-12=5h+2h-6-12$

$=7h-18$

27) $5x+3-12x-8=5x-12x+3-8$

$=-7x-5$

29) $\dfrac{3}{7}y+5-\dfrac{6}{7}y=\dfrac{3}{7}y-\dfrac{6}{7}y+5$

$=-\dfrac{3}{7}y+5$

31) $\dfrac{2}{5}y-\dfrac{2}{3}y-23=\dfrac{6}{15}y-\dfrac{10}{15}y-23$

$=-\dfrac{4}{15}y-23$

33) $10x-4v-3.6x+5v=10x-3.6x-4v+5v$

$=6.4x+v$

35) a) $t+f+t$

b)   turtles, they are like animals

b) $2t+f$

37) $4(3x)=12x$

39) $4z\cdot8=32z$

41) $4(y+6)=4y+24$

43) $6(3x-6)=18x-36$

45) $-8(3x+2)=-24x-16$

47) $-8(4x-1)=-32x+8$

49) $4(2z)-z+4=8z-z+4$
$\qquad\qquad\qquad =7z+4$

51) $\dfrac{3h}{3}+2h-17=h+2h-17$
$\qquad\qquad\qquad\quad =3h-17$

53) $-11x+3y-y-23=-11x+2y-23$

55) $10t-45-3v+5(4v)=10t-45-3v+20v$
$\qquad\qquad\qquad\qquad\quad =10t+17v-45$

57) $7n+4(n+1)=7n+4n+4$
$\qquad\qquad\qquad\quad =11n+4$

59) $9x-5(x+6)=9x-5x-30$
$\qquad\qquad\qquad\quad =4x-30$

61) $6z-3(4z-7)=6z-12z+21$
$\qquad\qquad\qquad\quad =-6z+21$

63) $5(6)+3m-5(m-2)=30+3m-5m+10$
$\qquad\qquad\qquad\qquad\quad =3m-5m+30+10$
$\qquad\qquad\qquad\qquad\quad =-2m+40$

65) $-7(t-3)+6(2t-5)=-7t+21+12t-30$
$\qquad\qquad\qquad\qquad\quad =-7t+12t+21-30$
$\qquad\qquad\qquad\qquad\quad =5t-9$

67) $6(x+1)-6(2+x)=6x+6-12-6x$
$\qquad\qquad\qquad\qquad =6x-6x+6-12$
$\qquad\qquad\qquad\qquad =-6$

69) $P=\text{Side}_1+\text{Side}_2+\text{Side}_3+\text{Side}_4+\text{Side}_5$
$P=(x+6)+(3x-7)+3+(2x-15)+(7x+5)$
$P=x+6+3x-7+3+2x-15+7x+5$
$P=x+3x+2x+7x+6-7+3-15+5$
$P=13x-8$ units

71) a) $P=2\cdot l+2\cdot w$
$\qquad P=2(15y-2)+2(7)$
$\qquad P=30y-4+14$
$\qquad P=30y+10$ units

b) $A=\text{length}\cdot\text{width}$
$\qquad A=7(15y-2)$
$\qquad A=105y-14$ square units

73) $A=\dfrac{1}{2}bh$
$\qquad =\dfrac{1}{2}(2x+9)6$
$\qquad =3(2x+9)$
$\qquad =6x+27$ square units

75) a)

b) $\text{Area}=A_{\text{triangle}}-A_{\text{rectangle}}$
$\qquad =\dfrac{1}{2}\cdot(2x+12)\cdot10-6\cdot5$
$\qquad =5(2x+12)-30$
$\qquad =10x+60-30$
$\qquad =10x+30$ ft^2

10.3 Solving One-Step Equations

GUIDED PRACTICE

1) $x+10=15$
$\qquad(5)\overset{?}{+}10=15$
$\qquad\quad 15=15$
$\quad x=5$ is a solution.

2) $4x=24$
$\qquad 4(6)\overset{?}{=}24$
$\qquad\quad 24=24$
Since $24=24$ is true, 6 is a solution.

3) $2x-4=x-8$
$\qquad 2(-3)-4\overset{?}{=}(-3)-8$
$\qquad\quad -6-4\overset{?}{=}-3-8$
$\qquad\qquad -10=-11$
Since $-10=-11$ is false, -3 isn't a solution.

4) a) 5 is being added to y.

b) Addition of 5 is undone by subtraction of 5.

5) a) y is being multiplied by -10.

   b) Multiplication by -10 is undone with division by -10.

6) a) z is being divided by 13.

   b) Division by 13 is undone with multiplication by 13.

7) $5 = 5$
   $5 + 2 = 5 + 2$
   $7 = 7$

   Add 2 to both sides, and the equation remains true.

8) $y + 5 = 15$
   $y + 5 - 5 = 15 - 5$
   $y = 10$

   Subtracting 5 from both sides of the equation isolates the variable and solves the equation.

9) $-23 = x - 8$
   $-23 + 8 = x - 8 + 8$
   $-15 = x$

   Check: $-23 = x - 8$

   $-23 \overset{?}{=} (-15) - 8$

   $-23 = -23$

   The solution checks.

10) $x + \dfrac{1}{3} = \dfrac{1}{2}$

   $x + \dfrac{1}{3} - \dfrac{1}{3} = \dfrac{1}{2} - \dfrac{1}{3}$

   $x = \dfrac{3}{6} - \dfrac{2}{6}$

   $x = \dfrac{1}{6}$

   Check: $x + \dfrac{1}{3} = \dfrac{1}{2}$

   $\left(\dfrac{1}{6}\right) + \dfrac{1}{3} \overset{?}{=} \dfrac{1}{2}$

   $\dfrac{1}{6} + \dfrac{2}{6} \overset{?}{=} \dfrac{1}{2}$

   $\dfrac{3}{6} \overset{?}{=} \dfrac{1}{2}$

   $\dfrac{1}{2} = \dfrac{1}{2}$

   The solution checks.

11) $6 = 6$
   $6 \cdot 4 = 6 \cdot 4$
   $24 = 24$

   Multiply both sides by 4, and the equation remains true.

12) $5x = 20$

   $\dfrac{5x}{5} = \dfrac{20}{5}$

   $\dfrac{\cancel{5}x}{\cancel{5}} = \dfrac{4 \cdot \cancel{5}}{\cancel{5}}$

   $x = 4$

   Dividing both sides of the equation by 5 isolates the variable and solves the equation.

13) $-9z = 90$

   $\dfrac{\cancel{-9}z}{\cancel{-9}} = \dfrac{90}{-9}$

   $z = -10$

   Check: $-9z = 90$

   $-9(-10) \overset{?}{=} 90$

   $90 = 90$

   The solution checks.

14) $\dfrac{x}{7} = 6$

   $7 \cdot \dfrac{x}{7} = 7 \cdot 6$

   $\dfrac{7}{1} \cdot \dfrac{x}{7} = 7 \cdot 6$

   $\dfrac{\cancel{7}}{1} \cdot \dfrac{x}{\cancel{7}} = 42$

   $x = 42$

   Check: $\dfrac{x}{7} = 6$

   $\dfrac{(42)}{7} \overset{?}{=} 6$

   $6 = 6$

   The solution checks.

15) $6x = 0$

   $\dfrac{\cancel{6}x}{\cancel{6}} = \dfrac{0}{6}$

   $x = 0$

   Check: $6x = 0$

   $6(0) \overset{?}{=} 0$

   $0 = 0$

   The solution checks.

16) $\dfrac{2}{3}x = \dfrac{10}{3}$

$\dfrac{3}{2} \cdot \dfrac{2}{3}x = \dfrac{3}{2} \cdot \dfrac{10}{3}$

$\dfrac{\cancel{3}}{\cancel{2}} \cdot \dfrac{\cancel{2}}{\cancel{3}}x = \dfrac{\cancel{3}}{\cancel{2}} \cdot \dfrac{\cancel{2} \cdot 5}{\cancel{3}}$

$x = 5$

Check: $\dfrac{2}{3}x = \dfrac{10}{3}$

$\dfrac{2}{3}(5) \overset{?}{=} \dfrac{10}{3}$

$\dfrac{2 \cdot 5}{3 \cdot 1} \overset{?}{=} \dfrac{10}{3}$

$\dfrac{10}{3} = \dfrac{10}{3}$

The solution checks.

CONCEPT CHECKS AND PRACTICE EXERCISES

A1)   Answers may vary. Example: A solution is the numerical value that makes the equation a true statement.

A2)   An equation is a statement where one expression is equal to another expression.

A3) $x - 16 = 10$

$(26) - 16 \overset{?}{=} 10$

$10 = 10$

Since $10 = 10$ is true, $x = 26$ is a solution.

A4) $24 = z + 14$

$24 \overset{?}{=} (10) + 14$

$24 = 24$

Since $24 = 24$ is true, $z = 10$ is a solution.

A5) $3y = 18$

$3(5) \overset{?}{=} 18$

$15 = 18$

Since $15 = 18$ is false, $y = 5$ is not a solution.

A6) $27 = 9y$

$27 \overset{?}{=} 9(4)$

$27 = 36$

Since $27 = 36$ is false, $y = 4$ is not a solution.

A7) $-53 = 8x + 3$

$-53 \overset{?}{=} 8(-7) + 3$

$-53 \overset{?}{=} -56 + 3$

$-53 = -53$

Since $-53 = -53$ is true, $x = -7$ is a solution.

A8) $-112 = 12x - 4$

$-112 \overset{?}{=} 12(-9) - 4$

$-112 \overset{?}{=} -108 - 4$

$-112 = -112$

Since $-112 = -112$ is true, $z = -9$ is a solution.

A9) $\dfrac{y}{5} + 1 = 9$

$\dfrac{(40)}{5} + 1 \overset{?}{=} 9$

$8 + 1 \overset{?}{=} 9$

$9 = 9$

Since $9 = 9$ is true, $y = 40$ is a solution.

A10) $11 = \dfrac{y}{10} + 3$

$11 \overset{?}{=} \dfrac{(80)}{10} + 3$

$11 \overset{?}{=} 8 + 3$

$11 = 11$

Since $11 = 11$ is true, $y = 80$ is a solution.

B1)   Addition and subtraction are inverse operations. Multiplication and division are inverse operations.

B2)   a) $3 + 4 - 4 = 3$

b) $2 \cdot 6 \div 6 = 2$

c) $4 \div 2 \cdot 2 = 4$

d) $8 - 5 + 5 = 8$

B3)   a)   17 is being added to y.

b)   Addition of 17 is undone by subtraction of 17.

B4)   a)   12 is being subtracted from x.

b)   Subtraction of 12 is undone by addition of 12.

B5)   a) y is being divided by 5.

b)   Division by 5 is undone with multiplication by 5.

B6) a) z is being divided by -2.

   b) Division by -2 is undone with multiplication by -2.

B7) a) -12 is being added to x. (Or, equivalently, 12 is being subtracted from x.)

   b) Addition of -12 is undone by subtraction of -12, (or, equivalently, addition of 12).

B8) a) -15 is being added to z. (Or, equivalently, 15 is being subtracted from z.)

   b) Addition of -15 is undone by subtraction of -15, (or equivalently, addition of 15).

B9) a) y is being multiplied by -9.

   b) Multiplication by -9 is undone with division by -9.

B10) a) x is being multiplied by -1.

   b) Multiplication by -1 is undone with division by -1.

C1) Answers may vary. Example: Whatever you do to one side of an equation, you must also do to the other side of the equation in order to keep the equation in balance.

C2) a) 4 is being subtracted from x.

   b) Subtraction of 4 is undone by addition of 4.

C3) Answers may vary. Example: To isolate the variable, apply inverse operations to get the variable alone on one side of the equal sign.

C4) To isolate the variable, $\frac{1}{8}$ must be subtracted from each side of the equation.

C5) $$y+10=14$$
$$y+10-10=14-10$$
$$y=4$$

C6) $$z+24=29$$
$$z+24-24=29-24$$
$$z=5$$

C7) $$-14=t-25$$
$$-14+25=t-25+25$$
$$11=t$$

C8) $$-16=z+24$$
$$-16-24=z+24-24$$
$$-40=z$$

C9) $$x-10=-10$$
$$x-10+10=-10+10$$
$$x=0$$

C10) $$x-30=-30$$
$$x-30+30=-30+30$$
$$x=0$$

C11) $$-4=t-16$$
$$-4+16=t-16+16$$
$$12=t$$

C12) $$7=t-11$$
$$7+11=t-11+11$$
$$18=t$$

C13) $$x-\frac{1}{2}=6$$
$$x-\frac{1}{2}+\frac{1}{2}=6+\frac{1}{2}$$
$$x=\frac{12}{2}+\frac{1}{2}$$
$$x=\frac{13}{2}$$

C14) $$y+\frac{1}{5}=4$$
$$y+\frac{1}{5}-\frac{1}{5}=4-\frac{1}{5}$$
$$y=\frac{20}{5}-\frac{1}{5}$$
$$y=\frac{19}{5}$$

C15) $$z+\frac{1}{3}=\frac{3}{4}$$
$$z+\frac{1}{3}-\frac{1}{3}=\frac{3}{4}-\frac{1}{3}$$
$$z=\frac{9}{12}-\frac{4}{12}$$
$$z=\frac{5}{12}$$

C16) $\quad x - \dfrac{1}{6} = \dfrac{1}{2}$

$$x - \dfrac{1}{6} + \dfrac{1}{6} = \dfrac{1}{2} + \dfrac{1}{6}$$

$$x = \dfrac{3}{6} + \dfrac{1}{6}$$

$$x = \dfrac{4}{6}$$

$$x = \dfrac{2}{3}$$

D1) Answers may vary. Example:

$$100 = 100$$

$$\dfrac{100}{2} = \dfrac{100}{2}$$

$$50 = 50$$

D2) To isolate the variable, both sides of the equation must be multiplied by $\dfrac{8}{3}$.

D3) $\quad 4x = 32$

$$\dfrac{\cancel{4}x}{\cancel{4}} = \dfrac{32}{4}$$

$$x = 8$$

D4) $\quad 5y = 45$

$$\dfrac{\cancel{5}y}{\cancel{5}} = \dfrac{45}{5}$$

$$y = 9$$

D5) $\quad -4z = 20$

$$\dfrac{\cancel{-4}z}{\cancel{-4}} = \dfrac{20}{-4}$$

$$z = -5$$

D6) $\quad -7y = 28$

$$\dfrac{\cancel{-7}y}{\cancel{-7}} = \dfrac{28}{-7}$$

$$y = -4$$

D7) $\quad -55 = -5t$

$$\dfrac{-55}{-5} = \dfrac{-5t}{-5}$$

$$11 = t$$

D8) $\quad 48 = -6t$

$$\dfrac{48}{-6} = \dfrac{-6t}{-6}$$

$$-8 = t$$

D9) $\quad 5x = 0$

$$\dfrac{\cancel{5}x}{\cancel{5}} = \dfrac{0}{5}$$

$$x = 0$$

D10) $\quad 7z = 0$

$$\dfrac{\cancel{7}z}{\cancel{7}} = \dfrac{0}{7}$$

$$z = 0$$

D11) $\quad \dfrac{x}{2} = -3$

$$2 \cdot \dfrac{x}{2} = 2 \cdot (-3)$$

$$\dfrac{\cancel{2}}{1} \cdot \dfrac{x}{\cancel{2}} = -6$$

$$x = -6$$

D12) $\quad \dfrac{y}{3} = -7$

$$3 \cdot \dfrac{y}{3} = 3 \cdot (-7)$$

$$\dfrac{\cancel{3}}{1} \cdot \dfrac{y}{\cancel{3}} = -21$$

$$y = -21$$

D13) $\quad \dfrac{4}{5}y = \dfrac{12}{5}$

$$\dfrac{5}{4} \cdot \dfrac{4}{5}y = \dfrac{5}{4} \cdot \dfrac{12}{5}$$

$$\dfrac{\cancel{5}}{\cancel{4}} \cdot \dfrac{\cancel{4}}{\cancel{5}}y = \dfrac{\cancel{5}}{\cancel{4}} \cdot \dfrac{3 \cdot \cancel{4}}{\cancel{5}}$$

$$y = 3$$

D14) $\quad \dfrac{3}{4}x = \dfrac{9}{4}$

$$\dfrac{4}{3} \cdot \dfrac{3}{4}x = \dfrac{4}{3} \cdot \dfrac{9}{4}$$

$$\dfrac{\cancel{4}}{\cancel{3}} \cdot \dfrac{\cancel{3}}{\cancel{4}}x = \dfrac{\cancel{4}}{\cancel{3}} \cdot \dfrac{\cancel{3} \cdot 3}{\cancel{4}}$$

$$x = 3$$

SECTION 10.3   EXERCISES

For 1–12, refer to Concept Checks and Practice Exercises.

13) solve an equation

15) isolate a variable

17) multiplication property of equations

19) equation

21) $y - 14 = 5$

$(19) - 14 \overset{?}{=} 5$

$5 = 5$

Since $5 = 5$ is true, $y = 19$ is a solution.

23) $35 = 6t$

$35 \overset{?}{=} 6(5)$

$35 = 30$

Since $35 = 30$ is false, $t = 5$ is not a solution.

25) $16 = 8x - 8$

$16 \overset{?}{=} 8(3) - 8$

$16 \overset{?}{=} 24 - 8$

$16 = 16$

Since $16 = 16$ is true, $x = 3$ is a solution.

27) $\dfrac{y}{2} - 10 = 8$

$\dfrac{(30)}{2} - 10 \overset{?}{=} 8$

$15 - 10 \overset{?}{=} 8$

$5 = 8$

Since $5 = 8$ is false, $y = 30$ is not a solution.

29)   a) 7 is being added to t.

b) Addition of 7 is undone by subtraction of 7.

31)   a) z is being divided by 15.

b) Division by 15 is undone with multiplication by 15.

33)   a) $\dfrac{2}{3}$ is being added to x.

b) Addition of $\dfrac{2}{3}$ is undone by subtraction

of $\dfrac{2}{3}$.

35)   a) y is being multiplied by -2.

b) Multiplication by -2 is undone with division by -2.

37) $y + 12 = 23$

$y + 12 - 12 = 23 - 12$

$y = 11$

39) $-18 = t - 14$

$-18 + 14 = t - 14 + 14$

$-4 = t$

41) $x - 15 = -15$

$x - 15 + 15 = -15 + 15$

$x = 0$

43) $-9 = t - 12$

$-9 + 12 = t - 12 + 12$

$3 = t$

45) $18 + v = -4$

$18 - 18 + v = -4 - 18$

$v = -22$

47) $x - \dfrac{1}{3} = 2$

$x - \dfrac{1}{3} + \dfrac{1}{3} = 2 + \dfrac{1}{3}$

$x = \dfrac{6}{3} + \dfrac{1}{3}$

$x = \dfrac{7}{3}$

49) $\dfrac{3}{4} = z + \dfrac{2}{3}$

$\dfrac{3}{4} - \dfrac{2}{3} = z + \dfrac{2}{3} - \dfrac{2}{3}$

$\dfrac{9}{12} - \dfrac{8}{12} = z$

$\dfrac{1}{12} = z$

51) $x - 4 = 5.8$

$x - 4 + 4 = 5.8 + 4$

$x = 9.8$

53) $7z = 56$

$\dfrac{\cancel{7}z}{\cancel{7}} = \dfrac{56}{7}$

$z = 8$

55) $-3z = -21$

$\dfrac{\cancel{-3}z}{\cancel{-3}} = \dfrac{-21}{-3}$

$z = 7$

57) $3 = -2x$

$\dfrac{3}{-2} = \dfrac{\cancel{-2}x}{\cancel{-2}}$

$-\dfrac{3}{2} = x$

59) $\dfrac{v}{4} = 10$

$4 \cdot \dfrac{v}{4} = 4 \cdot 10$

$\dfrac{\cancel{4}}{1} \cdot \dfrac{v}{\cancel{4}} = 40$

$v = 40$

61) $-25 = -5t$

$\dfrac{-25}{-5} = \dfrac{\cancel{-5}t}{\cancel{-5}}$

$5 = t$

63) $0 = 3x$

$\dfrac{0}{3} = \dfrac{\cancel{3}x}{\cancel{3}}$

$0 = x$

65) $\dfrac{2}{5}y = \dfrac{1}{2}$

$\dfrac{5}{2} \cdot \dfrac{2}{5}y = \dfrac{5}{2} \cdot \dfrac{1}{2}$

$\dfrac{\cancel{5}}{\cancel{2}} \cdot \dfrac{\cancel{2}}{\cancel{5}}y = \dfrac{5}{2} \cdot \dfrac{1}{2}$

$y = \dfrac{5}{4}$

67) $15 = -\dfrac{x}{3}$

$15 \cdot (-3) = -\dfrac{x}{3} \cdot (-3)$

$-45 = \dfrac{x}{\cancel{3}} \cdot (\cancel{-3})$

$-45 = x$

69) a) $A = l \cdot w$

$60 = 12 \cdot w$

$60 = 12w$

b) $60 = 12w$

$\dfrac{60}{12} = \dfrac{\cancel{12}w}{\cancel{12}}$

$5 = w$

The width of the rectangle is 5 meters.

71) a) $A = \dfrac{1}{2}b \cdot h$

$20 = \dfrac{1}{2} \cdot 8 \cdot h$

b) $20 = \dfrac{1}{2} \cdot 8 \cdot h$

$20 = 4h$

$\dfrac{20}{4} = \dfrac{4h}{4}$

$5 = h$

The height of the triangle is 5 inches.

73) a) The mistake was dividing by 4 instead of subtracting 4.

b) $20 = y + 4$

$20 - 4 = y + 4 - 4$

$16 = y$

75) a) There was a division fact error.

b) $-5z = 10$

$\dfrac{-5z}{-5} = \dfrac{10}{-5}$

$z = -2$

10.4 Solving Multistep Equations

GUIDED PRACTICE

1) 8 is less than 9. Move $8x$ to the right by subtracting $8x$ from both sides.

Then, move 4 to the other side by subtracting 4 from both sides.

$8x + 1 = 9x + 4$

$8x - 8x + 1 = 9x - 8x + 4$

$1 = x + 4$

$1 - 4 = x + 4 - 4$

$-3 = x$

2) -4 is less than -3. Move $-4x$ to the left by adding $4x$ to both sides.

Then, move -3 to the other side by adding 3 to both sides.

$-3x - 3 = -4x + 1$

$-3x + 4x - 3 = -4x + 4x + 1$

$x - 3 = 1$

$x - 3 + 3 = 1 + 3$

$x = 4$

3) Move $6x$ to the left side.

Move 2 to the other side.

Divide both sides by 3 to undo the multiplication.

$$9x + 2 = 6x + 5$$
$$9x - 6x + 2 = 6x - 6x + 5$$
$$3x + 2 = 5$$
$$3x + 2 - 2 = 5 - 2$$
$$3x = 3$$
$$\frac{3x}{3} = \frac{3}{3}$$
$$x = 1$$

4) Move $-6x$ to the right side.

Move -2 to the left side.

Divide both sides by 10 to undo the multiplication.

$$-6x + 18 = 4x - 2$$
$$-6x + 6x + 18 = 4x + 6x - 2$$
$$18 = 10x - 2$$
$$18 + 2 = 10x - 2 + 2$$
$$20 = 10x$$
$$\frac{20}{10} = \frac{10x}{10}$$
$$2 = x$$

5) $$3(2x + 5) = -3x + 5x + 1$$
$$6x + 15 = -3x + 5x + 1$$
$$6x + 15 = 2x + 1$$
$$6x - 2x + 15 = 2x - 2x + 1$$
$$4x + 15 = 1$$
$$4x + 15 - 15 = 1 - 15$$
$$4x = -14$$
$$\frac{4x}{4} = \frac{-14}{4}$$
$$x = -\frac{7}{2}$$

6) $$4 + 8t - 5 = 3(t + 8)$$
$$4 + 8t - 5 = 3t + 24$$
$$8t - 1 = 3t + 24$$
$$8t - 3t - 1 = 3t - 3t + 24$$
$$5t - 1 = 24$$
$$5t - 1 + 1 = 24 + 1$$
$$5t = 25$$
$$\frac{5t}{5} = \frac{25}{5}$$
$$t = 5$$

7) $$F = \frac{9}{5}C + 32$$
$$-4 = \frac{9}{5}C + 32$$
$$-4 - 32 = \frac{9}{5}C + 32 - 32$$
$$-36 = \frac{9}{5}C$$
$$\frac{5}{9} \cdot \frac{-36}{1} = \frac{5}{9} \cdot \frac{9}{5}C$$
$$\frac{5 \cdot -4 \cdot \cancel{9}}{\cancel{9}} = C$$
$$-20 = C$$

$-4°\,F$ is the same as $-20°\,C$.

CONCEPT CHECKS AND PRACTICE EXERCISES

A1)   Addition and subtraction are used to gather variable and constant terms on opposite sides.

A2)   When there are variable terms on both sides of an equation, moving the term with the smaller coefficient often makes the exercise easier because the result is a positive coefficient on the variable that is being isolated.

A3) $$2x + 4 = 3x - 5$$
$$2x - 2x + 4 = 3x - 2x - 5$$
$$4 = x - 5$$
$$4 + 5 = x - 5 + 5$$
$$9 = x$$

A4) $$5x - 3 = 4x + 2$$
$$5x - 4x - 3 = 4x - 4x + 2$$
$$x - 3 = 2$$
$$x - 3 + 3 = 2 + 3$$
$$x = 5$$

A5) $$9x + 5 = 8x - 1$$
$$9x - 8x + 5 = 8x - 8x - 1$$
$$x + 5 = -1$$
$$x + 5 - 5 = -1 - 5$$
$$x = -6$$

A6) $$6x - 13 = 5x - 2$$
$$6x - 5x - 13 = 5x - 5x - 2$$
$$x - 13 = -2$$
$$x - 13 + 13 = -2 + 13$$
$$x = 11$$

A7) $-3x - 3 = -2x + 4$
$-3x + 3x - 3 = -2x + 3x + 4$
$-3 = x + 4$
$-3 - 4 = x + 4 - 4$
$-7 = x$

A8) $-6x + 2 = -5x - 4$
$-6x + 6x + 2 = -5x + 6x - 4$
$2 = x - 4$
$2 + 4 = x - 4 + 4$
$6 = x$

A9) $-17x + 3 = -16x + 1$
$-17x + 17x + 3 = -16x + 17x + 1$
$3 = x + 1$
$3 - 1 = x + 1 - 1$
$2 = x$

A10) $-9x + 7 = -8x + 2$
$-9x + 9x + 7 = -8x + 9x + 2$
$7 = x + 2$
$7 - 2 = x + 2 - 2$
$5 = x$

B1) a) Subtract $2x$ from both sides.

b) Subtract 4 from both sides.

c) Add $7x$ to both sides.

B2) The "doing" operation is multiplication. Multiplication is "undone" using division, not addition.

B3) $6x + 5 = 2x - 3$
$6x - 2x + 5 = 2x - 2x - 3$
$4x + 5 = -3$
$4x + 5 - 5 = -3 - 5$
$4x = -8$
$\dfrac{4x}{4} = \dfrac{-8}{4}$
$x = -2$

B4) $4x + 7 = 8x - 5$
$4x - 4x + 7 = 8x - 4x - 5$
$7 = 4x - 5$
$7 + 5 = 4x - 5 + 5$
$12 = 4x$
$\dfrac{12}{4} = \dfrac{4x}{4}$
$3 = x$

B5) $y + 6 = 4y - 3$
$y - y + 6 = 4y - y - 3$
$6 = 3y - 3$
$6 + 3 = 3y - 3 + 3$
$9 = 3y$
$\dfrac{9}{3} = \dfrac{3y}{3}$
$3 = y$

B6) $7y - 16 = y + 8$
$7y - y - 16 = y - y + 8$
$6y - 16 = 8$
$6y - 16 + 16 = 8 + 16$
$6y = 24$
$\dfrac{6y}{6} = \dfrac{24}{6}$
$y = 4$

B7) $4x - 12 = 3x - 12$
$4x - 3x - 12 = 3x - 3x - 12$
$x - 12 = -12$
$x - 12 + 12 = -12 + 12$
$x = 0$

B8) $8x - 22 = 5x - 22$
$8x - 5x - 22 = 5x - 5x - 22$
$3x - 22 = -22$
$3x - 22 + 22 = -22 + 22$
$3x = 0$
$\dfrac{3x}{3} = \dfrac{0}{3}$
$x = 0$

B9) $-2z + 8 = z + 5$
$-2z + 2z + 8 = z + 2z + 5$
$8 = 3z + 5$
$8 - 5 = 3z + 5 - 5$
$3 = 3z$
$\dfrac{3}{3} = \dfrac{3z}{3}$
$1 = z$

B10) $4x - 10 = -3x + 4$
$4x + 3x - 10 = -3x + 3x + 4$
$7x - 10 = 4$
$7x - 10 + 10 = 4 + 10$
$7x = 14$
$\dfrac{7x}{7} = \dfrac{14}{7}$
$x = 2$

C1) Simplify each side, gather variable terms and constant terms to opposite sides, isolate the variable.

C2) Answers may vary. Example: Tell her to write $-2(x-5)$ as the sum of two multiplication problems, $-2(x)+(-2)(-5)$.

C3)
$$6(x+5)=9+2x-3$$
$$6x+30=9+2x-3$$
$$6x+30=6+2x$$
$$6x-2x+30=6+2x-2x$$
$$4x+30=6$$
$$4x+30-30=6-30$$
$$4x=-24$$
$$\frac{4x}{4}=\frac{-24}{4}$$
$$x=-6$$

C4)
$$5(x+7)=10+3x-3$$
$$5x+35=10+3x-3$$
$$5x+35=7+3x$$
$$5x-3x+35=7+3x-3x$$
$$2x+35=7$$
$$2x+35-35=7-35$$
$$2x=-28$$
$$\frac{2x}{2}=\frac{-28}{2}$$
$$x=-14$$

C5)
$$6=4(y-1)+y$$
$$6=4y-4+y$$
$$6=5y-4$$
$$6+4=5y-4+4$$
$$10=5y$$
$$\frac{10}{5}=\frac{5y}{5}$$
$$2=y$$

C6)
$$2=8(z-2)+z$$
$$2=8z-16+z$$
$$2=9z-16$$
$$2+16=9z-16+16$$
$$18=9z$$
$$\frac{18}{9}=\frac{9z}{9}$$
$$2=z$$

C7)
$$4x-1-2x=3x-12$$
$$2x-1=3x-12$$
$$2x-2x-1=3x-2x-12$$
$$-1=x-12$$
$$-1+12=x-12+12$$
$$11=x$$

C8)
$$3x+14=3x-1-x$$
$$3x+14=2x-1$$
$$3x-2x+14=2x-2x-1$$
$$x+14=-1$$
$$x+14-14=-1-14$$
$$x=-15$$

C9)
$$3-2(x-5)=4x+1$$
$$3-2x+10=4x+1$$
$$13-2x=4x+1$$
$$13-2x+2x=4x+2x+1$$
$$13=6x+1$$
$$13-1=6x+1-1$$
$$12=6x$$
$$\frac{12}{6}=\frac{6x}{6}$$
$$2=x$$

C10)
$$4-3(t-8)=2t-7$$
$$4-3t+24=2t-7$$
$$28-3t=2t-7$$
$$28-3t+3t=2t+3t-7$$
$$28=5t-7$$
$$28+7=5t-7+7$$
$$35=5t$$
$$\frac{35}{5}=\frac{5t}{5}$$
$$7=t$$

SECTION 10.4   EXERCISES

For 1–9, refer to Concept Checks and Practice Exercises.

11) constant term

13) variable term

15)
$$5x-7=6x+2$$
$$5x-5x-7=6x-5x+2$$
$$-7=x+2$$
$$-7-2=x+2-2$$
$$-9=x$$

17)
$$4x - 11 = 5x + 2$$
$$4x - 4x - 11 = 5x - 4x + 2$$
$$-11 = x + 2$$
$$-11 - 2 = x + 2 - 2$$
$$-13 = x$$

19)
$$-4x - 3 = -5x - 4$$
$$-4x + 5x - 3 = -5x + 5x - 4$$
$$x - 3 = -4$$
$$x - 3 + 3 = -4 + 3$$
$$x = -1$$

21)
$$-7x + 5 = -8x + 3$$
$$-7x + 8x + 5 = -8x + 8x + 3$$
$$x + 5 = 3$$
$$x + 5 - 5 = 3 - 5$$
$$x = -2$$

23)
$$6x - 18 = 24$$
$$6x - 18 + 18 = 24 + 18$$
$$6x = 42$$
$$\frac{6x}{6} = \frac{42}{6}$$
$$x = 7$$

25)
$$5 = 8 - 3x$$
$$5 - 8 = 8 - 8 - 3x$$
$$-3 = -3x$$
$$\frac{-3}{-3} = \frac{-3x}{-3}$$
$$1 = x$$

27)
$$\frac{c}{4} - 6 = 4$$
$$\frac{c}{4} - 6 + 6 = 4 + 6$$
$$\frac{c}{4} = 10$$
$$4 \cdot \frac{c}{4} = 4 \cdot 10$$
$$c = 40$$

29)
$$20 - 5v = 20 - 3v$$
$$20 - 5v + 5v = 20 - 3v + 5v$$
$$20 = 20 + 2v$$
$$20 - 20 = 20 - 20 + 2v$$
$$0 = 2v$$
$$\frac{0}{2} = \frac{2v}{2}$$
$$0 = v$$

31)
$$6x - 25 = -7x + 1$$
$$6x + 7x - 25 = -7x + 7x + 1$$
$$13x - 25 = 1$$
$$13x - 25 + 25 = 1 + 25$$
$$13x = 26$$
$$\frac{13x}{13} = \frac{26}{13}$$
$$x = 2$$

33)
$$\frac{3}{4}v - 6 = -\frac{1}{4}v + 5$$
$$\frac{3}{4}v + \frac{1}{4}v - 6 = -\frac{1}{4}v + \frac{1}{4}v + 5$$
$$\frac{4}{4}v - 6 = 5$$
$$v - 6 + 6 = 5 + 6$$
$$v = 11$$

35)
$$5(x + 7) = 10$$
$$5x + 35 = 10$$
$$5x + 35 - 35 = 10 - 35$$
$$5x = -25$$
$$\frac{5x}{5} = \frac{-25}{5}$$
$$x = -5$$

37)
$$-3x + 19 = 3x - 1 - x$$
$$-3x + 19 = 2x - 1$$
$$-3x + 3x + 19 = 2x + 3x - 1$$
$$19 = 5x - 1$$
$$19 + 1 = 5x - 1 + 1$$
$$20 = 5x$$
$$\frac{20}{5} = \frac{5x}{5}$$
$$4 = x$$

39)
$$10 = 3(z - 2) + z$$
$$10 = 3z - 6 + z$$
$$10 = 4z - 6$$
$$10 + 6 = 4z - 6 + 6$$
$$16 = 4z$$
$$\frac{16}{4} = \frac{4z}{4}$$
$$4 = z$$

41) $-2c+12+4c=4(c+6)$
$-2c+12+4c=4c+24$
$2c+12=4c+24$
$2c-2c+12=4c-2c+24$
$12=2c+24$
$12-24=2c+24-24$
$-12=2c$
$\dfrac{-12}{2}=\dfrac{2c}{2}$
$-6=c$

43) $4-3(t-8)=37$
$4-3t+24=37$
$28-3t=37$
$28-28-3t=37-28$
$-3t=9$
$\dfrac{-3t}{-3}=\dfrac{9}{-3}$
$t=-3$

45) $-5(c-9)=-9(c-5)$
$-5c+45=-9c+45$
$-5c+9c+45=-9c+9c+45$
$4c+45=45$
$4c+45-45=45-45$
$4c=0$
$\dfrac{4c}{4}=\dfrac{0}{4}$
$c=0$

47) $4x=-2$
$\dfrac{4x}{4}=\dfrac{-2}{4}$
$x=-\dfrac{1}{2}$

49) $3(t-8)=7t+4$
$3t-24=7t+4$
$3t-3t-24=7t-3t+4$
$-24=4t+4$
$-24-4=4t+4-4$
$-28=4t$
$\dfrac{-28}{4}=\dfrac{4t}{4}$
$-7=t$

51) $4(y+2)=5y-3(y+4)$
$4y+8=5y-3y-12$
$4y+8=2y-12$
$4y-2y+8=2y-2y-12$
$2y+8=-12$
$2y+8-8=-12-8$
$2y=-20$
$\dfrac{2y}{2}=\dfrac{-20}{2}$
$y=-10$

53) $y-32=5y+8$
$y-y-32=5y-y+8$
$-32=4y+8$
$-32-8=4y+8-8$
$-40=4y$
$\dfrac{-40}{4}=\dfrac{4y}{4}$
$-10=y$

55) $9x-2+x=3+2(x+1)$
$9x-2+x=3+2x+2$
$10x-2=5+2x$
$10x-2x-2=5+2x-2x$
$8x-2=5$
$8x-2+2=5+2$
$8x=7$
$\dfrac{8x}{8}=\dfrac{7}{8}$
$x=\dfrac{7}{8}$

57) $3x-(2+2x)=7+2x$
$3x-2-2x=7+2x$
$x-2=7+2x$
$x-x-2=7+2x-x$
$-2=7+x$
$-2-7=7-7+x$
$-9=x$

59) $F=\dfrac{9}{5}C+32$
$F=\dfrac{9}{5}(21)+32$
$F=\dfrac{189}{5}+\dfrac{160}{5}$
$F=\dfrac{349}{5}$
$F=69.8$

21°C is equal to 69.8°F.

61)
$$F = \frac{9}{5}C + 32$$
$$84 = \frac{9}{5}C + 32$$
$$84 - 32 = \frac{9}{5}C + 32 - 32$$
$$52 = \frac{9}{5}C$$
$$\frac{5}{9} \cdot (52) = \frac{5}{9} \cdot \frac{9}{5}C$$
$$28\frac{8}{9} = C$$

$84°F$ is equal to $28\frac{8}{9}°C$ or about $28.9°C$.

63) a)
$$60 = 2l + 2w$$
$$60 = 2l + 2(12)$$
$$60 = 2l + 24$$
$$60 - 24 = 2l + 24 - 24$$
$$36 = 2l$$
$$\frac{36}{2} = \frac{2l}{2}$$
$$18 = l$$
The length of the rectangle is 18 inches.

b)
$$A = l \cdot w$$
$$A = 18 \cdot 12$$
$$A = 216$$
The area of the rectangle is 216 square inches.

65)
$$T = 3f + p$$
$$94 = 3f + 31$$
$$94 - 31 = 3f + 31 - 31$$
$$63 = 3f$$
$$\frac{63}{3} = \frac{3f}{3}$$
$$21 = f$$
Bryant kicked 21 field goals.

67)
$$T = 3f + p$$
$$90 = 3(20) + p$$
$$90 = 60 + p$$
$$90 - 60 = 60 - 60 + p$$
$$30 = p$$
Janikowski kicked 30 extra points.

69) Axis Rental
$$C = 20d + 0.32m$$
$$C = 20(1) + 0.32(190)$$
$$C = 20 + 60.8$$
$$C = 80.8$$
$$C = \$80.80$$

Harts Rental
$$C = 40d + 0.18m$$
$$C = 40(1) + 0.18(190)$$
$$C = 40 + 34.2$$
$$C = 74.2$$
$$C = \$74.20$$

Harts Rental has the better deal.

71) Axis Rental
$$C = 20d + 0.32m$$
$$C = 20(2) + 0.32(120)$$
$$C = 40 + 38.4$$
$$C = 78.4$$
$$C = \$78.40$$

Harts Rental
$$C = 40d + 0.18m$$
$$C = 40(2) + 0.18(120)$$
$$C = 80 + 21.6$$
$$C = 101.6$$
$$C = \$101.60$$

Yes. Andrea can afford the plan from Axis Rental.

73) a) Did not distribute -2 correctly.

b)
$$3x - 2(x + 4) = 2x$$
$$3x - 2x - 8 = 2x$$
$$x - 8 = 2x$$
$$x - x - 8 = 2x - x$$
$$-8 = x$$

75) a) Divided by 3 instead of by -3.

b)
$$-5x + 2(x + 1) = 8$$
$$-5x + 2x + 2 = 8$$
$$-3x + 2 = 8$$
$$-3x + 2 - 2 = 8 - 2$$
$$-3x = 6$$
$$\frac{-3x}{-3} = \frac{6}{-3}$$
$$x = -2$$

77) $x+(x+1)+(x+2)=168$
$x+x+1+x+2=168$
$3x+3=168$
$3x+3-3=168-3$
$3x=165$
$\dfrac{3x}{3}=\dfrac{165}{3}$
$x=55$

$x=55$
$x+1=56$
$x+2=57$

79) $x+(x+2)+(x+4)=252$
$x+x+2+x+4=252$
$3x+6=252$
$3x+6-6=252-6$
$3x=246$
$\dfrac{3x}{3}=\dfrac{246}{3}$
$x=82$

$x=82$
$x+2=84$
$x+4=86$

CHAPTER 10   REVIEW EXERCISES

1) Let b = the number of candy bars in one box.

Number of candy bars $=2b+2$

2) Let b = the number of lollipops in one box.

Number of lollipops $=b+5$

3) Let b = the number of baseball caps in one box.

Number of baseball caps $=5b+15$

4) Let b = the number of DVDs in one box.

Number of DVDs $=10b+14$

5) The two types of terms are variable terms and constant terms.

6) The numerical factor of a term is called its coefficient.

7) Like terms have identical variable parts.

8) a) There are four terms.

    b) The constant term is 3.

c) The variable terms are $12x$, $-4y$, and $-1x$.

d) The numerical coefficients are $12, -4, -1,$ and 3.

e) $12x$ and $-1x$ are like terms.

9) a) There are four terms.

    b) The constant term is 5.

    c) The variable terms are $3z^2$, $-5y$, and $-6r^2$.

    d) The numerical coefficients are $3, -5, -6,$ and 5.

    e) There are no like terms.

10) a) There are four terms.

    b) The constant term is -17.

    c) The variable terms are $3y^3$, $-y^2$, and $-14y^3$.

    d) The numerical coefficients are $-17, 3, -1,$ and -14.

    e) $3y^3$ and $-14y^3$ are like terms.

11) a) There are four terms.

    b) The constant term is 35.

    c) The variable terms are $3z^2$, $9z$, and $-2z^2$.

    d) The numerical coefficients are $3, 9, -2,$ and 35.

    e) $3z^2$ and $-2z^2$ are like terms.

12) $d=rt$
$=50\cdot 3$
$=150$ mi
Harvey traveled 150 miles.

13) $d=rt$
$=6\cdot 1.5$
$=9$ mi
Karel jogged 9 miles.

14) $5-2z+7=-2z+5+7$
$=-2z+12$

15) $3x+5-2x+6=3x-2x+5+6$
$=x+11$

16) $10z^2 + 8 - 4z - 5z^2 = 10z^2 - 5z^2 - 4z + 8$
$$= 5z^2 - 4z + 8$$

17) $-9x - 12 - 7x^2 - 11x = -7x^2 - 9x - 11x - 12$
$$= -7x^2 - 20x - 12$$

18) $3(4x) = 12x$

19) $5(2y) = 10y$

20) Answers may vary. Example: When adding things together, there must be a way to describe the combined thing. Adding carrots to carrots results in carrots. Adding x to x results in x.

21) Answers may vary. In order to add things together, there must be a way to describe the combined thing. There is not a name for the combination of a carrot and a potato. Also there is not a name for the addition of x and y.

22) $5(x + 7) = 5x + 35$

23) $7(y - 8) = 7y - 56$

24) $-4(z + 8) = -4z - 32$

25) $-2(y - 3) = -2y + 6$

26) $3(4x) + 5x = 12x + 5x$
$$= 17x$$

27) $\dfrac{5h}{5} + 7h = h + 7h$
$$= 8h$$

28) $7n + 2(n + 1) = 7n + 2n + 2$
$$= 9n + 2$$

29) $5t - 3(t + 3) = 5t - 3t - 9$
$$= 2t - 9$$

30) $5(x - 1) + 3(2x - 2) = 5x - 5 + 6x - 6$
$$= 5x + 6x - 5 - 6$$
$$= 11x - 11$$

31) $6(x + 4) - 3(x + 4) = 6x + 24 - 3x - 12$
$$= 6x - 3x + 24 - 12$$
$$= 3x + 12$$

32) $3y = y + 20$
$$3(10) \overset{?}{=} 10 + 20$$
$$30 = 30$$
Since $30 = 30$ is true, $y = 10$ is a solution.

33) $5x - 16 = 4x$
$$5(12) - 16 \overset{?}{=} 4(12)$$
$$60 - 16 \overset{?}{=} 48$$
$$44 = 48$$
Since $44 = 48$ is false, $x = 12$ isn't a solution.

34) Answers may vary. Example: A solution is a value that, when substituted for the variable, creates a true statement.

35) $x - 6 = 10$
$$x - 6 + 6 = 10 + 6$$
$$x = 16$$

36) $15 = y + 7$
$$15 - 7 = y + 7 - 7$$
$$8 = y$$

37) $-12 = z + 5$
$$-12 - 5 = z + 5 - 5$$
$$-17 = z$$

38) $x + \dfrac{1}{3} = \dfrac{3}{5}$
$$x + \dfrac{1}{3} - \dfrac{1}{3} = \dfrac{3}{5} - \dfrac{1}{3}$$
$$x = \dfrac{9}{15} - \dfrac{5}{15}$$
$$x = \dfrac{4}{15}$$

39) $56 = 8y$
$$\dfrac{56}{8} = \dfrac{8y}{8}$$
$$7 = y$$

40) $-5x = 55$
$$\dfrac{-5x}{-5} = \dfrac{55}{-5}$$
$$x = -11$$

41) $\dfrac{x}{5} = 9$
$$5 \cdot \dfrac{x}{5} = 5 \cdot 9$$
$$x = 45$$

42) $\dfrac{2}{3}x = \dfrac{1}{2}$

$\dfrac{3}{2} \cdot \dfrac{2}{3}x = \dfrac{3}{2} \cdot \dfrac{1}{2}$

$x = \dfrac{3}{4}$

43) a) 3 should have been added to both sides of the equation.

b) $x - 3 = 10$

$x - 3 + 3 = 10 + 3$

$x = 13$

44) a) Both sides should have been divided by −2.

b) $12 = -2x$

$\dfrac{12}{-2} = \dfrac{-2x}{-2}$

$-6 = x$

45) a) Both sides should have been divided by 4.

b) $4 = 4x$

$\dfrac{4}{4} = \dfrac{4x}{4}$

$1 = x$

46) Answers may vary. Example: An equal sign indicates that both sides are in balance. To be an equation, both sides must be in balance. Adding to or subtracting from only one side will cause the equation to become unbalanced.

47) $5x + 3 = 4x - 7$

$5x - 4x + 3 = 4x - 4x - 7$

$x + 3 = -7$

$x + 3 - 3 = -7 - 3$

$x = -10$

48) $-2x + 4 = -x + 6$

$-2x + 2x + 4 = -x + 2x + 6$

$4 = x + 6$

$4 - 6 = x + 6 - 6$

$-2 = x$

49) $5x + 2 = 3x - 8$

$5x - 3x + 2 = 3x - 3x - 8$

$2x + 2 = -8$

$2x + 2 - 2 = -8 - 2$

$2x = -10$

$\dfrac{2x}{2} = \dfrac{-10}{2}$

$x = -5$

50) $-7x - 8 = 4x - 8$

$-7x + 7x - 8 = 4x + 7x - 8$

$-8 = 11x - 8$

$-8 + 8 = 11x - 8 + 8$

$0 = 11x$

$\dfrac{0}{11} = \dfrac{11x}{11}$

$0 = x$

51) $3(x + 4) = 5x - 3x$

$3x + 12 = 5x - 3x$

$3x + 12 = 2x$

$3x - 3x + 12 = 2x - 3x$

$12 = -x$

$\dfrac{12}{-1} = \dfrac{-x}{-1}$

$-12 = x$

52) $6(x - 2) = 10 + 2x + 2$

$6x - 12 = 10 + 2x + 2$

$6x - 12 = 2x + 12$

$6x - 2x - 12 = 2x - 2x + 12$

$4x - 12 = 12$

$4x - 12 + 12 = 12 + 12$

$4x = 24$

$\dfrac{4x}{4} = \dfrac{24}{4}$

$x = 6$

53) $-15x = 45$

$\dfrac{-15x}{-15} = \dfrac{45}{-15}$

$x = -3$

54) $\dfrac{x}{7} = 9$

$7 \cdot \dfrac{x}{7} = 7 \cdot 9$

$x = 63$

55) $-\dfrac{2}{3}y = \dfrac{1}{4}$

$\left(-\dfrac{3}{2}\right) \cdot \left(-\dfrac{2}{3}\right)y = \left(-\dfrac{3}{2}\right) \cdot \left(\dfrac{1}{4}\right)$

$y = -\dfrac{3}{8}$

56) $10y + 3 = 9y - 17$

$10y - 9y + 3 = 9y - 9y - 17$

$y + 3 = -17$

$y + 3 - 3 = -17 - 3$

$y = -20$

57)
$$x + 45 = 45 + 3x$$
$$x - x + 45 = 45 + 3x - x$$
$$45 = 45 + 2x$$
$$45 - 45 = 45 - 45 + 2x$$
$$0 = 2x$$
$$\frac{0}{2} = \frac{2x}{2}$$
$$0 = x$$

58)
$$-3x + 13 = 4x - 8$$
$$-3x + 3x + 13 = 4x + 3x - 8$$
$$13 = 7x - 8$$
$$13 + 8 = 7x - 8 + 8$$
$$21 = 7x$$
$$\frac{21}{7} = \frac{7x}{7}$$
$$3 = x$$

59)
$$-5x = 2$$
$$\frac{-5x}{-5} = \frac{2}{-5}$$
$$x = -\frac{2}{5}$$

60)
$$7(x - 2) = 8 + 2x + 3$$
$$7x - 14 = 8 + 2x + 3$$
$$7x - 14 = 2x + 11$$
$$7x - 2x - 14 = 2x - 2x + 11$$
$$5x - 14 = 11$$
$$5x - 14 + 14 = 11 + 14$$
$$5x = 25$$
$$\frac{5x}{5} = \frac{25}{5}$$
$$x = 5$$

61)
$$4 - 2(x + 8) = 2x$$
$$4 - 2x - 16 = 2x$$
$$-2x - 12 = 2x$$
$$-2x + 2x - 12 = 2x + 2x$$
$$-12 = 4x$$
$$\frac{-12}{4} = \frac{4x}{4}$$
$$-3 = x$$

62)
$$6 - 3(x + 6) = 3x$$
$$6 - 3x - 18 = 3x$$
$$-12 - 3x = 3x$$
$$-12 - 3x + 3x = 3x + 3x$$
$$-12 = 6x$$
$$\frac{-12}{6} = \frac{6x}{6}$$
$$-2 = x$$

63)
$$9y = 5$$
$$\frac{9y}{9} = \frac{5}{9}$$
$$y = \frac{5}{9}$$

64)
$$5x + 10 = 4x + 13$$
$$5x - 4x + 10 = 4x - 4x + 13$$
$$x + 10 = 13$$
$$x + 10 - 10 = 13 - 10$$
$$x = 3$$

CHAPTER 10   TEST

1) Let $c =$ the number of cans of soda in one case.

Number of cans of soda $= 2c + 3$

2) Let $j =$ the number of jerseys in one box.

Number of jerseys $= 4j + 25$

3) When simplifying an expression, like terms can be combined.

4) Terms that do not contain a variable part are called constant terms.

5) a) There are four terms.

b) The constant terms are -5 and -14.

c) The variable terms are $4y$ and $-2y^2$.

d) The numerical coefficients are $-5, 4, -2,$ and -14.

e) -5 and -14 are like terms.

6) a) There are five terms.

b) The constant terms are 3 and -8.

c) The variable terms are $5z, 4z^2,$ and $-z$.

d) The numerical coefficients are $3, 5, 4, -1,$ and -8.

e) 3 and -8 are like terms.
$5z$ and $-z$ are like terms.

7) $d = rt$
$$= 70 \cdot 6$$
$$= 420 \text{ mi}$$
Mark traveled 420 miles.

8) $5x - 12 + 17x = 5x + 17x - 12$
$$= 22x - 12$$

9) $7(8x) = 56x$

10) $x^2 + 8x - 4 - x^2 = x^2 - x^2 + 8x - 4$
$$= 8x - 4$$

11) $4x - 7 - 2x - 12 = 4x - 2x - 7 - 12$
$$= 2x - 19$$

12) $6(x + 9) = 6x + 54$

13) $-5(y - 7) = -5y + 35$

14) $n + 9(n + 4) = n + 9n + 36$
$$= 10n + 36$$

15) $\dfrac{3x}{3} + 5x = x + 5x$
$$= 6x$$

16) $9y - 2(y + 3) = 9y - 2y - 6$
$$= 7y - 6$$

17) $2(x + 1) - 7(x + 2) = 2x + 2 - 7x - 14$
$$= 2x - 7x + 2 - 14$$
$$= -5x - 12$$

18) $2y = y + 20$
$$2(22) \overset{?}{=} (22) + 20$$
$$44 \overset{?}{=} 22 + 20$$
$$44 = 42$$
Since $44 = 42$ is false, $y = 22$ isn't a solution.

19) $x - 12 = 40$
$$x - 12 + 12 = 40 + 12$$
$$x = 52$$

20) $121 = 11y$
$$\dfrac{121}{11} = \dfrac{11y}{11}$$
$$11 = y$$

21) $\dfrac{2}{1} \cdot \dfrac{1}{2}x = \dfrac{2}{1} \cdot \dfrac{4}{5}$
$$x = \dfrac{8}{5}$$

22) $y + \dfrac{1}{4} = \dfrac{4}{5}$
$$y + \dfrac{1}{4} - \dfrac{1}{4} = \dfrac{4}{5} - \dfrac{1}{4}$$
$$y = \dfrac{16}{20} - \dfrac{5}{20}$$
$$y = \dfrac{11}{20}$$

23) Add 10 to both sides of the equation.

24) Multiply both sides of the equation by -6.

25) Divide both sides of the equation by -7.

26) Subtract $\dfrac{2}{3}$ from both sides of the equation.

27) Answers may vary.
Goal 1: Simplify both sides of the equation.
Goal 2: Gather variable terms together on one side of the equation. Gather constant terms together on the other side of the equation.
Goal 3: Isolate the variable.

28) $5(x + 2) = 4x - 2x + 25$
$$5x + 10 = 2x + 25$$
$$5x - 2x + 10 = 2x - 2x + 25$$
$$3x + 10 = 25$$
$$3x + 10 - 10 = 25 - 10$$
$$3x = 15$$
$$\dfrac{3x}{3} = \dfrac{15}{3}$$
$$x = 5$$

29) $20 = -10y$
$$\dfrac{20}{-10} = \dfrac{-10y}{-10}$$
$$-2 = y$$

30) $7y + 13 = 6y - 12$
$$7y - 6y + 13 = 6y - 6y - 12$$
$$y + 13 = -12$$
$$y + 13 - 13 = -12 - 13$$
$$y = -25$$

31) $2+2x+3=7(x-5)$
       $2+2x+3=7x-35$
       $2x+5=7x-35$
       $2x-2x+5=7x-2x-35$
       $5=5x-35$
       $5+35=5x-35+35$
       $40=5x$
       $\dfrac{40}{5}=\dfrac{5x}{5}$
       $8=x$

32) $\dfrac{x}{4}=8$

       $4\cdot\dfrac{x}{4}=4\cdot 8$

       $x=32$

33) $-10x+11=4x-17$
       $-10x+10x+11=4x+10x-17$
       $11=14x-17$
       $11+17=14x-17+17$
       $28=14x$
       $2=x$

34) $3y+14=2y+14$
       $3y-2y+14=2y-2y+14$
       $y+14=14$
       $y+14-14=14-14$
       $y=0$

35) $5x-4(2x-6)=x$
       $5x-8x+24=x$
       $-3x+24=x$
       $-3x+3x+24=x+3x$
       $24=4x$
       $\dfrac{24}{4}=\dfrac{4x}{4}$
       $6=x$

36) $8y-1=4$
       $8y-1+1=4+1$
       $8y=5$
       $\dfrac{8y}{8}=\dfrac{5}{8}$
       $y=\dfrac{5}{8}$

Appendix A Additional Practice and Review

Section 1.2   **Extra Practice Exercises; Addition Facts**

1)
$$\begin{array}{r} 23 \\ +11 \\ \hline 34 \end{array}$$

2)
$$\begin{array}{r} 21 \\ +48 \\ \hline 69 \end{array}$$

3)
$$\begin{array}{r} 52 \\ +33 \\ \hline 85 \end{array}$$

4)
$$\begin{array}{r} 55 \\ +44 \\ \hline 99 \end{array}$$

5)
$$\begin{array}{r} \overset{1}{8}3 \\ +18 \\ \hline 101 \end{array}$$

6)
$$\begin{array}{r} \overset{1}{4}8 \\ +35 \\ \hline 83 \end{array}$$

7)
$$\begin{array}{r} \overset{1}{6}7 \\ +73 \\ \hline 140 \end{array}$$

8)
$$\begin{array}{r} 93 \\ +66 \\ \hline 159 \end{array}$$

9)
$$\begin{array}{r} 6 \\ 7 \\ +5 \\ \hline 18 \end{array}$$

10)
$$\begin{array}{r} 9 \\ 5 \\ +3 \\ \hline 17 \end{array}$$

11)
$$\begin{array}{r} \overset{1}{1}7 \\ 14 \\ +13 \\ \hline 44 \end{array}$$

12)
$$\begin{array}{r} \overset{1}{1}5 \\ 11 \\ +17 \\ \hline 43 \end{array}$$

13)
$$\begin{array}{r} \overset{1}{4}53 \\ +\ 54 \\ \hline 507 \end{array}$$

14)
$$\begin{array}{r} \overset{1\,1}{6}67 \\ +\ 38 \\ \hline 705 \end{array}$$

15)
$$\begin{array}{r} \overset{1\,1}{6}52 \\ +\ 49 \\ \hline 701 \end{array}$$

16)
$$\begin{array}{r} \overset{1\,1}{6}45 \\ +\ 58 \\ \hline 703 \end{array}$$

17)
$$\begin{array}{r} \overset{1}{7}63 \\ +528 \\ \hline 1291 \end{array}$$

18)
$$\begin{array}{r} \overset{1}{8}78 \\ +381 \\ \hline 1259 \end{array}$$

19)
$$\begin{array}{r} \overset{1\,1}{6}75 \\ +378 \\ \hline 1053 \end{array}$$

20)
$$\begin{array}{r} \overset{1\,1}{9}83 \\ +658 \\ \hline 164\,1 \end{array}$$

21)
$$
\begin{array}{r}
25 \\
42 \\
+\,81 \\
\hline
148
\end{array}
$$

22)
$$
\begin{array}{r}
\overset{1}{2}7 \\
74 \\
+\,67 \\
\hline
168
\end{array}
$$

23)
$$
\begin{array}{r}
\overset{1}{7}5 \\
52 \\
+\,63 \\
\hline
190
\end{array}
$$

24)
$$
\begin{array}{r}
\overset{1}{8}7 \\
41 \\
+\,36 \\
\hline
164
\end{array}
$$

Section 1.3   Extra Practice Exercises; Subtraction Facts

1)
$$
\begin{array}{r}
23 \\
-\,11 \\
\hline
12
\end{array}
$$

2)
$$
\begin{array}{r}
26 \\
-\,15 \\
\hline
11
\end{array}
$$

3)
$$
\begin{array}{r}
\overset{4}{\cancel{5}}\,^{1}2 \\
-\,3\ 3 \\
\hline
1\ 9
\end{array}
$$

4)
$$
\begin{array}{r}
\overset{4}{\cancel{5}}\,^{1}3 \\
-\,4\ 4 \\
\hline
9
\end{array}
$$

5)
$$
\begin{array}{r}
\overset{7}{\cancel{8}}\,^{1}3 \\
-\,1\ 8 \\
\hline
6\ 5
\end{array}
$$

6)
$$
\begin{array}{r}
\overset{3}{\cancel{4}}\,^{1}1 \\
-\,3\ 5 \\
\hline
6
\end{array}
$$

7)
$$
\begin{array}{r}
\overset{6}{\cancel{7}}\,^{1}3 \\
-\,2\ 8 \\
\hline
4\ 5
\end{array}
$$

8)
$$
\begin{array}{r}
\overset{8}{\cancel{9}}\,^{1}3 \\
-\,6\ 6 \\
\hline
2\ 7
\end{array}
$$

9) $26 - 7 - 5 = 19 - 5$
   $= 14$

10) $19 - 5 - 3 = 14 - 3$
    $= 11$

11) $37 - 14 - 13 = 23 - 13$
    $= 10$

12) $45 - 11 - 17 = 34 - 17$
    $= 17$

13)
$$
\begin{array}{r}
\overset{3}{\cancel{4}}\,\overset{14}{\cancel{5}}\,^{1}3 \\
-\,\ \ 5\ 4 \\
\hline
3\ 9\ 9
\end{array}
$$

14)
$$
\begin{array}{r}
\overset{5}{\cancel{6}}\,\overset{16}{\cancel{7}}\,^{1}7 \\
-\,\ \ 7\ 8 \\
\hline
5\ 9\ 9
\end{array}
$$

15)
$$
\begin{array}{r}
\overset{5}{\cancel{6}}\,\overset{14}{\cancel{5}}\,^{1}2 \\
-\,\ \ 6\ 9 \\
\hline
5\ 8\ 3
\end{array}
$$

16)
$$
\begin{array}{r}
\overset{5}{\cancel{6}}\,\overset{13}{\cancel{4}}\,^{1}5 \\
-\,\ \ 5\ 8 \\
\hline
5\ 8\ 7
\end{array}
$$

17)
$$
\begin{array}{r}
\overset{6}{\cancel{7}}\,\overset{15}{\cancel{6}}\,^{1}3 \\
-\,5\ 7\ 8 \\
\hline
1\ 8\ 5
\end{array}
$$

18) $\overset{7}{\cancel{8}}\,\overset{\cancel{1}6}{\cancel{7}}\,^{1}8$
       $-\ 3\ 8\ 9$
       $\overline{\ \ \ 4\ 8\ 9}$

19) $\overset{5}{\cancel{6}}\,\overset{\cancel{1}6}{\cancel{7}}\,^{1}5$
       $-\ 3\ 7\ 8$
       $\overline{\ \ \ 2\ 9\ 7}$

20) $\overset{8}{\cancel{9}}\,\overset{\cancel{1}7}{\cancel{8}}\,^{1}3$
       $-\ 6\ 8\ 8$
       $\overline{\ \ \ 2\ 9\ 5}$

21) $\overset{1}{\cancel{2}}\,\overset{\cancel{1}4}{\cancel{5}}\,\overset{\cancel{9}}{\cancel{0}}\,\overset{\cancel{9}}{\cancel{0}}\,^{1}2$
       $-\ \ \ 8\ 1\ 4\ 5$
       $\overline{\ \ 1\ 6\ 8\ 5\ 7}$

22) $2\,\overset{6}{\cancel{7}}\,\overset{\cancel{9}}{\cancel{0}}\,^{1}0\ 6$
       $-\ \ \ \ 6\ 7\ 3\ 4$
       $\overline{\ \ 2\ 0\ 2\ 7\ 2}$

23) $\overset{3}{\cancel{4}}\,\overset{\cancel{1}4}{\cancel{5}}\,\overset{\cancel{1}5}{\cancel{6}}\,\overset{\cancel{9}}{\cancel{0}}\,^{1}0$
       $-\ \ \ \ 8\ 7\ 6\ 5$
       $\overline{\ \ 3\ 6\ 8\ 3\ 5}$

24) $8\,\overset{6}{\cancel{7}}\,\overset{\cancel{9}}{\cancel{0}}\,\overset{\cancel{9}}{\cancel{0}}\,^{1}0\ 3$
       $-\ \ \ \ \ \ 7\ 4\ 5\ 3$
       $\overline{\ 8\ 6\ 2\ 5\ 5\ 0}$

Section 1.4   Extra Practice Exercises; Multiplication Facts

1) $\overset{3}{4}5$
      $\times\ 6$
      $\overline{270}$

2) $\overset{1}{5}2$
      $\times\ 5$
      $\overline{260}$

3) $\overset{4}{6}7$
      $\times\ 7$
      $\overline{469}$

4) $\overset{2}{8}3$
      $\times\ 9$
      $\overline{747}$

5) $\overset{1}{5}26$
      $\times\ \ \ 3$
      $\overline{1578}$

6) $\overset{3\ 2}{9}87$
      $\times\ \ \ 4$
      $\overline{3948}$

7) $\overset{3\ 5\ 2}{2}463$
      $\times\ \ \ \ 8$
      $\overline{19,704}$

8) $\overset{2\ 3}{3}351$
      $\times\ \ \ \ 7$
      $\overline{23,457}$

9) 67
      $\times 10$
      $\overline{670}$

10) 92
      $\times 10$
      $\overline{920}$

11) 234
      $\times\ \ 1000$
      $\overline{234,000}$

12) 542
      $\times\ \ 1000$
      $\overline{542,000}$

13) $\overset{1}{4}5$
      $\times 30$
      $\overline{1350}$

14)
$$\overset{1}{93}$$
$$\times 40$$
$$\overline{3720}$$

15)
$$\overset{1}{6300}$$
$$\times\ 5000$$
$$\overline{31,500,000}$$

16)
$$\overset{4}{1800}$$
$$\times\ 6000$$
$$\overline{10,800,000}$$

17)
$$65$$
$$\times 28$$
$$\overline{520}$$
$$\underline{1300}$$
$$1820$$

18)
$$73$$
$$\times 45$$
$$\overline{365}$$
$$\underline{2920}$$
$$3285$$

19)
$$71$$
$$\times 64$$
$$\overline{284}$$
$$\underline{4260}$$
$$4544$$

20)
$$51$$
$$\times 76$$
$$\overline{306}$$
$$\underline{3570}$$
$$3876$$

21) $5 \cdot 3 \cdot 4 = 15 \cdot 4$
$= 60$

22) $6 \cdot 6 \cdot 5 = 36 \cdot 5$
$= 180$

23) $13 \cdot 14 \cdot 15 = 182 \cdot 15$
$= 2,730$

24) $12 \cdot 20 \cdot 14 = 240 \cdot 14$
$= 3,360$

25)
$$332$$
$$\times 92$$
$$\overline{664}$$
$$\underline{29880}$$
$$30544$$

26)
$$561$$
$$\times 85$$
$$\overline{2805}$$
$$\underline{44880}$$
$$47685$$

27)
$$643$$
$$\times 39$$
$$\overline{5787}$$
$$\underline{19290}$$
$$25077$$

28)
$$265$$
$$\times 74$$
$$\overline{1060}$$
$$\underline{18550}$$
$$19610$$

29)
$$537$$
$$\times 657$$
$$\overline{3759}$$
$$26850$$
$$\underline{322200}$$
$$352809$$

30)
$$210$$
$$\times 734$$
$$\overline{840}$$
$$6300$$
$$\underline{147000}$$
$$154140$$

31)
$$705$$
$$\times 408$$
$$\overline{5640}$$
$$000$$
$$\underline{282000}$$
$$287640$$

32)
$$803$$
$$\times 650$$
$$\overline{000}$$
$$40150$$
$$\underline{481800}$$
$$521950$$

Mid-Chapter 1 Review Exercises

1) Identify the digit in the thousands place.
3$\boxed{2}$,894; 32,894 will round to either 32,000 or
33,000. Since the hundreds digit is 8, we round
up. 32,894 rounds to 33,000.

2) Identify the digit in the tens place. 1$\boxed{2}$1; 121
will round to either 120 or 130. Since the ones
digit is 1, we round down. 121 rounds to 120.

3) Identify the digit in the hundreds place. $\boxed{3}$49;
349 will round to either 300 or 400. Since the
tens digit is 4, we round down. 349 rounds to
300.

4) Identify the digit in the ten thousands place.
1,3$\boxed{3}$2,984; 1,332,984 will round to either
1,330,000 or 1,340,000. Since the thousands
digit is 2, we round down. 1,332,984 rounds to
1,330,000.

5) Four thousand, nine hundred eight

6) Twenty-five thousand, forty-five

7)
```
   1 1
   3 9 8
 + 2 1 5
 ───────
   6 1 3
```

8)
```
   1
   4 6 7
 + 2 4 1
 ───────
   7 0 8
```

9)
```
  1 10
  2 1̶ ¹2
 −   7 8
 ───────
   1 3 4
```

10)
```
  3 14
  4̶ 5̶ ¹2
 − 3 8 3
 ───────
     6 9
```

11)
```
   1̶
   8̶
   4 8
 × 2 7
 ─────
   3 3 6
   9 6 0
 ───────
 1 2 9 6
```

12)
```
   8̶
   1̶
   5 6
 × 9 4
 ─────
   2 2 4
 5 0 4 0
 ───────
 5 2 6 4
```

13)
```
      1 9 R2
 8 ) 1 5 4
   − 8
   ─────
     7 4
   − 7 2
   ─────
       2
```

14)
```
      2 3 R6
 9 ) 2 1 3
   − 1 8
   ─────
     3 3
   − 2 7
   ─────
       6
```

15)
```
       2 6 R3
 1 4 ) 3 6 7
     − 2 8
     ─────
       8 7
     − 8 4
     ─────
         3
```

16)
```
       2 2 R12
 1 3 ) 2 9 8
     − 2 6
     ─────
       3 8
     − 2 6
     ─────
       1 2
```

17)
```
   5  9
   6̶ 0̶ ¹0
 − 1 2 6
 ───────
   4 7 4
```

18)
$$
\begin{array}{r}
{\scriptstyle 7\ \ 9} \\
{\scriptstyle 8\ \ \cancel{0}\ {}^{1}0} \\
-\ 1\ 3\ 5 \\
\hline
6\ 6\ 5
\end{array}
$$

19) 100 has two zeros.
$81 \cdot 100 = 8{,}100$

20) 1,000 has three zeros.
$45 \cdot 1{,}000 = 45{,}000$

21)
$$
\begin{array}{r}
{\scriptstyle 8} \\
7\ \cancel{9}\ {}^{1}2 \\
-\ 3\ 2\ 5 \\
\hline
4\ 6\ 7
\end{array}
$$
The new balance is $467.

22)
$$
\begin{array}{r}
3\ 5 \\
2\ 4\ )\overline{\ 8\ 4\ 0} \\
-\ 7\ 2 \\
\hline
1\ 2\ 0 \\
-\ 1\ 2\ 0 \\
\hline
0
\end{array}
$$
Each monthly payment is $35.

23)
$$
\begin{array}{r}
{\scriptstyle \cancel{1}} \\
{\scriptstyle \cancel{3}} \\
5\ 5 \\
\times\ 3\ 6 \\
\hline
3\ 3\ 0 \\
1\ 6\ 5\ 0 \\
\hline
1\ 9\ 8\ 0
\end{array}
$$
He will pay $1,980 in total.

24)
$$
\begin{array}{r}
{\scriptstyle 1\ \ 1} \\
6\ 3 \\
1\ 2\ 8 \\
+\ 1\ 6 \\
\hline
2\ 0\ 7
\end{array}
$$
He drove 207 miles is total.

25) $32 \approx 30$
$675 \approx 700$
$32 \cdot 675 \approx 30 \cdot 700$
$ = 21{,}000$

26) $84 \approx 80$
$512 \approx 500$
$84 \cdot 512 \approx 80 \cdot 500$
$ = 40{,}000$

27) $831 \approx 800$
$39 \approx 40$
$$
\begin{array}{r}
2\ 0 \\
4\ 0\ )\overline{8\ 0\ 0} \\
-8\ 0 \\
\hline
0\ 0
\end{array}
$$
$831 \div 39 \approx 20$

28) $976 \approx 1{,}000$
$45 \approx 50$
$$
\begin{array}{r}
2\ 0 \\
5\ 0\ )\overline{1\ 0\ 0\ 0} \\
-1\ 0\ 0 \\
\hline
0\ 0
\end{array}
$$
$976 \div 45 \approx 20$

29)
$$
\begin{array}{r}
{\scriptstyle 1} \\
\cancel{2}\ {}^{1}4\ 6 \\
-\ 1\ 8\ 6 \\
\hline
6\ 0
\end{array}
$$
60 invited guests did not attend.

30)
$$
\begin{array}{r}
6\ 4 \\
7\ )\overline{4\ 4\ 8} \\
-\ 4\ 2 \\
\hline
2\ 8 \\
-\ 2\ 8 \\
\hline
0
\end{array}
$$
They were traveling 64 miles per hour.

31) $A = \frac{1}{2} \cdot b \cdot h$
$ = \frac{1}{2} \cdot 32 \cdot 12$
$ = 16 \cdot 12$
$ = 192$
The area of the gable is 192 square feet.

32) $P = 32 + 20 + 20$
$P = 52 + 20$
$P = 72$
The perimeter of the gable is 72 feet.

33) $7 \cdot x = 42$
$ x = 6$
Basic Fact: $7 \cdot 6 = 42$

34) $ x + 18 = 53$
$x + 18 - 18 = 53 - 18$
$ x = 35$

35) $x \div 5 = 8$
Basic fact: $40 \div 5 = 8$
$x = 40$

36) $x - 14 = 7$
$x - 14 + 14 = 7 + 14$
$x = 21$

Mid-Chapter 2 Review Exercises

1) LCD $= 84$
$$\frac{1}{12} + \frac{1}{14} = \frac{1}{12} \cdot \frac{7}{7} + \frac{1}{14} \cdot \frac{6}{6}$$
$$= \frac{7}{84} + \frac{6}{84}$$
$$= \frac{13}{84}$$

2) LCD $= 162$
$$\frac{1}{18} + \frac{3}{81} = \frac{1}{18} \cdot \frac{9}{9} + \frac{3}{81} \cdot \frac{2}{2}$$
$$= \frac{9}{162} + \frac{6}{162}$$
$$= \frac{15}{162}$$
$$= \frac{\cancel{3} \cdot 5}{\cancel{3} \cdot 54}$$
$$= \frac{5}{54}$$

3) LCD $= 30$
$$\frac{3}{15} - \frac{1}{10} = \frac{3}{15} \cdot \frac{2}{2} - \frac{1}{10} \cdot \frac{3}{3}$$
$$= \frac{6}{30} - \frac{3}{30}$$
$$= \frac{3}{30}$$
$$= \frac{1 \cdot \cancel{3}}{\cancel{3} \cdot 10}$$
$$= \frac{1}{10}$$

4) LCD $= 56$
$$\frac{7}{8} - \frac{1}{14} = \frac{7}{8} \cdot \frac{7}{7} - \frac{1}{14} \cdot \frac{4}{4}$$
$$= \frac{49}{56} - \frac{4}{56}$$
$$= \frac{45}{56}$$

5) $$\frac{5}{4} \cdot \frac{3}{10} = \frac{5 \cdot 3}{4 \cdot 10}$$
$$= \frac{\cancel{5} \cdot 3}{4 \cdot 2 \cdot \cancel{5}}$$
$$= \frac{3}{8}$$

6) $$\frac{5}{12} \cdot \frac{4}{5} = \frac{5 \cdot 4}{12 \cdot 5}$$
$$= \frac{\cancel{5} \cdot \cancel{4}}{3 \cdot \cancel{4} \cdot \cancel{5}}$$
$$= \frac{1}{3}$$

7) $$\frac{13}{15} \div \frac{7}{10} = \frac{13}{15} \cdot \frac{10}{7}$$
$$= \frac{13 \cdot 2 \cdot \cancel{5}}{3 \cdot \cancel{5} \cdot 7}$$
$$= \frac{26}{21}$$

8) $$\frac{1}{24} \div \frac{1}{10} = \frac{1}{24} \cdot \frac{10}{1}$$
$$= \frac{\cancel{2} \cdot 5}{\cancel{2} \cdot 12}$$
$$= \frac{5}{12}$$

9) LCD $= 30$
$$\frac{3}{5} + \frac{3}{10} - \frac{1}{3} = \frac{3}{5} \cdot \frac{6}{6} + \frac{3}{10} \cdot \frac{3}{3} - \frac{1}{3} \cdot \frac{10}{10}$$
$$= \frac{18}{30} + \frac{9}{30} - \frac{10}{30}$$
$$= \frac{27}{30} - \frac{10}{30}$$
$$= \frac{17}{30}$$

10) LCD $= 40$
$$\frac{7}{8} + \frac{4}{5} - \frac{1}{10} = \frac{7}{8} \cdot \frac{5}{5} + \frac{4}{5} \cdot \frac{8}{8} - \frac{1}{10} \cdot \frac{4}{4}$$
$$= \frac{35}{40} + \frac{32}{40} - \frac{4}{40}$$
$$= \frac{67}{40} - \frac{4}{40}$$
$$= \frac{63}{40}$$

11) $LCD = 36$

$$\frac{5}{6} - \frac{1}{18} + \frac{4}{36} = \frac{5}{6} \cdot \frac{6}{6} - \frac{1}{18} \cdot \frac{2}{2} + \frac{4}{36}$$

$$= \frac{30}{36} - \frac{2}{36} + \frac{4}{36}$$

$$= \frac{28}{36} + \frac{4}{36}$$

$$= \frac{32}{36}$$

$$= \frac{\cancel{4} \cdot 8}{\cancel{4} \cdot 9}$$

$$= \frac{8}{9}$$

12) $LCD = 18$

$$\frac{3}{2} - \frac{2}{3} + \frac{1}{9} = \frac{3}{2} \cdot \frac{9}{9} - \frac{2}{3} \cdot \frac{6}{6} + \frac{1}{9} \cdot \frac{2}{2}$$

$$= \frac{27}{18} - \frac{12}{18} + \frac{2}{18}$$

$$= \frac{15}{18} + \frac{2}{18}$$

$$= \frac{17}{18}$$

13)

$$\frac{9}{10} \div \frac{5}{12} \cdot \frac{2}{4} = \left(\frac{9}{10} \cdot \frac{12}{5}\right) \cdot \frac{2}{4}$$

$$= \left(\frac{9 \cdot \cancel{2} \cdot 6}{\cancel{2} \cdot 5 \cdot 5}\right) \cdot \frac{2}{4}$$

$$= \frac{54}{25} \cdot \frac{2}{4}$$

$$= \frac{\cancel{2} \cdot 27 \cdot \cancel{2}}{25 \cdot \cancel{2} \cdot \cancel{2}}$$

$$= \frac{27}{25}$$

14)

$$\frac{4}{25} \div \frac{6}{15} \cdot \frac{15}{2} = \left(\frac{4}{25} \cdot \frac{15}{6}\right) \cdot \frac{15}{2}$$

$$= \left(\frac{\cancel{2} \cdot 2 \cdot \cancel{3} \cdot \cancel{5}}{\cancel{5} \cdot 5 \cdot \cancel{2} \cdot \cancel{3}}\right) \cdot \frac{15}{2}$$

$$= \frac{2}{5} \cdot \frac{15}{2}$$

$$= \frac{\cancel{2} \cdot 3 \cdot \cancel{5}}{\cancel{5} \cdot \cancel{2}}$$

$$= 3$$

15)

$$\frac{6}{25} \cdot \frac{15}{4} \div \frac{9}{10} = \left(\frac{\cancel{2} \cdot 3 \cdot 3 \cdot \cancel{5}}{\cancel{5} \cdot 5 \cdot \cancel{2} \cdot 2}\right) \div \frac{9}{10}$$

$$= \frac{9}{10} \div \frac{9}{10}$$

$$= 1$$

16)

$$\frac{9}{18} \cdot \frac{1}{24} \div \frac{9}{12} = \left(\frac{\cancel{9} \cdot \cancel{9} \cdot 1}{2 \cdot \cancel{9} \cdot \cancel{9} \cdot 24}\right) \div \frac{9}{12}$$

$$= \frac{1}{48} \div \frac{9}{12}$$

$$= \frac{1}{48} \cdot \frac{12}{9}$$

$$= \frac{1 \cdot \cancel{12}}{4 \cdot \cancel{12} \cdot 9}$$

$$= \frac{1}{36}$$

17) Addition and subtraction require you to have like fractions.

18) Multiplication and division can be performed with unlike fractions.

19) Answers may vary.

Example: $\frac{1}{5}, \frac{2}{5}$

20) Answers may vary.

Example: $\frac{1}{2}, \frac{2}{4}$

21) A fraction will be equal to one if its denominator is the same as the numerator.

22) b; $\frac{1}{2} \cdot \frac{1}{3} = \frac{1}{6}$

23) g; $\left(\frac{1}{3}\right)^2 = \frac{1}{9}$

24) h; $\frac{6}{7} - \frac{1}{3} = \frac{18}{21} - \frac{7}{21}$

25) d; $\frac{2+1}{8} \cdot \frac{6}{5} = \frac{3}{8} \cdot \frac{6}{5}$

26) c; $\frac{1}{9} \cdot \frac{1}{9} = \frac{1}{81}$

27) f; $\frac{3}{8} \div \frac{6}{5} = \frac{3}{8} \cdot \frac{5}{6}$

28) e; $\frac{1}{2} \div \frac{1}{3} = \frac{1}{2} \cdot \frac{3}{1}$

29) a; $\frac{1}{2} + 3 = \frac{1}{2} + \frac{3}{1}$

30) $LCD = 171$

$\frac{5}{9} \cdot \frac{19}{19} = \frac{95}{171}$ and $\frac{12}{19} \cdot \frac{9}{9} = \frac{108}{171}$

$\frac{95}{171} < \frac{108}{171}$, so $\frac{5}{9} < \frac{12}{19}$

31) The numerators are equal.

$8 < 13$, so $\frac{7}{8} > \frac{7}{13}$

32) $LCD = 10$

$\frac{3}{10} = \frac{3}{10}$ and $\frac{2}{5} \cdot \frac{2}{2} = \frac{4}{10}$

$\frac{3}{10} < \frac{4}{10}$, so $\frac{3}{10} < \frac{2}{5}$

33) The denominators are equal.

$5 < 7$, so $\frac{5}{12} < \frac{7}{12}$

34) $LCD = 21$

$\frac{3}{7} \cdot \frac{3}{3} = \frac{9}{21}$ and $\frac{10}{21} = \frac{10}{21}$

$\frac{9}{21} < \frac{10}{21}$, so $\frac{3}{7} < \frac{10}{21}$

35) $\frac{20}{39} \approx \frac{20}{40} = \frac{1}{2}$, but since $40 > 39$, then $\frac{20}{40} < \frac{20}{39}$

and $\frac{1}{2} < \frac{20}{39}$.

36) $LCD = 442$

$\frac{11}{34} \cdot \frac{13}{13} = \frac{143}{442}$ and $\frac{5}{13} \cdot \frac{34}{34} = \frac{170}{442}$

$\frac{143}{442} < \frac{170}{442}$, so $\frac{11}{34} < \frac{5}{13}$

37) $LCD = 80$

$\frac{13}{20} \cdot \frac{4}{4} = \frac{53}{80}$ and $\frac{51}{80} = \frac{51}{80}$

$\frac{53}{80} > \frac{51}{80}$, so $\frac{13}{20} > \frac{51}{80}$

38) $\frac{1}{2} + \frac{1}{2} = 1$ $\qquad$ $\frac{1}{2} \cdot \frac{5}{6} = \frac{5}{12}$

$1 = \frac{12}{12}$ $\qquad$ $\frac{5}{12} = \frac{5}{12}$

$\frac{12}{12} > \frac{5}{12}$, so $\frac{1}{2} + \frac{1}{2} > \frac{1}{2} \cdot \frac{5}{6}$

39) a) $\frac{7}{8} + \frac{3}{4} = \frac{7}{8} + \frac{6}{8} = \frac{13}{8}$ $\qquad$ $\frac{8}{7} \cdot \frac{5}{2} = \frac{\cancel{2} \cdot 4 \cdot 5}{7 \cdot \cancel{2}} = \frac{20}{7}$

b) $\frac{13}{8} \cdot \frac{7}{7} = \frac{91}{56}$ $\qquad$ $\frac{20}{7} \cdot \frac{8}{8} = \frac{160}{56}$

c) $\frac{91}{56} < \frac{160}{56}$, so $\frac{7}{8} + \frac{3}{4} < \frac{8}{7} \cdot \frac{5}{2}$

40) $\frac{15}{4} - \frac{5}{2} = \frac{15}{4} - \frac{10}{4} = \frac{5}{4}$ $\qquad$ $\frac{8}{3} \cdot \frac{5}{3} = \frac{40}{9}$

$\frac{5}{4} \cdot \frac{9}{9} = \frac{45}{36}$ $\qquad$ $\frac{40}{9} \cdot \frac{4}{4} = \frac{160}{36}$

$\frac{45}{36} < \frac{160}{36}$, so $\frac{15}{4} - \frac{5}{2} < \frac{8}{3} \cdot \frac{5}{3}$

41) $\frac{2}{3} + \frac{3}{4} = \frac{8}{12} + \frac{9}{12} = \frac{17}{12}$ $\qquad$ $\frac{5}{6} \cdot \frac{1}{13} = \frac{5}{78}$

$\frac{17}{12} \cdot \frac{13}{13} = \frac{221}{156}$ $\qquad$ $\frac{5}{78} \cdot \frac{2}{2} = \frac{10}{156}$

$\frac{221}{156} > \frac{10}{156}$, so $\frac{2}{3} + \frac{3}{4} > \frac{5}{6} \cdot \frac{1}{13}$

42) $\frac{15}{16} + \frac{7}{4} = \frac{15}{16} + \frac{28}{16} = \frac{43}{16}$ $\qquad$ $\frac{8}{3} \cdot \frac{6}{5} = \frac{48}{15}$

$\frac{43}{16} \cdot \frac{15}{15} = \frac{645}{240}$ $\qquad$ $\frac{48}{15} \cdot \frac{16}{16} = \frac{768}{240}$

$\frac{645}{240} < \frac{768}{240}$, so $\frac{15}{16} + \frac{7}{4} < \frac{8}{3} \cdot \frac{6}{5}$

43) $\frac{21}{5} - \frac{5}{2} = \frac{42}{10} - \frac{25}{10} = \frac{17}{10}$ $\qquad$ $\frac{2}{1} \cdot \frac{3}{5} = \frac{6}{5}$

$\frac{17}{10} = \frac{17}{10}$ $\qquad$ $\frac{6}{5} \cdot \frac{2}{2} = \frac{12}{10}$

$\frac{17}{10} > \frac{12}{10}$, so $\frac{21}{5} - \frac{5}{2} > \frac{2}{1} \cdot \frac{3}{5}$

44) 1 mile $= 5,280$ feet

$6 \cancel{\text{mi}} \cdot \left(\frac{5,280 \text{ ft}}{1 \cancel{\text{mi}}} \right) = 31,680$ ft

6 miles is equal to 31,680 feet.

45) 1 pound $= 16$ ounces

$80 \cancel{\text{oz}} \cdot \left(\frac{1 \text{ lb}}{16 \cancel{\text{oz.}}} \right) = 5$ lb

80 ounces is equal to 5 pounds.

46) 1 gallon = 8 pints

$$\frac{500 \ \cancel{\text{jelly beans}}}{1 \ \cancel{\text{pt}}} \cdot \left(\frac{8 \ \cancel{\text{pt}}}{1 \ \cancel{\text{gal}}}\right) \cdot \left(5 \ \cancel{\text{gal}}\right)$$

= 20,000 jelly beans

Mid-Chapter 9 Review Exercises

1) $-5, -3, 4, 7$

2) $-13, -8, 7, 12$

3) $|-8| = 8 \qquad -|-4| = -4$
$-(-3) = 3 \qquad -(7) = -7$
$-(7), -|-4|, -(-3), |-8|$

4) $-(-2) = 2 \qquad 0 = 0$
$|-9| = 9 \qquad -|-2| = -2$
$-|-2|, 0, -(-2), |-9|$

5) $-8 + (-16) = -24$

6) $-5 + 12 = 7$

7) $(-3)(-12) = 36$

8) $-32 \div 8 = -4$

9) $7 - (-8) = 7 + 8 = 15$

10) $-13 - 7 = -20$

11) $(-2)(-5)(7) = (10)(7)$
$= 70$

12) $(-12) \div 3 \cdot (-2) = -4 \cdot (-2)$
$= 8$

13) $-6 - 5 + 8 = -11 + 8$
$= -3$

14) $6 + (-15) - 4 = -9 - 4$
$= -13$

15) $-2^4 = -16$

16) $(-3)^2 = 9$

17) $\frac{-1}{5} + \frac{-7}{10} = \frac{-1}{5} \cdot \frac{2}{2} + \frac{-7}{10}$
$= \frac{-2}{10} + \frac{-7}{10}$
$= -\frac{9}{10}$

18) $\frac{1}{7} + \frac{-3}{14} = \frac{1}{7} \cdot \frac{2}{2} + \frac{-3}{14}$
$= \frac{2}{14} + \frac{-3}{14}$
$= -\frac{1}{14}$

19) $\frac{-3}{5} \div \frac{9}{10} = \frac{-3}{5} \cdot \frac{10}{9}$
$= -\frac{\cancel{3} \cdot \cancel{5} \cdot 2}{\cancel{5} \cdot \cancel{3} \cdot 3}$
$= -\frac{2}{3}$

20) $\frac{-12}{25} \cdot \frac{-35}{16} = \frac{3 \cdot \cancel{4} \cdot \cancel{5} \cdot 7}{5 \cdot \cancel{5} \cdot \cancel{4} \cdot 4}$
$= \frac{21}{20}$

21) $4 \cdot (-6) - 3 \cdot (5) = -24 - 3 \cdot (5)$
$= -24 - 15$
$= -39$

22) $8 - (5) \cdot (7) + (-21) = 8 - 35 + (-21)$
$= -27 + (-21)$
$= -48$

23) $8 - 10 + (4-6)^2 = 8 - 10 + (-2)^2$
$= 8 - 10 + 4$
$= -2 + 4$
$= 2$

24) $(2^3 - 3^2) \cdot 8 \div (-2) = (8-9) \cdot 8 \div (-2)$
$= (-1) \cdot 8 \div (-2)$
$= -8 \div (-2)$
$= 4$

25) $\frac{4 \cdot 5 - (-7)}{9 - 6} = \frac{20 - (-7)}{3}$
$= \frac{27}{3}$
$= 9$

26) $\dfrac{(-4)\cdot 3-(-2)}{-4+2}=\dfrac{-12+2}{-2}$

$\qquad\qquad\qquad =\dfrac{-10}{-2}$

$\qquad\qquad\qquad =5$

27) $\dfrac{1-8+4}{3}=\dfrac{-3}{3}=-1$

Their average margin of victory was -1.

28) $\dfrac{8-2.3-4.2+7+1}{5}=\dfrac{9.5}{5}=1.9$

The average rate of return is 1.9%.

29) List factors of -15:

$\begin{array}{ll} 1\cdot(-15)=-15 & 1+(-15)=-14 \\ 3\cdot(-5)=-15 & 3+(-5)=-2 \\ 5\cdot(-3)=-15 & 5+(-3)=2 \\ 15\cdot(-1)=-15 & 15+(-1)=14 \end{array}$

-3 and 5 have a product of -15 and a sum of 2.

30) List factors of 24:

$\begin{array}{ll} 1\cdot(24)=24 & 1+24=25 \\ 2\cdot(12)=24 & 2+12=14 \\ 3\cdot(8)=24 & 3+8=11 \\ 4\cdot(6)=24 & 4+6=10 \\ -1\cdot(-24)=24 & -1+(-24)=-25 \\ -2\cdot(-12)=24 & -2+(-12)=-14 \\ -3\cdot(-8)=24 & -3+(-8)=-11 \\ -4\cdot(-6)=24 & -4+(-6)=-10 \end{array}$

-8 and -3 have a product of 24 and a
sum of -11.

31) $100\cdot(-0.32)+150\cdot(0.80)=-32+120$

$\qquad\qquad\qquad\qquad\qquad\qquad =88$

The total change in value of the stocks is $88.

32) $\dfrac{36\ \text{ft}}{1\ \cancel{\text{sec}}}\cdot 60\ \cancel{\text{sec}}=2{,}160\ \text{ft}$

The elevator was 2,160 feet from the entrance
of the mine after 1 minute.